Proceedings of the Seventh International Colloquium on Differential Equations

Also available from VSP

Proceedings of the Sixth International Colloquium on Differential Equations
Plovdiv, Bulgaria, 18 - 23 August 1995

Proceedings of the Fifth International Colloquium on Differential Equations
Plovdiv, Bulgaria, 18 - 23 August 1994

Proceedings of the Fourth International Colloquium on Differential Equations
Plovdiv, Bulgaria, 18 - 22 August 1993

Proceedings of the Third International Colloquium on Differential Equations
Plovdiv, Bulgaria, 18 - 23 August 1992

Proceedings of the Third International Colloquium on Numerical Analysis
Plovdiv, Bulgaria, 13 - 17 August 1994

Proceedings of the Second International Colloquium on Numerical Analysis
Plovdiv, Bulgaria, 13 - 17 August 1993

Proceedings of the First International Colloquium on Numerical Analysis
Plovdiv, Bulgaria, 13 - 17 August 1992

Proceedings of the Seventh International Colloquium on Differential Equations

Plovdiv, Bulgaria, 18 - 23 August, 1996

Editor: D. Bainov

///VSP///
Utrecht, The Netherlands, 1997

VSP BV
P.O. Box 346
3700 AH Zeist
The Netherlands

First published in 1997

ISBN 90-6764-233-9

Printed in The Netherlands by Ridderprint bv, Ridderkerk.

CONTENTS

Preface

The Seventh International Colloquium on Differential Equations was organized by the Institute for Basic Science of Inha University, the International Federation of Nonlinear Analysts, the Mathematical Society of Japan, Pharmaceutical Faculty of the Medical University of Sofia, University of Catania and UNESCO, with the cooperation of the Association Suisse d'Informatique, the Canadian Mathematical Society, Kyushu University, the London Mathematical Society, Technical University of Plovdiv, and it was partially sponsored by the Bulgarian Ministry of Education, Science and Technologies under Grant MM-511.

The Colloquium was held August 18-23, 1996, in Plovdiv, Bulgaria. The subsequent colloqiua will take place each year from 18 to 23 August in Plovdiv, Bulgaria.

The publishing of this volume is fully supported by the Bulgarian Ministry of Education, Science and Technologies under Grant MM-511.

Organizing Committee:

H.Adeli, A.S.A. Al-Hammadi, D. Bainov (Chairman), Q.-H. Choi, J. Diblik, S. Hristova, J. Kajiwara (Vice Chairman), V. Lakshmikantham, M. Marino, A. Maugeri (Vice Chairman), E. Minchev (Secretary), K. Nakagawa, N. Popivanov, P. Popivanov, M. Robnik, H.M. Srivastava, Dj. Takaci, R.U. Verma

Scientific Committee:

D. Bainov (Bulgaria), C. Dafermos (USA), G. Da Prato (Italy), H. Fujita (Japan), R. Ivanov (Bulgaria), T. Kaczorek (Poland), J. Kajiwara (Japan), A. Kaneko (Japan), N. Kenmochi (Japan), V. Lakshmikantham (USA), A. Maugeri (Italy), N. Pavel (USA), N. Popivanov (Bulgaria), P. Popivanov (Bulgaria), M. Robnik (Slovenia), P. Sobolevskii (Israel), H.M. Srivastava (Canada), Fr. Stenger (USA), E. Titi (USA), M. Tsutsumi (Japan). R.U. Verma (USA).

The address of the Organizing Committee is:
Drumi Bainov, P.O. Box 45, 1504 Sofia, Bulgaria.

The Editor

7th Int. Coll. on Differential Equations, p.p. 1-7
D. Bainov (Ed)

Identifying the Jordan Canonical Form and Associated Non-Singular Transformation

S. Askarpour and T. J. Owens
Department of Electrical Engineering and Electronics, Brunel University, Uxbridge, Middlesex UB8 3PH, United Kingdom

Keywords: Jordan canonical form

INTRODUCTION

At present all existing methods for determining the Jordan canonical form of a matrix are ambiguous in identifying which chains of equations lead to the correct selection of characteristic vectors that achieve the Jordan canonical form through similarity transformation. Golub and Wilkinson [1] discuss several methods for computing the Jordan canonical form but do not discuss a numerically stable method.

It is easily shown that only some linear combinations of characteristic vectors can be employed in order to determine the Jordan canonical form leading to limited freedom in the selection of characteristic vectors. Gantmacher [2] has described some of the possible operations that can be performed on the transformation matrix without altering the shape of the Jordan form. However, the operations discussed by Gantmacher [2] are performed only on complete Jordan blocks and no proofs are given that the operations do not change the shape of the Jordan form.

In this paper six theorems are presented relating to which linear combinations of characteristic vectors can be employed to achieve the Jordan canonical form. From these theorems two rules are deduced concerning the freedom available in the choice of the transformation matrix for the Jordan canonical form. A seventh theorem is presented which provides conditions that enable a new method for determining the Jordan canonical form and associated non-singular transformation to be developed. The new method does not possess the ambiguities associated with existing methods. The method is demonstrated through a seventh order example.

STATEMENT OF THE PROBLEM

Square matrices of dimension n which have repeated eigenvalues may be transformed into the Jordan canonical form using a similarity transformation. That is there exists a transformation V, obtained by using the characteristic vectors of the original matrix, such that:

$$J = V^{-1}AV$$

where J has the Jordan form.

Construction of a matrix V is achieved by searching for linearly independent eigenvectors from the standard eigenvalue-eigenvector problem

$$(A - I\lambda_i)\nu_i = 0$$

which produces $\nu \leq n$ linearly independent vectors and completing the n dimensional basis by using generalised eigenvectors of Jordan chains. To clearly express the chaining process the following notation is introduced:

Let ν^l_{xyz} be an eigenvector or generalised eigenvector relating to eigenvalue λ_l where x shows the level of the vector in the chaining process e.g. ν^l_{0yz} denotes an eigenvector and ν^l_{2yz} denotes a second chained generalised eigenvector. The suffix y indicates the Jordan block with which the vector is associated. The maximum value that y takes is the geometric multiplicity of λ_l. The suffix z indicates the number of generalised eigenvectors chained off the vector. e.g. ν^l_{030} denotes the third eigenvector of λ_l which has no following chain and ν^l_{211} denotes the second chained generalised eigenvector of the first Jordan block which has one generalised eigenvector chained off it.

Now from the definition of eigenvectors and generalised eigenvectors we may write:

$$(A - I\lambda_l)\nu^l_{0yz} = 0$$
$$(A - I\lambda_l)\nu^l_{x+1yz-x-1} = \nu^l_{xyz-x}$$

and

$$(A - I\lambda_l)^{x+1}\nu^l_{xyz-x} = 0$$
$$(A - I\lambda_l)^{x}\nu^l_{xyz-x} = \nu^l_{0yz}$$

From the above it is clear that the nullity of $(A - I\lambda_l)$ is equal to the number of eigenvectors belonging to λ_l and difference between the nullity of $(A - I\lambda_l)^2$ and the nullity of $(A - I\lambda_l)$ is equal to the number of first chain generalised eigenvectors belonging to λ_l and so on. Thus, it is conceptually simple to determine the Jordan form. Determination of a transformation matrix for the Jordan form is, however, a challenging problem.

This presentation will address the construction of a transformation matrix for the Jordan form. Firstly, however, rules that enable many similarity transformations that achieve the Jordan form to be determined once one transformation is known will be developed.

NON-UNIQUENESS OF THE TRANSFORMATION MATRIX

It is well known that the Jordan canonical form is unique up to the size and the order of the Jordan blocks. The following six theorems determine which combinations of eigenvectors and generalised eigenvectors can be used in the transformation matrix without altering the Jordan form.

Theorem 1. *Any generalised eigenvector can only be chained off one eigenvector or generalised eigenvector.*

Theorem 2. *Any multiple of an eigenvector of a repeated eigenvalue added to any multiple of another eigenvector associated with the same repeated eigenvalue with a smaller or equal number of chained vectors can replace the latter without altering the Jordan form.*

Theorem 3. *No linear combination of an eigenvector of a repeated eigenvalue and an eigenvector associated with the same repeated eigenvalue with a longer following chain can replace the latter without altering the Jordan form.*

Theorem 4. *Any multiple of an eigenvector of a repeated eigenvalue added to a generalised eigenvector associated with the same eigenvalue with a shorter or equal length of chain can replace the latter without altering the Jordan form.*

Theorem 5. *No linear combination of a generalised eigenvector of a repeated eigenvalue and an eigenvector associated with the same repeated eigenvalue can replace the latter without altering the Jordan form.*

Theorem 6. *No linear combination of a generalised eigenvector of a repeated eigenvalue with another generalised eigenvector associated with the same repeated eigenvalue can replace the latter without altering the Jordan form.*

The six theorems can be summarised in two rules:

Rule 1. *No linear combination of a generalised eigenvector of a repeated eigenvalue and any other vector associated with the same repeated eigenvalue can replace the latter without altering the Jordan form. This also means that a generalised eigenvector can not be replaced by multiples of itself.*

Rule 2. *A multiple of an eigenvector of a repeated eigenvalue can only be added to a generalised eigenvector or multiple of an eigenvector associated with the same repeated eigenvalue to replace the latter without altering the Jordan block if the length of the chain of the former is greater than or equal to the length of the chain of the latter.*

DETERMINATION OF A TRANSFORMATION MATRIX

A condition is needed satisfaction of which ensures chaining can take place. Such a condition is:

Theorem 7. *Any vector ν^l_{xyz} satisfying the condition of a further z generalised eigenvectors to be chained from it must satisfy:*

$$\nu^l_{xyz} \in \Re(A - I\lambda_l)^z \cap \aleph(A - I\lambda_l)^{x+1}.$$

Recall that: x gives level of the vector in the chaining process, y indicates the Jordan block with which the vector is associated, z gives the number of generalised eigenvectors chained off the vector.

Remark 1. To develop a computational algorithm for determining a similarity transformation for the Jordan canonical form based on Theorem 7 the following propositions are required:

Proposition 1. *$\Re(A) \cap \Re(B)$ can be determined from $\aleph[A - B]$ where A and B are matrices with an equal number of rows.*

Proposition 2. *If $A\nu_2 = \nu_1$ and $\nu_2 \in \Re(B)$, where A and B are matrices with an equal number of rows and A is square, then all possible vectors ν_2 can be obtained from $\aleph[AB - \nu_1]$.*

Remark 2. Propositions 1 and 2 mean that all the subspaces needed to determine a similarity transformation to achieve the Jordan canonical form can be obtained from basis vectors of null spaces.

A COMPUTATIONAL ALGORITHM FOR DETERMINING A SIMILARITY TRANSFORMATION FOR THE JORDAN FORM

The following algorithm constructs the generalised eigenstructure associated with the Jordan blocks of a repeated eigenvalue λ_l. It enables the entire generalised eigenstructure of a matrix with repeated eigenvalues to be determined.

Step 1: Determine the nullity nl of $(A - I\lambda_l)$ and a set of basis vectors NL which span the nullspace of $(A - I\lambda_l)$ using the matlab null command. Determine successive higher powers of $(A - I\lambda_l)$ and the nullities of the powers up to the point where the nullity of the power is equal to the algebraic multiplicity of the eigenvalue.

Step 2: Determine the shape and size of the Jordan blocks of λ_l from the nullities of step 1.

Step 3: Determine an eigenvector ν^l_{0yz} belonging to a largest Jordan block of λ_l for which the eigenvector is unknown using the result of Proposition 1: Determine a vector $u^l_{0yz} \in \aleph([NL - (A - I\lambda_l)^z])$ then $\nu^l_{0yz} = NL * u^l_{0yz}(1 : nl)$ where $u^l_{0yz}(1 : nl)$ represents a vector containing the first nl elements of u^l_{0yz}.

Step 4: Determine the next generalised eigenvector in the chain using the result of Proposition 2 with $A = (A - I\lambda_l)$ and $B = (A - I\lambda_l)^z$:
Determine $u^l_{xyz} \in \aleph([(A - I\lambda_l)^{z+1} - \nu^l_{x-1yz+1}])$ and $w^l_{xyz} = u^l_{xyz}/\gamma$ where γ is the last element of u^l_{xyz}. Then $\nu^l_{xyz} = (A - I\lambda_l)^z * w^l_{xyz}(1 : n)$ where n is the dimension of A.

Step 5: Follow step 4 until all the vectors in the chain are determined.

Step 6: Follow steps 3 to 5 for the remaining blocks until all the vectors are determined remembering that all eigenvectors chosen in step 3 must be linearly independent.

EXAMPLE

Consider the 7th order matrix:

$$A = \begin{bmatrix} -1 & 1 & -1 & -1 & 0 & 1 & -1 \\ 0.5 & -0.5 & 0 & -2 & 0.5 & 0.5 & 0 \\ 0 & -1 & 0 & 3 & 0 & -1 & 1 \\ 0 & 0 & 0 & -1 & 0 & 0 & 0 \\ -1 & -1 & -1 & 0 & -2 & 0 & -1 \\ 0.5 & -0.5 & 2 & 5 & 0.5 & -2.5 & 2 \\ 0.5 & 0.5 & 1 & 1 & 0.5 & -0.5 & 0 \end{bmatrix}$$

with seven eigenvalues at -1 and Jordan form:

$$J = \begin{bmatrix} -1 & 1 & 0 & 0 & 0 & 0 & 0 \\ 0 & -1 & 1 & 0 & 0 & 0 & 0 \\ 0 & 0 & -1 & 1 & 0 & 0 & 0 \\ 0 & 0 & 0 & -1 & 0 & 0 & 0 \\ 0 & 0 & 0 & 0 & -1 & 1 & 0 \\ 0 & 0 & 0 & 0 & 0 & -1 & 0 \\ 0 & 0 & 0 & 0 & 0 & 0 & -1 \end{bmatrix}$$

The condition number of A in the sense of the ratio of the largest to the smallest singular value of A is 62.3144.

Applying the above algorithm to determine a similarity transformation for the Jordan form gives:

$V =$

Columns 1 through 3

9.965710984760412e−02	−6.106349859081298e+00	1.034450927331190e−01
−1.153122400305962e−16	9.965710984760381e−02	−6.206006968928888e+00
−9.965710984760394e−02	6.106349859081298e+00	9.586912696208919e−02
6.609734426920564e−17	0	0
−2.981555974335137e−18	−9.965710984760381e−02	1.211493659975772e+01
−9.965710984760383e−02	6.106349859081298e+00	−5.813060503866741e+00
−4.522478311980159e−17	0	−6.008586740676437e+00

Columns 4 through 6

−1.210516041365604e+01	4.309822056027898e−01	−1.771408807344095e+01
1.015511012903993e−01	−5.035309947468366e−16	3.281031771563433e−01
7.483295486981004e−02	−4.309822056027889e−01	2.343571382500273e+00
9.965710984760397e−02	2.909728334617533e−16	−1.026385132017130e−14
−1.273665645020327e−01	1.028790284464509e−01	1.086922180115206e+01
1.175903784002304e−01	−5.338612340492395e−01	6.516763095132537e+00
−5.879367775990322e−02	−1.028790284464512e−01	4.070312684185825e+00

Column 7

7.896658308256629e−01
−9.018596958641767e−16
−1.513142880156455e−01
4.964561747811125e−16
−4.488544555234243e−01
−3.408113753022363e−01
−1.894970872865920e−01

which transforms A to:

$J =$

Columns 1 through 3

−9.999999999973138e−01	9.999999998204316e−01	2.293010226139813e−11
3.913067548819219e−14	−1.000000000002642e+00	1.000000000000640e+00
7.104776836263623e−16	−4.485474491833230e−14	−9.999999999999921e−01
−1.618216441070205e−27	−6.632476534735134e−16	−7.595152795467902e−27
−3.357545837229238e−14	1.945887895260512e−12	3.944344850737025e−13
−6.700578002036070e−17	4.362482597386474e−14	3.191891195797325e−16
−5.683230036341138e−15	1.304512053934559e−12	−7.838174553853605e−14

Columns 4 through 6

−4.269400111223121e−11	1.229467916363802e−11	−1.276924788129463e−09
−7.339165082956289e−13	1.818202706449501e−13	−1.905840762983502e−11
9.999999999999996e−01	3.330249891558917e−15	−3.139154536175963e−13
−9.999999999999999e−01	−1.987387723286967e−20	−3.056296636131730e−15
−4.900363968773913e−13	−1.000000000000146e+00	1.000000000017093e+00
−3.005007267325999e−14	−4.923362065256676e−16	−9.999999999998127e−01
−1.146157821430016e−13	−2.572941859568800e−14	5.115963208623953e−12

Column 7

2.318017999769495e−11
3.324940149596056e−13
6.079984695933130e−15
4.942214969267663e−18
−3.316973432032633e−13
−3.935653886122381e−16
−1.000000000000035e+00

CONCLUDING REMARKS

The largest error in J is $-1.276924788129463e-09$ in element $(1,6)$. This error is large for such a well-conditioned A matrix. Much of it may be associated with the error in the inversion of V as the condition number of V is $6.6305e+07$. However, it might be possible to greatly reduce the error by using rules 1 and 2 to construct from the V delivered by the algorithm an "optimally conditioned" transformation matrix.

Alternatively, the error might be reduced by constraining the algorithm to deliver characteristic vectors that result in a relatively well-conditioned transformation matrix.

REFERENCES

[1] G. H. Golub and J. H. Wilkinson, Ill conditioned Eigensystems and the computation of the Jordan Canonical Form, *SIAM Review,* **18** (1976), 578–619.

[2] F. R. Gantmacher, *The Theory of Matrices,* Chelsea Publ., 1959.

7th Int. Coll. on Differential Equations, pp. 9-15
D. Bainov (Ed)

FORWARD DYNAMIC PROGRAMMING FOR THE NUMERICAL SOLUTION OF PARABOLIC CONTROL PROBLEMS

S. A. BELBAS
Mathematics Department, University of Alabama
Tuscaloosa, AL 35487-0350, USA

Abstract. The numerical solution of parabolic control problems is carried out by using an implicit or semi-implicit finite-difference discretization and a corresponding type of dynamic programming that deserves the name "forward dynamic programming". By contrast, the standard dynamic programming equations are solved backwards in time. We present here the essential features of forward dynamic programming.

1. INTRODUCTION

We are interested in optimal control problems for parabolic systems. A model problem consists of a controlled parabolic equation

$$\frac{\partial y(t,x)}{\partial t} = \mathcal{L}(t,x)y(t,x) + f(t,x,y(t,x),\nabla y(t,x)) + g(t,x,u(t)) \tag{1.1}$$

with appropriate initial and boundary conditions, where $\mathcal{L}(t,x)$ is a second-order uniformly elliptic operator with sufficiently smooth coefficients, y is an $\mathbb{R}^n$-valued function, and $u(t)$ is an $\mathbb{R}^m$-valued control function.

The spatial domain Ω is an open bounded set in $\mathbb{R}^3$ with sufficiently smooth boundary, and the temporal horizon is $(0,T]$. An optimal control problem associated with (1.1) concerns the minimization of a functional J of the form

$$J = \int_0^T \int_\Omega F(t,x,y(t,x),\nabla y(t,x),u(t))dtdx + \int_\Omega \Phi(T,x,y(T,x))dx \tag{1.2}$$

Of course, other more general functionals can also be considered.

Many applied problems lead to mathematical problems of the type (1.1)-(1.2). For example, problems of atmospheric pollution (transport and diffusion of aerosols), coastal pollution (transport and diffusion of hydrosols), and groundwater pollution (flow in porous media), lead to problems of the type (1.1)-(1.2); cf. [1, 2, 3, 4, 5].

The numerical solution of problem (1.1-2) relies on three steps: first, the discretization of the system (1.1) and the functional (1,2); second, an algorithm for the approximate solution of the discretized version of (1.1); and third, an algorithm for the approximate solution of the corresponding discretized optimal control problem. The first two steps have been extensively studied in the research literature; in our opinion, relatively less work has been done for the third step. In this paper, we are concerned primarily with the third step.

After a discretization in both space and time, we obtain a system in discrete time, with discrete time-variable still denoted by t, where now t takes nonnegative integer values. If $z(t)$ represents a discrete approximation to $y(t,x)$, $x \in \tilde{\Omega}$, where $\tilde{\Omega}$ is the set of interior grid points in Ω, so that z takes values in $\mathbb{R}^{2N}$, where N is the cardinality of $\tilde{\Omega}$.

If an explicit scheme is used, then the discrete system for z has the form

$$z(t+1) = g(t, z(t), u(t)) \tag{1.3}$$

If a fully implicit scheme is employed, then the corresponding dynamics for z are obtained as

$$z(t) = g(t+1, z(t+1), u(t+1)) \tag{1.4}$$

Finally, if a semi-implicit (e.g., Crank-Nicolson) scheme is used, the control system in discrete time can be written in either of the two forms below:

$$z(t) = \tilde{g}(t+1, z(t), z(t+1), u(t), u(t+1) \tag{1.5}$$

$$z(t+1) = \tilde{\gamma}(t, z(t), z(t+1), u(t), u(t+1)) \tag{1.6}$$

For the purpose of applying the ideas of dynamic programming to semi-implicit discretizations, it is useful to approximate (1.5) or (1.6) by (1.7) or (1.8) below, respectively:

$$z(t) = g(t+1, z(t+1), u(t), u(t+1)) \tag{1.7}$$

$$z(t+1) = \gamma(t, z(t), u(t), u(t+1)) \tag{1.8}$$

The functional J given by (1.2) is discretized as

$$\tilde{J} = \sum_{t=0}^{N} G(t, z(t), u(t)) + \Psi(T, z(T)) \tag{1.9}$$

2. BRIEF OVERVIEW OF CLASSICAL DYNAMIC PROGRAMMING FOR DETERMINISTIC DISCRETE-TIME SYSTEMS

The method of dynamic programming for discrete-time systems has been beautifully explained in the classical work of Boltyanskii [6], as well as in several other books and papers. Here, we shall give a very terse summary, entirely for the purpose of paving the way for the rest of this paper; our notation and terminology are different from [6].

Consider a discrete dynamical system (recursive formula)

$$y(t+1) = f(t, y(t), u(t)), y(0) = y_0 = \text{ given} \tag{2.1}$$

where t is discrete time-parameter, $t = 0, 1, 2, \ldots, N$. The function $u(t)$ is the **control**. The problem is to minimize

$$J \equiv \sum_{t=0}^{N} F(t, y(t), u(t)) + \Phi(y(N+1)) \tag{2.2}$$

Define a parametrized system

$$y_{sx}(t+1) = f(t, y(t), u(t)), y_{sx}(s) = x, t = s, s+1, s+2, \ldots, N. \tag{2.3}$$

Define

$$J_{sx} \equiv \sum_{t=s}^{N} F(t, y_{sx}(t), u(t)) + \Phi(y_{sx}(N+1)) \tag{2.4}$$

and

$$V(s, x) = \inf J_{sx} \tag{2.5}$$

the infimum being taken over all admissible control functions $u(\cdot)$. For simplicity, assume that u takes values in a compact set K, and then (assuming continuity of f, F, Φ)

$$V(s, x) = \min_{u(\cdot)} J_{sx} \tag{2.6}$$

Compare $V(s, x)$ to the value of J_{sx} obtained by using an arbitrary value $a \in K$ of the control $u(s)$, and using an optimal u from time $s+1$ until the final time N, starting with the value $y_{sx}(s+1) = f(s, x, a)$:

$$V(s, x) \leq F(s, x, a) + V(s+1, y_{sx}(s+1)) \tag{2.7}$$

or

$$V(s, x) \leq F(s, x, a) + V(s+1, f(s, x, a)). \tag{2.8}$$

If a happens to be the value of an optimal control u^* at time s, i.e., if $a = a^* \equiv u^*(s)$ where $u^*(\cdot)$ is a minimizer of J_{sx}, then

$$V(s, x) = F(s, x, a^*) + V(s+1, f(s, x, a^*)) \tag{2.9}$$

The formulae (1) and (2) can be combined into

$$V(s, x) = \min_{a \in K}\{V(s+1, f(s, x, a)) + F(s, x, a)\}; \tag{2.10}$$

$$a^* = \arg\min_{a}\{V(s+1, f(s, x, a)) + F(s, x, a)\} \tag{2.11}$$

where "$\arg\min_a$" denotes a value of a that minimizes the expression in brackets. In addition, from the definition of V and J_{sx}, we have

$$V(N+1, x) = \Phi(x) \tag{2.12}$$

because, for $s = N+1$,

$$J_{N+1,x} = \Phi(y_{N+1,x}(N+1)) \equiv \Phi(x) \tag{2.13}$$

using the convention that a sum over an empty set of indices is zero, so that, for $s = N+1$, the sum part of J_{sx} vanishes.

The system of (2.10-11) is a backward recursive relation (starting from time $N+1$, and moving, at each step, from $s+1$ to s). This yields the values of an optimal control.

The advantage of dynamic programming is this: suppose K is a finite set with k elements. In the minimization problem in (2.10), the variable a takes k values. It is readily seen from the definition of J that the variable of minimization is $\{u(0), u(1), \ldots, u(N)\}$, which takes k^{N+1} values. Thus, by dynamic programming, a minimization over a k^{N+1}-dimensional variable is reduced to $N+1$ minimizations, each with respect to a k-dimensional variable. This implies serious reduction in the amount of computations. This reduction is possible because of the particular structure of the dynamical system and the functional J.

3. THE FORWARD DYNAMIC PROGRAMMING EQUATIONS AND RELATED EQUATIONS

When the discrete-time dynamical system has the form (1.4) and the functional $\tilde{J}$ has the form (1.9), then we define a parameterized functional

$$\tilde{J}_{tx} \equiv \sum_{s=0}^{t} G(s, z_{tx}(s), u(s)), \qquad \text{for } 0 \le t \le N; \tag{3.1}$$

$$\tilde{J}_{N+1,x} \equiv \sum_{s=0}^{N} G(s, z_{N+1,x}(s), u(s)) + \Psi(N+1, x) \tag{3.2}$$

provided z_{tx} solves (1.4) with $z_{tx}(t) = x$, and the associated Hamilton-Jacobi function

$$V(t+1, x) \equiv \inf_{u(\cdot)} J_{tx} \tag{3.3}$$

Then a standard argument of dynamic programming shows that V must satisfy the discrete dynamic programming equation

$$V(t+1, x) = \inf_{a}\{V(t, g(t, x, a)) + G(t, x, a)\} \tag{3.4}$$

with initial conditions

$$V(0, x) = \inf_{a} G(0, x, a) \tag{3.5}$$

and with the convention $G(N+1, x, a) \equiv \Psi(N+1, x)$.

As in the case of ordinary (backward-in-time) dynamic programming, we can show that, under minimal smoothness condiutions on the various functions involved, the solution of (3.3-4) gives the optimal cost for the problem (1.4, 1.9). Still, the numerical solution of (3.3,4) requires another approximation. The complication arises from the simple fact that, if the spatial domain of all possible values of the variable x in (3.3) is approximated by a discrete grid $\mathcal{G}$, then, in general, $g(t, x, a)$ is not an element of $\mathcal{G}$. The description of the approximation is facilitated by introducting the following notation: for every $t, 0 \le t \le N+1$, for every value a of the control $u(\cdot)$, for every $x \notin \mathcal{G}$, we define a set $\omega(t, x, a)$, the set of neighbors of $g(t, x, a)$ in the grid $\mathcal{G}$ (the set $\omega(t, x, a)$ depends on the intended interpolation scheme for the values of V on $\mathcal{G}$); also, we define functions $\lambda_y(t, x, a)$, for each $y \in \omega(t, x, a)$, by the following requirements:

$$\lambda_y(t, x, a) \ge 0; \sum_{y \in \omega(t,x,a)} \lambda_y(t, x, a) = 1; f(t, x, a) = \sum_{y \in \omega(t,x,a)} \lambda_y(t, x, a) y \tag{3.6}$$

Then the function V is approximated by $\tilde{V}$, where $\tilde{V}$ solves

$$\tilde{V}(t+1, x) = \inf_{a}\{G(t, x, a) + \sum_{y \in \omega(t,x,a)} \lambda_y(t, x, a) \tilde{V}(t, y)\} \tag{3.7}$$

and $\tilde{V}$ has the same initial conditions as V. Estimates for the error $|V - \tilde{V}|$ are provided in the next section.

Let $\tilde{N}$ be the dimension of $\tilde{x}$. In order to be able to find λ's satisfying (3.5), $\omega(t, x, a)$ should contain at least $\tilde{n} + 1$ points; on the other hand, if $\omega(t, x, a)$ contains at least $\tilde{N} + 1$ points, then the determination of the λ's is generally not unique. For computational convenience, particularly with the prospect of parallel implementation, it is useful to have a grid $\mathcal{G}$ and corresponding sets $\omega(t, x, a)$ of the following type: points y of $\mathcal{G}$ have the form $x_0 + h_i e_i$, $1 \leq i \leq \tilde{N}$, where h_i is the discretization mesh in the i-th coordinate, and e_i is the corresponding unit $\tilde{N}$-dimensional vector; the neighbors of $x \notin \mathcal{G}$ are all those points $y \in \mathcal{G}$ that satisfy $y_i = [\frac{x_i}{h_i}]h_i$ or $y_i = [\frac{x_i}{h_i}]h_i + h_i$. Here, x_i, y_i, $1 \leq i \leq \tilde{N}$, are the components of x, y, respectively.

One way to obtain the coefficients λ_y is this: let $w_k = \frac{x_k}{h_k}, i_k = [w_k], \tilde{\omega}(x) =$ set of neighbors of x, y=generic element of $\tilde{\omega}(x), j_k = \frac{y_k}{h_k}$, so that j_k is either i_k or i_k+1; for each $y \in \tilde{\omega}(x)$, define $\lambda_k = \frac{i_k+1-w_k}{i_k}$, and $\overline{\lambda}_k = \lambda_k$ if $j_k = i_k, \overline{\lambda}_k = 1 - \lambda_k$ if $j_k = i_k + 1, \overline{\lambda}_y(x) = \Pi_{k=1}^{\tilde{N}} \overline{\lambda}_k$; then $\lambda_y(t, x, a) = \overline{\lambda}_y(g(t, x, a))$.

When a semi-implicit scheme is used for the original controlled partial differential equation, we assume that we utilize (1.7) or (1.8) as the discrete approximation to the continuous controlled distributed system. We present here the dynamic programming equations for (1.7) with functional (1.9) and the convention made immediately after (3.4). In this case, we define

$$J_{txv} \equiv \sum_{s=0}^{t} G(s, z_{txv}(s), u(s)), \text{ for } 0 \leq t \leq N+1 \tag{3.8}$$

where z_{txv} solves (1.7) with $z_{txv}(t) = x$ and $u(t) = v$. Also, we define

$$V(t, x, v) \equiv \inf_{u(\cdot)} J_{txv} \tag{3.9}$$

Then V satisfies

$$V(t+1, x, b) = \inf_a \{G(t, x, b) + V(t, g(t, x, a, b), a)\} \tag{3.10}$$

with initial conditions (3.4).

A similar treatment can be given for a system of the form (1.8). We can show, under minimal assumptions, that the system (3.8-3.4) is necessary and sufficient for optimality.

4. ERROR ESTIMATES

Convergence results and error estimates for the difference between the solution of an optimal control problem and the exact solution of a finite-difference discretization have been obtained by many authors, and, among many papers, we mention [7,8]. We shall not be concerned with that type of error estimate in this paper.

We have seen that the optimal control problem for the discrete problem that approximates the original (continuous) problem needs a further approximation, namely the approximation of the exact dynamic programming equation for the discrete system by an approximate (with interpolated right-hand side) dynamic programming equation. We now examine the error of this type of approximation. For definiteness, we compare (3.5) and (3.3).

Set $\tilde{h} = |h|$, for some vector norm $|\cdot|$ on $\mathbb{R}^{\tilde{N}}$. Assume $V(t, \cdot)$ is uniformly Lipschitz in x with

$$|V(t, x_1) - V(t, x_2)| \leq \eta(t)|x_1 - x_2| \tag{4.1}$$

If we have an error estimate

$$\sup_{x \in \mathcal{G}} |V(t, x) - \tilde{V}(t, x)| \leq \epsilon(t) \tag{4.2}$$

then we also obtain

$$\sup_{x\in\mathcal{G}} |V(t+1,x) - \tilde{V}(t+1,x)| \leq \epsilon(t+1) \tag{4.3}$$

where

$$\epsilon(t+1) = \eta(t)\tilde{h} + \epsilon(t) \tag{4.4}$$

Indeed, from (3. - 3.) we get

$$\begin{aligned}
|V(t+1,x) - \tilde{V}(t+1,x)| &\leq \\
&\leq \sup_a |V(t,g(t,x,a)) - \sum_{y\in\omega(t,x,a)} \lambda_y It,x,a)\tilde{V}(t,y)| \leq \\
&\leq \sup_a \sum_{y\in\omega(t,x,a)} \lambda_y(t,x,a)|V(t,g(t,x,a)) - \tilde{V}(t,y)| \leq \\
&\leq \sup_a \sum_{y\in\omega(t,x,a)} \lambda_y(t,x,a)\{|V(t,g(t,x,a)) - V(t,y)| + \\
&+ |V(t,y) - \tilde{V}(t,y)|\} \leq \\
&\leq \sup_a \eta(t)|g(t,x,a) - y| + \epsilon(t) \leq
\end{aligned}$$

$$\leq \eta(t)\tilde{h} + \epsilon(t) \tag{4.5}$$

Now, if we start with $\epsilon(0) = 0$, then (4.5) shows that

$$|V(t,x) - \tilde{V}(t,x)| \leq \sum_{s=0}^{t-1} \eta(s)\tilde{h} \tag{4.6}$$

and the lost inequality in particular implies the convergence of V to $\tilde{V}$ as $\tilde{h} \to 0^+$.

The Lipschitz condition on V follows from Lipschitz conditions on G and g. If we assume that

$$|G(t,x,a) - G(t,y,a)| \leq L_G(t)|x-y|; \tag{4.7}$$

$$|g(t,x,a) - g(t,y,a)| \leq L_g(t)|x-y| \tag{4.8}$$

and if

$$|V(t,x) - V(t,y)| \leq \eta(t)|x-y| \tag{4.9}$$

then

$$|V(t+1,x) - V(t+1,y)| \leq \eta(t+1)|x-y| \tag{4.10}$$

where

$$\eta(t+1) = L_G(t) + \eta(t)L_g(t) \tag{4.11}$$

The proof of (4.9)-(4.10) follows the same type of argumentation as the induction for $\epsilon(t)$.

REFERENCES

1. G.I. Marchuk, Mathematical modelling in environmental problems, (in Russian), Nauka, Moscow (1982).

2. G.I. Marchuk, Adjoint equations and analysis of complicated systems (in Russian), Nauka, Moscow, 1992.

3. G.I. Marchuk, V.I. Agoshkov, V.P. Shutyaev, Adjoint equations and perturbation methods in nonlinear problems of mathematical physics (in Russian), Nauka, Moscow, 1993.

4. Z. Zlatev, I. Dimov, K. Georgiev, Comput. Sci. Eng., Fall 1994, 45-52 (1994).

5. J. Bear, Dynamics of fluids in porous media, Elsevier, New York, 1972.

6. V. Boltyanskii, Optimal control of discrete systems (in Russian), Nauka, Moscow (1973).

7. T.P. Evseenko, in Approximate solution of optimal control problems for distributed parameter systems (in Russian), pp. 31-65, Ilim, Frunze (1976).

8. B.M. Budak, E.M. Berkovich, Yu. L. Gaponenko, Zh. Vychislit. Matem. i Matem. Fiz., 9, 2, 86-99 (1969).

7th Int. Coll. on Differential Equations, pp. 17-21
D. Bainov (Ed)

OPTIMAL CONTROL OF GOURSAT SYSTEMS OVER VARYING DOMAINS

S. A. BELBAS
Mathematics Department, University of Alabama
Tuscaloosa, AL 35487-0350, USA

Abstract. We present necessary conditions for optimal control of Goursat systems over a domain with variable end-points and final state constraints, and also for problems over a certain class of non-rectangular domains.

1. INTRODUCTION

The topic of optimal control of Goursat systems is an interesting generalization of the classical problems of optimal control for systems governed by ordinary differential equations. A classical problem of control of O.D.E. systems consists of a state equation

$$\frac{dy(t)}{dt} = f(t, y(t), u(t)), y(0) = y_0 \tag{1.1}$$

where u is the control function (usually piecewise continuous or bounded measurable), and a cost functional

$$J = \int_0^T F(t, y(t), y(t))dt + F_0(T, y(T)) \tag{1.2}$$

Such problems may contain other ingredients, e.g., constraints on the control and/or the state, final-time constraints, etc. A controlled Goursat system involves in "2-dimensional time" (s, t), according to the state equation

$$\begin{aligned} \frac{\partial^2 y(s,t)}{\partial s \partial t} &= f(s, t, y(s,t), y_s(s,t), y_t(s,t), u(s,t)); \\ & y(s,0) = y_1(s), y(0,t) = y_2(t) \end{aligned} \tag{1.3}$$

and the optimal control problem is the minimization of a functional

$$\begin{aligned} J &= \int_0^A \int_0^B F(s, t, y(s,t), y_s(s,,t), y_t(s,t), u(s,t))dtds + \\ &+ \int_0^A F_1(s, B, y(s,B), y_s(s,B), y_t(s,B))ds + \\ &+ \int_0^B F_2(A, t, y(A,t), y_s(A,t), y_t(A,t))dt \\ &+ F_0(A, B, y(A,B), y_s(A,B), y_t(A,B)) \end{aligned} \tag{1.4}$$

For the case in which F_1 is independent of y_t, F_2 is independent of y_s, and F_0 is independent from both y_s and y_t, the problem of minimizing (1.4) subject to (1.3) has been studied by several authors, including the first (to the best of our knowledge) work on this subject by Egorov [1], the monographs [2,3], and several other papers including works by this author. Under appropriate natural assumptions, necessary conditions have been proved in a form akin to the maximum principle of Pontryagin and Boltyanskii. We denote the components of y, f and u by y^i, f^i, u^j $(1 \leq i \leq n, 1 \leq j \leq m)$, the derivatives y^i_s and y^i_t by p^i and q^i, the components of the co-state $\lambda(s,t)$ by $\lambda_i(s,t)$, $1 \leq i \leq n$, and the operators of total differentiation by $\frac{D}{Ds}, \frac{D}{Dt}$, where

$$\begin{aligned} \frac{D}{Ds} &= y^i_s\frac{\partial}{\partial y^i} + y^i_{ss}\frac{\partial}{\partial p^i} + f^i\frac{\partial}{\partial q^i} + u^i_s\frac{\partial}{\partial u^i} + \frac{\partial}{\partial s}; \\ \frac{D}{Dt} &= y^i_t\frac{\partial}{\partial y^i} + f^i\frac{\partial}{\partial p^i} + y^i_{tt}\frac{\partial}{\partial q^i} + u^i_t\frac{\partial}{\partial u^i} + \frac{\partial}{\partial t} \end{aligned} \tag{1.5}$$

The Hamiltonian $H(s,t,y,p,q,\lambda,u)$ is defined by

$$H = \lambda Pif^i(s,t,y,p,q,u) + F(s,t,y,p,q,u) \tag{1.6}$$

The Hamilton equations for λ are

$$\frac{\partial^2\lambda_i(s,t)}{\partial s\partial t} = \frac{\partial H(s,t,\ldots)}{\partial y^i} - \frac{D}{Ds}\frac{\partial H(s,t,\ldots)}{\partial p^i} \tag{1.7}$$

$$- \frac{D}{Dt}\frac{\partial H(s,t,\ldots)}{\partial q^i} \tag{1.8}$$

with conditions along the final boundary $\{(s,B) : 0 < s \leq A\} \bigcup \{(A,t) : 0 < t \leq B\}$:

$$\begin{aligned} \frac{\partial \lambda_i(s,B)}{\partial s} &+ \frac{\partial H(s,B,\ldots)}{\partial q^i} - \frac{D}{Ds}\frac{\partial F_1(s,B,\ldots)}{\partial p^i} \\ &+ \frac{\partial F_1(s,B,\ldots)}{\partial y^i} = 0; \\ \frac{\partial \lambda_i(A,t)}{\partial t} &+ \frac{\partial H(A,t,\ldots)}{\partial p^i} - \frac{D}{Dt}\frac{\partial F_2(A,t,\ldots)}{\partial q^i} \\ &+ \frac{\partial F_2(A,t,\ldots)}{\partial y^i} = 0; \\ &- \lambda_i(A,B) + \frac{\partial F_1(A,B,\ldots)}{\partial p^i} + \frac{\partial F_2(A,B,\ldots)}{\partial q^i} \\ &+ \frac{\partial F_0(A,B,\ldots)}{\partial y^i} = 0 \end{aligned} \tag{1.9}$$

The associated maximum principle asserts that, if $u^*(s,t)$ is an optimal' control with associated state trajectory $y^*(s,t)$ and co-state $\lambda^*(s,t)$, then, for almost all $s,t)(0 \leq s \leq A, 0 \leq t \leq B)$,

$$\begin{aligned} & H(s,t,y^*(s,t),y^*_s(s,t),y^*_t(s,t),u^*(s,t),\lambda^*(s,t)) = \\ = \; & \max_u H(s,t,y^*(s,t),y^*_s(s,t),y^*_t(s,t),u,\lambda^*(s,t)) \end{aligned} \tag{1.10}$$

In this paper, we address 2 problems related to the standard Goursat control problem presented above: the problem of optimal control with variable end-points (A,B), and the problem of optimal control over a non-rectangular domain on the st-plane.

2. OPTIMALITY CONDITIONS WITH VARIATION OF THE END-POINTS

We consider the optimal control problem (1.1) with variable end-points (A, B), and side-conditions

$$\begin{aligned} y^i(s,B) &= \phi_0^i(s,A,B), y_s^i(s,B) = \phi_1^i(s,A,B), y_t^i(s,B) = \phi_2^i(s,A,B), \\ y^i(A,T) &= \psi_0^i(t,A,B), y_s^i(A,t) = \psi_1^i(t,A,B), y_t^i(A,t) = \psi_2^i(t,A,B), \\ y^i(A,B) &= \omega_0^2(A,B), y_s^i(A,B) = \omega_1^i(A,B), y_t^i(A,B) = \omega_2^i(A,B) \end{aligned} \tag{2.1}$$

where the functions $\phi_j^i, \psi_j^i, \omega_j^i$ satisfy appropriate consistency conditions. Under the assumptions of *reachability* (roughly, existence of a control such that the corresponding state $y(s,t)$ satisfies (2.1) for an open set of values of (A, B)) and *localization of an optimal control* (roughly, sufficiently small local variations of an optimal control exist so that the corresponding state still satisfies the constraints (2.1)), we can evaluate the variations of J with respect to the end-points (A, B). We use the symbol $\frac{\partial}{\partial A}$ for partial derivatives with respect to the explicit dependence of an expression relative to A, and $\frac{\theta}{\theta A}$ for the corresponding implicit dependence.

After some calculations, using also the Hamiltonian equations (1.7), we obtain a splitting of $\frac{DJ}{DA}$ as

$$\frac{DJ}{DA} = D_1 J + D_2 J \tag{2.2}$$

where $D_1 J$ contains only single integrals, and $D_2 J$ is given by

$$\begin{aligned} D_2 J &= \int_0^A \int_0^B \{\frac{\partial^2 \lambda_i}{\partial s \partial t} - \lambda_j \frac{\partial f^j}{\partial y^i} + \\ &+ \frac{D}{Ds}(\lambda_j \frac{\partial f^j}{\partial p^i}) + \frac{D}{Dt}\left(\lambda_i \frac{\partial f^j}{\partial q^i}\right)\}. \\ \frac{\theta y^i}{\theta A} dt ds &- \int_0^\Delta \int_0^B \lambda_j \frac{\partial f^j}{\partial u^i} \frac{\theta u^i}{\theta A} dt ds \end{aligned} \tag{2.3}$$

The original system (1.3) yields

$$\begin{aligned} \frac{\theta \partial^2 y^i}{\theta A \partial s \partial t} &= \frac{\partial f^i}{\partial y^j}\frac{\theta y^j}{\theta A} + \frac{\partial f^i}{\partial p^j}\frac{\theta y_s^j}{\theta A} \\ &+ \frac{\partial f^i}{\partial q^j}\frac{\theta y_t^i}{\theta A} + \frac{\partial f^i}{\partial u^j}\frac{\theta u^j}{\theta A} \end{aligned} \tag{2.4}$$

We use the simple fact

$$\begin{aligned} \frac{\partial^2}{\partial s \partial t}(\lambda_j \frac{\theta y^j}{\theta A}) &= \frac{\partial^2 \lambda_j}{\partial s \partial t}\frac{\theta y^j}{\theta A} + \frac{\partial \lambda_j}{\partial s}\frac{\theta y_t^i}{\theta A} \\ &+ \frac{\partial \lambda_j}{\partial t}\frac{\theta y_s^j}{\theta A} + \lambda_j \frac{\theta y_{st}^i}{\theta A} \end{aligned} \tag{2.5}$$

Now, we substitute the right-hand side of (2.4) in lieu of $\frac{\theta y_{st}^i}{\theta A}$ on the right-hand side of (2.5), and, after a concatenation of calculations, we find

$$-\int_0^A \int_0^B \frac{\partial^2}{\partial s \partial t}[\lambda_j(s,t)\frac{\theta y^j(s,t)}{\theta A}] dt ds \tag{2.6}$$

$$
\begin{aligned}
&= D_2 J - \int_0^A \{\lambda_j(s,B)\frac{\partial f^j(s,B,\ldots)}{\partial p^i} + \frac{\partial \lambda_i(s,B)}{\partial s}\}.\\
&\quad \cdot \frac{\theta y^i(s,B)}{\theta A} ds\\
&\quad - \int_0^B \{\lambda_j(A,t)\frac{\partial f^j(A,t,\ldots)}{\partial q^i} + \frac{\partial \lambda_i(A,t)}{\partial t}\}.\\
&\quad \cdot \frac{\theta y^i(A,t)}{\theta A} dt
\end{aligned} \tag{2.7}
$$

Since the double integral on the left-hand side of (2.6) equals $-\lambda_j(A,B)\frac{\theta y^j(A,B)}{\theta A}$ (the fact that $\lambda_j(A,0) = \lambda_j(0,B) = \lambda_j(0,0) = 0$ follows from separate arguments), we finally obtain an expression for the total derivative $\frac{DJ}{DA}$ where $\frac{D}{DA} \equiv \frac{\partial}{\partial A} + \frac{\theta}{\theta A}$ denotes total derivative with respect to the parameter A:

$$
\begin{aligned}
\frac{DJ}{DA} &= \{\frac{\partial F_1(A,B,\ldots)}{\partial p^i} - \lambda_i(A,B)\}\frac{\theta y^2(A,B)}{\theta A}\\
&+ \int_0^A \{\lambda_i(s,B)\frac{\partial f^j(s,B,\ldots)}{\partial p^i} + \frac{\partial \lambda_i(s,B)}{\partial s}\\
&+ \frac{\partial F_1(s,B,\ldots)}{\partial y^i} - \frac{D}{Ds}\frac{\partial F_1(s,B,\ldots)}{\partial p^i}\\
&+ \frac{\partial F_1(s,B,\ldots)}{\partial q^i}\frac{\theta y_t^i(s,B)}{\theta A}\} ds\\
&+ \int_0^B \{[\lambda_j(A,t)\frac{\partial f^j(A,t,\ldots)}{\partial q^i}\\
&+ \frac{\partial \lambda_i(A,t)}{\partial t} - \frac{\partial F(A,t),\ldots)}{\partial p^i} - \frac{\partial F(A,t,\ldots)}{\partial q^i}]\frac{\theta y^i(A,t)}{\theta A}\\
&+ [\frac{\partial F_2(A,t,\ldots)}{\partial y^i} - \frac{D}{Dt}\frac{\partial F_2(A,t,\ldots)}{\partial p^i}]\frac{Dy^i(A,t)}{DA}\\
&+ \frac{\partial F_2(A,t,\ldots)}{\partial A} + \frac{\partial F_2(A,t,\ldots)}{\partial p^i}\frac{Dy_s^i(A,t)}{DA}\} dt\\
&+ \frac{\partial F_1(A,B,\ldots)}{\partial q^i}\frac{Dy^i(A,B)}{DA} + \frac{\partial F_0(A,B,\ldots)}{\partial A}\\
&+ \frac{\partial F_0(A,B,\ldots)}{\partial y^i}\frac{Dy^i(A,B)}{DA} + \frac{\partial F_0(A,B,\ldots)}{\partial p^i}\frac{Dy_s^i(A,B)}{DA}\\
&+ \frac{\partial F_0(A,B,\ldots)}{\partial q^i}\frac{Dy^i(A,B)}{DA}
\end{aligned} \tag{2.8}
$$

The significant feature of (2.7) is that it involves values of λ and y only on the final backward characteristics through the point (A,B). The values of y and its derivatives, along the final backward characteristics, are evaluated from the constraints (2.1) and substituted into (2.7) to yield boundary conditions for λ at the final boundary. Of course, differentiation of J with respect to B yields another final boundary condition for λ.

3. OPTIMALITY CONDITIONS FOR A NON-RECTANGULAR DOMAIN

We consider the system (1.1) over a domain G on the st-plane bounded by the segments $[0,A]$ and $[0,B]$ on the s- and t-axes, respectively, and by a nonincreasing curve (γ) from point $(A, 0)$ to point $(0, B)$. The curve is continuous and piecewise C^2, and a collection of points $P_0 \equiv *A, 0)$, $P_1, P_2, \dots, P_N, P_{N+1} \equiv (0, B)$ is given on (γ); this collection includes all points of non-smoothness of (γ). For each point on $(\gamma), \vec{n} = (n_1, n_2)$ is the unit outward normal to (γ), and $\vec{\ell} = (-n_2, n_1)$ is the unit tangent vector in the direction of orientation of (γ) from P_0 to P_{N+1}. For a piecewise continuous function ψ on (γ), we set $\Omega_j(\Psi) = \lim_{P \to P_j^+} \Psi(P) - \lim_{P \to P_j^-} \Psi(P)$ for $1 \leq j \leq N$.

We consider the problem of minimizing a functional K given by

$$\begin{aligned} K &= \int\int_G F(s,t,y(s,t),y_s(s,t),y_t(s,t),u(s,t))dA(s,t) \\ &+ \int_{(\gamma)} F_1(s,t,y(s,t),y_\ell(s,t))d\ell(s,t) + \sum_{j=0}^{N+1} F_0(P_j, y(P_j)) \end{aligned} \tag{3.1}$$

where $dA, d\ell$ are the elements of area on G and arc length on (γ), respectively.

For this problem, with sufficiently smooth functions F, F_1, F_0, we have proved that the co-state λ satisfies the same Hamiltonian equations (1.7) and the extremum condition (1.9) holds for an optimal control, but the final boundary conditions for λ are

$$\begin{aligned} R\frac{\partial \lambda_k}{\partial n} &+ 2T\frac{\partial \lambda_k}{\partial \ell} + \lambda_k \frac{\partial T}{\partial \ell} + n_1 \frac{\partial H}{\partial p^k} \\ &+ n_2 \frac{\partial H}{\partial q^k} + \frac{\partial F_1}{\partial y^k} - \frac{D}{D\ell}\frac{\partial F_1}{\partial \eta^k} = 0, \ on\ (\tilde{\gamma}); \\ \frac{\partial F_0(P_j, y(P_j))}{\partial y^k} &+ \Omega_j(T\lambda_k - \frac{\partial F_1}{\partial \eta^k}) = 0 \\ & for\ 1 \leq j \leq N \end{aligned} \tag{3.2}$$

where $\eta^k \equiv \frac{\partial y^k}{\partial \ell}$, $(\tilde{\gamma})$ is the part of (γ) consisting of the open arcs $P_i P_{i+1}, 0 \leq i \leq N, R \equiv n_1, n_2, T \equiv \frac{1}{2}(n_1^2 - n_2^2)$.

Our proof the validity of the Hamiltonian equations, including the conditions (3.2), for the problem of minimizing K, is quite lengthy and it is omitted from the present paper.

REFERENCES

1. A.I. Egorov, Prilk. Matem. Mekh., No. 4, 688-696 (1963).
2. O.V. Vasilyev, V.A. Srochko, V.V. Terletskii, Methods of Optimization and their Applications, Part 2: Optimal Control (in Russian), Nauka, Novosibirsk (1990).
3. O.V. Vasilyev (general editor), Extremum conditions and constructive solution methods in optimization problems for hyperbolic systems (in Russian), Nauka, Novosibirsk (1993).

7th Int. Coll. on Differential Equations, pp. 23-30
D. Bainov (Ed)

On Newton's problem under side constraints

Marino Belloni
Dipartimento di Matematica
Universitá di Parma
I-43100 Parma (Italy)
belloni@prmat.math.unipr.it

Bernd Kawohl
Matematisches Institut
Universität zu Köln
D-50923 Köln (Germany)
kawohl@mi.uni-koeln.de

Abstract: *We give a survey of some results on Newton's problem of minimal resistance. In particular we present our results from* [1], *but there are also new results, e.g. Remark 3.1.*

KEYWORDS: Optimal solutions belonging to restricted classes, Variational methods

1. Introduction

In the late 17th century, I.Newton [7] investigated a model for the resistance of a three dimensional rotational body which moves vertically upwards in direction of its axis of rotation through a rarefied medium. Assume that the fluid consists of small particles, independent from each other and equally distributed with constant mass and velocity. If we suppose that a particle hits the body in a perfectly elastic manner, its momentum can be divided into a tangential part and a normal part. Only the normal part has influence on the resistance.

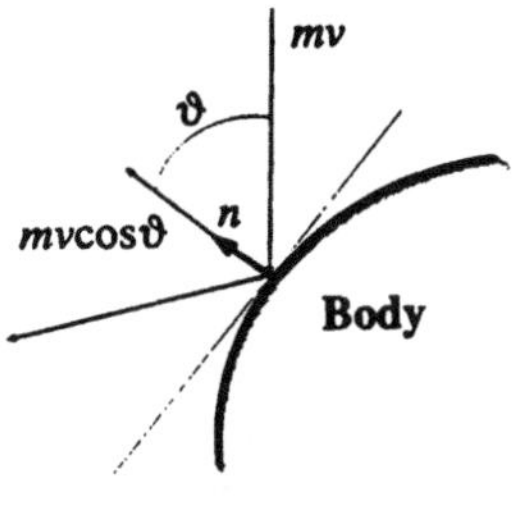

Figure 1: Elastic collision

But the normal part can again be divided into a horizontal part and a vertical part. If mv is the momentum of the hitting particle and θ is the angle between the normal to the surface and the incoming path of the particle, we have that the vertical part is $mv\cos^2\theta$. The horizontal part has no contribution, because we assume rotational symmetry and therefore the total contribution due to the horizontal part is zero (see Figure 1).

If $u(x)$ describes the upper boundary of the rotational body and if the plane $\{z = 0\}$ describes its lower boundary, we have $\cos\theta = \left(\sqrt{1+|Du|^2}\right)^{-1}$. Therefore the total resistance is proportional by to

$$R(u) := \int_\Omega \frac{1}{1+|Du|^2}\,dx. \tag{1}$$

Since u is supposed to be radial, if $\Omega = B(0,R) \subseteq \mathbb{R}^2$, in polar coordinates the functional reduces to

$$R(u) := 2\pi \int_0^R \frac{r}{1+{u_r}^2(r)}\,dr, \tag{2}$$

with $u \geq 0$, $u(R) = 0$ and $u_r = \partial u/\partial r$. The so called Newton's problem is to find a function u wich minimize the resistance. Newton did not specify the class of admissible functions; he integrated the Euler-Lagrange equation for bodies of revolution of prescribed height. His solution is well-known (see e.g. Tonelli [9]): on an initial interval $[0,a]$ it is constant, at a its derivative from the right equals minus one, and between a and 1 the solution u is concave. Finally, at radius 1 it has boundary value zero (see Figure 2).

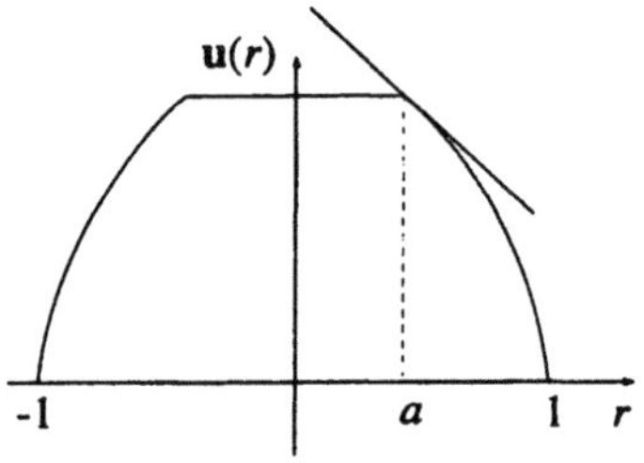

Figure 2: Newton's solution

A more recent discussion of several classes of admissible functions can be found in the article of Buttazzo, Ferone and Kawohl [3]. Notice that the functional R is neither convex nor coercive, so that standard existence proofs fail.

It should be noted also that, when the admissible class is not specified, the solution in general does not exist. Let us consider the Newton problem

$$\min_{v\in X} R(v) \text{ on an admissible set } X \subseteq W^{1,1}(\Omega;\mathbb{R}). \tag{3}$$

Along the sequence $u_n(x) = n\,\mathrm{dist}\,(x,\partial\Omega)$ the functional R converges to its infimum zero, but this infimum is not a minimum (see Figure 3).

This effect can be ruled out by requiring that X is bounded in L^∞.

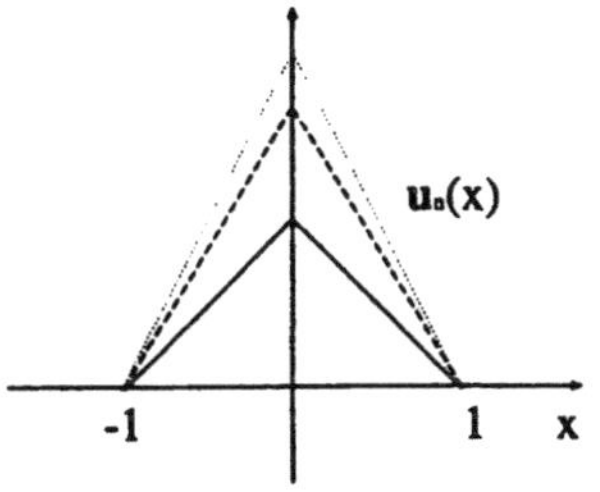

Figure 3: Unbounded sequence

Boundedness of a minimizing sequence is necessary, but not *sufficient*. In fact, as pointed out by Legendre [6], boundedness cannot prevent oscillatory phenomena. Arguing as before, we get again non-existence of the solution by using the sequence $u_n(x) = M \sin^2(h|x|)$ of bounded functions (see Figure 4).

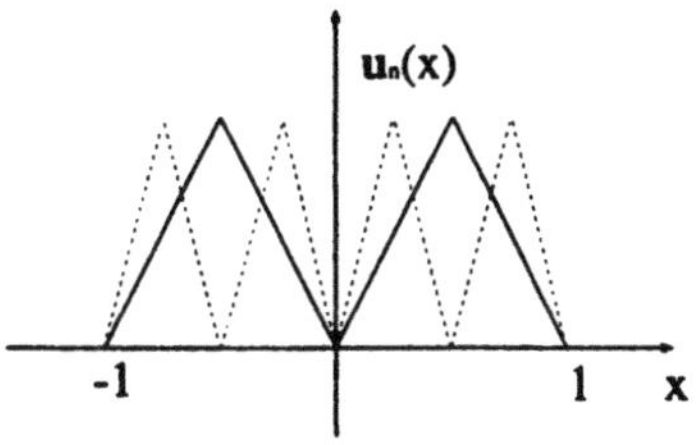

Figure 4: Oscillating sequence

It is now obvious that we have to further restrict the admissible class in order to have existence for (1). A recent result in this direction is the article of Buttazzo, Ferone and Kawohl [3], where some admissible classes of functions are considered. In particular, they show an existence result for the following minimum problem

$$\min\{R(u) \;:\; 0 \le u \le M,\; u \textit{ concave}\}. \tag{4}$$

They also consider constraints different from concavity, in particular a constraint which guarantees single collision between the particles and the body. It must be pointed out that $R(u)$ measures only the resistance due to the first collision, and concavity is sufficient (but not *necessary*) to assure single collision (see [3] for more details). In the following sections we investigate two additional classes of admissible functions, one of which can be found in a paper of Legendre [6]. We consider the case that $\Omega \subset \mathbb{R}^n$ is the unit ball, $n \in \mathbb{N}$ and that $u : \Omega \mapsto \mathbb{R}_0^+$ is a radial function. Of course the physically relevant case is $n = 2$. First we minimize R on the set of functions for which the arclength

$$l = \int_0^1 \sqrt{1 + u_r^2}\, dr \tag{5}$$

is prescribed, and prove existence of this problem in Theorem 2.1. A careful look at Legendre's paper reveals that he sought to minimize R over a slightly different class, namely over those functions for which

$$L = \int_0^1 \sqrt{1 + u_r^2}\ \chi_{\{u_r \neq 0\}}\, dr \tag{6}$$

is prescribed. The existence result for functions in this class is treated in Theorem 2.2. It should be pointed out that L does not measure the length of horizontal segments of u. The two classes of functions which are (loosely) described by (5) and (6) have enough compactness properties to obtain existence of solutions. In

section 3 we further investigate the shape of solutions. First we discuss the case $n = 1$, then we consider the physically relevant case $n = 2$ and we show that the solutions, in case (6), must be concave, bounded and satisfy $|u_r(\rho)| \notin (0, \sqrt{2})$ for a.e. $\rho \in (0, 1)$. By integrating the Euler-Lagrange equation for a special case we recover Legendre's special solution.

2. Existence results

Let us consider Newton's functional (1), where $\Omega \subseteq \mathbb{R}^n$ is the n-dimensional ball of radius 1.

We want to solve the minimization problem

$$\min_{v \in Y_l} R(v) \tag{7}$$

where the set Y_l is defined as

$$Y_l = \left\{ u \in W^{1,1}(\Omega) \mid u \geq 0,\ u \text{ radial},\ \int_0^1 \sqrt{1 + u_r^2(r)}\, dr = l \right\}. \tag{8}$$

Theorem 2.1: *(see [1]) For every $l \geq 1$ there exists a solution $u \in Y_l$ to the minimum problem* (7). *Moreover $u \in Y_l \cap C_l$, where*

$$C_l = \{u \in W^{1,1}(\Omega) \mid u \geq 0,\ u \textit{ radial},\ :\ \|u\|_\infty \leq l\ ,\ u \textit{ concave}\}.$$

We can also consider another arc-length constraint different from the previous one introduced in (8), considered first by Legendre [6]:

$$X_L = \left\{ u \in W^{1,1}(\Omega) \mid u \geq 0,\ u \text{ radial}, \int_0^1 \sqrt{1 + u_r^2}\ \chi_{\{u_r \neq 0\}}\, dr = L \right\} \tag{9}$$

and the related minimization problem

$$\min_{v \in X_L} R(v) \ . \tag{10}$$

Theorem 2.2: *For every $L > 0$ there exists a solution $u \in X_L$ to the minimum problem* (10). *Moreover $u \in C_L$, where C_L is defined in Theorem 2.1.*

Two ideas are essential for the proofs of Theorem 2.1 and 2.2. The first is a rearrangement of u_r which makes minimizers concave. The second is a compactness lemma for sequences in C_l or C_L, see [8, p.248].

3. More on the solutions

Remark 3.1: There is at least one situation in which we can calculate a solution of (7). If $n = 1$ set $g(r) = \sqrt{1 + u_r^2}$ and note that the average of g on $(0, 1)$ is l. Moreover $R(u) = \in_0^1 g^{-2}(r)\, dr$ is a convex function of g. Upon taking averages of g the functional R will decrease. Therefore g must be constant if u is a minimizers. Consequently cones of slopes $\sqrt{l^2 - 1}$ are solutions of (7) if $n = 1$. This observation is due to [4]. For $n = 2$ the trick fails because (2) contains a factor r. In fact, for $n = 2$, conical slopes are no longer optimal, because there are frustum cones in $Y_l \cap C_l$ with smaller resistance, see [2].

For the remainder of the paper we shall consider problem (10). Let us consider the case $n = 1$. In this case the Euler equation becomes autonomous and reads

$$-\frac{u_r}{(1 + u_r^2)^2} + \lambda \frac{u_r}{\sqrt{1 + u_r^2}} = C \tag{11}$$

and, as we shall see below, any solution of problem (10) must be a piecewise linear function, e.g. of trapezoidal shape. Now it is easily seen, that one can also have nonconcave solutions, see Figure 5.

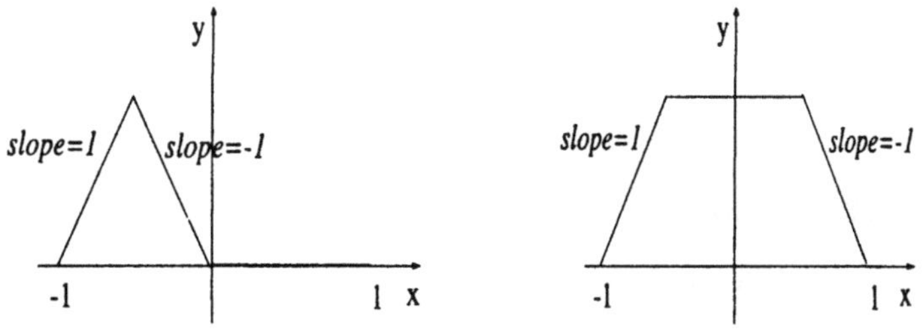

Figure 5: Two different solutions of Problem (10) for $n = 1$

First of all, we have that the multiplier λ is positive. This follows by a classical argument (see Hestenes [5, p.36]), replacing the Legendre arc-lenght constraint by

$$\int_0^1 \sqrt{1+u_r^2}\,\chi_{\{u_r\neq 0\}}\,dr \leq L.$$

The left hand side of (11) is decreasing in u_r for $u_r \leq -1$. Therefore there is at most one (constant) solution $u_r = -K$ of (11), with $K \geq 1$, for every choice of $\lambda > 0$ and C. Hence every concave solution of the minimum problem (10) must be a trapezoid. But more can be said.

Theorem 3.1: *(see [1]) Let $n = 1$ and u be a concave solution of the minimum problem (10). Then we have*

$$|u_r(r)| \notin (0, \sqrt{2}) \quad \text{for } a.e.\ r \in (0,1). \tag{12}$$

Moreover, if $L < \sqrt{3}$, then

$$u(r) = \begin{cases} L\sqrt{\frac{2}{3}} & \text{if } r \in [0, 1-\frac{L}{\sqrt{3}}) \\ -\sqrt{2}(1-r) & \text{if } r \in [1-\frac{L}{\sqrt{3}}, 1] \end{cases}$$

and if $L \geq \sqrt{3}$, then $u(r) = \sqrt{L^2-1}(1-r)$ for every $r \in [0,1]$.

For the moment, we only know that there exists "at least" one concave solution of (10). In fact, if $n = 1$, we have given examples of minimizers that are nonconcave. But if we restrict our attention to higher dimensions we have the following result.

Corollary 3.2: *(see [1]) For $n \geq 2$ every solution of (10) must be concave.*

This follows from an ispection of an inequality due to Riesz.

When $n > 1$, a result analogous to Theorem 3.1 holds as follows.

Theorem 3.3: *Let u be a radial solution of the minimum problem (10), with $n > 1$, and suppose that u is piecewise of class C^1.*

Then u must be concave and we have

$$|\nabla u(x)| \notin (0, \sqrt{2}) \quad \text{for } a.e.\ x \in \Omega$$

and $u_r \leq -\sqrt{2}$ at the free boundary.

Can we compute a solution of (10) if $n = 2$? Now the Euler-Lagrange equation reads

$$-r^{n-1}\frac{u_r}{(1+u_r^2)^2} + \lambda\frac{u_r}{\sqrt{1+u_r^2}} = c\ , \tag{13}$$

where λ is a Lagrange multiplier and c is an undetermined constant. Note that this equation holds (in the weak sense) for every r for which $u_r(r) \neq 0$, because then we can calculate the first variation of

$$\int_0^1 \frac{r}{1+u_r^2}\, dr + \lambda \left(\int_0^1 \sqrt{1+u_r^2}\ \chi_{\{u_r \neq 0\}}\, dr - L \right).$$

Since u is concave, equation (13) holds on an interval from a to 1, where a may be zero or positive. If (3.10) holds in 0 and u is of class C^1, then $c = 0$. This might serve as a justification to choose $c = 0$. In fact, as noted by Legendre in the case $n = 2$, choosing $c = 0$ simplifies further calculations, because then the equation becomes

$$\frac{r u_r}{(1+u_r^2)^2} = \lambda \frac{u_r}{\sqrt{1+u_r^2}}.$$

At this point Legendre stopped his calculation. We continue by noting that $\lambda > 0$ and

$$u_r = -\sqrt{\left(\frac{r}{\lambda}\right)^{2/3} - 1}. \tag{14}$$

From (14) we have

$$u(r) - u(1) = \int_r^1 \sqrt{\left(\frac{r}{\lambda}\right)^{2/3} - 1\, dr}$$

and noting that $u(1) = 0$, we obtain

$$u(r) = 12\lambda \left[2\,(2z-1)\left(z^2 - z\right)^{1/2} - \log\left(2\sqrt{z^2 - z} + 2z - 1\right) \right]_{(r/\lambda)^{2/3}}^{(1/\lambda)^{2/3}} \tag{15}$$

and obviously the solution depends on λ. Since by Theorem 3.3 $u_r(a) = -\sqrt{2}$, from (14) we have that $a = \lambda\sqrt{27}$. But from the arc-length constraint we obtain

$$L = \int_a^1 \sqrt{1 + u_r^2(r)}\, dr = \frac{3}{4}\left(\frac{1}{\lambda}\right)^{1/3}\left(1 - a^{4/3}\right)$$

so

$$L = \frac{3}{4}\left(\lambda^{-1/3} - 9\lambda\right).$$

Since L is a monotone decreasing function of λ, it is invertible as $\lambda = \lambda(L)$. Substituting in (15) we have the explicit form of Legendre's solutions. So, if for instance $L = 2$, then $\lambda = 1/27$ and $a = 1/\sqrt{27}$ (see Figure 6).

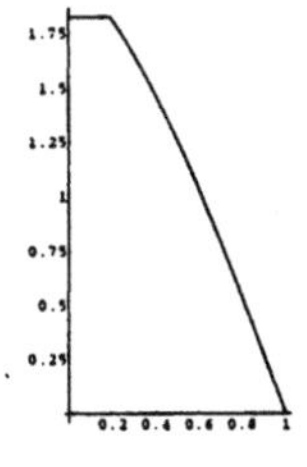

Figure 6: Legendre's solution in the case $L = 2$

Remark 3.4: It is interesting to note, that for every $L < \infty$ there exists a solution to (10) given by (15) with a nonempty flat region. This is in contrast to the case $n = 1$.

References

[1] M.Belloni & B.Kawohl, Forum Math., *to appear.*
[2] M.Belloni & B.Kawohl, *in preparation.*
[3] Buttazzo,G., V.Ferone & B.Kawohl, Mathem. Nachrichten, 173, pp.71–89 (1995).
[4] V.Ferone & B.Kawohl, *in preparation.*
[5] Hestenes, M.R., *Calculus of variations and optimal control theory,* John Wiley & Sons, Inc., New York - London - Sydney (1966).
[6] Legendre,A.M., Mémoires de L'Académie Royale des Sciences, Année MDC-CLXXVI - Paris, pp. 7-37 (1788).
[7] Newton,I., *Philosophiae Naturalis Principia Mathematica* (1686).
[8] R.T.Rockafellar, *Convex analysis,* Princeton Univ. Press, Princeton (1970).
[9] Tonelli,L. *Fondamenti di Calcolo delle Variazioni*, Zanichelli, Bologna (1923).

7th Int. Coll. on Differential Equations, pp. 31-38
D. Bainov (Ed)

Parameters estimation of some electro-elasticity models

NIKOLAI BOTKIN

Institute of Applied Mathematics and Statistics of Technical University of Munich, Dachauer str. 9a, D-80335 Munich, Germany

ABSTRACT

An algorithm for the estimation of unknown parameters of a nonlinear electro-elasticity model is proposed. Distributions of the displacement observed with an error are input data of the algorithm. Finite element approximations of the input data and of the model are used. The algorithm implements the idea of the direct minimization of the residual of equations [1, 2, 3, 4]. The paper is illustrated by computer simulations.

Keywords: *Nonlinear thin plates, Piezoelectric actuators, Estimation of parameters.*

1. PROBLEM SETTING

We consider a nonlinear system describing oscillations of thin plates excited by patches made of piezoelectric ceramics

$$\begin{gathered}\rho\xi_{tt} + \operatorname{div}\nabla\left(\gamma\Delta\xi\right) - \mu\Delta\xi_t - \frac{\partial}{\partial x_\alpha}\left(\tau_{\alpha\beta}\xi_{x_\beta}\right) = v(t)K\Delta\left(I_{S_P}\right) - v(t)G\frac{\partial}{\partial x_\alpha}\left(\xi_{x_\alpha}I_{S_P}\right),\\ \rho u_{\alpha tt} - \frac{\partial}{\partial x_\beta}\tau_{\alpha\beta} = -v(t)G\frac{\partial}{\partial x_\alpha}I_{S_P}, \quad \alpha = 1,2.\end{gathered} \tag{1}$$

Here ξ and u_α, $\alpha = 1, 2$, are vertical and longitudinal displacements, Δ is the Laplace operator, ∇ is the operator of the gradient, $v(t)$ is the voltage applied to the piezoelectric patch, I_{S_P} is the indicator function of S_P. The summation over repeating indices is assumed in (1). The coefficients μ, K, and G are constant; ρ, γ, and $\tau_{\alpha\beta}$ are discontinuous functions defined as follows:

$$\rho = \begin{cases} \rho_P, & (x_1, x_1) \in S_P, \\ \rho_B, & (x_1, x_2) \in S_B, \end{cases} \qquad \gamma = \begin{cases} \gamma_P, & (x_1, x_2) \in S_P, \\ \gamma_B, & (x_1, x_2) \in S_B, \end{cases} \qquad \tau_{\alpha\beta} = \begin{cases} \tau^P_{\alpha\beta}, & (x_1, x_2) \in S_P, \\ \tau^B_{\alpha\beta}, & (x_1, x_2) \in S_B, \end{cases}$$

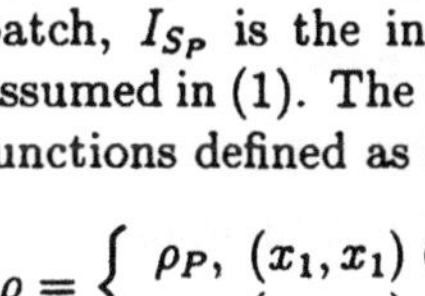

$S = S_P \cup S_B$

$$\tau^P_{11} = \frac{R_P}{1-\sigma_P^2}(d_{11} + \sigma_P d_{22}), \quad \tau^B_{11} = \frac{R_B}{1-\sigma_B^2}(d_{11} + \sigma_B d_{22}),$$

$$\tau^P_{22} = \frac{R_P}{1-\sigma_P^2}(d_{22} + \sigma_P d_{11}), \quad \tau^B_{22} = \frac{R_B}{1-\sigma_B^2}(d_{22} + \sigma_B d_{11}),$$

$$\tau^P_{12} = \frac{R_P}{1+\sigma_P}d_{12}. \qquad \tau^B_{12} = \frac{R_B}{1+\sigma_B}d_{12},$$

$$d_{\alpha\beta} = \frac{1}{2}\left(\frac{\partial u_\alpha}{\partial x_\beta} + \frac{\partial u_\beta}{\partial x_\alpha} + \frac{\partial\xi}{\partial x_\alpha}\frac{\partial\xi}{\partial x_\beta}\right), \quad \alpha, \beta = 1, 2.$$

Here $d_{\alpha\beta}$ are components of the plain strain tensor; $\tau^P_{\alpha\beta}$ and $\tau^B_{\alpha\beta}$ are components of the plain stress tensors; R_P and R_B are the elastic modules; σ_P and σ_B are the Poisson numbers. Indices P and B point to the piezoelectric and basis materials, respectively.

The system (1) is derived using the technique described in [5] and the theory of electro-elasticity developed in [6, 7, 8]. Note that the system (1) should be supplied with some interface conditions that hold on the boundary between S_P and S_B because of the integration by parts when deriving (1) from a weak formulation. We do not write down these conditions because we shall go back to the weak formulation when introducing a residual whose minimization will yield the estimation of parameters.

The following smoothness of solutions of (1) is expected (see [9, 10]):

$$\xi \in L_2(0,T; H_0^2(S)) \cap H^1(0,T; H_0^1(S)),$$

$$u_i \in L_2(0,T; H_0^1(S)) \cap H^1(0,T; L_2(S)), \ i = 1,2.$$

The internal smoothness is infinite due to the damping term $\mu\Delta\xi_t$:

$$\xi,\ u_1,\ u_2 \in C^\infty(S' \times (0,T)), \quad \forall\, S' \subset \operatorname{int}(S_P) \cup \operatorname{int}(S_B).$$

Denote

$$a_1^p = \frac{R_P}{1-\sigma_P^2},\ a_2^p = \frac{\sigma_P R_P}{1-\sigma_P^2},\ a_3^p = \frac{R_P}{1+\sigma_P},\quad a_1^b = \frac{R_B}{1-\sigma_B^2},\ a_2^b = \frac{\sigma_B R_B}{1-\sigma_B^2},\ a_3^b = \frac{R_B}{1+\sigma_B}.$$

We estimate the parameters

$$\gamma_P,\ \gamma_B,\ \mu,\ K,\ G,\ a_1^p,\ a_2^p,\ a_3^p,\ a_1^b,\ a_2^b,\ a_3^b$$

on the basis of a solution observed with an error η at points (x_1^i, x_2^i), $i = 1...N$, of a grid on some subregion $S' \subset \operatorname{int}(S_P) \cup \operatorname{int}(S_B)$ at times t^j, $j = 0...M$. We set $t^0 = 0$ and $t^M = T$ for definiteness.

2. Estimation procedure

Let us introduce vector notations:

$$\begin{aligned}
&\pi = (\gamma_P,\ \gamma_B,\ \mu,\ K,\ G,\ a_1^p,\ a_2^p,\ a_3^p,\ a_1^b,\ a_2^b,\ a_3^b) && \text{is the vector of parameters,}\\
&u = (\xi,\ u_1,\ u_2) && \text{is the solution vector,}\\
&\psi = (\varphi,\ \psi_1,\ \psi_2) && \text{is the vector of test functions,}\\
&V = H_0^2(S') \times H_0^1(S') \times H_0^1(S') && \text{is the space of test functions.}
\end{aligned}$$

An a priori constraint $\pi \in \Pi \in \operatorname{comp}(R^{11})$ is assumed. Let us define the residual:

$$\ell(A(u),\ L\psi,\ \pi, S') =$$

$$\iint\limits_{S_P \cap S'} \rho_P \xi_{tt}\varphi + \gamma_P \Delta\xi\Delta\varphi + \mu\nabla\xi_t\nabla\varphi + \tau^P_{\alpha\beta}\xi_{x_\beta}\varphi_{x_\alpha} - v(t)K\Delta\varphi + v(t)G\xi_{x_\alpha}\varphi_{x_\alpha} + \tag{2}$$

$$\iint\limits_{S_B \cap S'} \rho_B \xi_{tt}\varphi + \gamma_B \Delta\xi\Delta\varphi + \mu\nabla\xi_t\nabla\varphi + \tau^B_{\alpha\beta}\xi_{x_\beta}\varphi_{x_\alpha} +$$

$$\iint\limits_{S_P \cap S'} \rho_P u_{1tt}\psi_1 + \tau^P_{1\beta}\psi_{1x_\beta} + v(t)G\psi_{1x_1} + \iint\limits_{S_B \cap S'} \rho_B u_{1tt}\psi_1 + \tau^B_{1\beta}\psi_{1x_\beta} +$$

$$\iint\limits_{S_P \cap S'} \rho_P u_{2tt}\psi_2 + \tau^P_{2\beta}\psi_{2x_\beta} + v(t)G\psi_{2x_2} + \iint\limits_{S_B \cap S'} \rho_B u_{2tt}\psi_2 + \tau^B_{2\beta}\psi_{2x_\beta}.$$

This expression is precisely the left-hand-side of the weak formulation of (1) restricted to the region S'. The test functions appear linearly here, which is reflected by writing $L\psi$, where L denotes the corresponding linear operator. The solution u is nonlinearly involved into the definition of ℓ, which is depicted by the notation $A(u)$, where A is the corresponding nonlinear operator.

Obviously, it holds

$$\int_0^T \max_{\psi \in V^1} [\ell(A(u), L\psi, \pi^0, S')]^2 dt = 0, \qquad V^1 = \{\psi \in V : \|\psi\|_V \leq 1\}$$

for the set π^0 of parameters compatible with the exact solution u. Therefore,

$$\pi^0 = \arg\min_{\pi \in \Pi} \int_0^T \max_{\psi \in V^1} [\ell(A(u), L\psi, \pi, S')]^2 dt. \tag{3}$$

We have observations $u^\eta = (\xi^\eta, u_1^\eta, u_2^\eta)$ and two parameters more:

$$h = \operatorname{diam}\{(x_1^i, x_2^i),\ i = 1 \ldots N\}, \quad \tau = \operatorname{diam}\{t^j,\ j = 1 \ldots M\}.$$

Denote $\varepsilon = (\eta, \tau, h)$ and agree that

$$\varepsilon \to 0 \Longleftrightarrow \eta,\ \tau,\ h,\ \eta/h^2,\ \eta/\tau^2 \to 0.$$

One can construct (on the basis of u^η) a piecewise constant with respect to t function w^ε such that $w^\varepsilon \to A(u)$ in $[L_2((0,T) \times S')]^q$ (q is an appropriate integer) as $\varepsilon \to 0$. Some analysis shows that one should be able to approximate functions of the form

$$\xi_{tt},\ \xi_{x_\alpha t},\ \Delta\xi,\ \xi_{x_\alpha}\xi_{x_\beta},\ \xi_{x_\alpha}\xi_{x_\beta}\xi_{x_\gamma},\ u_{\alpha tt},\ u_{\alpha x_\beta},\ u_{\alpha x_\beta}\xi_{x_\gamma},\quad \alpha, \beta, \gamma = 1, 2, \tag{4}$$

to construct the desired approximation w^ε. Approximations of the functions in (4) are constructed using finite differences with respect to the time variable and finite elements with respect to the spatial variables. A space V_h of finite elements with the property $V_h \subset V = H_0^2(S') \times H_0^1(S') \times H_0^1(S')$ is proposed. The interpolation operator corresponding to V_h has three components because V is the product of three spaces. The first component is similar to the interpolation operator related to the Bogner-Fox-Schmit finite elements (see [11]). The difference is that the modified operator uses values of functions without any use of their derivatives. The other two components are interpolation operators defined with the use of linear finite elements.

Substituting w^ε in place of $A(u)$ and replacing V^1 by the unit ball $V_h^1 \subset V_h$ in (3), we obtain the following estimation of π^0:

$$\pi^\varepsilon = \arg\min_{\pi \in \Pi} \int_0^T \max_{\psi \in V_h^1} [\ell(w^\varepsilon, L\psi, \pi, S')]^2 dt. \tag{5}$$

The maximum in (5) can be computed with the use of an auxiliary function $\psi^{t,\varepsilon} \in V_h$ defined by the relation

$$(\psi^{t,\varepsilon}, \psi)_{V_h} = \ell(w^\varepsilon, L\psi, \pi, S'), \ \forall \psi \in V_h,$$

so that we have

$$\max_{\psi \in V_h^1} [\ell(w^\varepsilon, L\psi, \pi, S')]^2 = \ell(w^\varepsilon, L\psi^{t,\varepsilon}, \pi, S'), \quad \forall t, \varepsilon.$$

Thus, we arrive at the following estimation:

$$\pi^\varepsilon = \arg\min_{\pi \in \Pi} \int_0^T \ell(w^\varepsilon, L\psi^{t,\varepsilon}, \pi, S')dt. \tag{6}$$

Because of the linearity of ℓ with respect to π (the auxiliary function depends linearly on π, too), the integral in (6) is a quadratic form:

$$\int_0^T \ell(w^\varepsilon, L\psi^{t,\varepsilon}, \pi, S')dt = \pi' Q_\varepsilon \pi - 2q_\varepsilon' \pi + d_\varepsilon =: G^\varepsilon_{S'}(\pi). \tag{7}$$

Here the matrix Q_ε has the following structure:

$$Q_\varepsilon = \int_0^T R_\varepsilon'(t) D_h^{-1} R_\varepsilon(t) dt,$$

where R_ε is some matrix and D_h^{-1} is some positive definite matrix. Let

$$\Pi^0 = \{\pi^0 \in \Pi : \pi^0 \text{ is compatible with the exact solution } u\}.$$

We assume in addition that S' depends on h and may shrink with h.

Suppose that the function $G^\varepsilon_{S'_h}(\pi)$ defined by (7) has the following properties:

G1. There is a unique minimizing element $\pi^\varepsilon \in \Pi$ of the function $G^\varepsilon_{S'_h}(\cdot)$.

G2. There is $\nu > 0$ and a positive function $X(S'_h, \varepsilon)$ such that, for any $\pi \in \Pi$,

$$G^\varepsilon_{S'_h}(\pi) - G^\varepsilon_{S'_h}(\pi^\varepsilon) \geq X(S'_h, \varepsilon)\|\pi - \pi^\varepsilon\|^\nu.$$

One can prove the following property:

G3. There is a function $Z(\varepsilon)$ such that $Z(\varepsilon) \to 0$ as $\varepsilon \to 0$ and

$$\max_{\pi \in \Pi^0} |G^\varepsilon_{S'_h}(\pi) - G^\varepsilon_{S'_h}(\pi^\varepsilon)| \leq Z(\varepsilon).$$

Theorem. If $Z(\varepsilon)/X(S'_h, \varepsilon) \to 0$ as $\varepsilon \to 0$, then Π^0 consists of a unique element π^0 and $\pi^\varepsilon \to \pi^0$ as $\varepsilon \to 0$.

Remark. Properties G1, G2 express the positive definiteness of the matrix Q_ε (this matrix depends on the domain S'_h). As a rule, the function X decreases with S'_h. Therefore, if S'_h decreases not too quick with h, one can expect the conditions of the Theorem are satisfied.

3. Simulation

The simulation was done for the following data

$$\gamma_P = 5,\ \gamma_B = 3,\ \mu = 0.1,\ K = 5,\ G = 5,\ v(t) = 5\sin(5t),$$

$$a_1^p = 2.38,\ a_2^p = 0.714,\ a_3^p = 0.952,\ a_1^b = 1.041,\ a_2^b = 0.416,\ a_3^b = 0.208.$$

The results are shown in Figures 1–3. The horizontal lines show the nominal values. The horizontal axes stand for the duration T of the observation. The error η was equal to 0, 0.00001, and 0.0001 for the results shown in Figures 1,2, and 3, respectively.

It should be noted that the estimation of K is not possible if $S' \cap \partial S_P = \emptyset$ because the coefficient $\int\int_{S_P \cap S'} v(t)\Delta\varphi$ preceding K in (2) is identically equal to zero for any φ with $\operatorname{supp}(\varphi) \subset S_P \cap S'$. To estimate K, a region S' with $S' \cap \partial S_P \neq \emptyset$ was used. Theoretically, this should not work because of the insufficient smoothness of solutions of (1) in such a region: the convergence $w^\varepsilon \to A(u)$ is not guaranteed. Nevertheless, the simulation shows that the procedure still works.

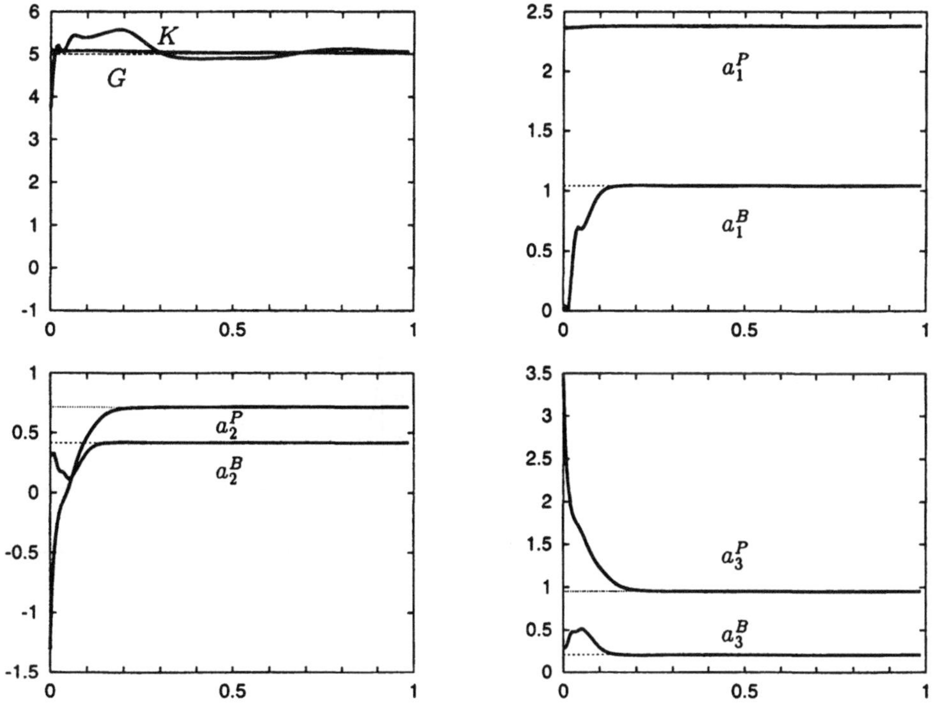

Figure 1. Error $\eta = 0$.

Here γ_P, γ_B, and μ are not shown because they are exactly equal to their nominal values.

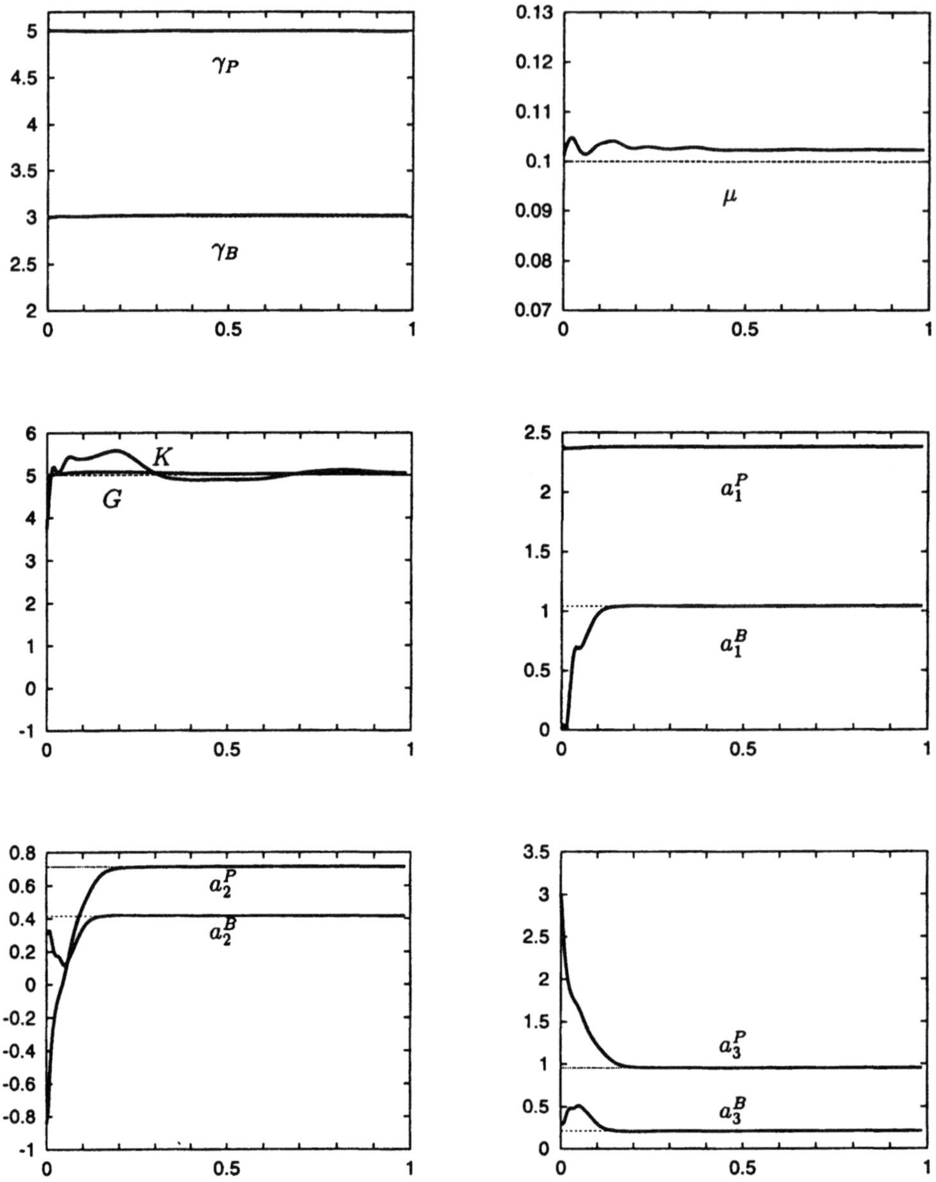

Figure 2. Error $\eta = 0.00001$.

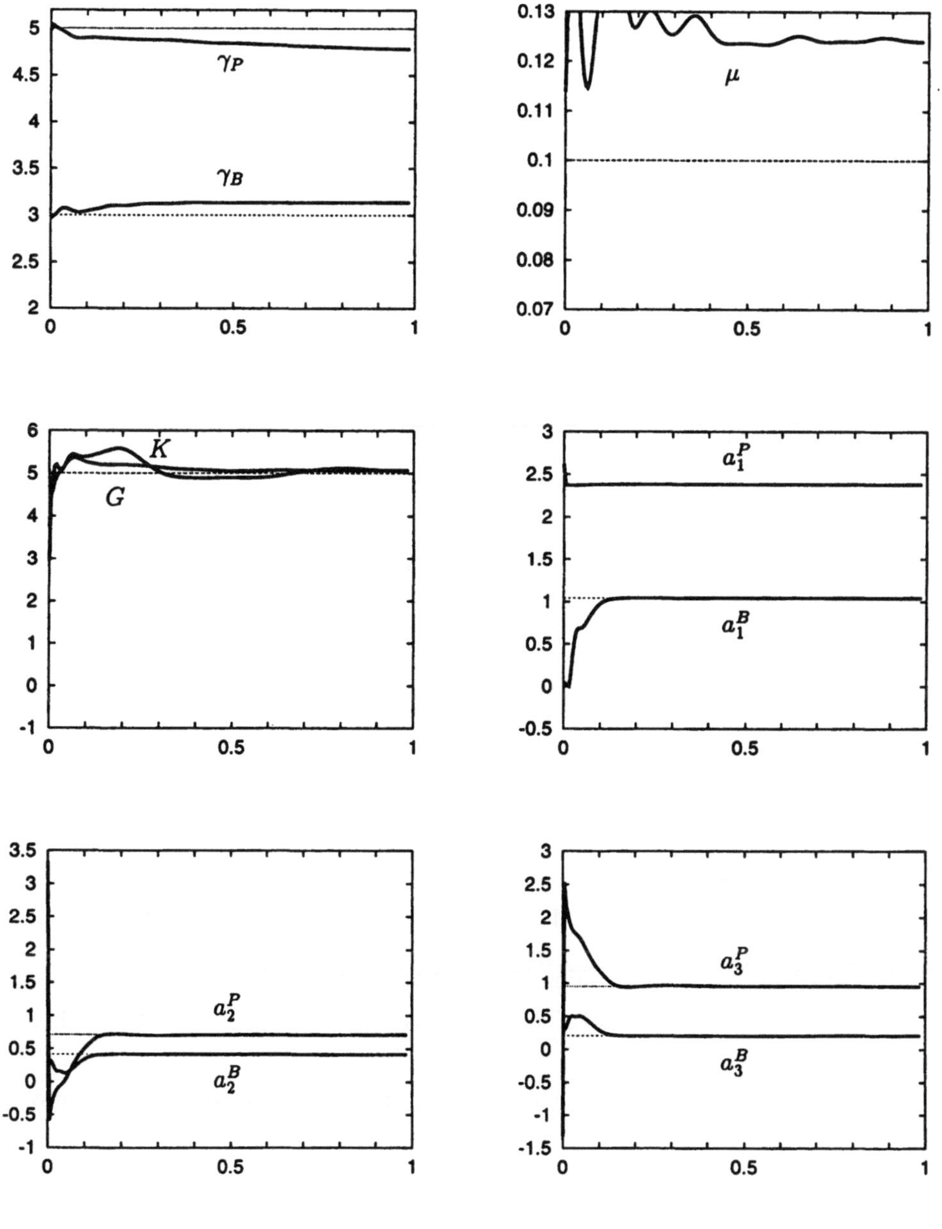

Figure 3. Error $\eta = 0.0001$.

CONCLUSION

The algorithm proposed can be extended to other systems of partial differential equations practically independently from their complexity. Two requirements must be satisfied: Solutions must be sufficiently smooth for them to be approximated along with all their derivatives involved in the weak formulation of systems. Observations of solutions must be done with sufficiently high precision because the algorithm is very sensitive to errors in the input data.

ACKNOWLEDGEMENTS

The author thanks Prof. Dr. K.-H. Hoffmann for the setting this problem and for discussions throughout the work.

REFERENCES

1. A.V.Kryazhimskii and Yu.S.Osipov, Proc. of the Steklov Institute of Mathematics, (2), 143–164 (1990).
2. H.T.Banks and K.Kunisch, Estimation Techniques for Distributed Parameter Systems, Birkhäuser, Boston (1989).
3. K.-H.Hoffmann and J.Sprekels, Numer. Funct. Anal. and Optimiz., 7(2 & 3), 157-178, (1984/1985).
4. N.D.Botkin, Numer. Funct. Anal. and Optimiz., 16(5 & 6), 583–599 (1995).
5. L.D.Landau and E.M.Lifschitz, Elastizitätstheorie, Akademie - Verlag, Berlin (1975).
6. G.A.Maugin, Continuum mechanics of electromagnetic solids, North-Holland series in Applied Mathematics and Mechanics, 33,J.D.Achenbach, B.Budiansky, W.T.Koiter, H.A.Lauwerier, and L.Van Wijngaarden, (Eds.), North-Holland, Amsterdam-New York-Oxford-Tokyo (1987).
7. H.F.Tiersten, Int. J. Engng Sci., (9), 587–604 (1971).
8. J.Zelenka, Piezoelectric Resonators and their Applications, Studies in Electrical and Electronic Engineering, 24, Elsevier, Amsterdam-Oxford-New York-Tokyo (1986).
9. J-P.Puel and M.Tucsnak, SIAM J. Control and Optimiz., 33(1), 255–273 (1995).
10. J-P.Puel and M.Tucsnak, Appl. Math. Optimiz., 34, 139–160 (1996).
11. P.G.Ciarlet, The finite element method for elliptic problems, North-Holland, Amsterdam (1978).
12. V.Ya.Tikhonov and A.N.Arsenin, Solution of ill-posed problems, Winston-Wiley, New York (1977).

7th Int. Coll. on Differential Equations, pp. 39-48
D. Bainov (Ed)

A Nonlinear Suspension Bridge Equation

Q-Heung Choi[1] and Tacksun Jung[2]

Department of Mathematics, Inha University, Incheon 402-751, Korea

Department of Mathematics, Kunsan National University, Kunsan 573-360, Korea

Abstract : Let L be the differential operator, $Lu = u_{tt}+u_{xxxx}$. We consider a nonlinear suspension bridge equation $Lu + bu^+ = f$ in H, where H is the Hilbert space spanned by eigenfunctions of L. We reveal a relation between multiplicity of solutions and source terms in the nonlinear suspension bridge equation.

Keywords : Eigenvalue, Palais-Smale condition, critical point theory, degree theory.

0. Introduction

In this paper we investigate a relation between the multiplicity of solutions and source terms in a nonlinear suspension bridge equation in an interval $\left(-\frac{\pi}{2}, \frac{\pi}{2}\right)$, with nonconstant load

$$u_{tt} + u_{xxxx} + bu^+ = f(x) \quad \text{in} \quad (-\frac{\pi}{2}, \frac{\pi}{2}) \times R, \tag{0.1}$$

$$u(\pm\frac{\pi}{2}, t) = u_{xx}(\pm\frac{\pi}{2}, t) = 0, \tag{0.2}$$

$$u \textit{ is } \pi-\textit{periodic in } t \textit{ and even in } x \textit{ and } t, \tag{0.3}$$

where the nonlinearity $-(bu^+)$ crosses an eigenvalue λ_{10}. This equation represents a bending beam supported by cables under a load f. The constant b represents the restoring force if the cables stretch. The nonlinearity u^+ models the fact that cables resist expansion but do not resist compression.

Let L be the differential operator, $Lu = u_{tt} + u_{xxxx}$. Then the eigenvalue problem for $u(x,t)$

$$Lu = \lambda u \quad \text{in} \quad (-\frac{\pi}{2}, \frac{\pi}{2}) \times R$$

---

[0]1991 AMS Subject Classification: 35B10, 35Q40

[1]Research supported in part by GARC-KOSEF and Inha University Reserach Fund.

[2]Research supported in part by BSRI Program, Ministry of Education BSRI-96-1436.

with (0.2) and (0.3) has infinitely many eigenvalues

$$\lambda_{mn} = (2n+1)^4 - 4m^2 \qquad (m, n = 0, 1, 2, \cdots)$$

and corresponding eigenfunctions $\phi_{mn}(m, n \geq 0)$ given by

$$\begin{aligned} \phi_{0n} &= \frac{\sqrt{2}}{\pi}\cos(2n+1)x \quad \text{for} \quad n \geq 0, \\ \phi_{mn} &= \frac{2}{\pi}\cos 2mt \cos(2n+1)x \quad \text{for} \quad m \geq 1, n \geq 0. \end{aligned}$$

We note that all eigenvalues in the interval $(-19, 45)$ are given by

$$\lambda_{20} = -15 < \lambda_{10} = -3 < \lambda_{00} = 1 < \lambda_{41} = 17.$$

Let Q be the square $\left[-\frac{\pi}{2}, \frac{\pi}{2}\right] \times \left[-\frac{\pi}{2}, \frac{\pi}{2}\right]$ and H the Hilbert space defined by

$$H = \left\{u \in L^2(Q): \quad u \text{ is even in } x \text{ and } t \quad \right\}.$$

Then the set of eigenfunctions $\{\phi_{mn}\}$ is an orthonormal base in H. Hence equation (0.1) with (0.2) and (0.3) is equivalent to

$$Lu + bu^+ = f \quad \text{in} \quad H. \tag{0.4}$$

In this paper, we suppose that $3 < b < 15$ and f is generated by ϕ_{00} and ϕ_{10}. Our goal is to reveal a relation between multiplicity of solutions and source terms in equation (0.4) when f belongs to the two dimensional subspace space of H spanned by ϕ_{00} and ϕ_{10}.

1. A Variational Reduction

In this section, we suppose $3 < b < 15$. Under this assumption, we have a concern with the multiplicity of solutions of a nonlinear suspension bridge equation

$$Lu + bu^+ = f \quad \text{in} \quad H. \tag{1.1}$$

Here we suppose that f is generated by two eigenfunctions ϕ_{00} and ϕ_{10}, that is, $f = s_1\phi_{00} + s_2\phi_{10} (s_1, s_2 \in R)$.

To study equation (1.1), we use the contraction mapping theorem to reduce the problem from an infinite dimensional one in H to a finite dimensional one.

Let V be the two dimensional subspace of H spanned by $\{\phi_{00}, \phi_{10}\}$ and W be the orthogonal complement of V in H. Let P be an orthogonal projection H onto V. Then every element $u \in H$ is expressed by

$$u = v + w,$$

where $v = Pu$, $w = (I - P)u$. Hence equation (1.1) is equivalent to a system

$$Lw + (I - P)(b(v + w)^+) = 0, \tag{1.2}$$

$$Lv + P(b(v + w)^+) = s_1\phi_{00} + s_2\phi_{10}. \tag{1.3}$$

Here we look on (1.2) and (1.3) as a system of two equations in the two unknowns v and w.

LEMMA 1.1. *[4] For fixed $v \in V$, (1.2) has a unique solution $w = \theta(v)$. Furthermore, $\theta(v)$ is Lipschitz continuous (with respect to the L^2 norm) in terms of v.*

By Lemma 1.1, the study of the multiplicity of solutions of (1.1) is reduced to the study of the multiplicity of solutions of an equivalent problem

$$Lv + P[(b(v + \theta(v))^+] = s_1\phi_{00} + s_2\phi_{10} \tag{1.5}$$

defined on the two dimensional subspace V spanned by $\{\phi_{00}, \phi_{10}\}$.

While one feels instinctively that (1.5) ought to be easier to solve, there is the disadvantage of an implicitly defined term $\theta(v)$ in equation (1.5). However, in our case, it turns out that we know $\theta(v)$ for special v's.

If $v \geq 0$ or $v \leq 0$, then $\theta(v) \equiv 0$. For example, let us take $v \geq 0$ and $\theta(v) = 0$. Then equation (1.2) reduces to

$$L0 + (I - P)(bv^+) = 0$$

which is satisfied because $v^+ = v$ and $(I - P)v = 0$, since $v \in V$.

Since the subspace V is spanned by $\{\phi_{00}, \phi_{10}\}$, there exists a cone C_1 defined by

$$C_1 = \{v = c_1\phi_{00} + c_2\phi_{10} \mid c_1 \geq 0, |c_2| \leq c_1\}$$

so that $v \geq 0$ for all $v \in C_1$ and a cone C_3 defined by

$$C_3 = \{v = c_1\phi_{00} + c_2\phi_{10} \mid c_1 \leq 0, |c_2| \leq |c_1|\}$$

so that $v \leq 0$ for all $v \in C_3$.

Thus, even if we do not know $\theta(v)$ for all $v \in V$, we know $\theta(v) \equiv 0$ for $v \in C_1 \cup C_3$.

Now, we define a map $\Phi : V \to V$ given by

$$\Phi(v) = Lv + P(b(v + \theta(v))^+), \quad v \in V. \tag{1.6}$$

Then Φ is continuous on V, since θ is continuous on V and we have the following lemma.

LEMMA 1.2.[5] *$\Phi(cv) = c\Phi(v)$ for $c \geq 0$.*

Lemma 1.2 implies that Φ maps a cone with vertex 0 onto a cone with vertex 0. Now we investigate the image of the cones C_1 and C_3 under Φ. First we consider the image of the cone C_1. If $v = c_1\phi_{00} + c_2\phi_{10} \geq 0$, we have

$$\begin{aligned} \Phi(v) &= L(v) + P(b(v + \theta(v))^+) \\ &= c_1\lambda_{00}\phi_{00} + c_2\lambda_{10}\phi_{10} + b(c_1\phi_{00} + c_2\phi_{10}) \\ &= c_1(b + \lambda_{00})\phi_{00} + c_2(b + \lambda_{10})\phi_{10}. \end{aligned}$$

Thus the images of the rays $c_1\phi_{00} \pm c_1\phi_{10} (c_1 \geq 0)$ can be explicitly calculated and they are

$$c_1(b + \lambda_{00})\phi_{00} \pm c_1(b + \lambda_{10})\phi_{10} \qquad (c_1 \geq 0).$$

Therefore Φ maps C_1 onto the cone

$$R_1 = \left\{ d_1\phi_{00} + d_2\phi_{10} \mid d_1 \geq 0, |d_2| \leq \left(\frac{b + \lambda_{10}}{b + \lambda_{00}}\right) d_1 \right\}.$$

Here the restriction $\Phi|_{C_1} : C_1 \to R_1$ is bijective. Second we consider the image of the cone C_3. If

$$v = -c_1\phi_{00} + c_2\phi_{10} \leq 0 \quad (c_1 \geq 0,\ |c_2| \leq c_1),$$

we have

$$\begin{aligned} \Phi(v) &= L(v) + P(b(v + \theta(v))^+) \\ &= Lv + P(0) \\ &= -c_1\lambda_{00}\phi_{00} + c_2\lambda_{10}\phi_{10}. \end{aligned}$$

Thus the images of the rays $-c_1\phi_{00} \pm c_1\phi_{10}$ $(c_1 \geq 0)$ can be explicitly calculated and they are

$$-c_1\lambda_{00}\phi_{00} \pm c_1\lambda_{10}\phi_{10} \qquad (c_1 \geq 0).$$

Thus Φ maps the cone C_3 onto the cone

$$R_3 = \left\{ d_1\phi_{00} + d_2\phi_{10} \mid d_1 \leq 0,\ d_2 \leq |\frac{\lambda_{10}}{\lambda_{00}}||d_1| \right\}.$$

Here the restriction $\Phi|_{C_1} : C_3 \to R_3$ is bijective. We note that R_1 is in the right half plane and R_3 is in the left half plane.

THEOREM 1.1. *(i) If f belongs to R_1, then equation (1.1) has a positive solution and no negative solution. (ii) If f belongs to R_3, then equation (1.1) has a negative solution.*

Now we set

$$C_2 = \{c_1\phi_{00} + c_2\phi_{10} \mid c_2 \geq 0,\ c_2 \geq |c_1|\},$$

$$C_4 = \{c_1\phi_{00} + c_2\phi_{10} \mid c_2 \leq 0,\ c_2 \leq -|c_1|\}.$$

Then the union of C_1, C_2, C_3, C_4 is the space V.

Lemma 1.2 means that the images $\Phi(C_2)$ and $\Phi(C_4)$ are the cones in the plane V. Before we investigate the images $\Phi(C_2)$ and $\Phi(C_4)$, we set

$$R_2' = \left\{d_1\phi_{00} + d_2\phi_{10} \mid d_2 \geq 0, -|\frac{\lambda_{00}}{\lambda_{10}}|d_2 \leq d_1 \leq \left|\frac{b+\lambda_{00}}{b+\lambda_{10}}\right| d_2\right\},$$

$$R_4' = \left\{d_1\phi_{00} + d_2\phi_{10} \mid d_2 \leq 0, |\frac{\lambda_{00}}{\lambda_{10}}|d_2 \leq d_1 \leq \left(\frac{b+\lambda_{00}}{b+\lambda_{10}}\right) d_2\right\}.$$

Then the union of R_1, R_2', R_3, R_4' is the plane V.

To investigate a relation between the multiplicity of solutions and source terms in a nonlinear suspension bridge equation

$$Lu + bu^+ = f \quad \text{in} \quad H \tag{1.6}$$

we consider the restrictions $\Phi|_{C_i}(1 \leq i \leq 4)$ of Φ to the cones C_i. Let $\Phi_i = \Phi|_{C_i}$, i.e.,

$$\Phi_i : C_i \rightarrow V.$$

For $i = 1, 3$, the image of Φ_i is R_i and $\Phi_i : C_i \rightarrow R_i$ is bijective. From now on, our goal is to find the image of C_i under Φ_i for $i = 2, 4$.

Suppose that γ is a simple path in C_2 without meeting the origin, and end points (initial and terminal) of γ lie on the boundary ray of C_2 and they are on each other boundary ray. Then the image of one end point of γ under Φ is on the ray $c_1(b+\lambda_{00})\phi_{00} + c_1(b+\lambda_{10})\phi_{10}, c_1 \geq 0$ (a boundary ray of R_1) and the image of the other end point of γ under Φ is on the ray $-c_1\lambda_{00}\phi_{00} + c_1\lambda_{10}, c_1 \geq 0$ (a boundary ray of R_3). Since Φ is continuous, $\Phi(\gamma)$ is a path in V. By Lemma 1.2, $\Phi(\gamma)$ does not meet the origin. Hence the path $\Phi(\gamma)$ meets all rays (starting from the origin) in $R_1 \cup R_4'$ or all rays (starting from the origin) in $R_2' \cup R_3$.

Therefore it follows from Lemma 1.2 that the image $\Phi(C_2)$ of C_2 contains one of sets $R_1 \cup R_4'$ and $R_2' \cup R_3$.

Similarly, we have that the image $\Phi(C_4)$ of C_4 contains one of sets $R_1 \cup R_2'$ and $R_4' \cup R_3$.

LEMMA 1.3.[5] *Let A be one of the sets $R_1 \cup R_4'$ and $R_2' \cup R_3$ such that it is contained in $\Phi(C_2)$. Let γ be any simple path in A with end points on ∂A, where each ray (starting from the origin) in A intersect only one point of γ. Then the inverse image $\Phi_2^{-1}(\gamma)$ of γ is a simple path in C_2 with end points on ∂A, where any ray (starting from the origin) in C_2 intersects only one point of this path.*

Similarly, we have the following lemma.

LEMMA 1.3'.[5] *Let A be one of the sets $R_1 \cup R_2'$ and $R_4' \cup R_3$ such that it is contained in $\Phi(C_4)$. Let γ be any simple path in A with end points on ∂A, where each ray (starting from the origin) in A intersect only one point of γ. Then the inverse image $\Phi_4^{-1}(\gamma)$ of γ is a simple path in C_4 with end points on ∂A, where any ray (starting from the origin) in C_4 intersects only one point of this path.*

With Lemma 1.3 and Lemma 1.3', we have the following theorem, which is very important to investigate a relation between the multiplicity of solutions and source terms in a nonlinear suspension bridge equation.

THEOREM 1.2. *For $i = 2, 4$, if we let $\Phi_i(C_i) = R_i$, then R_2 is one of sets $R_1 \cup R_4'$, $R_2' \cup R_3$ and R_4 is one of sets $R_3 \cup R_4'$, $R_1 \cup R_2'$. Furthermore, for each $1 \leq i \leq 4$, the restriction Φ_i maps C_i onto R_i. In particular, Φ_1 and Φ_3 are bijective.*

To determine the images R_2 and R_4, we shall investigate the nonlinear suspension bridge equation

$$Lu + bu^+ = s\phi_{00} \quad \text{in} \quad H,$$

where we $3 < b < 15$ and s is real.

2. MULTIPLICITY OF SOLUTIONS AND SOURCE TERMS

In this section we reveal the relation between multiplicity of solutions and source terms in the nonlinear suspension bridge equation (1.1). Now, we remember the map $\Phi : V \to V$ given by

$$\Phi(v) = Lv + P(b(v + \theta(v))^+), \quad v \in V,$$

where $3 < b < 15$, $\theta(v)$ is a solution of (1.2), and V is the same space as in section 1. The map Φ is continuous on V, since θ is continuous on V. For $1 \leq i \leq 4$, let C_i be the same cone, in V, as in section 1.

For $f \in V$, we establish an *a priori* bound for solutions of

$$Lv + P(b(v+\theta(v))^+) = f \quad \text{in} \quad V. \tag{3.1}$$

LEMMA 2.1. *Let $-1 < b < 15$ and $k(\geq b+1)$ be fixed. Let $f \in V$ with $\|f\| = k$ and $\alpha > 0$ be given. Then there exists $R_0 > 0$ (depending only on k and α) such that for all b with $-1+\alpha \leq b \leq 15-\alpha$ the solutions of (3.1) satisfy $\|v\| < R_0$.*

Proof. We note that Lemma 2.2 holds when $-1 < b < 15$. Suppose Lemma 3.1 does not hold. Then there is a sequence (b_n, v_n) with $b_n \in [-1+\alpha, 15-\alpha]$, $\|v_n\| \to +\infty$ such that

$$v_n = L^{-1}(f - P(b(v_n + \theta(v_n))^+) \quad \text{in} \quad V.$$

Let $u_n = v_n + \theta(v_n)$. Then the sequence (b_n, u_n) with $b_n \in [-1+\alpha, 15-\alpha]$ satisfies $\|u_n\| \to +\infty$ and

$$u_n = L^{-1}(f - bu_n^+) \quad \text{in} \quad H_0.$$

Put $w_n = \frac{u_n}{\|u_n\|}$. Then we have

$$w_n = L^{-1}(\frac{f}{\|u_n\|} - bw_n^+).$$

The operator L^{-1} is compact. Therefore we may assume that $w_n \to w_0$ and $b_n \to b_0 \in (-1, 15)$. Since $\|w_n\| = 1$ for all n, $\|w_0\| = 1$ and w_0 satisfies

$$w_0 = L^{-1}(-bw_0^+) \quad \text{in} \quad H_0.$$

This contradicts the fact (cf. [7]) that $Lu + bu^+ = 0$ $(-1 < b < 15)$ has only the trivial solution. ∎

LEMMA 2.2. *Let $-1 < b < 15$ and $k(\geq b+1)$ be fixed. Let $f \in V$ with $\|f\| = k$. Then we have*

$$d(v - L^{-1}(f - P(b(v+\theta(v))^+)), B_R, 0) = 1$$

for all $R \geq R_0$.

Proof. Let $b = 0$. Then we have

$$d(v - L^{-1}(f), B_R, 0) = 1,$$

since the map is simply a translation of the identity and since $\|L^{-1}(f)\| < R_0$ by Lemma 2.1.

In case $b \neq 0(-1 < b < 15)$, the result follows in the usual way by invariance under homotopy, since all solutions are in the open ball B_{R_0} (cf. [7]). ∎

LEMMA 2.3. *Let $3 < b < 15$ and $f = (b+1)\phi_{00}$. Then equation (2.1) has a positive solution in* IntC_1, *at least one sign changing solution in* IntC_2, *and at least one sign changing solution in* IntC_4.

Proof. First we compute the degree $(R > R_0)$

$$\begin{aligned} &d(v - L^{-1}(f - P(b(v+\theta(v))^+)), B_R \cap C_1, 0) \\ &= d(v - L^{-1}(f - bv), B_R \cap C_1, 0) = -1, \end{aligned}$$

since $v - L^{-1}(f - bv) = 0$ has a unique solution in IntC_1 and $1 + \frac{b}{\lambda_{00}} > 0$, $1 + \frac{b}{\lambda_{10}} < 0$. Since $\tilde{I}_b(v)$ has a critical point in Int$(C_2 \cap C_4)$ (Theorem 2.1 of [5]), equation (2.1) has a solution in Int$(C_2 \cap C_4)$.

Suppose that (2.1) has a solution in IntC_2. Then $\Phi(C_2) \cap R_1 \neq \phi$ and hence $R_2 = \Phi(C_2) = R_1 \cup R_4'$ by Theorem 1.2. Let $B : V \to V$ be a linear map, where the matrix B is given by

$$\begin{pmatrix} \frac{b+2}{2} & \frac{b}{2} \\ \frac{b}{2} & \frac{b-6}{2} \end{pmatrix}.$$

Then $B(C_2) = R_2 = \Phi(C_2)$ and $Bv = \Phi(v)$ for all $v \in \partial C_2$. Now we may assume that the solution of $Bv = f$ is in B_{R_0}. Hence if $0 \leq t \leq 1$ and $R \geq 0$, then we have

$$tBv + (1-t)\Phi(v) \neq f, \quad v \in \partial(B_R \cap C_2).$$

So we have

$$\begin{aligned} &d(v - L^{-1}(f - P(b(v+\theta(v))^+)), B_R \cap C_2, 0) \\ &= d(v - L^{-1}(f - Bv + Lv), B_R \cap C_2, 0) = 1, \end{aligned}$$

since $Bv = f$ has a unique solution in IntC_2 and $\det(L^{-1}B) > 0$. Since $d(v - L^{-1}(f - P(b(v+\theta(v))^+)), B_R, 0) = 1$ and $d(v - L^{-1}(f - P(b(v+\theta(v))^+)), B_R \cap C_3, 0) = 0$,

$$d(v - L^{-1}(f - P(b(v+\theta(v))^+)), B_R \cap C_4, 0) = 1.$$

Therefore (2.1) has at least one solution in IntC_4.

Similarly, if we assume that (2.1) has a solution in IntC_4, then $d(v - L^{-1}(f - P(b(v+\theta(v))^+)), B_R \cap C_4, 0) = 1$ and hence we get $d(v - L^{-1}(f - P(b(v+\theta(v))^+)), B_R \cap C_2, 0) = 1$. Therefore (2.1) has at least one solution in IntC_2. ∎

With Theorem 1.2, Lemma 2.3, we get the following.

LEMMA 2.4. *Let $3 < b < 15$. For $1 \leq i \leq 4$, let $\Phi(C_i) = R_i$. Then $R_2 = R_1 \cup R_4'$ and $R_4 = R_1 \cup R_2'$, where R_2', R_4' are the same cones as in section 1.*

Proof. Let $3 < b < 15$ and $f \in V$. Then it follows from Lemma 2.3 that $R_2 \cap R_1 \neq \phi$. Since R_2 is one of sets $R_1 \cup R_4'$, $R_3 \cup R_2'$ (Theorem 1.2), the image R_2 of C_2 under Φ must be $R_1 \cup R_4'$.

On the other hand, it follows from Lemma 2.3 that $R_4 \cap R_1 \neq \phi$. Since R_4 is one of sets $R_1 \cup R_2'$, $R_3 \cup R_4'$ (Theorem 1.2), the image R_4 of C_4 under Φ must be $R_1 \cup R_2'$. ∎

If a solution of (1.5) is in C_1, then it is positive. If a solution of (1.5) is in C_3, then it is negative. If a solution of (1.5) is in $\text{Int}(C_2 \cup C_4)$, then it has both signs. Therefore we have the main theorem of this paper with aid of Theorem 1.1, Theorem 1.2, and Lemma 2.4.

THEOREM 2.1. *Let $3 < b < 15$. Then we have the following.*
(i) If $f \in \text{Int } R_1$, then equation (1.1) has a positive solution and at least two sign changing solutions.
(ii) If $f \in \partial R_1$, then equation (1.1) has a positive solution and at least one sign changing solution.
(iii) If $f \in \text{Int } R_i'(i = 2, 4)$, then equation (1.1) has at least one sign changing solution.
(iv) If $f \in \text{Int } R_3$, then equation (1.1) has only the negative solution.
(v) If $f \in \partial R_3$, then equation (1.1) has a negative solution.

REFERENCES

[1] **H. Amann**, *Saddle points and multiple solutions of differential equations.* Math. Z. (1979), 127-166.

[2] **A. Ambrosetti** and **P.H. Rabinowitz**, *Dual variational methods in critical point theory and applications.* J. funct. Analysis, **14** (1973), 349-381.

[3] **A. Castro** and **A.C. Lazer**, *Applications of a max-min principle.* Rev. Colombian Mat., **10** (1979), 141-149.

[4] **Q.H. Choi, T. Jung,** and **P.J. McKenna**, *The study of a nonlinear suspension bridge equation by a variational reduction method.* Applicable Analysis, **50** (1993), 71-90.

[5] **Q.H.Choi** and **T. Jung,** *A nonlinear suspension bridge equation with nonconstant load*, Preprint.

[6] **A.C. Lazer** and **P.J. McKenna,** *Critical points theory and boundary value problems with nonlinearities crossing multiple eigenvalues II.* Comm.in P.D.E., **11** (1986), 1653-1676.

[7] **P.J. McKenna** and **W. Walter,** *Nonlinear Oscillations in a Suspension Bridge.* Archive for Rational Mechanics and Analysis, **98** (1987), 167-177.

[8] **P.H. Rabinowitz,** *Minimax methods in critical point theory with applications to differential equations.* Conference board of the mathematical sciences regional conference series in mathematics, A.M.S., No.65 (1988).

7th Int. Coll. on Differential Equations, pp. 49-56
D. Bainov (Ed)

UNIQUENESS IN THE CAUCHY PROBLEM FOR THE HEAT EQUATION

SOON-YEONG CHUNG

Department of Mathematics
Sogang University,
Seoul 121-742, Korea
e-mail: sychung@ccs.sogang.ac.kr

ABSTRACT. We relax the growth condition in time for uniqueness of solutions of the Cauchy problem for the heat equation as follows: Let $u(x,t)$ be a continuous function on $\mathbb{R}^n \times [0,T]$ satisfying the heat equation in $\mathbb{R}^n \times (0,t)$ and the following:

(i) There exist constants $a > 0, 0 < \alpha < 1$, and $C > 0$ such that

$$|u(x,t)| \le C \exp\left[\left(\frac{a}{t}\right)^\alpha + a|x|^2\right] \quad \text{in} \quad \mathbb{R}^n \times (0,T).$$

(ii) $u(x,0) = 0$ for $x \in \mathbb{R}^n$.

Then $u(x,t) \equiv 0$ on $\mathbb{R}^n \times [0,T]$.

We also prove that the condition $0 < \alpha < 1$ is optimal.

1. Introduction

Here we deal with uniqueness of solutions of the Cauchy problem for the heat equation

$$\begin{cases} (\partial_t - \Delta)u(x,t) = f & \text{in} \quad \mathbb{R}^n \times (0,T) \quad (T > 0), \\ u(x,0) = \varphi(x) & \text{for} \quad x \in \mathbb{R}^n. \end{cases}$$

It is well known that the temperature of the infinite rod is not uniquely determined by its initial temperature (see [1, 3, 8, 6]).

With additional growth conditions there are uniqueness theorems. The following is the famous uniqueness theorem which was originally given by Tychonoff.

Theorem A ([8]). *Let $u(x,t)$ be a continuous function on $\mathbb{R}^n \times [0,T]$ satisfying that*

$$(\partial_t - \Delta)u(x,t) = 0 \quad \textit{in} \quad \mathbb{R}^n \times (0,T)$$

and for some constants $a > 0$ and $C > 0$

$$|u(x,t)| \le C \exp a|x|^2 \quad \textit{on} \quad \mathbb{R}^n \times [0,T].$$

Then $u(x,0) = 0$ *implies* $u(x,t) \equiv 0$ *on* $\mathbb{R}^n \times [0,T]$.

For most uniqueness theorems the solution $u(x,t)$ must be uniformly bounded with respect to t variable. However, sometimes we need a uniqueness theorem with a milder condition on time.

Some authors have had interests in this direction and have relaxed the condition on t. For example, Shapiro [7] showed:

Theorem B ([7]). *Let* $u(x,t)$ *be a solution of the heat equation in the strip* $0 < t < c$ *and bounded in every substrip of the form* $0 < t_0 \leq t < c$. *Suppose that*

(i) $\|u(x,t)\|_\infty = o(t^{-1})$ as $t \to 0$;
(ii) $\lim_{t\to 0} u(x,t) = 0$ *except possibly for a countable set* E;
(iii) $\liminf_{t\to 0} t^{1/2} u(x,t) = 0$ *for every* x *in* E.

Then $u(x,t) \equiv 0$ *in the strip* $0 < t < c$.

On the other hand, Chung and Kim [2] showed the following:

Theorem C ([2]). *Let* $u(x,t)$ *be a solution of the heat equation in* $\mathbb{R}^n \times (0,T)$ *satisfying:*

(i) *There exist constants* $k > 0$ *and* $C > 0$ *such that*

$$|u(x,t)| \leq C \exp k(|x|^2 + \frac{1}{t}), \quad 0 < t < T;$$

(ii) $\lim_{t\to 0+} \int u(x,t)\phi(x)dx = 0$ *for every* C^∞ *function* $\phi(x)$ *such that for every* $h > 0$,

$$\sup_{\substack{x\in\mathbb{R}^n \\ \alpha}} \frac{|\partial^\alpha \phi(x)| \exp 2k|x|}{h^{|\alpha|}\alpha!} < \infty.$$

Then $u(x,t) \equiv 0$ *on* $\mathbb{R}^n \times [0,T]$.

In the above two theorems they relaxed the growth condition on time so that the uniqueness classes determined by them are larger than that of Theorem A. In fact, Theorem C gives a much larger uniquness class than others. But, nevertheless, the hypotheses are not so natural that one can apply them effectively. The hypothesis (ii) is stronger than $u(x,0) = 0$, since there exists a nonzero temperature function satisfying (i) and $u(x,0) = 0$ (see the section 4).

The purpose of this lecture is to give the following theorem:

Theorem 3.1. *Let* $u(x,t)$ *be a continuous function on* $\mathbb{R}^n \times [0,T]$ *satisfying the heat equation in* $\mathbb{R}^n \times (0,T)$ *and the following:*

(i) *There exist constants* $a > 0$, $0 < \alpha < 1$, *and* $C > 0$ *such that*

$$|u(x,t)| \leq C \exp\left[\left(\frac{a}{t}\right)^\alpha + a|x|^2\right] \quad \text{in} \quad \mathbb{R}^n \times (0,T).$$

(ii) $u(x,0) = 0$ *for* $x \in \mathbb{R}^n$.

Then $u(x,t) \equiv 0$ *on* $\mathbb{R}^n \times [0,T]$.

Moreover, the uniqueness of the solution with weaker initial condition, not imposing the continuity of $u(x,t)$ at $t=0$ will be given here.

Besides these results the condition $0<\alpha<1$ in the above can be seen optimal with a counterexample showing that if we take $\alpha=1$ then the theorem is no longer true.

2. Notations and basic results.

We introduce briefly the ultradistributions of Gevrey type which will be very useful later. See [5] for more details.

Definition 2.1. Let Ω be an open subset of $\mathbb{R}^n$ and $\varphi \in C^\infty(\Omega)$. Then we say that φ belongs to $\mathcal{E}_{(s)}(\Omega)$ for $s>1$ if for any compact subset K of Ω and for every $h>0$ there exists a constant $C=C(K,h)>0$ such that

$$\sup_{x\in K}|\partial^\alpha \varphi(x)| \le C h^{|\alpha|}\alpha!^s, \quad \alpha \in \mathbb{N}_0^n$$

where $\mathbb{N}_0$ is the set of nonnegative integers, and we use the multi-index notations $|\alpha| = \alpha_1 + \cdots + \alpha_n$ and $\partial^\alpha = \partial_1^{\alpha_1}\partial_2^{\alpha_2}\cdots\partial_n^{\alpha_n}$, $\partial_j = \frac{\partial}{\partial x_j}$, $j=1,2,\cdots,n$, for $\alpha = (\alpha_1,\alpha_2,\cdots,\alpha_n)$.

We denote by $\mathcal{D}_{(s)}(\Omega)$ the subspace of $\mathcal{E}_{(s)}(\Omega)$ which consists of functions with compact support in Ω. The topologies of these spaces are defined as follows:

(i) $\{\phi_j(x)\} \in \mathcal{E}_{(s)}(\Omega)$ converges to zero in $\mathcal{E}_{(s)}(\Omega), s>1$, if for any compact subset K of Ω and for every $h>0$,

$$\sup_{\substack{x\in K\\ \alpha\in\mathbb{N}_0^n}} \frac{|\partial^\alpha \phi_j(x)|}{h^{|\alpha|}\alpha!^s} \to 0 \quad \text{as} \quad j\to\infty.$$

(ii) $\{\phi_j(x)\} \in \mathcal{D}_{(s)}(\Omega)$ converges to zero in $\mathcal{D}_{(s)}, s>1$, if there is a compact set K of Ω such that $\mathrm{supp}\phi_j \subset K, j=1,2,\cdots$, and $\phi_j \to 0$ in $\mathcal{E}_{(s)}(\Omega)$.

As usual, we denote by $\mathcal{D}'_{(s)}(\Omega)$ (by $\mathcal{E}'_{(s)}(\Omega)$) the strong dual space of $\mathcal{D}_{(s)}(\Omega)$ (of $\mathcal{E}_{(s)}(\Omega)$, respectively) and we call its elements the ultradistributions of Gevrey type.

We have the inclusion

$$\mathcal{D}'(\Omega) \subset \mathcal{D}'_{(s)}(\Omega), \quad \mathcal{E}'(\Omega) \subset \mathcal{E}'_{(s)}(\Omega), \quad s>1,$$

where $\mathcal{D}'(\Omega)$ and $\mathcal{E}'(\Omega)$ are the space of Schwartz distributions and the space of Schwartz distributions with compact support respectively. It is well known that $\mathcal{E}'_{(s)}(\Omega)$ consists of the ultradistributions in $\mathcal{D}'_{(s)}(\Omega)$ with compact support in Ω like the space $\mathcal{E}'(\Omega)$ in $\mathcal{D}'(\Omega)$. In fact, since there exist cut off functions and partitions of unity in $\mathcal{D}_{(s)}(\Omega), s>1$, the properties of the ultradistributions in $\mathcal{D}'_{(s)}(\Omega)$ is very similar to those of Schwartz distributions. In particular, the concepts of the support, convolution, etc. are defined very similarly and naturally.

We note here that in view of the topology on $\mathcal{D}_{(s)}(\Omega), u$ belongs to $\mathcal{D}'_{(s)}(\Omega)$ if and only if for every compact subset K of Ω there exist constants $h > 0$ and $C > 0$ such that

$$|u(\phi)| \le C \sup_{\substack{x\in K \\ \alpha\in\mathbb{N}^n}} \frac{|\partial^\alpha \phi(x)|}{h^{|\alpha|}\alpha!^s}$$

for every $\phi \in \mathcal{D}_{(s)}(\Omega)$ with support in K.

On the other hand, define the partial differential operators of class (s) of infinite order as follows :

$$p(\partial) = \sum_\alpha a_\alpha \partial^\alpha, \quad |a_\alpha| \le CL^{|\alpha|}/\alpha!^s$$

for some $L > 0$ and $C > 0$. Then the mappings

$$p(\partial) : \mathcal{D}_{(s)}(\Omega) \to \mathcal{D}_{(s)}(\Omega), \quad \mathcal{E}_{(s)}(\Omega) \to \mathcal{E}_{(s)}(\Omega)$$
$$p(\partial) : \mathcal{D}'_{(s)}(\Omega) \to \mathcal{D}'_{(s)}(\Omega), \quad \mathcal{E}'_{(s)}(\Omega) \to \mathcal{E}'_{(s)}(\Omega)$$

are continuous.

From now on we denote by $E(x,t)$ the fundamental solution of the heat equation:

$$E(x,t) = \begin{cases} (4\pi t)^{-n/2} \exp(-|x|^2/4t), & t > 0 \\ 0, & t < 0. \end{cases}$$

Then the following can be obtained by some tedious calculation.

Proposition 2.2. *Let $g(x)$ be a continuous function on $\mathbb{R}^n$ satisfying that for some constants $a > 0$ and $C > 0$*

$$|g(x)| \le C \exp a|x|^2, \quad x \in \mathbb{R}^n.$$

Then $G(x,t) = g(x) * E(x,t)$ is a well defined C^∞ function in $\mathbb{R}^n \times (0, 1/4a)$ and satisfies

(i) $(\partial_t - \Delta)G(x,t) = 0, \quad 0 < t < 1/4a$

(ii) $|G(x,t)| \le C\exp(2a|x|^2), \quad 0 < t < 1/6a,$

(iii) *$G(x,t) \to g(x)$ uniformly on each compact subset of $\mathbb{R}^n$ as $t \to 0+$.*

Here, $*$ denotes the convolution with respect to x variable.

The following lemma is the main tool used to prove the main theorem.

Lemma 2.3. *For any $L > 0$ and $\varepsilon > 0$ there exist functions $v(t)$, $w(t) \in C_0^\infty(\mathbb{R})$ and a differential operator $p(d/dt)$ of infinite order such that*

$$p(d/dt)v(t) = \delta(t) + w(t); \tag{2.1}$$

$$\operatorname{supp} v \subset [0, \varepsilon], \quad \operatorname{supp} w \subset [\varepsilon/2, \varepsilon]; \tag{2.2}$$

$$p(d/dt) = \sum_{k=0}^\infty a_k (d/dt)^k, \quad |a_k| \le CL^k/k!^s, \quad s > 1; \tag{2.3}$$

$$|v(t)| \le C\exp\left[-(L/t)^{1/(s-1)}\right]. \tag{2.4}$$

where δ is the Dirac measure.

3. Main Theorems

We give here a much better uniqueness theorem than those in [2, 3, 4, 7].

Theorem 3.1. *Let $u(x,t)$ be a continuous function on $\mathbb{R}^n \times [0,T]$ satisfying*

(i) $(\partial_t - \Delta)u(x,t) = 0$ *in* $\mathbb{R}^n \times (0,T]$

(ii) $|u(x,t)| \leq C\exp\left[(a/t)^\alpha + a|x|^2\right]$ *in* $\mathbb{R}^n \times (0,T]$ *for some constants* $a > 0, 0 < \alpha < 1$, *and* $C > 0$.

(iii) $u(x,0) = 0$ *on* $\mathbb{R}^n$.

Then $u(x,t)$ is identically zero on $\mathbb{R}^n \times [0,T]$. Here T may be ∞.

Sketch of Proof. In view of Theorem A given by Tychonoff in the introduction we have only to show that $u \equiv 0$ on $\mathbb{R} \times [0,T_0]$ for sufficiently small $T_0 > 0$.

Let $s = \frac{1}{2}\left(1 + \frac{1}{\alpha}\right) > 1$. Then according to Lemma 2.3 we can choose C_0^∞ functions $v(t), w(t)$ on $\mathbb{R}$ and a differential operater $p(d/dt)$ of infinite order such that

$$p(d/dt)v(t) = \delta(t) + w(t);$$

$$\operatorname{supp} v \subset [0,T_0], \quad \operatorname{supp} w \subset [T_0/2, T_0];$$

$$p(d/dt) = \sum_{k=1}^{\infty} a_k (d/dt)^k, \quad |a_k| \leq C(2a)^k/k!^{2s};$$

$$|v(t)| \leq C\exp\left[-(2a/t)^{1/(2s-1)}\right],$$

where $T_0 > 0$ is a small number such that $2T_0 < \min(T, 1/16a)$.

Define two functions $G(x,t)$ and $H(x,t)$ by

$$G(x,t) = \int_0^T u(x,t+\tau)v(\tau)d\tau,$$

and

$$H(x,t) = -\int_0^T u(x,t+\tau)w(\tau)d\tau.$$

Then it follows that the integrals converge and are continuous functions on $\mathbb{R}^n \times [0,T_0]$. Moreover, they satisfy the heat equation and the growth condition

$$|G(x,t)| \leq C\exp a|x|^2$$

and

$$|H(x,t)| \leq C\exp a|x|^2$$

for all $(x,t) \in \mathbb{R}^n \times [0,T_0]$. If we take $g(x) = G(x,0)$ and $h(x) = H(x,0)$ then Proposition 2.2 implies that $g * E$ and $h * E$ are continuous functions on $\mathbb{R}^n \times [0,T_0]$ which converge to $g(x)$ and $h(x)$ respectively as $t \to 0+$. In view of Theorem A in the introduction and the growth conditions it follows that

$$G(x,t) = g(x) * E(x,t), \quad H(x,t) = h(x) * E(x,t)$$

on $\mathbb{R}^n \times [0,T_0]$.

Thus, if we define

$$u = p(-\Delta)g(x) + h(x),$$

then u belongs to $\mathcal{D}'_{(s)}(\mathbb{R}^n)$.

Applying the differential operator $p(\partial/\partial t)$ we have

$$p(\partial/\partial t)G(x,t) = p(-\Delta)G(x,t) = u(x,t) - H(x,t)$$

in $\mathbb{R}^n \times (0,T_0)$. Then we note that $u = p(-\Delta)g + h$ is a $\mathcal{D}'_{(s)}$-limit (or weak limit) of $p(-\Delta)G(x,t) + H(x,t)$, since $G(x,t) \to g(x)$ and $H(x,t) \to h(x)$ uniformly on each compact subset of $\mathbb{R}^n$. Therefore, we have

$$u = \lim_{t\to 0+} u(x,t) \quad \text{in} \quad \mathcal{D}'_{(s)}(\mathbb{R}^n).$$

From this we obtain that

$$u(\varphi) = \lim_{t\to 0} \int u(x,t)\varphi(x)dx = 0, \qquad \varphi \in \mathcal{D}_{(s)}(\mathbb{R}^n),$$

which implies that $u = 0$ as an element of $\mathcal{D}'_{(s)}(\mathbb{R}^n)$.

Then it follows that

$$\begin{aligned} u(x,t) &= p(-\Delta)G(x,t) + H(x,t) \\ &= p(-\Delta)g * E + h * E \\ &= [p(-\Delta)g + h] * E \\ &= u * E \equiv 0. \end{aligned}$$

In the above the continuity of $u(x,t)$ at $t = 0$ can be weakened.

Theorem 3.2. *If $u(x,t)$ satisfies the conditions in $\mathbb{R}^n \times (0,T)$:*

(i) $(\partial_t - \Delta)u(x,t) = 0$,
(ii) $|u(x,t)| \le C\exp\left[(a/t)^\alpha + a|x|^2\right]$
for some constants $a > 0$, $0 < \alpha < 1$ and $C > 0$
(iii) $\lim_{t\to 0+} \int u(x,t)\varphi(x)dx = 0, \quad \varphi \in \mathcal{D}_{(s)}(\mathbb{R}^n)$ *where* $s = \frac{1}{2}\left(1 + \frac{1}{\alpha}\right)$, *then* $u(x,t) \equiv 0$ *in* $\mathbb{R}^n \times [0,T)$.

4. Example of the nonuniqueness

In this section we show that the uniqueness class given in the previous section is the optimal one. In particular, it will be shown that the growth condition on the time variable is optimal.

For the space variable it is well known (see [3]) that for every $\varepsilon > 0$ there exists a C^∞ function $u(x,t) \not\equiv 0$ satisfying that

(i) $(\partial_t - \Delta)u(x,t) = 0$ in $\mathbb{R}^n \times (0,T)$.
(ii) u is continuous on $\mathbb{R}^n \times [0,T)$.
(iii) $|u(x,t)| \le C_\varepsilon \exp|x|^{2+\varepsilon}$ on $\mathbb{R}^n \times (0,T)$.
(iv) $u(x,0) = 0$ on $\mathbb{R}^n$.

This shows that Theorem 3.1 is not true if in that condition we replace $\exp(a|x|^2)$ by $\exp(a|x|^{2+\varepsilon}), \varepsilon > 0$.

Now we will show that theorem 3.1 is also no longer true if we replace the condition $0 < \alpha < 1$ by $\alpha = 1$. To see this let D_N be a domain in the complex plane $\mathbb{C}$ given by

$$D_N = \{z \in \mathbb{C} \,|\, z = x + yi, \quad x \geq N, -\pi \leq y \leq \pi\}, \quad N \geq 0$$

and C_N be the boundary of D_N. Define a function $u(x,t)$ on $\mathbb{R} \times (0,\infty)$ by

$$u(x,t) = \frac{1}{2\pi i} \int_{C_N} E(x-\zeta, t) \exp(e^\zeta) d\zeta,$$

where the integral is taken counterclockwise.

Then it follows that for some constant $a > 0$

$$|u(x,t)| \leq C(N) \exp\left(\frac{a}{t}\right) \exp\left[-\frac{d(x, D_N)^2}{4t}\right]$$

where d is the Euclidean distance.

Thus we have

$$|u(x,t)| \leq C(N) \exp\left(\frac{a}{t}\right) \quad \text{in} \quad \mathbb{R}^n \times (0,\infty).$$

Let $r > 0$ and $x \leq r$. Since the integral is independent of N we may choose a sufficiently large $N > 0$ so that $4a < (N-r)^2$. Then we obtain

$$\sup_{x \leq r} |u(x,t)| \leq C(N) \exp\left[\frac{4a - (N-r)^2}{4t}\right], \quad t > 0.$$

The right hand side converges to 0 as $t \to 0+$ and $u(x,t)$ converges uniformly to 0 as $t \to 0+$ in every half-line $(-\infty, r]$, $r > 0$. Therefore, we can conclude that $u(x,t)$ is continous on $\mathbb{R}^n \times [0,\infty)$ and $u(x,0) = 0$.

Now it remains to show that $u(x,t) \not\equiv 0$.

To do this we suppose that $u(x,t) \equiv 0$ in $\mathbb{R} \times [0,\infty)$. Then we obtain that

$$\int_{C_N} \exp\left[-\frac{(x-\zeta)^2}{4t}\right] \cdot \exp(e^\zeta) d\zeta \equiv 0$$

in $\mathbb{R}^n \times [0,\infty)$. Applying the Lebesgue dominated convergence theorem we can see that

$$\int_{C_N} \exp(e^\zeta) d\zeta = 0.$$

Since the integral does not depend on $N \geq 0$ we may choose $N = 0$. Then it can be written as

$$\begin{aligned} 0 = &- \int_0^\infty \exp(-e^\zeta) d\zeta - \int_{-\pi}^{\pi} \exp(e^{yi}) dy \\ &+ \int_0^\infty \exp(-e^\zeta) d\zeta \\ = &- 2i \int_0^\pi e^{\cos y} \cos(\sin y) dy. \end{aligned}$$

But the integral $e^{\cos y} \cos(\sin y) > 0$ on $[0,\pi]$, which leads a contradiction. Thus we can conclude that $u(x,t) \not\equiv 0$.

REFERENCES

[1] S.-Y. Chung, D. Kim,An example of nonuniqueness of the Cauchy problem for the heat equation, *Comm. Partial Differential Equations* **19** (1994), 1257-1261.

[2] S.-Y. Chung, D. Kim, Uniqueness for the Cauchy problem of the heat equation without uniform condition on time, *J. Korean Math. Soc.* **31** (1994), 245-254.

[3] A. Friedman, "Partial differential equations of parabolic type ", Englewood Cliffs, N. J.: Prentice Hall, Inc. 1964.

[4] R. M. Hayne, Uniqueness in the Cauchy problem for the parabolic equations, *Trans. Amer. Math. Soc.* **241** (1978), 373-399.

[5] H. Komatsu, "Introduction to the theory of hyperfunctions", Tokyo: Iwanami 1978 (In Japanese).

[6] H. Komatsu, Ultradistributions I; Structure theorems and a characterization, J. Fac. Sci. Univ. Tokyo, Sect IA, **20** (1973), 25-105.

[7] P. C. Rosenbloom, D. V. Widder, A temperature function which vanishes initially, *Amer. Math. Monthly* **65** (1958), 607-609.

[8] V. L. Shapiro, The uniqueness of solutions of the heat equation in an infinite strip, *Trans. Amer. Math. Soc.* **125** (1966), 326-361.

[9] D. V. Widder, "The heat equation", New York and London: Academic Press 1975.

7th Int. Coll. on Differential Equations, pp. 57-65
D. Bainov (Ed)

EXISTENCE AND UNIQUENESS OF SOLUTIONS FOR AN CLASS OF ABSTRACT NONLINEAR WAVE EQUATION

Marcondes R. Clark
UFPB - CCT - DME
mclark@dme.ufpb.br
Cep 58.109-970 - C. Grande-Pb - Brazil

Abstract

In this paper we study the Existence and Uniqueness of solutions for a wide class of abstract nonlinear evolution equation of the type

$$(A_2(t)u'(t))' + A_1(t)u'(t) + A(t)u(t) + M(u(t)) = f(t), \quad t \in (0,T) \qquad (I)$$

in a Hilbert space H where $\{A_1(t), t \geq 0\}, \{A_2(t), t \geq 0\}$ are families of bounded linear operators in H, $\{A(t), t \geq 0\}$ is a family of selfadjoint unbounded positive definite operators, M is a non-linear application on H and f is a function from $(0,T)$ in H. We prove the existence and uniqueness of global weak solutions for equation (I) where we use the Faedo-Galerkin method. We illustrate our results with examples of this family of hyperbolic-parabolic equations.

Keywork: Existence and Uniquesness of Solutions; Evolution Equation; Faedo-Galerkin method.

AMS Subject Classifications: 35L15

1 INTRODUCTION

Let $T \geq 0$ be a real number and Ω be a bounded open set of $\mathbb{R}^n$,with smooth boundary Γ. In the cylinder $Q = \Omega \times (0,T)$, Bensoussan, Lions and Papanicolaou (see [01]), have studied the homogeneization for the following Cauchy problem:

$$\begin{aligned} k_2(x)u'' + k_1(x)u' - \Delta u = f \quad &\text{in } \Omega, u(x,0) = u_0(x) \\ \text{and} \quad k_2(x)u'(x,0) = k_2^{1/2}&(x)u_1(x), \ x \in \Omega. \end{aligned} \qquad \text{(II)}$$

Many authors have been investigating the solvability of solutions for non-linear equations associated with problem (II). See Clark, [03] and [04]; Larkin, [05]; Lima, [07] ;Medeiros, [11]-[13] ; Maciel, [10]; Neves, [14]; Strauss, [15]; Vagrov, [16].

In Section 2, we establish some notations for the function spaces, make assumptions on the families of operators of H; $\{A(t); t \geq 0\}, \{A_1(t); t \geq 0\}, \{A_2(t); t \geq 0\}, M$ and on the function f of equation (I). In section 3, we establish and prove our main results. In the final Section we make applications of the Cauchy problem associated with (I).

2 NOTATIONS AND MAIN RESULTS.

Let H be a real abstract Hilbert space, with the inner product and the norm denoted by $(.,.)$ and $|.|$, respectively.

We consider here the following assumptions on the families of linear operators on $H, \{A(t); t \geq 0\}$; $\{A_1(t); t \geq 0\}$; $\{A_2(t); t \geq 0\}$.

i) $A_1(t),\ A_2(t) : H \to H$ are symmetric, strongly continuously differentiable linear operators, satisfying: $A_2(t),\ A_2'(t)$ are positive for each $t \geq 0$;

$$((A_1(t)u, u) \geq \beta|u|^2,\ 0 < \beta \in I\!R, \text{ for all } u \in H,\ \forall t \geq 0.$$

ii) $A(t)$ is a self-adjoint linear operator on H, such that there exists a constant $\alpha > 0$, satisfying $(A(t)u, u) \geq \alpha|u|^2$ for all $u \in D(A(t))$, where we assume that the domain $D(A(t))$ is constant, i.e. $D(A(t)) = D(A(s))$ for all $t, s \geq 0$, and dense in H.Then, $D(A(t)) \subseteq D(A^{\frac{1}{2}}(t))$ and $D(A^{\frac{1}{2}}(t))$ is constant,(see [04]).

Let $V_t = D(A^{\frac{1}{2}}(t))$ be a Hilbert space dense in H with the inner product $((.,.))$ and associated norm $\|.\|_t$. Then $\|u\|_t^2 = |A^{\frac{1}{2}}(t)u|^2 \geq \alpha|u|^2$ and so that $V_t \hookrightarrow H$. Besides this, we assume that $V_t \hookrightarrow V_0 \hookrightarrow H$, with the embedding $V_0 \hookrightarrow H$ compact. Therefore we have that the spectrum of the operator $A(t)$ is discrete.

iii) $A(t)$ is strongly continuously differentiable and for $u \in D(A(t))$, and exists a real $\eta \geq 0$, independent of t, such that:$|(A'(t)u, u)| \leq \eta|u|^2$, for all $t \in [0, T]$.

Identifying, by the Theorem of Representation of Riesz, H with its dual H', we have: $V_0 \hookrightarrow H \hookrightarrow V_0'$; where each space is dense in the following one.

iv) For the operator $M : V_0 \longrightarrow H$, we assume the same hypothesis of [04] or [07], i.e.

1) M is monotone, hemi-continuous and limited.

2) There is a constant $\sigma > 0$ such that

$$\int_0^T (M(u(s)), u'(s))ds \geq -\sigma, \quad \forall\, t\ \in [0, T] \quad \text{and} \ \forall\ u \in E_C$$

where E_C is the set $\{u \in L^\infty(0, T; V_0);\ u' \in L^2(0, T; H) \quad \text{and} \quad \|u(0)\|_0 \leq C\}$

We establish the main results:

Theorem 1 *Under the assumptions (i)-(iii) and considering*

$$f \in L^2(0, T; H); \quad u_0 \in V_0; \quad u_1 \in H \tag{1.1}$$

then there exists a function $u : (0, T) \longrightarrow V_0$ *in the class:*

$$u \in L^\infty(0, T; V_0); \quad u' \in L^2(0, T; H) \tag{1.2}$$

besides this, u is solution of (I) in the following sense in the following sense:

$$\begin{aligned}&-\int_0^T(A_2(t)u'(t),\phi'(t)v)dt+\int_0^T(A_1(t)u',\phi(t)v)dt+\int_0^T(A^{\frac{1}{2}}(t)u,A^{\frac{1}{2}}(t)\phi(t)v)dt\\&+\int_0^T(M(u(t)),\phi(t)v)dt=\int_0^T(f(t),\phi(t)v)dt;\ \forall\phi\in C_0^1(0,T)\ \text{and}\ \forall v\in V_0\end{aligned} \tag{1.3}$$

plus initial conditions:

$$(u(0),v)=(u_0,v)\ \text{and}\ (A_2(0)u'(0),v)=(A_2^{\frac{1}{2}}(0)u_1,v),\ \forall v\in V \tag{1.4}$$

For the uniqueness we need the following condition on A_1' and M :
v)There exists $\rho>0$ such that $|(A_1'(t)u,v)|\le\rho|u|\,|v|$ and given $C>0$, there exists $\tau>0$, which depends on C such that:

$$|M(u)-M(v)|\le\tau|u-v|\ \forall u,v\in V \quad \text{whenever } \|u\|_0\le C,\ \|v\|_0\le C.$$

Theorem 2 *Suppose that the operators $A_1(t)$, $A_2(t)$, $A(t)$ and M satisfy the conditions (i)-(v) and map functions of $L^\infty(0,T;V_0)$ into functions of $L^2(0,T;H)$. Then there exists at most one function u in the class*

$$u\in L^\infty(0,T;V_0);\quad u'\in L^2(0,T;H),$$

such that u is a solution of problem (I) in the sense (1.7)-(1.9) of Theorem-1.

3 PROOF OF THEOREMS

We shall use the Faedo-Galerkin method , perturbing the equation (I) with the term $\lambda u''$, $\lambda>0$. For the uniqueness we use the method of Visik-Ladyzhenskaja, see [07] for example. In this part we use the following result:
Lemma 1. *Let $u\in L^2(0,T;H)$, $u'\in L^2(0,T;V_0)$ and v, $v'\in L^2(0,T;V_0)$. Then*

$$\frac{d}{dt}<u,v>=<u',v>+(u,v').$$

For the proof of this lemma see [07]. Let (w_ν) be a base of $D(A(0))$ that it is a base of H, by density. From (i), we have that $(A_2\lambda^{\frac{1}{2}}(0)w_\nu)$ is a base of H too; where $0<\lambda$ is a constant and $A_{2\lambda}(t)=A_2(t)+\lambda I$. Let $V_m(0)=[w_1,\dots,w_m]$ be a subspace of $D(A(0))$ and $V_m^\lambda(0)=[A_{2\lambda}^{\frac{1}{2}}(0)w_1,\dots,A_{2\lambda}^{\frac{1}{2}}(0)w_m]$ a subspace of H.

We denote $u_{\lambda m}(t)=\sum_{i=1}^m g_{i\lambda m}(t)w_i$ a solution of the approximated perturbed problem:

$$((A_{2\lambda}(t)u_{\lambda m}'(t))'+A_1(t)u_{\lambda m}'(t)+A(t)u_{\lambda m}(t)+M(u_{\lambda m}(t)),v)=(f(t),v), \tag{1.5}$$

for all $v\in V_m(0)$.

$$u_{\lambda m}(0)=u_{0m};\quad u_{\lambda m}'(0)=u_{1\lambda m} \tag{1.6}$$

where $u_{0m} = \sum_{i=1}^{m} \alpha_{im} w_i \longrightarrow u_0$ strongly in V_0; $u_{1\lambda m} = \sum_{i=1}^{m} \beta_{i\lambda m} w_i$ and the coefficient $\beta_{i\lambda m}$ is the coordinates of the vector $P_{\lambda m} u_1$, the orthogonal projections of vector u_1 upon the subspace $V_m^{\lambda}(0)$ in relation to the base $(A_{2\lambda}^{\frac{1}{2}}(0) w_\nu)$, such that:

$$P_{\lambda m} u_1 = \sum_{i=1}^{m} \beta_{i\lambda m} A_{2\lambda m}^{\frac{1}{2}}(0) w_i \longrightarrow u_1$$

strongly in H and satisfying,

$$|P_{\lambda m} u_1| \le |u_1|, \quad \forall m \in \mathbb{N} \text{ and } \forall \lambda > 0. \tag{1.7}$$

The system (1.5)-(1.7) has, by the theorem of Carathéodory (see [02]), a solution $u_{\lambda m}(t)$ defined in $[0, t_{\lambda m})$, where $0 < t_{\lambda m} \le T$.In the following step, we are going to obtain the a priori estimates which shows, in particular, that $t_{\lambda m} = T$.

(i) A Priori Estimates.

In (1.5) taking $v = 2u'_{\lambda m}(t)$ we obtain

$$\begin{aligned} &\tfrac{d}{dt}(A_{2\lambda}(t) u'_{\lambda m}(t), u'_{\lambda m}(t)) + ((A(t) u_{\lambda m}(t), u_{\lambda m}(t)) + \\ &((A'_2(t) + 2A(t)) u'_{\lambda m}(t), u'_{\lambda m}(t)) + 2(M(u_{\lambda m}(t)), u'_{\lambda m}(t)) = \\ &2(f(t), u'_{\lambda m}(t)) + (A'(t) u_{\lambda m}(t), u_{\lambda m}(t)) \end{aligned} \tag{1.8}$$

Integrating the last above identity from 0 to t, $0 < t < t_{\lambda m}$, and using the hypotheses (i) and (iii) we obtain:

$$\begin{aligned} &|A_{2\lambda}^{\frac{1}{2}}(t) u'_{\lambda m}(t)|^2 + \beta \textstyle\int_0^t |u'_{\lambda m}(s)|^2 ds + \|u_{\lambda m}(t)\|_t^2 \le \\ &|P_{\lambda m} u_1|^2 + \|u_{0m}\|_0^2 + \tfrac{1}{\beta} \textstyle\int_0^T |f(s)|^2 ds + \eta \int_0^t |u_{\lambda m}(s)|^2 ds \end{aligned} \tag{1.9}$$

From (1.1), (1.6) , (1.7), from embedding $V_t \hookrightarrow V_0 \hookrightarrow H$ and from Gronwall's inequality, we conclude from (1.9) that the there exists a constant $C > 0$ independent of m and λ such that

$$|A_{2\lambda}^{\frac{1}{2}}(t) u'_{\lambda m}(t)|^2 + \beta \int_0^t |u'_{\lambda m}(s)|^2 ds + \|u_{\lambda m}(t)\|_t^2 \le C + \eta \int_0^t |u_{\lambda m}(s)|^2 ds \tag{1.10}$$

From the embedding $V_t \hookrightarrow V_0 \hookrightarrow H$ and from Gronwall's inequality, we conclude from (1.4) that:

$$|A_{2\lambda}^{\frac{1}{2}}(t) u'_{\lambda m}(t)|^2 + \beta \int_0^t |u'_{\lambda m}(s)| ds + \|u_{\lambda m}(t)\|_t^2 \le C, \tag{1.11}$$

independently from $0 < \lambda \in \mathbb{R}$, $m \in \mathbb{N}$ and of $t \in [0, t_m)$.

Therefore, (1.11) and by Carathéodory Theorem (see [02]) there is a solution in all interval $[0, T]$. From this, we obtain the following estimates:

$$\|u_{\lambda m}\|_{L^\infty(0,T;V_0)} \le C, \ \forall \lambda > 0, \ m \in \mathbb{N} \tag{1.12}$$

$$\|u'_{\lambda m}\|_{L^2(0,T;H)} \le C, \ \forall \lambda > 0, \ m \in \mathbb{N}. \tag{1.13}$$

By (1.12) and noting that M is bounded in V_0, it follows that

$$\|M(u_{\lambda m})\|_{L^\infty(0.T;H)} \le C, \ \forall \lambda > 0, \ m \in \mathbb{N} \tag{1.14}$$

From (1.12)-(1.14), there exists a subsequence of the $(u_{\lambda m})$, still denoted by $(u_{\lambda m})$ and a function u_λ such that:

$$u_{\lambda m} \longrightarrow u_\lambda \text{ weak-star in } L^\infty(0,T;V_0) \tag{1.15}$$

$$u'_{\lambda m} \longrightarrow u'_\lambda \text{ weak in } L^2(0,T;H) \tag{1.16}$$

$$A^{\frac{1}{2}}(t)u_{\lambda m} \longrightarrow A^{\frac{1}{2}}(t)u_\lambda \text{ weak-star in } L^\infty(0.T;H) \tag{1.17}$$

$$A_{2\lambda}(t)u'_{\lambda m} \longrightarrow A_{2\lambda}(t)u'_\lambda \text{ weak in } L^2(0,T;H) \tag{1.18}$$

$$A_1(t)u'_{\lambda m} \longrightarrow A_1(t)u'_\lambda \text{ weak in } L^2(0,T;H) \tag{1.19}$$

$$M(u_{\lambda m}) \longrightarrow \Psi \text{ weak-star in } L^\infty(0,T;H) \tag{1.20}$$

Note that in (1.17)-(1.19) we use the fact that the operators $A^{\frac{1}{2}}(t)$, $A_1(t)$, $A_2(t)$ are weakly closed in $L^2(0,T;H)$.

The proof that $M(u_\lambda) = \Psi$ is made exactly as in M. Clark [04] or O. Lima [07]. There it is used of the hypoteses: $V_0 \hookrightarrow H \hookrightarrow V'_0$, the Lemma of Aubins-Lions (see J. L. Lions [09]) and the monotonicity of M.

By multiplying both sides of (1.5) by $\phi \in C_0^1(0,T)$, integrating from $t=0$ to $t=T$, passing to the limit in $n \to \infty$, using the convergencies (1.15)-(1.20) and noting that V_m are dense in $D(A(0))$, we obtain:

$$-\int_0^T (A_{2\lambda}(t)u_\lambda, \phi' v)dt + \int_0^T (A_1(t)u'_\lambda, \phi v)dt + \int_0^T (A^{\frac{1}{2}}(t)u_\lambda, A^{\frac{1}{2}}(t)\phi v)dt$$

$$+\int_0^T (M(u_\lambda(t)), \phi v)dt = \int_0^T (f(t), \phi v)dt; \forall \phi \in C_0^1(0,T) \text{ and } \forall v \in V_0. \tag{1.21}$$

From this it follows that

$$(A_{2\lambda}(t)u'_\lambda)' = -A_1 u'_\lambda - A(t)u_\lambda - M(u_\lambda) + f \in L^2(0,T;V'_0) \tag{1.22}$$

Noting that the estimates (1.12), (1.13) and (1.14) are independent of $\lambda > 0$, we obtain the same convergencies (1.15)-(1.20) and the equality $M(u) = \Psi$ by changing $u_{\lambda m}$ by u_λ and u_λ by u. In particular

$$A_{2\lambda}(t)u'_\lambda \longrightarrow A_2(t)u' \text{ in } L^2(0,T;H).$$

So, by the above arguments, taking the limit in (1.21) we have that u satisfies (1.2) and (1.3). From (1.3) it follows that

$$(A_2(t)u')' + A_1(t)u' + A(t)u + M(u) = f \quad \text{in} \quad L^2(0,T;V'_0) \tag{1.23}$$

Therefore

$$(A_2(t)u')' \in L^2(0,T;V'_0) \tag{1.24}$$

(ii) The Initial Conditions.

Changing $u_{\lambda m}$ by u_λ and u_λ by u the initial condition (1.4) is obtained by the convergencies (1.15), (1.16) and using Lemma 1.

2-Proof of Theorem 2.

If u and v satisfy Theorem-1, then $w = u - v$ satisfies:

$$(A_2(t)w')' + A_1(t)w' + A(t)w + M(u) - M(v) = 0 \quad \text{in } L^2(0,T;V_0') \tag{1.25}$$

$$w(0) = 0; \quad A_2(0)w'(0) = 0.$$

Will prove that $w = 0$ in $[0,T]$.

Being the solution $u'(t) \in H$ and $(A_2(t)u'(t))' \in V_0'$, it does not make sense the duality between these vectors. In this case, will use the method introduced by Visik and Ladyzenskaja [06].

For each s with $0 < s < T$, will consider the function $z(t)$ given by:

$$z(t) = \begin{cases} -\int_t^s w(\xi)d\xi & \text{if } 0 \le t \le s \\ 0 \text{ if } s < t \le T. \end{cases} \tag{1.26}$$

Therefore $z(s) = 0, z'(t) = w(t)$ for $0 \le t \le s$ and $z(t) \in V_0$ for each $t \in [0,T]$.

Let us consider the function $w_1(t)$ given by, $w_1(t) = \int_0^t w(\tau)d\tau$. Then, $z(t) = w_1(t) - w_1(s)$, $0 \le t \le s$.

Taking the duality of (1.25) with (1.26) and integrating from $t = 0$ to $t = T$, we obtain:

$$\begin{aligned} &\int_0^T < (A_2(t)w')', z > dt + \int_0^T (A_1(t)w', z)dt + \int_0^T < A(t)w, z > dt + \\ &\int_0^T (M(u) - M(v), z)dt = 0 \end{aligned} \tag{1.27}$$

We use Lemma 1 to obtain:

$$\int_0^T < (A_2(t)w')', z > dt = -\frac{1}{2}|A_2^{\frac{1}{2}}(s)w(s)|^2 + \frac{1}{2}\int_0^s (A_2'(t)w, w)dt.$$

Also by simple calculations one has:

$$\int_0^T (A_1(t)w', z)dt = -\int_0^s (A_1'(t)w, z)dt - \int_0^s (A_1(t)w, w)dt$$

$$\int_0^T < A(t)w, z > dt = -\frac{1}{2}|A^{\frac{1}{2}}(0)w_1(s)|^2 - \frac{1}{2}\int_0^s (A'(t)z, z)dt$$

Substituting the above equalities in (1.27) and using hypotheses (i),(ii),(iii) and (v) we obtain:

$$\frac{1}{2}|A_2^{\frac{1}{2}}(s)w(s)|^2 + \beta\int_0^s |w(t)|^2dt + \frac{\alpha}{2}|w_1(s)|^2 \le \gamma\int_0^s |w(s)||z(t)|dt + \frac{\eta}{2}\int_0^s |z(t)|^2dt;$$

where $\gamma = \tau + \rho$.

Being $z(t) = w_1(t) - w_1(s)$.From the elementary inequality $ab \le \frac{\lambda a^2}{2} + \frac{b^2}{2\lambda}$; $\lambda > 0$, the above inequality it comes:

$$(\beta - \gamma^2\lambda)\int_0^s |w(t)|^2dt + [\frac{\alpha}{2} - (\frac{1}{2\lambda} + \eta)s]|w_1(s)|^2 \le (\frac{1}{2\lambda} + \eta)\int_0^s |w_1(t)|^2dt,$$

$\forall\lambda > 0$ such that $\beta - \gamma^2\lambda > 0$ and $\frac{\alpha}{2} - (\frac{1}{2\lambda} + \eta)s > 0$.

Taking $\lambda > 0$ such that $\beta - \gamma^2\lambda = \frac{\beta}{2}$, that is, $\lambda = \frac{\gamma}{2\gamma^2}$, and s_0 such that $\frac{\alpha}{2} - (\frac{1}{2\lambda} + \eta)s_0 = \frac{\alpha}{4}$, that is, $s_0 = \frac{\alpha\lambda}{1+2\lambda\eta}$, we obtain from the above equality:

$$\frac{\beta}{2}\int_0^s |w(t)|^2 dt + \frac{\alpha}{4}|w_1(s)|^2 \leq (\frac{1}{2\lambda} + \eta)\int_0^s |w_1(t)|^2 dt, \quad \forall s \in [0, s_0].$$

Applying Gronwall's inequality it implies that $w_1(s) = 0 \ \forall s \in [0, s_0]$, and consequently $w(t) = 0, \ \forall t \in [0, s_0]$.

As $w(s_0) = 0$, we have from (1.25) that: $(A_2(t)w')' + A_1(t)w' = 0$, a. e. in $[0, s_0]$. Then $A_2(t)w' + A_1(t)w$ is constant a. e. in $[0, s_0]$. As $w(0) = A_2(0)w'(0) = 0$, it implies that $A_2(t)w' + A_1(t)w = 0, \quad \forall t \in [0, s_0]$. Then $A_2(s_0)w'(s_0) = 0$.

Applying the same arguments in $[s_0, 2s_0]$ to the Cauchy problem:

$$(A_2(t)w')' + A_1(t)w' + A(t)w + M(u) - M(v) = 0, \ w(s_0) = 0; \qquad A_2(s_0)w'(s_0) = 0,$$

we obtain that $w(t) = 0$, for all $t \in [0, 2s_0]$. Therefore, after a finite number of steps we conclude that $w = 0$ in $[0, T]$ and the proof of the Theorem 2 is completed.

4 EXAMPLES

I) Let Ω be a regular bounded open subset of $\mathbb{R}^n$, $0 < T \in \mathbb{R}$, and $H = L^2(\Omega), V = \dot{H}_0^1(\Omega)$. We denote by $(.,.)$, $|.|$ and $((.,.))$, $\|.\|$ the inner product and norm or $L^2(\Omega)$ and $H_0^1(\Omega)$ respectively.

Let us define the functions k_1 and k_2 such that: i) $k_1, \ k_2, \ k_2' \in L^\infty(0,T;H)$ ii) $k_2(x,t) \geq 0$; $k_2'(x,t) \geq 0$ a. e. in $Q = \Omega \times (0,T)$.

Will define the operators $A_1(t)$, $A_2(t)$, of H by:

$$(A_1(t)u)(x) = k_1(x,t)u(x); (A_2(t)u)(x) = k_2(x,t)u(x)$$

such that

$$((A_1(t)u, u) \geq \beta|u|^2, \ 0 < \beta \in \mathbb{R}, \forall u \in H; \forall t \in [0,T].$$

Note that $(A_2(t)u, u) \geq 0, \forall u \in H, \ t \geq 0$.

There exists functions k_1 and k_2 satisfying (i) and (ii) above. For example: $k_1(x,t) = 1 + \frac{1}{2}K(x)t^\alpha; k_2(x,t) = \frac{1}{\alpha+1}K(x)t^{\alpha+1}$, where $K \in L^\infty(\Omega)$ such that $K(x) \geq 0$ in Ω and $\alpha > -1$.

We consider

$$A(t) = -\sum_{ij=1}^n \frac{\partial}{\partial x_j}(a_{ij}(x,t)\frac{\partial}{\partial x_i})$$

defined in the domain $D(A(t)) = H^2(\Omega) \cap H_0^1(\Omega)$ that is dense in $H = L^2(\Omega)$; where $a_{ij} = a_{ji}$ and $a_{ij}, a_{ij}' = \frac{\partial}{\partial t}a_{ij} \in L^\infty(Q)$ and $\frac{\partial}{\partial t}$ is classic derivative.

Then $A(t)$ is a family of self-adjoint operators and, if we assume that there exists $\rho > 0$ such that

$$\sum_{i,j=1}^n a_{ij}(x,t)\xi_i\xi_j \leq \lambda(|\xi_1|^2 + \cdots + |\xi_n|^2)$$
$$\sum_{i,j=1}^n \frac{\partial}{\partial t}a_{ij}(x,t)\xi_i\xi_j \geq \mu(|\xi_1|^2 + \cdots + |\xi_n|^2);$$

$(x,t) \in Q$,and $\xi = (\xi_1, \cdots, \xi_n) \in \mathbb{R}$. Then, the Poincaré-Friedrichs inequality implies that $(A(t)u, u) \geq \alpha|u|^2$, $\forall u \in D(A(t))$ and α is a positive constant. It is easy to see that $A(t)$ is strongly continuously differentiable in H and that

$$|(A'(t)u, u)| \leq \eta|u|^2, \quad \forall t \in [0, T].$$

Let $F : \mathbb{R} \longrightarrow \mathbb{R}$ be the function defined by $F(s) = s^3$, and a operator $M : H_0^1(\Omega) \longrightarrow L^2(\Omega)$ defined by $(Mu)(x) = F(u(x))$.Then, from the properties of F it follows that M is monotone, bounded , hemicontinuous and Lipschitzian in the norm $L^2(\Omega)$, and

$$\int_0^T (M(u(s)), u'(s))ds > -\sigma, \forall t \in [0, T], \forall u \in E_C$$

where E_C is the set $\{u \in L^\infty(0, T; H_0^1(\Omega)), u' \in L^2(0, T; L^2(\Omega))$and $\|u(0)\| < C\}$. The constant σ depends on C (see [04]).

If $G : \mathbb{R} \longrightarrow \mathbb{R}$ is a function defined by

$$G(s) = \begin{cases} F(s), & \text{if} \quad |s| \leq a \in \mathbb{R} \\ a_0, & \text{if} \quad |s| \geq a, \end{cases}$$

we have that $|G'(s)| \leq C$, $\forall\ s \in \mathbb{R}$. Therefore M is Lipschitzian.

We are in the conditions of the Theorems-1 and 2. Therefore one has studied the existence and uniqueness of solutions of the mixed problem for the equation

$$(k_2(x,t)u')' + k_1(x,t)u' + A(t)u + u^3 = f.$$

II) Let Ω be a regular bounded open subset of $\mathbb{R}^3$; $H = L^2(\Omega)$ and $V = L^2(\Omega)$. In the same scheme we have analogous results for the equation

$$(k_2(x,t)u')' + k_1(x,t)u' + A(t)u + M(u) = f$$

where k_1, k_2 functions in the conditions i) and ii) of the example (I);$(Mu)(x) = F(u(x))$ where $F(s)$, defined in (Lima, O. [07]) as,

$$F(s) = \begin{cases} \text{linear} & \text{if} \quad s < -s_0 \\ s^3 \quad \text{if} & -s_0 \leq s \leq s_0 \\ \text{linear} & \text{if} \quad s > s_0 \end{cases}$$

$s_0 > 0$ is real number and $M : H_0^1(\Omega) \longrightarrow L^2(\Omega)$ is defined by $(Mu)(x) = F(u(x))$. Note that $H_0^1(\Omega) \hookrightarrow L^6(\Omega)$.

REFERENCES

[1] A. BENSOUSSAN,; J. L. LIONS and G. PAPANICOLAU, <u>*Perturbations et aumentatiom des conditions initiales*</u>, Singular Perturbation and Boundary Layer Theory, Springer Verlarg, Lion (1976),10-29.

[2] CODDINGTON, E.& LEVISON, N.-*Theory of ordinary differential equations*, McGraw-Hill,New York (1955).

[3] CLARK, M.R.-*Uma equação hiperbólica-parabólica não linear* , Proceeding of XXIV Bras. Sem. of Analysis (1986).

[4] CLARK, M. R. - *Uma Equação Hiperbólica-Parabólica abstrata não Linear, existência e unicidade de soluções Fracas*, Thesis of Doctoral, Institute of Mathematics of the Federal University of Rio the Janeiro, Brazil(1987).

[5] LARKIN, N. A. -*Boundary problems in the large for a class of hyperbolic equations* - Siberian Math. Journal, vol. 18, n^0 6, (1977), 1003 - 1006.

[6] LADYZENSKAJA, O.A.- VISIK, M.I.-*On boundary value problems for PDE and certain class of operators equations,* AMS Transl. 10 (2), (1958).

[7] LIMA,O.A.-*Existence and uniqueness of solutios for an abstract nonlinear hyperbolic - parabolic equation*, Applicable Analysis, vol. 24.(1987), 101-116.

[8] LIONS, J. L. - *Quelques Méthodes de Resolution des Problemes aux Limites non Linéaires* - Dunod, Gauthier-Villars,Paris,(1969).

[9] LIONS, J. L.: MAGENES, E.- *Problémes aux Limites non Homogenes et Applications*, Vol-I Dunod, Paris .(1968).

[10] MACIEL, A.- *On a hyperbolic-parabolic equation with a continuos non-linearity.* Thesis of Doctor, IM-UFRJ, Rio de Janeiro (1986).

[11] MEDEIROS, L.A. -*Non-linear Hyperbolic-Parabolic Partial Differential Equations*- Funkciala, Ekvacioj, n^0 23, (1978), 151-158.

[12] MEDEIROS, L. A.-*Remarks on Hyperbolic-Parabolic Partial Differential Equations* - Proccedings of XI Bras. Sem. of Analysis, Rio de Janeiro - (1978), 1-12.

[13] MEDEIROS. L. A.- *Semilinear wave Equations*, Lectures Notes in Mathematics, Springer Verlag - N. Y.,(1975).

[14] NEVES, B.P.- *Perturbações e aumento das condições iniciais de equações hiperbólica-parabólicas não Lineares*- Procceding of IX Bras. Sem. of Analysis, Rio de Janeiro, (1979), 109-123.

[15] STRAUSS, W.A.- *The Energy Method in Nonlinear Pertial Differential Equations* - Notas de Matemática, n° 47,IMPA ,(1969).

[16] VAGROV, V.N.- *On the Cauchy problem for some parabolic-hyperbolic equations* - Soviet. Math. Dokl, vol. 14-5, (1973), 1396-1400.

7th Int. Coll. on Differential Equations, pp. 67-74
D. Bainov (Ed)

Asymptotic expansions for some classical operators and their use in approximation theory

*Francesco COSTABILE, *Maria Italia GUALTIERI and **Stefano SERRA
*Dipartimento di Matematica, Università della Calabria. (E-mail:Costabil@unical.it)
**Dipartimento di Energetica, Università di Firenze. (E-mail:Serra@morse.dm.unipi.it)

Abstract. In this paper we review some results recently introduced by the authors about the use of extrapolation techniques in approximation theory.

We used very classical and well-known operators, i.e., the Bernstein one $B_n(\cdot)$ and that one associated with the De La Vallée Poussin integral $V_n(\cdot)$ [3, 4] (see also [5] for general exponential type operators). When the function to be approximated f is continuous, the standard polynomials $B_n(f)$ converge uniformly to the given function f; it is also true that the same property holds in the case where f is a continuous, 2π-periodic and we approximate it by means of the trigonometric polynomials $V_n(f)$. In addition these operators allow to approximate simultaneously the first k derivatives under the only assumption of continuity of $f^{(k)}$. These very good qualitative features are, unfortunately, joint with a dramatic slowness in the order of the approximation. Actually, in general, we cannot expect an order of approximation asymptotically better than $O(n^{-1})$ [16]. Therefore these polynomials, interesting from an analytical point of view, did not look attractive from a computational point of view.

In earlier works [3, 4] we proved two asymptotic expansions of the polynomials $B_n(f)$, $V_n(f)$ with respect to the grid "parameter" $h = n^{-1}$. These formulas used in connection with extrapolation techniques have shown to be effective in approximating the given function f, at the same time, fast and efficiently.

The main advantages, with respect to the orthogonal polynomial based approximation procedures (the best ones for the order of the approximation), are the use of evaluations of f only on an uniform grid and the simultaneous approximation, the last property being inherited by the basic operators.

Finally several numerical experiments performed in [3, 4] have fully confirmed the theoretical analysis.
Keywords: Bernstein operator, De La Vallée Poussin operator, extrapolation.
AMS SC: 65D15, 65B05.

1 INTRODUCTION

In this paper, we consider the use of classical linear operators, such as the De La Vallée Poussin integrals $V_n(f)$ [13]

$$V_n(f) = \frac{(2n)!!}{(2n-1)!!}\,\frac{1}{2\pi}\int_{-\pi}^{\pi} f(t)\,\cos^{2n}\frac{(t-x)}{2}\,dt \tag{1.1}$$

and the Bernstein polynomials $B_n(f)$ [16]

$$B_n(f) = \sum_{k=0}^{n} \binom{n}{k} f\left(\frac{k}{n}\right) x^k (1-x)^{n-k} \tag{1.2}$$

in order to approximate a given function belonging to the linear space of the continuous 2π functions $C_{2\pi}$ or to the space of the continuous function over $[0, 1]$, respectively. We remark that, in the first case, the approximation is obtained by means of trigonometric polynomials while, in the other case, we use standard polynomials.

The theorem of Korovkin (see [11], pages 14-17) asserts that the convergence of positive linear operators generically defined in C_{b-a} on itself, to the identity, depends only on its convergence on three test functions. Moreover, the related rate of convergence depends also on the specific convergence speed with regard to these test functions. We point out that the operators defined by (1.1) and (1.2) have the previous feature [13] and, consequently, they give a *theoretical* scheme to approximate the given function f.

Unfortunately, the magnitude of the convergence is computationally not satisfactory and improves with the smoothness of the functions up to $C^2_{2\pi}$ (or $f \in C^2[0, 1]$), whereas it does not improve with smoother class of functions. More precisely, if $f \in C^2_{2\pi}$ or $f \in C^2[0, 1]$ we find [13, 16]

$$\|V_n(f) - f\|_\infty = O(n^{-1}), \quad \|B_n(f) - f\|_\infty = O(n^{-1}). \tag{1.3}$$

As shown in [8], to improve the order of the approximation we must consider nonpositive operators. Actually, in [2, 12, 7, 8] the authors also define special linear combinations of few terms $B_n(f)$ and $V_n(f)$, respectively, with different choices of n and they prove approximation results finer than (1.3).

In the sequel, we fully generalise this idea by following two main stages (see also [3, 4]):

- We demonstrate an asymptotic expansion for $B_n(f)$ and $V_n(f)$ by proving that, under some regularity conditions, there exist constant values $S_i(f, x)$ and $Z_i(f, x)$ independent of n such that the following relations hold

$$B_n(f) = f(x) + \sum_{i=1}^{k} S_i(f, x)h^i + s_k(f, x, h)o(h^k),$$

$$V_n(f) = f(x) + \sum_{i=1}^{k} Z_i(f, x)h^i + z_k(f, x, h)o(h^k).$$

  Here $h = n^{-1}$ and s_k, z_k are uniformly bounded with the parameter h.

- We apply an extrapolation algorithm and, by virtue of the previous expansion, we obtain an error of order h^k for the extrapolated expression (of course, as remarked in [8], the new operator is a nonpositive one).

Observe that the resulting expressions are linear combinations of the Bernstein and De La Vallée Poussin terms; actually, by using this procedure we can easily obtain also the linear combinations considered in [2, 12, 7, 8]. For the approximation by linear combinations of $V_n(f)$, from a computational point of view, the situation is a little bit more complicate: actually, each trigonometric polynomial $V_n(f)$ is not given explicitly, but is implicitly defined by using an integral. Consequently, since we are interested in computational algorithms, we do not calculate the terms $V_n(f)$, but we use an extended trapezoidal rule in order to evaluate the given integral. In addition, by calling $T_N(f_n)$

the extended trapezoidal rule applied to $f_n(t) = f(t)\cos^{2n}\frac{(t-x)}{2}$, we prove another asymptotic expansion for $T_N(f_n)$. More precisely, putting $H = N^{-1}$ and t_n the constant defined in the definition 2.3, we have

$$t_n T_N(f_n) = f(x) + \sum_{i=1}^{k} S_i(f,x)h^i + z_k(f,x,h)o(h^k) + w_k(f,x,H)H^{2k}$$

with z_k and w_k uniformly bounded with regard to h and H, respectively. We notice that, in this way, we require the knowledge of the function f only on a uniform grid, since we apply the extrapolation methods, not to the theoretic function $V_n(f)$, but to the values $t_n T_N(f_n)$.

The paper is organised as follows. In the section 2, we give the asymptotic expansions for $B_n(f)$ and $V_n(f)$ proved in [3, 4] while, in section 3, we discuss the application of the extrapolation procedures, we exhibit a new asymptotic expansion for $T_N(f_n)$ and we indicate how to choose, in an "suitable" way, the parameter N. Finally, in the last section 4, we report few numerical experiments which confirm the correctness of the theoretical analysis.

2 MAIN RESULTS

We start by recalling some classical preliminary definitions and results.

Definition 2.1 *Let $f \in C[0,1]$, then the n-th Bernstein polynomial is defined as follows*

$$B_n(f) = \sum_{k=0}^{n} \binom{n}{k} f\left(\frac{k}{n}\right) x^k(1-x)^{n-k}. \tag{2.1}$$

Definition 2.2 *If $f \in C_{2\pi}$, then we define as the De La Vallée-Poussin integral the trigonometric polynomial of degree n given by*

$$V_n(x) = \frac{(2n)!!}{(2n-1)!!}\frac{1}{2\pi}\int_{-\pi}^{\pi} f(t)\cos^{2n}\frac{(t-x)}{2}\,dt \tag{2.2}$$

where the symbol $n!!$ denotes the product of all the positive natural numbers not exceeding n and even or uneven according as n is even or uneven.

In addition, by setting

$$t_n^{-1} = \int_0^{\pi/2} \cos^{2n} t\,dt = \frac{(2n-1)!!}{(2n)!!}\frac{\pi}{2} \tag{2.3}$$

we have

$$V_n(x) = t_n \frac{1}{\pi}\int_0^{\pi/2} [f(x+2t) + f(x-2t)]\cos^{2n} t\,dt. \tag{2.4}$$

Finally, the following very classical approximation results hold true

Theorem 2.1 [16, 13] *if $f \in C[0,1]$, then for all $x \in [0,1]$*

$$\lim_{n\to\infty} B_n(x) = f(x) \tag{2.5}$$

Otherwise, if $f \in C_{2\pi}$, then for all $x \in R$

$$\lim_{n\to\infty} V_n(x) = f(x) \tag{2.6}$$

Now we give the main results, which allows one to use extrapolation algorithms in order to speed up the convergence of linear combinations of Bernstein's and De La Vallée Pussin's terms respectively. We remark that the following theorems are a strong refinement of the results resumed in theorem 2.1.

Theorem 2.2 [3] *If $f \in C^{2k}[0,1]$ then the following expansion holds true*

$$B_n(f) = f(x) + S_1(f,x)h + S_2(f,x)h^2 + \ldots + S_k(f,x)h^k + s_k(f,x,h)o(h^k) \qquad (2.7)$$

with $S_i(f,x)$ independent of $h = \frac{1}{n}$ and $s_k(f,x,h)$ uniformly bounded with respect to x, with h.

Theorem 2.3 [4] *If $f \in C^{2k}_{2\pi}$ then the following expansion holds true*

$$V_n(f) = f(x) + Z_1(f,x)h + Z_2(f,x)h^2 + \ldots + Z_k(f,x)h^k + z_k(f,x,h)o(h^k) \qquad (2.8)$$

with $Z_i(f,x)$ independent of $h = \frac{1}{n}$ and $z_k(f,x,h)$ uniformly bounded with respect to x, with h.

3 THE USE OF THE POLYNOMIAL EXTRAPOLATION

As first step we apply a polynomial procedure of extrapolation to the terms $B_n(f)$ and $V_n(f)$ by obtaining, as a very special case, the nonpositive operators found in [2, 12, 9, 8].

Theorem 3.1 [15] *Let $f \in C[0,1]$ (or $f \in C^{2k}_{2\pi}$), let $\{n_0 = n < n_1 < n_2 < \cdots\}$ be an increasing sequence of positive integers and let us define $h_i = n_i^{-1}$. We set $\{X_q^{(i)}\}$ a sequence of values constructed in the following way*

$$\begin{cases} X_0^{(i)} & := \quad B_{n_i}(f) \ \text{ or } \ (V_{n_i}(f)), \\ X_q^{(i)} & := \quad \frac{h_{i+q}X_{q-1}^{(i)} - h_i X_{q-1}^{(i+1)}}{h_{i+q}-h_i}, \quad q = 1,2,\ldots,k-1. \end{cases}$$

Then, for i fixed positive number, we have

$$\lim_{h_i \to 0} X_q^{(i)} = f(x), \qquad (3.1)$$

while the error takes the form

$$X_q^{(i)} - f(x) = h_i h_{i+1} \cdots h_q (-1)^q (S_{q+1}(f,x) + O(h_i)). \qquad (3.2)$$

3.1 Comparison with the literature

3.1.1 The Bernstein case

In his paper May [12] defined a general class of linear combinations of Bernstein polynomials which contains, for instance, the classical combinations considered by Butzer [2]. Now we recall the definition of these polynomials and we show that our technique is much more general in the sense that we can obtain all the May polynomials as a special case of extrapolated values $X_q^{(i)}$.

Let $\Delta = \{d_0, \ldots, d_k\}$ be a set of $k+1$ arbitrary but fixed and increasingly ordered distinct positive integers. The operator $S_n(f, \Delta)$ is a linear combination of Bernstein polynomials $B_{nd_j}(f)$ defined as

$$S_n(f,\Delta) = \sum_{j=0}^{k} c(j,k) B_{nd_j}(f)$$

with

$$c(j,k) = \prod_{i=0,\ i\neq j}^{k} \frac{d_j}{d_j - d_i}, \quad k \neq 0, \quad c(0,0) = 1.$$

By pointing out that the values $c(j,k)$ are the Lagrange polynomials calculated in $h = 0$ (see [3] for the details), it is trivial to conclude that $S_n(f,\Delta) = X_k^{(0)}$ with respect to the sequence $\{n_0 = n < n_1 < n_2 < \cdots\}$ with $n_j = n \cdot d_j$. In particular, for the choice of $d_j = 2^{j-1}$, we find the Butzer operator [2].

Moreover, from this remark we deduce also that our approach is more general. In fact, for instance, all the sets of the form $n_j = n + z(j)$, with $z(j)$ increasing function not depending on n, cannot be obtained following the scheme given by May.

Finally our class contains a special operator which has an optimality property with regard to the approximation of a given function f: if $f \in C^{2k}[0.1]$ and $n_j = n + j$, then the polynomial $X_k^{(0)}$ obtained as combination of $k+1$ distinct Bernstein polynomials is the combination of **smallest degree** among all the polynomials satisfying the subsequent properties:

- $\|f - p\|_\infty \leq \frac{M}{n^k}$, M independent of n.
- p is a linear combination of $k+1$ distinct Bernstein polynomials of degree greater or equal to n.

3.1.2 The De La Vallée Poussin case

We start by observing that the De La Vallée linear combinations have received less attention than the linear combinations of Bernstein's polynomials. However, in this paragraph we observe that, if we choose the Romberg sequence $n_i = n\rho^i$ with $\rho = 2$ and we apply the procedure given in the theorem 3.1 by assuming $q = 1, 2$, we obtain

$$X_1^{(0)} = 2V_{2n}(f) - V_n(f), \quad (n_0 = n),$$

$$X_2^{(0)} = \frac{8}{3}V_{4n}(f) - 2V_n(f) + \frac{1}{3}V_n(f), \quad (n_0 = n).$$

It is worth noting that this two operators have been considered as linear combinations of the De La Vallée Poussin terms in [8]; according to the previous theorem, if the function f is regular enough then we have an error asymptotic to $\frac{1}{n^2}$ and $\frac{1}{n^3}$, respectively. Finally, we remark that by means of this extrapolation procedure, we can construct, in a general and uniform way, several examples of linear combinations of the terms $V_n(f)$ which well approximate the given function f. Even in this case the choice of $n_j = n + j$ possesses the same optimality property stated in the previous paragraph for the Bernstein case.

3.2 Another asymptotic expansion for the term $V_n(f)$

Since we are interested in computational methods and the trigonometric polynomials $V_n(f)$ are defined by an integral, in the following we consider the use of an integration rule.

First we recall a classical result about the extended trapezoidal rule for a regular function.

Theorem 3.2 *Let $f \in C^{2k}[a,b]$, let us indicate with $T_N(f)$ the extended trapezoidal rule applied to f with stepsize $H = N^{-1}$. Then the following expansion holds*

$$T_N(f) = \tau_0 + \sum_{j=1}^{k-1} \tau_j H^{2j} + \tau_k(H)H^{2k}$$

where $\tau_0 = \int_a^b f(x)dx$, $\tau_k(H)$ is uniformly bounded with respect to H and has the following expression

$$\tau_k = \frac{-1}{(2k)!}\int_a^b f^{(2k)}(x)(b)\left[S_{2k}\left(\frac{x-a}{h}\right) - S_{2k}(0)\right] dx; \tag{3.3}$$

here the symbol S_j denotes the Bernouilli polynomials periodically extended on the whole real axis. The other parameters τ_j have the form

$$\tau_j = \frac{(-1)^{j+1}B_j}{(2j)!}\left[f^{(2j-1)}(b) - f^{(2j-1)}(a)\right] \tag{3.4}$$

with B_j being the Bernouilli numbers [15].

Proof For a detailed proof see [15], pp. 106-111. •

Observe that if f is also periodic it follows that $f^{(2j-1)}(b) = f^{(2j-1)}(a)$ for $j = 1, \dots, k$ and therefore the extended trapezoidal rule becomes very precise. In particular we have

$$T_N(f) = \int_a^b f(x)dx + \tau_k(H)H^{2k}.$$

Now, we want to link this kind of asymptotic expansion with that one found for the De La Vallée Poussin polynomials in the previous section. The idea is motivated by computational reasons; since each term $V_n(f)$ is not easily evaluable, we will use a trapezoidal rule in order to have an approximation of it. More precisely, we consider

$$V_n(f) = t_n \int_{-\pi}^{\pi} f_n(x,t)dt$$

where $f_n(x,t) = f(t)\cos^{2n}\dfrac{t-x}{2}$ and we point out that $f_n(x,t)$ has the same regularity of $f(t)$ and is also 2π-periodic. Therefore we obtain

$$t_n T_N(f_n) = V_n(f) + t_n\tau_k(H)H^{2k}$$

where, in the light of theorem 2.1, the quantity $V_n(f)$ is asymptotic to $f(x)$. Finally we combine the two expansions and we apply directly the extrapolation procedures to the terms $T_N(f_n)$.

Of course, in order to obtain a convergent scheme and to evaluate the related error, we must consider only one parameter. Here we choose $H = h^{1+\delta} \cdot \sqrt{h}$, $\delta > 1/(4k)$ and we observe that $f_n^{(j)} = f(x)(n^j \cos^{2n-j}((x-t)/2) + g_n(x,t))$ where g_n is a trigonometric polynomial with $(x-t)/2$ whose coefficients are standard polynomial of degree strictly less than j. In this way, by observing that $t_n \sim \left(\frac{n}{\pi}\right)^{1/2}$ (see the Wallis formula [1]), we obtain a new asymptotic expansion of the form [4]

$$t_n T_{N(n)}(f_n) = f(x) + \sum_{j=1}^{k} S_j(f,x)h^j + o(h^k). \tag{3.5}$$

To conclude, in the following theorem [15], we summarise the definition of a scheme of polynomial extrapolation and the behaviour of the related error.

Theorem 3.3 *Let $f \in C_{2\pi}^{2k}$, let $\{n_0 = n < n_1 < n_2 < \cdots\}$ be an increasing sequence of positive integers, let us define $h_i = n_i^{-1}$ and $N(n_i) = \left\lceil n_i^{1+\delta} \cdot \sqrt{n_i}\right\rceil$, $\delta > 1/(4k)$. We set $\{T_q^{(i)}\}$ a sequence of values constructed in the following way*

$$\begin{cases} T_0^{(i)} & := \quad t_{n_i} T_{N(n_i)}(f_{n_i}), \\ T_q^{(i)} & := \quad \frac{h_{i+q}T_{q-1}^{(i)} - h_i T_{q-1}^{(i+1)}}{h_{i+q}-h_i}, \quad q = 1,2,\ldots,k-1. \end{cases}$$

Then, for i fixed positive number, we have

$$\lim_{h_i \to 0} T_q^{(i)} = f(x), \tag{3.6}$$

while the error takes the form

$$T_q^{(i)} - f(x) = h_i h_{i+1} \cdots h_q (-1)^q (S_{q+1}(f,x) + O(h_i)). \tag{3.7}$$

We can consider several choices of the sequences $\{n_i\}$: the Romberg one $n_i = n\rho^i$ (typically $\rho = 2$) or the choice $n_i = n \cdot i$ or the choice $n_i = n + i$. Observe that the last choice minimises the number of evaluation points in the calculation of $T_{N(n_i)}(f_{n_i})$ as n_i grows as slow as possible. However, it is interesting to point out that, in view of the preceding theorem, the error $T_q^{(i)} - f(x)$ is asymptotical to $\frac{1}{n^k}$ in all the considered case.

4 NUMERICAL RESULTS

In order to conclude, we perform two numerical tests for the problem considered here: the approximation by polynomials of a given function. It should be noted that, if f has a fixed order of smoothness, then, as shown in theorems 3.1 and 3.3, the convergence rate of $T_q^{(0)}(f)$ to f (or $X_q^{(0)}(f)$ to f) is the same for all the considered choices of the sequences n_i.

Consequently we perform all the numerical tests by using the choice $n_i = n + i$ which guarantees the smallest number of functional evaluations of the function f.

CASE 1.

In the first example we consider the 2π-periodic function $f(x) = sin(3x) + cos(x) \in C_{2\pi}^{\infty}$. By taking $n = 10$ and $q = 10$ we obtain

$$\left\| f - T_q^{(0)}(f) \right\|_{\infty} \le 1.64\,10^{-7}.$$

CASE 2.

In the second example we consider the function $f(x) = x^{5.5} \in C^5[0,1]$ and we apply the extrapolation to the Bernstein polynomials. By taking $n = 1$ and $q = 10$ we obtain

$$\left\| f - X_q^{(0)}(f) \right\|_{\infty} \le 0.183\,10^{-9}.$$

REFERENCES

[1] M. Abramowitz, I.A. Stegun, Editors, Handbook of Mathematical Functions, National Bureau of Standards, US Gov. Printing Office 1964.

[2] P.L. Butzer, Can. J. Math., 5, pp. 559-567 (1953).

[3] F. Costabile, M.I. Gualtieri, S. Serra, BIT, 36, pp. 40-51 (1996).

[4] F. Costabile, M.I. Gualtieri, S. Serra, J. Approx. Th., submitted.

[5] F. Costabile, M.I. Gualtieri, S. Serra, Proc. third Int. Conf. on "Functional Analysis and Approximation Theory", Maratea - Italy (1996), submitted.

[6] R. De Vore, G. Lorentz, Constructive Approximation, Springer-Verlag, Berlin (1991).

[7] Z. Dizian, J. Approx. Th., 26, pp. 277-292 (1976).

[8] Z. Dizian, G. Freud, J. Approx. Th., 12, pp. 23-31 (1974).

[9] M. Frentiu, St. Univ. Babes-Boljai Set. Math. Mech., 15, pp. 63-68 (1970).

[10] D. Jackson, The Theory of Approximation, AMS Coll. Publ., 11, New York (1930).

[11] P.P. Korovkin, Linear Operators and Approximation Theory (English translation), Hindustan Publishing Co., Delhi (1960).

[12] C.P. May, Canad. J. Math., 28, pp. 1224-1250 (1976).

[13] I.P. Natanson, Constructive Function Theory, 1, Frederick Ungar Publishing Co., New York (1964).

[14] M.J.D. Powell, Comput. J., 9, pp. 404-407 (1966).

[15] J. Stoer, R. Bulirsh, Introduction to Numerical Analysis, 1, Springer-Verlag, Berlin (1970).

[16] E. Voronowskaja, C.R. Acad. Sci. USSR, pp. 79-85 (1932).

7th Int. Coll. on Differential Equations, pp. 75-80
D. Bainov and V. Covachev (Eds)

Some Over-Determined Boundary-Value Problems for Hyperbolic Equations

D. E. EDMUNDS
Center for Mathematical Analysis and its Applications, University of Sussex, Brighton BN1 9QH, ENGLAND
E-mail: mmfb7@sussex.ac.uk

N. I. POPIVANOV
Department of Mathematics and Informatics, University of Sofia, 1164 Sofia, BULGARIA
E-mail: nedyu@fmi.uni-sofia.bg

Abstract. Some 3-dimensional analogoues of the plane Darboux-problems for hyperbolic equation with degeneracy are investigated. In 1954, Protter initiated the study of such 3-dimensional problems, and it is now well known, that for an infinite number of smooth right-hand sides these problems have solutions with a strong power-type singularity on the characteristic cone. This effect appear even for small perturbations of certain C_0^∞ right-hand sides. Using Friedrichs' theory of symmetric positive operators, we find and investigate a non-local problem which is a regulariser, in some sence, of these ill-posed problems. To describe an appropriate regularization for that improperly posed problems we shall discuss here.

Keywords: ill-posed problems, regularization methods, boundary value problems, non-local problems, degenerate hyperbolic equations.

1. Introduction and statement of the problem

Let we denote points of $\mathbb{R}^3$ by $x = (x_1, x_2, t)$ and put $\rho = \sqrt{x_1^2 + x_2^2}$, $\varphi = \arctan(x_2/x_1)$. Let $K : [0, \infty) \to \mathbb{R}$ be of class C^1 and such that $K(0) = 0, K'(0) > 0$ with $K(t) > 0$ and $K'(t) > 0$ if $t > 0$. Let G be the domain

$$G = \left\{x \in \mathbb{R}^3 : \int_0^t \sqrt{K(\tau)}d\tau < \rho < 1 - \int_0^t \sqrt{K(\tau)}d\tau\right\},$$

The boundary of G is $\partial G = S_0 \cup S_1 \cup S_2$, where S_0 is the disc $S_0 = \{x : t = 0,\ 0 \le \rho \le 1\}$ and

$$S_1 = \left\{\rho = 1 - \int_0^t \sqrt{K(\tau)}d\tau\right\}, S_2 = \left\{\rho = \int_0^t \sqrt{K(\tau)}d\tau\right\}.$$

We shall consider the equation

$$(1) \qquad Lu := K(t)\left(u_{x_1x_1} + u_{x_2x_2}\right) - u_{tt} \equiv K(t)\left\{\rho^{-1}\left(\rho u_\rho\right)_\rho + \rho^{-2}u_{\varphi\varphi}\right\} - u_{tt} = f,$$

where f is a prescribed function; S_1 and S_2 are characteristics of (1).

Problem P. Find a solution of (1) in G, which satisfies the condition

$$u = 0 \quad \text{on} \quad S_0 \cup S_1. \tag{2}$$

Problem P*. Find a solution of (1) in G, which satisfies the condition

$$u = 0 \quad \text{on} \quad S_0 \cup S_2. \tag{3}$$

Protter [1] formulated these adjoint problems as multi-dimensional analogues of the Darboux problem in the plane. He worked with the wave equation, corresponding to $K(t) = 1$; and also investigated equation (1) in a domain which contained G in its hyperbolic part and contained an other set G' in which (1) is elliptic. For equation (1), which is of changing type in $G \cup G' \cup S_0$, he formulated certain other problems, which are three-dimensional analogues of a plane problem examined by Morawetz [2] and Lax and Phillips [3]. When equation (1) is of changing type, Protter's problem given in Protter [1] was studied by Aziz and Schneider [4], Bicadze [5], Didenko [6] and others. Problems P and P* for (1) in the domain G were considered by Didenko [6] in the case of the Tricomi equation (in which $K(t) \equiv t$). In the same case, after the paper of Khe Kan Cher [7], Popivanov shown in 1986, that the homogeneous Problem P* has infinitely many classical solutions v_n $(n = 4, 5, \ldots)$, where

$$v_n(t, \rho, \varphi) = t\rho^{-n} \left(\rho^2 - 4t^3/9\right)^{n-4/3} (a_n \cos n\varphi + b_n \sin n\varphi), \tag{4}$$

and a_n, b_n are arbitrary constants. This corresponds to the result of Tong Kwang-Chang [8] for the wave equation, and implies that for classical solvability of Problem P an infinite number of conditions of the form $f \perp v_n$ $(n = 4, 5, \ldots)$ are necessary. For further work on the problem (1), (2), see Aldashev [9], Khe Kan Cher [10], Popivanov and Schneider [11,12] and the references cited in these works.

Popivanov and Schneider [11] proceeded in another way, being interested in the question: why does Problem P not have a classical solution when $f = v_n$ (v_n as in (4))? They introduced a new class of 'generalised solutions' of Problem P and proved that some kind of 'generalised solution' exists and is unique, but that it is unstable and has a very strong singularity on the characteristic cone S_2. More precisely, they showed that given any $\ell \in \mathbb{N}$, there is a function $f_\ell \in C^\ell\left(\overline{G}\right)$ such that the corresponding 'generalised solution' $u_\ell \in C\left(\overline{G} \setminus S_2\right)$ of Problem P exists, is unique and satisfies the estimate

$$\int_{S_{2,\varepsilon}} |u_\ell| \, ds \geq \varepsilon^{-\ell}, \quad 0 < \varepsilon < 1, \tag{5}$$

where $S_{2,\varepsilon} = \left\{x \in G : \rho = \varepsilon + \int_0^t \sqrt{K(\tau)} d\tau\right\}$.

This situation can be interpreted in terms of improperly posed (or ill-posed) problems: we recall that these are problems which fail to have a unique global solution which depends continuously on the data. For investigations of such problems for partial differential equations and for numerous regularisation methods we refer to the papers of Ames, Levine and Payne [13], Ames [14], as well as to the many references cited there.

In Problem P, the position is the following:

(i) According to the results of Popivanov and Schneider cited above, there are infinitely many distinct right-hand sides f_n $(n \in \mathbb{N})$ of (1) for which there is a generalised solution with a strong singularity, of at least power-type (see (5)).

(ii) Let $u_0 \in C_0^\infty(G)$ be fixed and suppose that $K(t) \equiv t$. Then $f_0 := Lu_0 \in C_0^\infty(G)$ and for any right-hand side

$$f_{n,\delta} := f_0 + \delta f_n \qquad (n \in \mathbb{N},\ \delta \neq 0) \tag{6}$$

there is a generalised solution with a strong singularity (see (5)) and there is no classical solution. This shows that the Problem P for the Tricomi equation is very unstable, even though $f_0 \in C_0^\infty(G)$, because a small perturbation (6) in an infinite number of directions has such a strong effect.

With this in mind, and having regard to the work on ill-posed problems which was mentioned earlier, it is appropriate to consider a new problem which regularises Protter's problem. This new problem should be such that its solutions are free of the singularity, typified by (5), which appears on the characteristic cone S_2. This suggests connecting points from G with ones on the cone S_2, and so to investigate Problems P and P* we introduce a new, non-local problem. Let α be a small positive parameter. Given any t_0, ρ_0 and with $C = \{t_0 + \alpha\rho_0/(\alpha+1)\}\,\rho_0^\alpha$, let $(p_\alpha(t_0,\rho), q_\alpha(t_0,\rho_0))$ be the point of intersection of the curves

$$\rho = \int_0^t \sqrt{K(\tau)}d\tau, \quad t = C\rho^{-\alpha} - \alpha\rho/(\alpha+1).$$

Problem A. Find a solution $\omega(t,\rho,\varphi)$ of the equation

$$(L\omega)(t,\rho,\varphi) - \rho^{-2}K(t)\omega_{\varphi\varphi}(p_\alpha(t,\rho), q_\alpha(t,\rho), \varphi) = f(t,\rho,\phi) \tag{7}$$

in G which satisfies the boundary conditions (2), i.e. $\omega = 0$ on $S_0 \cup S_1$.

Equation (7) is non-local because it involves points with coordinates (t,ρ,φ) and $(p_\alpha(t,\rho), q_\alpha(t,\rho), \varphi)$. We remark here merely that in our non-local Problem A, in the additional term $\rho^{-2}K(t)\omega_{\varphi\varphi}(p_\alpha(t,\rho), q_\alpha(t,\rho), \varphi)$ of (7), the point $(p_\alpha, q_\alpha, \varphi)$ lies just on the characteristic cone S_2, where the big singularity shown by (5) appears in the 'generalised solution' of the original Problem P. The derivative $\omega_{\varphi\varphi}$ is tangential to S_2 at that point.

Unlike [11], where the 'generalised solution' belongs to a weighted space of smooth functions we work here in a weighted Sobolev space. Following the work of Morawetz [2] and Lax and Phillips [3] in the two-dimensional case and Sorokina [15] in the multi-dimensional case, we introduce the weighted Sobolev space

$$\widetilde{W}_2^1(G) := \left\{\omega : \|\omega\|_{\widetilde{W}_2^1(G)} = \left(\int_G \left(\omega^2 + \omega_t^2 + r\left(\omega_{x_1}^2 + \omega_{x_2}^2\right)\right)dx\right)^{1/2} < \infty\right\}, \tag{8}$$

where $r = \sqrt{x_1^2 + x_2^2 + t^2}$. In this space we establish existence, uniqueness and an a priori estimate of a generalised solution of Problem A for every $f \in L_2(G)$. We also prove the infinite smoothness of the solution with respect to φ.

2. The nonlocal problem : existence, uniqueness and smoothness of a generalised solution

Definition 1 *A function* $\omega(t,\rho,\varphi)$ *is called a generalised solution of Problem A if: (a)* ω, $\frac{\partial\omega}{\partial t}$, $\sqrt{r}\frac{\partial\omega}{\partial\rho} \in L_2(G)$; *(b)* $\sqrt{r}\rho^{-1}\frac{\partial\omega^1}{\partial\varphi} \in L_2(G)$, *where* $\omega^1(t,\rho,\varphi) := \omega(t,\rho,\varphi) - \omega\left(p_\alpha(t,\rho), q_\alpha(t,\rho), \varphi\right)$; *(c)* $\omega = 0$ *on* $S_0 \cup S_1$; *(d) for any function* $v \in C^* := \{v \in C^1(\overline{G}) : v = 0$ *on* $S_0 \cup S_2$ *and in some neighbourhood of* $(0,0,0)\}$, *the equality*

$$\int_G \left(\omega_t v_t - K(t)\omega_\rho v_\rho - K(t)\rho^{-2}\omega^1_\varphi v_\varphi - fv\right)\rho\, d\rho\, d\varphi\, dt = 0 \tag{9}$$

holds.

Theorem 2.1 *Suppose that the parameter* $\alpha > 0$ *is so small that conditions:*

$$\rho - \alpha(\rho+t)\sqrt{K(t)} > 0 \ \text{ in } \overline{G}\backslash\{(0,0,0)\}, \quad \alpha K(t) < 2+\alpha \qquad \text{in G,} \tag{E1}$$

are satisfied. Let $f \in L_2(G)$ be such that $\frac{\partial^k f}{\partial\varphi^k} \in L_2(G)$ for $k = 1,\dots,\ell$, for some $\ell \in \mathbb{N}$. .Then there is a unique generalised solution ω of Problem A for which

$$\frac{\partial^k\omega}{\partial\varphi^k} \in \widetilde{W}^1_2(G)\ (k=0,\dots\ell-1),\ \frac{\partial^{\ell+1}\omega}{\partial t\partial\varphi^\ell} \in L_2(G),\ \sqrt{r}\frac{\partial^{\ell+1}\omega}{\partial\rho\partial\varphi^\ell} \in L_2(G) \tag{10}$$

and

$$\sum_{k=0}^{\ell-1}\left\|\frac{\partial^k\omega}{\partial\varphi^k}\right\|_{\widetilde{W}^1_2(G)} + \left\|\frac{\partial^{\ell+1}\omega}{\partial t\partial\varphi^\ell}\right\|_{L_2(G)} + \left\|\sqrt{r}\frac{\partial^{\ell+1}\omega}{\partial\rho\partial\varphi^\ell}\right\|_{L_2(G)} \le C_\alpha\sum_{k=0}^{\ell}\left\|\frac{\partial^k f}{\partial\varphi^k}\right\|_{L_2(G)}. \tag{11}$$

Remark. It easy to see, that near the point $(0,0,0)$ the first of conditions (E1) is equivalent to $0 < \alpha < 2/3$. Note that (E1) is satisfied for every sufficiently small $\alpha > 0$. For example, if $K(t) = t$, then it holds if $0 < \alpha < 2/\left(3+\sqrt[3]{6}\right)$.

3. Local and non-local problems, and their connection

Definition 2 *A function* $u \in \tilde{W}^1_2(G)$ *is called a generalised solution of Problem P if, and only if,* $u = 0$ *on* $S_0 \cup S_1$ *and*

$$\int_G \left\{u_t v_t - K(t)u_\rho v_\rho - K(t)\rho^{-2}u_\varphi v_\varphi - fv\right\}dx = 0 \tag{12}$$

holds for all $v \in C^*$.

We shall also consider the following problem:

Problem $\mathbf{P}_\varphi$. Is there a generalised solution of Problem P which satisfies the extra condition

$$u_\varphi = 0 \text{ on } S_2 \text{ in a weak sense?} \tag{13}$$

For the Tricomi equation condition (13) was investigated by Didenko [6].

Denote by H the set of all $f \in L_2(G)$ for which there exists a generalised solution u_f of Problem P; denote by H_φ the subspace of functions $f \in H$ for which the solution u_f satisfies (13). We know that in the case of the Tricomi equation, where $K(t) = t$, $\dim(L_2(G)/H) = \infty$, since all the functions v_n given in (4) are orthogonal to H.

We shall examine the uniqueness and continuous dependence on the data of solutions of Problem P_φ. The non-local Problem A will give some information about this.

Theorem 3.2 *Let $u \in \tilde{W}_2^1(G)$ be a generalised solution of Problem P_φ. Then it is also a generalised solution of Problem A, and so is unique and satisfies the a priori estimate*

$$\|u\|_{\tilde{W}_2^1(G)} \leq C\,\|f\|_{L_2(G)}\,. \tag{14}$$

Corollary 3.2 *If $f \in H_\varphi$, then a generalised solution u_f of Problem P_φ exists and coincides with the generalised solution ω_f of Problem A.*

Corollary 3.2 *If $f \in H_\varphi$, and $\partial^k f/\partial\varphi^k \in L_2(G)$, for $k = 0, ..., l$, then the conclusions of Theorem 2.1 hold for a generalised solution u_f of Problem P_φ.*

Of course, for $f \in H\backslash H_\varphi$ a generalised solution of Problem P need not coincide with the generalised solution ω_f of Problem A; and if $f \in L_2(G)\backslash H$, there is no generalised solution of Problem P. Thus the non-local Problem A (for which Theorem 3.2 holds, ensuring uniqueness, existence and differentiability with respect to φ) is, so to speak, a 'regular continuation' of the strongly over-determined Problem P_φ when $f \in L_2(G)\backslash H_\varphi$. In this sense, we may regard Problem A to be a 'non-local regularisation' of Problems P and P_φ. All this suggests the following procedure for tackling the ill-posed Problem P. For the given function $f \in L_2(G)$ we first try to find a solution ω_f for the nonlocal Problem A. Then we check the value of the derivative $(\omega_f)_{\varphi\varphi}$ on the characteristic cone S_2 and if that value is very small, we might conclude that the solution u_f of the Problem P exists and is very close to the function ω_f already found.

Acknowlegements. This research was partially supported by the Bulgarian NSF under Grant MM-512/95.

REFERENCES

[1] M.H. Protter: J. Rat. Mech. Anal., 3(1954), pp.435-446.

[2] C.S. Morawetz: Comm. Pure Appl. Math., 11(1958), pp.315-331.

[3] P.D. Lax; R. Phyllips: Comm. Pure Appl. Math., 13(1960), pp.427-455.

[4] A.K. Aziz; M. Schneider: SIAM J. Math. Anal., 10(1979), pp.913-921.

[5] A.V. Bitzadze: Doklady ANSSR , 110(1956), pp.1021-1024.

[6] V.P. Didenko: Diff. Equationst , 11(1975), pp.25-28.

[7] Khe Kan Cher: Siberian Math. J. , 26(1985), pp.286-288.

[8] Tong Kwang-Chang: Science Record, New Series, 1, 1-3 (1957).

[9] S.A. Aldashev: Diff. Equations, 29(1993), pp.1829–1834.

[10] Khe Kan Cher: Soviet. Math.Dokl., 43, 887-891 (1991).

[11] N.I. Popivanov, M. Schneider: J. Math. Anal. Appl., 175, 537-579 (1993).

[12] N.I. Popivanov, M. Schneider: J. Math. Anal. Appl., 194, 50-77 (1995).

[13] K. Ames, H.A. Levine and L.E. Payne, in: Inverse and Ill-Posed Problems, H.W. Engl and C.W. Groetsch (Ed.), Mathematics in sciences and enginering, 4, Academic Press, 433-450 (1987).

[14] K.A. Ames, in: Nonlinear Equations in the Applied Sciences, W.F. Ames and C. Rogers (Ed.), Math. in sci. and eng., 185, Acad. Press (1992), pp.1-28.

[15] N.G. Sorokina: Diff. Equations, 13(1977), pp.128-134.

7th Int. Coll. on Differential Equations, pp. 81-83
D. Bainov (Ed)

On Asymptotic Properties in a Neighbourhood of Infinity of Solutions of Elliptic Equations

A.S. ABD EL-RADY
Department of Mathematics, Faculty of Science, South Valley University, Qena, EGYPT

Abstract. We shall consider the generalized, 1-periodical (w.r.t.y) solution of the Dirichlet or Neumann Problem for the second order homogeneous elliptic equation

$$L(u) \equiv \sum_{i,j=1}^{n+k} \frac{\partial}{\partial z_i}\left(a^{ij}(z)\frac{\partial u}{\partial z_j}\right) = 0, \quad \text{in } G \tag{1}$$

where $z = (x_1, \dots, x_n; y_1, \dots, y_k)$, G an unbounded domain.

It will be supposed that the generalized solution $u(z)$ and $a^{ij}(z)$ are 1-periodical w.r.t. $y_1, \dots, y_k$.

We study the behavior and the uniqueness of the solution in a neighbourhood of infinity under certain condition

$$\sup_{z\in\Omega} |u(z)| < \infty.$$

Keywords: Elliptic equation, asymptotic behavior, uniqueness.

INTRODUCTION

Let $z = (x, y)$, $x = (x_1, \dots, x_n) \in \mathbb{R}^n_x$, $y = (y_1, \dots, y_k) \in \mathbb{R}^k_y$, $n \geq 2$, $k \geq 0$, $z = (z_1, \dots, z_{n+k})$, $g \subset \{x : |x| < \ell\}$, $G = \{z : x \in \mathbb{R}^n_x \backslash \overline{g}, \quad y \in \mathbb{R}^k_y\}$, $G_\rho = G \cap \{z : |x| < \rho\}$.

Define

$$\Pi = \{y : \quad 0 < y_j < 1\},$$
$$\Omega = \{z : \quad x \in \mathbb{R}^n_x \setminus \overline{g}, \quad y \in \Pi\},$$
$$\Omega_\rho = \Omega \cap \{z : \quad |x| < \rho\},$$
$$S_\rho = \{z : \quad |x| = \rho, \quad y \in \Pi\}.$$

In [1] the behavior of the solutions of the non-homogeneous equation $L(u) = f$, in a neighbourhood of infinity under certain conditions on f was considered.

In [1], [2] the condition $\int_\Omega |\nabla_z u|^2 dz < \infty$ was considered.

Lemma 1. *If* $u(z)$ *is the generalized solution of* (1) *in* G *and* $\sup_{z\in\Omega} |u(z)| < \infty$, *then* $\exists \lim_{z\to\infty} u(z) < \infty$.

Proof. Since $\sup_{z\in\Omega}|u(z)| < \infty$, then $\liminf_{z\to\infty} u(z) = \alpha$. Let $u_1(z) = u(z) - \alpha$, then $u_1(z)$ is also a solution of (1) and

$$\liminf_{z\to\infty} u_1(z) = 0. \tag{2}$$

Let

$$u_{1-}(z) = \begin{cases} u_1 & \Big|\ u_1 < 0, \\ 0 & \Big|\ u_1 \geq 0. \end{cases}$$

So

$$\lim_{z\to\infty} u_{1-}(z) = 0. \tag{3}$$

Let

$$\Omega_\lambda: \quad z \in \Pi, \quad \frac{\lambda}{z} < |x| < 2\lambda, \quad \lambda > 1,$$

$$G_\lambda: \quad z \in \Pi, \quad \frac{3}{4}\lambda < |x| < \lambda,$$

$u(z) = u_1(z) + \left|\min_{\Omega_\lambda} u_1(z)\right|$, then

$$u \geq 0, \quad z \in \Omega_\lambda, \quad G_\lambda \subset \Omega_\lambda, \qquad \exists \ \ c \ \ \text{s.t.}$$

$$u(\overline{z}) \leq cu(\overline{\overline{z}}), \qquad \forall \ \ \overline{z},\ \overline{\overline{z}} \in G_\lambda$$

(Harnack's inequality), and from this and from the definition of (2), (3) we can get

$$\lim_{z\to\infty} u_1(z) = 0. \quad \square$$

Theorem 1. *Let* $n \geq 3$, $u(z)$ *generalized* 1*-periodical (w.r.t.* y*) solution of* (1), *satisfying* $\sup_{z\in\Omega}|u(z)| < \infty$.

Then $\exists$ *constant* M *s.t.*

$$\left|u(x,y) - M\right| \leq c|x|^{2-n}, \qquad c = \text{const} > 0.$$

Proof. Using Lemma 1, we get $\lim_{z\to\infty} u(z) = M$, and that is which was obtained in the proof of Theorem 3 p. 157 of [2] when the condition $\int\limits_\Omega \left|\nabla_z u\right|^2 dz < \infty$ was used, and the result of the Theorem follows immediately. $\square$

Lemma 2. *If* $\sup_{z\in\Omega}\left|u(z)\right| < \infty$ *then* $\int\limits_\Omega \left|\nabla_z u\right|^2 dz < \infty$.

Proof. Let $V = u - M$,

$$\phi(x) = \begin{cases} 0, & |x| > 2\ell, \\ 1, & |x| < \ell, \\ \dfrac{2\ell - |x|}{\ell}, & \ell < |x| < 2\ell. \end{cases}$$

Then from (1) we get

$$\int_{x\in\Omega,\ |x|<2\ell} \phi^2 V \sum_{i,j=1}^{n+k} \frac{\partial}{\partial z_i} a^{ij}(z) \frac{\partial V}{\partial z_j} dz = 0,$$

$$0 = -\sum_{i,j=1}^{n+k} \int_{\Omega_{2\ell}} \frac{\partial}{\partial z_i}(\phi^2 V) a^{ij}(z) \frac{\partial V}{\partial z_j} dz + \int_{\Gamma} \phi^2 V \frac{\partial V}{\partial \mu} ds,$$

where $\dfrac{\partial}{\partial \mu} \equiv \sum_{i,j=1}^{n+k} a^{ij}(z)\cos(\nu, z_i)\dfrac{\partial}{\partial z_j}$, $\nu = (\nu_1, \dots, \nu_{n+k})$ is the outward unit conormal vector on ∂G, and

$$\Gamma = \partial\Omega \cup \left\{z : \quad |x| = 2\ell, \quad z \in \Pi\right\}.$$

Then, using $|v| \le c|x|^{2-n}$, we can get

$$\int_{\Omega} \left|\nabla_2 V\right|^2 dz < c, \qquad \text{i.e.} \qquad \int_{\Omega} \left|\nabla_2 u\right|^2 dz < c. \quad \Box$$

The uniqueness Theorem of the problem in the case of the condition $\sup_{z\in\Omega} |u(z)| < \infty$, follows as a result of last Lemma 2 and [2] in which the condition $\int_{\Omega} \left|\nabla_2 u\right|^2 dz < \infty$ was used.

ACKNOWLEDGEMENT

The author would like to thank professor V.A. Kondrat'ev in Moscow State University, Faculty of Mechanics and Mathematics, USSR, for suggestions and valuable comments.

REFERENCES

[1] A.S. Abd El-Rady, Fifth International Colloquium on Differential Equations, Invited Lectures and Short Communications, Eds: D. Bainov and A. Dishliev, Vol. 1, Plovdiv, Bulgaria, August 18–23, 1994, Science, Culture, Technology, Publ., Singapore, 1–5.

[2] V.A. Kondrat'ev and O.A. Oleinik, Trudy Sem. Petrovski, **12** (1987), 149–163 (in Russian).

7th Int. Coll. on Differential Equations, pp. 85-92
D. Bainov (Ed)

Further Development of Minimum Curvature Multi-step Quasi-Newton Methods

J.A. FORD* and I.A. MOGHRABI†

*Department of Computer Science, University of Essex, Wivenhoe Park, Colchester, Essex, United Kingdom
†Natural Science Division, Lebanese American University, P.O. Box 13-5053, Beirut, Lebanon

ABSTRACT
Multi-step quasi-Newton methods for unconstrained optimization were introduced in [1, 2, 3], where it was shown how an interpolation in the variable-space could be used to generalize the Secant Equation. For the case of two-step methods, there is a free parameter used in the interpolation which is at our disposal. In [4], the problem of choosing this parameter to yield a "smooth" interpolation was discussed. It was shown, there, how this "minimum curvature" problem could be solved efficiently at each iteration. In this paper, we develop these concepts by considering some variations of the "minimum curvature" approach. The results of numerical experiments on the new methods are reported and compared with those obtained from existing methods.

Keywords: Unconstrained optimization, quasi-Newton methods, multi-step methods.

AMS Subject Classification: 65K10.

1 INTRODUCTION

We consider two-step quasi-Newton methods for the unconstrained optimization problem

$$\min f(\underline{x}) \quad (\underline{x} \in R^n).$$

If we denote the gradient and Hessian of f by $\underline{g}$ and G, respectively, then such methods closely resemble standard quasi-Newton methods, except that the approximation B_{i+1} to the Hessian $G(\underline{x}_{i+1})$ is required to satisfy a condition of the form

$$B_{i+1}(\underline{s}_i - \gamma_i \underline{s}_{i-1}) = (\underline{y}_i - \gamma_i \underline{y}_{i-1}), \tag{1}$$

in place of the more commonly-employed condition

$$B_{i+1}\underline{s}_i = \underline{y}_i, \tag{2}$$

known as the Secant Equation. (In equations (1) and (2), the vectors $\underline{s}_i$ and $\underline{y}_i$ are given by

$$\underline{s}_i \stackrel{\text{def}}{=} \underline{x}_{i+1} - \underline{x}_i; \quad \underline{y}_i \stackrel{\text{def}}{=} \underline{g}(\underline{x}_{i+1}) - \underline{g}(\underline{x}_i), \tag{3}$$

where $\{\underline{x}_j\}$ are the successive iterates produced by the method.)

The derivation of (1) is described by Ford and Moghrabi [1, 2]. Briefly, quadratic curves $\underline{x}(\tau)$ and $\hat{\underline{g}}(\tau)$ (where $\tau \in R$) are constructed in R^n which interpolate, respectively, the three most recent iterates $\underline{x}_{i-1}$, $\underline{x}_i$ and $\underline{x}_{i+1}$, and the three (assumed known) associated gradient evaluations. The derivatives of these two curves (at $\tau = \tau_2$, where τ_2 is the value of τ corresponding, on the respective curves, to $\underline{x}_{i+1}$ and $\underline{g}(\underline{x}_{i+1})$) are then substituted into the relation (derived from the Chain Rule applied to the function $\underline{g}(\underline{x}(\tau))$):

$$G(\underline{x}_{i+1})\underline{x}'(\tau_2) = \underline{g}'(\underline{x}(\tau_2)), \tag{4}$$

where primes denote differentiation with respect to τ. (We note that

$$\underline{w}_i \stackrel{\text{def}}{=} \hat{\underline{g}}'(\tau_2) \tag{5}$$

will, in general, only be an *approximation* to the vector $\underline{g}'(\underline{x}(\tau_2))$ required in (4), whereas

$$\underline{r}_i \stackrel{\text{def}}{=} \underline{x}'(\tau_2) \tag{6}$$

may be computed exactly.) On making these substitutions into (4) and removing a common scaling factor, a relation of the form (1) for $B_{i+1} \approx G(\underline{x}_{i+1})$ is obtained.

If the values of τ corresponding to $\underline{x}_{i-1}/\underline{g}(\underline{x}_{i-1})$ and $\underline{x}_i/\underline{g}(\underline{x}_i)$ are denoted by τ_0 and τ_1 respectively, then Ford and Moghrabi [4] observed that, without loss of generality, the values $\tau_0 = 0$ and $\tau_2 = 1$ could be specified, leaving the remaining value τ_1 to be chosen according to suitable criteria. Defining the related quantity

$$\delta = (\tau_2 - \tau_1)/(\tau_1 - \tau_0) = (1 - \tau_1)/\tau_1 \Rightarrow \tau_1 = (1 + \delta)^{-1}, \tag{7}$$

Ford and Moghrabi [4] then discussed choosing δ to minimize the norm $\| \underline{x}''(\tau) \|_M$, where

$$\| \underline{z} \|_M \stackrel{\text{def}}{=} \{\underline{z}^T M \underline{z}\}^{1/2} \tag{8}$$

and where M is a given symmetric positive-definite matrix, thus producing an interpolating curve which is the "smoothest" in the sense of the norm $\| \cdot \|_M$. It was shown in [4] that this proposal leads to the requirement, at each iteration, of solving the cubic polynomial

$$\psi(\delta) \equiv \delta^3 - \mu\delta^2 + \mu\delta - \sigma, \tag{9}$$

where

$$\sigma_j \stackrel{\text{def}}{=} \| \underline{s}_j \|_M^2; \quad \mu_j \stackrel{\text{def}}{=} \underline{s}_j^T M \underline{s}_{j-1}; \tag{10}$$

$$\sigma \stackrel{\text{def}}{=} \sigma_i/\sigma_{i-1}; \quad \mu \stackrel{\text{def}}{=} \mu_i/\sigma_{i-1}. \tag{11}$$

In [4], the properties of the cubic ψ were analyzed and it was shown how to determine the zeroes efficiently, under which circumstances ψ possesses more than one real zero and which zero to choose, in that case, in order to give a path $\underline{x}(\tau)$ with "minimum curvature". An algorithm based on this approach (using $M = I$) was then compared, in a series of numerical experiments, with the standard (one-step) BFGS (Broyden [5], Fletcher [6], Goldfarb [7], Shanno [8])) method and was shown to exhibit superior performance (in terms of evaluations, iterations and execution time), with the superiority becoming more marked as the dimension increased. In the next section, we will discuss three variants of this approach. For convenience, we will refer to the original version of the minimum curvature algorithm (as described in [4] and outlined above) by the symbol **A**.

2 OTHER MINIMUM CURVATURE METHODS

2.1 Smooth gradient approximation

Our first alternative method is obtained by applying the "minimum curvature" concept to the interpolating gradient approximation $\hat{\underline{g}}(\tau)$, instead of to $\underline{x}(\tau)$. Since we wish to use the quantity $\hat{\underline{g}}'(\tau_2)$ as an approximation to the vector $\underline{g}'(\underline{x}(\tau_2))$ required in (4), there would appear to be some merit in constructing a "smooth" approximation $\hat{\underline{g}}(\tau)$ for which the derivative $\hat{\underline{g}}'(\tau)$ is, therefore, not varying rapidly. In practice, the only modifications required in the original algorithm are to define the quantities σ_j and μ_j (and, hence, σ and μ) in terms of $\underline{y}_{j-1}$ and $\underline{y}_j$, instead of $\underline{s}_{j-1}$ and $\underline{s}_j$ (see equations (10) and (11)). This algorithm, using the weighting matrix $M = I$ again, is denoted by **G**.

2.2 Optimal ordering of the iterates

The approach developed in [4] and outlined in Section 1 is open to criticism when ψ has three real zeroes. In this case, it was shown that ψ has one positive zero (denoted by δ_*, say) and two negative ones (denoted by δ_1 and δ_2, say). Restricting consideration to the case when $\sigma < 1$ (the case "$\sigma = 1$" is trivial and the case "$\sigma > 1$" may be converted to the case "$\sigma < 1$" by 'inverting' the problem), it was further shown in [4] that

$$\delta_* < 1; \quad -\delta_* \leq \delta_1, \delta_2 < 0. \tag{12}$$

If we designate the (numerically) smaller of the two negative zeroes as δ_2, then it was proved in [4] that δ_2 yields a smaller value of $\| \underline{x}''(\tau) \|_M$ than does δ_*. Hence the algorithm developed in [4] used δ_2 in order to construct $\underline{x}(\tau)$. However, since $-1 < \delta_2 < 0$, it follows (see (7)) that $\tau_1 > 1$. Thus, the normalization originally introduced into the problem by specifying the range of the variable τ to be the interval $[0, 1]$ has now effectively been over-ridden. In other words, the reduction in the value of $\| \underline{x}''(\tau) \|_M$ has been achieved (at least in part) by 'stretching' the length of the interval from its original value of unity to $(1 + \delta_2)^{-1} > 1$. We therefore propose a variant of the original algorithm in which the three essentially different orderings of the iterates $\underline{x}_{i-1}$, $\underline{x}_i$ and $\underline{x}_{i+1}$ are treated in turn, in each case (in order to maintain the specified ordering and the normalization of the interval) only searching for the *positive* zero δ_* and, hence, a value of τ_1 lying within $(0, 1)$. (It is clear that ψ always has a positive zero, because σ is positive.) Then, the three minimum values of the norm (corresponding to the three orderings) are compared and the smallest is chosen. We denote this algorithm (which again uses $M = I$) by **A3**.

2.3 An alternative norm

In their study of other types of two-step methods, Ford and Moghrabi [1] found that a significant improvement in numerical performance could be obtained by using a metric based upon the current Hessian approximation B_i. It would therefore appear to be an attractive proposition to employ the same technique in constructing the norm required in the "minimum curvature" approach. However, we note (as was done in [1]) that computation of the values (as they currently stand) of σ_i, σ_{i-1} and μ_i would be too expensive, because of the matrix-vector products required. This difficulty can be alleviated, in part, by implementing the algorithm with a line-search along the quasi-Newton search direction

$$\underline{p}_i = -B_i^{-1}\underline{g}(\underline{x}_i), \tag{13}$$

so that there will exist a scalar $t_i > 0$ such that

$$\underline{s}_i = t_i\underline{p}_i \Rightarrow B_i\underline{s}_i = -t_i\underline{g}_i. \tag{14}$$

However, there still remains the difficulty of computing the product $B_i\underline{s}_{i-1}$ efficiently. In [1], this was achieved by approximation, but that option is not open to us here, because the theory developed in [4] concerning the zeroes of ψ depends upon the inequality $\mu^2 \leq \sigma$ being satisfied, and this may fail if μ_i and σ_{i-1} are only estimated. We therefore propose an algorithm in which ordinary single-step BFGS iterations are alternated with iterations involving the two-step "minimum curvature" construction (using B_i to define the norm). Then, on iterations when we apply the minimum curvature approach, B_i will satisfy

$$B_i\underline{s}_{i-1} = \underline{y}_{i-1}, \tag{15}$$

by virtue of the update performed on the previous ('single-step') iteration. The computation of the relevant quantities σ and μ required for the "minimum curvature" approach is now straightforward. We denote this algorithm by **B**.

3 NUMERICAL EXPERIMENTS

The algorithms **G**, **A3** and **B** developed in Section 2 were compared with each other, in our first set of experiments. All the multi-step algorithms tested in these and the following experiments employed the BFGS formula to update the inverse Hessian approximations $H_i \stackrel{\text{def}}{=} B_i^{-1}$, but with the usual vectors $\underline{s}_i$ and $\underline{y}_i$ replaced by the forms of $\underline{r}_i$ and $\underline{w}_i$ (see (5) and (6)) appropriate to that algorithm:

$$H_{i+1} = H_i + \left(1 + \frac{\underline{w}_i^T H_i \underline{w}_i}{\underline{r}_i^T \underline{w}_i}\right) \frac{\underline{r}_i \underline{r}_i^T}{\underline{r}_i^T \underline{w}_i} - \left(\frac{H_i \underline{w}_i \underline{r}_i^T + \underline{r}_i \underline{w}_i^T H_i}{\underline{r}_i^T \underline{w}_i}\right). \tag{16}$$

The suite of test functions used is as described in Ford and Moghrabi [2], with a small number of modifications to starting-points and convergence criteria. The suite comprises sixty functions, in all, and was chosen from standard sets described in the literature, such as the article by Moré, Garbow and Hillstrom [9]. For each function, four starting-points were used, giving a total of 240 test problems. For convenience, the functions were notionally classified into those of 'low' ($2 \leq n \leq 15$), 'medium' ($16 \leq n \leq 45$) and 'high' ($46 \leq n \leq 80$) dimension. In total, there were 10 functions in the 'low' set and 25 functions in each of the 'medium' and 'high' sets, giving, respectively, 40, 100 and 100 test problems in each of the sets. Information on the functions and starting-points used, together with details concerning the actual implementation of such algorithms as these, may be found in [2].

Table 1: Comparison of three new minimum curvature methods

	G	A3	B	Problem set
Totals	4866 (3657)	4885 (3682)	4746 (3568)	
Ratios	95.0% (93.1%)	95.4% (93.7%)	92.7% (90.8%)	**Low**
Scores	14	13	17	
Totals	17403 (15183)	17249 (15053)	16750 (14686)	
Ratios	83.4% (80.3%)	82.6% (79.6%)	80.3% (77.7%)	**Medium**
Scores	12	29	63	
Totals	15058 (13932)	14523 (13438)	14005 (12860)	
Ratios	81.1% (78.7%)	78.2% (75.9%)	75.4% (72.7%)	**High**
Scores	12	31	61	
Totals	37327 (32772)	36657 (32173)	35501 (31114)	
Ratios	83.8% (80.9%)	82.2% (79.4%)	79.7% (76.8%)	**Combined**
Scores	38	73	141	

Table 2: Comparison of **B** with **F2** and **A**

	A	B	F2	Problem set
Totals	4928 (3764)	4746 (3568)	5224 (3954)	
Ratios	96.2% (95.8%)	92.7% (90.8%)	102.0% (100.7%)	**Low**
Scores	13	17	12	
Totals	17205 (15019)	16750 (14686)	17203 (14760)	
Ratios	82.4% (79.4%)	80.3% (77.7%)	82.4% (78.1%)	**Medium**
Scores	13	51	42	
Totals	14560 (13462)	14005 (12860)	13487 (12340)	
Ratios	78.4% (76.1%)	75.4% (72.7%)	72.6% (69.7%)	**High**
Scores	13	43	46	
Totals	36693 (32245)	35501 (31114)	35914 (31054)	
Ratios	82.3% (79.6%)	79.7% (76.8%)	80.6% (76.6%)	**Combined**
Scores	39	111	100	

Results from these first experiments are summarized in Table 1. (Lack of space precludes the presentation of more detailed results, which may be found in [10].) Each of the tables which we will present is divided into four sections, three of which correspond to the three classes of functions referred to above, while the last refers to the complete set. In the row labelled 'Totals' in each section of a table, the total number (for each method) of function / gradient evaluations required to solve all the problems in the specified set is given, followed by the total number of iterations (in brackets). The entries in the 'Ratios' row give the proportions of evaluations and (in brackets) iterations for that method, expressed as percentages of the corresponding figures for the BFGS method. A 'best performance' for each problem was decided on the basis of the lowest number of evaluations, with ties resolved by the number of iterations. The row labelled 'Scores' shows the number of best performances by each method for the relevant set.

In their original paper on minimum curvature methods, Ford and Moghrabi [4] compared the method **A** with the 'fixed-point' method **F2**, which was the most successful of the methods developed in [1], finding that **F2** produced a better performance than **A** for functions of higher dimension. The next set of experiments conducted, therefore, was to compare the most successful method introduced in the present paper (namely, the alternating method **B**) with **F2** and with the original method **A**. The results of these tests are presented in Table 2.

Finally, as shown by the authors [11] recently, the technique of alternating a two-step method with the standard BFGS method (used in Section 2 to construct the method **B**) may also be applied to **F2**, and produces a further improvement in the numerical performance of that algorithm. This alternating method is denoted by **F21**. Hence, in our final set of experiments, we compared the performance of **B** with that of **F21** and (for completeness) of the standard single-step BFGS method, and the results of these tests are given in Table 3.

4 CONCLUSIONS

Comparison of the results presented in Table 1 with those for the method **A** in Table 2 indicates that only the third of the variant algorithms discussed in Section 2 (that is, **B**, using the Hessian approximation B_i to define the norm in which the curvature of the interpolation is 'measured') leads to any discernible improvement over the original "minimum

Table 3: Comparison of **B** with **F21** and **BFGS**

	BFGS	B	F21	Problem set
Totals	5122 (3928)	4746 (3568)	4840 (3658)	
Ratios	100.0% (100.0%)	92.7% (90.8%)	94.5% (93.1%)	**Low**
Scores	10	16	16	
Totals	20871 (18910)	16750 (14686)	15777 (13333)	
Ratios	100.0% (100.0%)	80.3% (77.7%)	75.6% (70.5%)	**Medium**
Scores	8	27	74	
Totals	18575 (17694)	14005 (12860)	12365 (10580)	
Ratios	100.0% (100.0%)	75.4% (72.7%)	66.6% (59.8%)	**High**
Scores	10	9	86	
Totals	44568 (40532)	35501 (31114)	32982 (27571)	
Ratios	100.0% (100.0%)	79.7% (76.8%)	74.0% (68.0%)	**Combined**
Scores	28	52	176	

curvature" method. Moreover, Table 2, in addition to showing that the performance of **B** is better than that of **A**. also indicates that **B** is broadly comparable with **F2**, with, perhaps, a slight balance in favour of **F2** for problems of higher dimension. Finally, however, it is clear from Table 3 that the method **F21** (which, like **B**, is an alternating method, interleaving single-step BFGS updates with the relevant two-step update) is significantly better than **B**′ in terms of both evaluations and iterations, for all but problems of the lowest dimension.

References

[1] J.A. Ford and I.A. Moghrabi, Optim. Meth. Software, 2, 357-370 (1993).

[2] J.A. Ford and I.A. Moghrabi, J. Comput. Appl. Math., 50, 305-323 (1994).

[3] J.A. Ford and I.A. Moghrabi, Scientia Iranica, 1, 327-334 (1995).

[4] J.A. Ford and I.A. Moghrabi, Computers Math. Applic., 31, 179-186 (1996).

[5] C.G. Broyden, J. Inst. Math. Appl., 6, 76-90 and 222-231 (1970).

[6] R. Fletcher, Computer J., 13, 317-322 (1970).

[7] D. Goldfarb, Maths. Comp., 24. 23-26 (1970).

[8] D.F. Shanno, Maths. Comp., 24, 647-656 (1970).

[9] J.J. Moré, B.S. Garbow and K.E. Hillstrom, TOMS, 7, 17-41 (1981).

[10] J.A. Ford and I.A. Moghrabi, Department of Computer Science (University of Essex) Technical Report CSM-271 (1996).

[11] J.A. Ford and I.A. Moghrabi, Department of Computer Science (University of Essex) Technical Report CSM-270 (1996).

7th Int. Coll. on Differential Equations, pp. 93-100
D. Bainov (Ed)

Some results about boundary conditions for a discrete model in elasticity [1]

M. Frontini L. Gotusso

Dipartimento di Matematica, Politecnico di Milano, Italia

Abstract

A discrete model which simulates the motion of a vibrating rod with different boundary constrains (hinged, clamped or free) is considered. The stability conditions, in the different cases, are studied. Numerical results, obtained by the discrete model, are compared with the ones obtained by using classical formulas of continuum mechanics.

Keywords: Differential and difference equations, discrete models, vibrating rod

1 Introduction

Some discrete models for the study of the vibrating string or rod have been recently suggested and studied in [1, 2, 3, 4, 5, 6]. It has been shown that the difference systems connected with these discrete models may be seen as finite difference schemes of partial differential equations which model the motion of a vibrating string or rod.

In this paper, after a short review, in §2, of previous results, we study, in §3, how different boundary conditions modify the discrete model of the vibrating rod. We give linear stability conditions for the following cases: rod hinged or clamped in both end points, rod clamped in the first end point and hinged or free to move in the last one.

Some numerical results are given in §4, where it is remarked that the stability conditions may lead, in particular in non linear cases, to have long calculation time. This remark suggests the use of unconditionally stable schemes which are being studied by the authors.

2 Discrete string and discrete rod

A discrete string (or rod) may be modelled as a set of $n+1$ particles $P_0, P_1, \dots, P_n$ free to move vertically only in the xy plane (transversal motion). Particles $P_1, P_2, \dots, P_{n-1}$ have mass m, while P_0 and P_n have mass $\frac{m}{2}$. Let $P_{i,k} = (x_i, y_{i,k})$ be the position of the center of P_i at time t_k ($t_k = k\Delta t, \quad k = 0, 1, \dots$). Defining $\Delta x = x_i - x_{i-1}$, $i = 1, 2, \dots, n$, if $v_{i,k}$ and $a_{i,k}$ are the velocity and the acceleration of P_i at time t_k, we use the following *leap-frog* formulas

[1] Work supported by italian M.U.R.S.T.

$$\begin{cases} v_{i,\frac{1}{2}} & = v_{i,0} + \frac{\Delta t}{2} a_{i,0} \\ v_{i,k+\frac{1}{2}} & = v_{i,k-\frac{1}{2}} + \Delta t a_{i,k} \\ y_{i,k+1} & = y_{i,k} + \Delta t v_{i,k+\frac{1}{2}} \end{cases} \tag{1}$$

in order to link position, velocity and acceleration of the particle P_i. Given the initial and boundary conditions, the motion of the system is determined by the total force $F_{i,k}$ acting on each particle P_i at time t_k according to the classical relation

$$F_{i,k} = m a_{i,k}. \tag{2}$$

In the case of a string we consider as internal forces the tension only while for the rod we consider the tension and the bending force. As external forces we consider the gravity and the viscosity. According to [3, 4] we define the preceding forces in the following way.

Tension force

$$F_{i,k} = T_{i,k}^{+} - T_{i,k}^{-} = KL\frac{\sqrt{\Delta x^2 + (y_{i+1,k} - y_{i,k})^2}}{\Delta x} - KL\frac{\sqrt{\Delta x^2 + (y_{i,k} - y_{i-1,k})^2}}{\Delta x} \tag{3}$$

where $L = n\Delta x$ is the length of the string (or rod) and $K = \frac{T_0}{M}$ is a physical constant depending on the tension T_0 and the total mass of the rod $M = nm$.

Bending force

$$F_{i,k} = F_{i,k}^{-} - F_{i,k}^{+-} + F_{i,k}^{+} - F_{i,k}^{-+} \tag{4}$$

where

$$\begin{cases} F_{i,k}^{-+} & = \frac{LC}{\Delta x}\frac{\theta_{i+1,k} - \theta_{i,k}}{\sqrt{\Delta x^2 + (y_{i+1,k} - y_{i,k})^2}} \\ F_{i,k}^{+-} & = \frac{LC}{\Delta x}\frac{\theta_{i-1,k} - \theta_{i-2,k}}{\sqrt{\Delta x^2 + (y_{i,k} - y_{i-1,k})^2}} \\ F_{i,k}^{+} & = \frac{LC}{\Delta x}\frac{\theta_{i,k} - \theta_{i-1,k}}{\sqrt{\Delta x^2 + (y_{i+1,k} - y_{i,k})^2}} \\ F_{i,k}^{-} & = \frac{LC}{\Delta x}\frac{\theta_{i,k} - \theta_{i-1,k}}{\sqrt{\Delta x^2 + (y_{i,k} - y_{i-1,k})^2}} \end{cases} \tag{5}$$

Gravity and viscosity

$$F_{i,k} = -mg, \qquad F_{i,k} = -\rho v_{i,k}, \tag{6}$$

where g is the gravity constant and ρ is the viscosity coefficient.

Following [4], if we assume that the deformations are sufficiently small so that

$$1 + \left(\frac{y_{i,k} - yi - 1, k}{\Delta x}\right) \simeq 1 \quad \forall i, \tag{7}$$

from (1),(2),(3),(4),(5),(6) we obtain (linear case):

$$\begin{aligned} \frac{y_{i,k+1} - 2y_{i,k} + y_{i,k-1}}{\Delta t^2} &= \frac{T_0 L}{M\Delta x^2}[y_{i+1,k} - 2y_{i,k} + y_{i-1,k}] - \frac{L^2 C}{M\Delta x^4}[y_{i+2,k} - 4y_{i+1,k} + \\ &+ 6y_{i,k} - 4y_{i-1,k} + y_{i-2,k}] - g - \frac{\rho}{m}\left(\frac{y_{i,k+1} - y_{i,k-1}}{2\Delta t}\right) \end{aligned} \tag{8}$$

where C is a physical constant and we have assumed $2v_{i,k} = v_{i,k+\frac{1}{2}} + v_{i,k-\frac{1}{2}}$. System (8) represents, in the linear case, the discrete model of the rod if $C \neq 0$ and of the string if $C = 0$ [4].

If the deformations are not sufficiently small, so that (7) does not hold, from the previous relations we have [4] (non linear case):

$$\begin{aligned}\frac{y_{i,k+1}-2y_{i,k}+y_{i,k-1}}{\Delta t^2} &= \frac{T_0 L}{M\Delta x^2}[y_{i+1,k}-2y_{i,k}+y_{i-1,k}]+\\ &-\frac{L^2C}{M\Delta x^3}\left\{\frac{1}{1+((y_{i+1,k}-y_{i,k})/\Delta x)^2}\left[\arctan(\frac{y_{i+2,k}-y_{i+1,k}}{\Delta x})+\right.\right.\\ &\left.-2\arctan(\frac{y_{i+1,k}-y_{i,k}}{\Delta x})+\arctan(\frac{y_{i,k}-y_{i-1,k}}{\Delta x})\right]+\\ &-\frac{1}{1+((y_{i,k}-y_{i-1,k})/\Delta x)^2}\left[\arctan(\frac{y_{i+1,k}-y_{i,k}}{\Delta x})+\right.\\ &\left.\left.-2\arctan(\frac{y_{i,k}-y_{i-1,k}}{\Delta x})+\arctan(\frac{y_{i-1,k}-y_{i-2,k}}{\Delta x})\right]\right\}+\\ &-g-\frac{\rho}{m}(\frac{y_{i,k+1}-y_{i,k-1}}{2\Delta t})\end{aligned} \qquad (9)$$

3 Initial and boundary conditions

As initial conditions we have to give the configuration of the string (or rod) at time t_0. It is easy to show that in both cases the initial conditions do not affect the stability conditions of the difference system.

As boundary conditions it seems natural, for the vibrating string, to assume the string fixed at the end points ($y_{0,k} = y_{n,k} = 0, \quad \forall k$), and the stability conditions for the difference system has been given in [2]. For the vibrating rod we can consider, as natural constraints, four different cases which lead to four different difference systems each of which must be studied from the stability point of view. In this study we will neglect the viscosity as this is the worst case for stability [4].

In our model the action of the constraints may by simulated by modifying the forces acting on the first two and last two particles. We consider here only the first two particles P_0 and P_1 (the same considerations hold, in a symmetric way, for the last two particles P_n and P_{n-1}).

As shown in figure 1, we remark that the forces $F^+_{i,k}$ and $F^-_{i,k}$ try to align the particles P_{i-1}, P_i and P_{i+1}, while the forces $-F^{+-}_{i,k}$ and $-F^{-+}_{i,k}$ try to align the particles P_{i-2}, P_{i-1}, P_i and the particles P_i, P_{i+1}, P_{i+2} respectively.

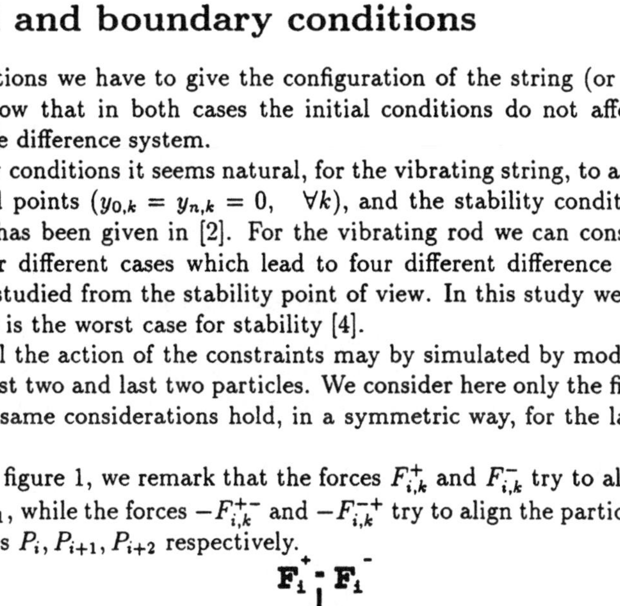

Figure 1

In the cases of hinged or clamped rod we have obviously $y_0 = 0$. Moreover for the hinged rod the masses P_{-1}, P_0, P_1 (where P_{-1} is a fictitious particle with $x_{-1} = -\Delta x$) are always aligned due to the presence of the hinge, so that $F^{+-}_{0,k} = 0$ which implies $y_{-1} = -y_1$.

In the case of the clamped rod the motion of P_1, due to gravity force, breaks the natural horizontal alignment of the masses P_{-1}, P_0, P_1; in order to equilibrate the bending force which arises we assume $y_{-1} = y_1$. Being $y_0 = 0$, this is equivalent to doubling the force $F_{0,k}^{+-}$.

When the mass P_0 is free to move it is natural to assume equal to zero all the forces due to the fictitious masses on the left of P_0. This is equivalent to assuming $F_{0,k}^{-} = F_{0,k}^{+} = F_{0,k}^{+-} = T_{0,k}^{-} = 0$ and $F_{1,k}^{+-} = 0$. We remark that, as P_0 has mass $\frac{m}{2}$, the only force different from zero, $F_{0,k}^{-+}$ defined in (5), must be halved according to (2), while the tension $T_{0,k}^{+}$ is unchanged according to the action and reaction principle.

To summarize the previous considerations, by taking into account the boundary constrains, we have, in the linear case:

a) for the rod hinged at both end points the resulting $(n-1) \times (n-1)$ difference system is given by equation (8), with $\rho = 0$, for $i = 2, \ldots, n-2$ and

$$\begin{aligned} \frac{y_{1,k+1}-2y_{1,k}+y_{1,k-1}}{\Delta t^2} &= \frac{T_0 L}{M\Delta x^2}[y_{2,k} - 2y_{1,k}]+ \\ &-\frac{L^2 C}{M\Delta x^4}[y_{3,k} - 4y_{2,k} + 5y_{1,k}] - g, \end{aligned} \tag{10}$$

$$\begin{aligned} \frac{y_{n-1,k+1}-2y_{n-1,k}+y_{n-1,k-1}}{\Delta t^2} &= \frac{T_0 L}{M\Delta x^2}[-2y_{n-1,k} + y_{n-2,k}]+ \\ &-\frac{L^2 C}{M\Delta x^4}[5y_{n-1,k} - 4y_{n-2,k} + y_{n-3,k}] - g; \end{aligned} \tag{11}$$

b) for the rod clamped at both end points we have a $(n-1) \times (n-1)$ difference system given by equation (8), with $\rho = 0$, for $i = 2, \ldots, n-2$ and

$$\begin{aligned} \frac{y_{1,k+1}-2y_{1,k}+y_{1,k-1}}{\Delta t^2} &= \frac{T_0 L}{M\Delta x^2}[y_{2,k} - 2y_{1,k}]+ \\ &-\frac{L^2 C}{M\Delta x^4}[y_{3,k} - 4y_{2,k} + 7y_{1,k}] - g, \end{aligned} \tag{12}$$

$$\begin{aligned} \frac{y_{n-1,k+1}-2y_{n-1,k}+y_{n-1,k-1}}{\Delta t^2} &= \frac{T_0 L}{M\Delta x^2}[-2y_{n-1,k} + y_{n-2,k}]+ \\ &-\frac{L^2 C}{M\Delta x^4}[7y_{n-1,k} - 4y_{n-2,k} + y_{n-3,k}] - g; \end{aligned} \tag{13}$$

c) for the rod clamped in $x = 0$ and hinged in $x = L$ we have a $(n-1) \times (n-1)$ difference system given by equation (8), with $\rho = 0$, for $i = 2, \ldots, n-2$, and equations (12) and (11);

d) for the rod clamped in $x = 0$ and free to move in $x = L$ we have a $n \times n$ difference system given by equation (8), with $\rho = 0$, for $i = 2, \ldots, n-2$, equation (12) and

$$\begin{aligned} \frac{y_{n-1,k+1}-2y_{n-1,k}+y_{n-1,k-1}}{\Delta t^2} &= \frac{T_0 L}{M\Delta x^2}[y_{n,k} - 2y_{n-1,k} + y_{n-2,k}]+ \\ &-\frac{L^2 C}{M\Delta x^4}[-2y_{n,k} + 5y_{n-1,k} - 4y_{n-2,k} + y_{n-3,k}] - g, \end{aligned} \tag{14}$$

$$\begin{aligned} \frac{y_{n,k+1}-2y_{n,k}+y_{n,k-1}}{\Delta t^2} &= \frac{T_0 L}{M\Delta x^2}[-2y_{n,k} + y_{n-1,k}]+ \\ &-\frac{L^2 C}{M\Delta x^4}[y_{n,k} - 2y_{n-1,k} + y_{n-2,k}] - g. \end{aligned} \tag{15}$$

The above four systems may be written in the matrix form (see [4]) $\underline{z}_{k+1} = \Pi \underline{z}_k$ where

$$\Pi = \begin{bmatrix} A & -I \\ I & 0 \end{bmatrix} \qquad \underline{z}_k = \begin{bmatrix} \underline{y}_k \\ \underline{y}_{k-1} \end{bmatrix}$$

and, setting $\gamma_1 = \frac{LT_0\Delta t^2}{M\Delta x^2}$, $\gamma_2 = \frac{L^2 C\Delta t^2}{M\Delta x^2}$, $\gamma_3 = 2 - 2\gamma_1$,
in case a) A is the $(n-1) \times (n-1)$ 5 diagonal matrix:

$$A = \begin{bmatrix} \gamma_3 - 5\gamma_2 & \gamma_1 + 4\gamma_2 & -\gamma_2 & 0 & 0 & 0 & \cdots \\ \gamma_1 + 4\gamma_2 & \gamma_3 - 6\gamma_2 & \gamma_1 + 4\gamma_2 & -\gamma_2 & 0 & 0 & \cdots \\ \cdots & \cdots & \cdots & \cdots & \cdots & \cdots & \cdots \\ \cdots & -\gamma_2 & \gamma_1 + 4\gamma_2 & \gamma_3 - 6\gamma_2 & \gamma_1 + 4\gamma_2 & -\gamma_2 & \cdots \\ \cdots & \cdots & \cdots & \cdots & \cdots & \cdots & \cdots \\ \cdots & 0 & 0 & -\gamma_2 & \gamma_1 + 4\gamma_2 & \gamma_3 - 6\gamma_2 & \gamma_1 + 4\gamma_2 \\ \cdots & 0 & 0 & 0 & -\gamma_2 & \gamma_1 + 4\gamma_2 & \gamma_3 - 5\gamma_2 \end{bmatrix} \tag{16}$$

in case b) A is the $(n-1) \times (n-1)$ 5 diagonal matrix

$$A = \begin{bmatrix} \gamma_3 - 7\gamma_2 & \gamma_1 + 4\gamma_2 & -\gamma_2 & 0 & 0 & 0 & \cdots \\ \gamma_1 + 4\gamma_2 & \gamma_3 - 6\gamma_2 & \gamma_1 + 4\gamma_2 & -\gamma_2 & 0 & 0 & \cdots \\ \cdots & \cdots & \cdots & \cdots & \cdots & \cdots & \cdots \\ \cdots & -\gamma_2 & \gamma_1 + 4\gamma_2 & \gamma_3 - 6\gamma_2 & \gamma_1 + 4\gamma_2 & -\gamma_2 & \cdots \\ \cdots & \cdots & \cdots & \cdots & \cdots & \cdots & \cdots \\ \cdots & 0 & 0 & -\gamma_2 & \gamma_1 + 4\gamma_2 & \gamma_3 - 6\gamma_2 & \gamma_1 + 4\gamma_2 \\ \cdots & 0 & 0 & 0 & -\gamma_2 & \gamma_1 + 4\gamma_2 & \gamma_3 - 7\gamma_2 \end{bmatrix} \tag{17}$$

in case c) A is the $(n-1) \times (n-1)$ 5 diagonal matrix

$$A = \begin{bmatrix} \gamma_3 - 7\gamma_2 & \gamma_1 + 4\gamma_2 & -\gamma_2 & 0 & 0 & 0 & \cdots \\ \gamma_1 + 4\gamma_2 & \gamma_3 - 6\gamma_2 & \gamma_1 + 4\gamma_2 & -\gamma_2 & 0 & 0 & \cdots \\ \cdots & \cdots & \cdots & \cdots & \cdots & \cdots & \cdots \\ \cdots & -\gamma_2 & \gamma_1 + 4\gamma_2 & \gamma_3 - 6\gamma_2 & \gamma_1 + 4\gamma_2 & -\gamma_2 & \cdots \\ \cdots & \cdots & \cdots & \cdots & \cdots & \cdots & \cdots \\ \cdots & 0 & 0 & -\gamma_2 & \gamma_1 + 4\gamma_2 & \gamma_3 - 6\gamma_2 & \gamma_1 + 4\gamma_2 \\ \cdots & 0 & 0 & 0 & -\gamma_2 & \gamma_1 + 4\gamma_2 & \gamma_3 - 5\gamma_2 \end{bmatrix} \tag{18}$$

in case d) A is the $n \times n$ 5 diagonal matrix

$$A = \begin{bmatrix} \gamma_3 - 7\gamma_2 & \gamma_1 + 4\gamma_2 & -\gamma_2 & 0 & 0 & 0 & \cdots \\ \gamma_1 + 4\gamma_2 & \gamma_3 - 6\gamma_2 & \gamma_1 + 4\gamma_2 & -\gamma_2 & 0 & 0 & \cdots \\ \cdots & \cdots & \cdots & \cdots & \cdots & \cdots & \cdots \\ \cdots & -\gamma_2 & \gamma_1 + 4\gamma_2 & \gamma_3 - 6\gamma_2 & \gamma_1 + 4\gamma_2 & -\gamma_2 & \cdots \\ \cdots & \cdots & \cdots & \cdots & \cdots & \cdots & \cdots \\ \cdots & 0 & 0 & -\gamma_2 & \gamma_1 + 4\gamma_2 & \gamma_3 - 5\gamma_2 & \gamma_1 + 2\gamma_2 \\ \cdots & 0 & 0 & 0 & -\gamma_2 & \gamma_1 + 2\gamma_2 & \gamma_3 - \gamma_2 \end{bmatrix} \tag{19}$$

The stability condition implies that in each case the eigenvalues λ_i of the corresponding matrix A must verify the condition $|\lambda_i| \leq 1$. In [4] has been proved that the eigenvalues of the matrix A of case a (or of cases b and c), neglecting the first and the last row and column, are less than one if

$$\Delta t \leq \left[\frac{LT_0}{M\Delta x^2} + \frac{4L^2C}{M\Delta x^4}\right]^{-\frac{1}{2}}. \tag{20}$$

By using this result and standard techniques of linear algebra [7] it is possible to prove that, if condition (20) holds, than also the eigenvalues λ_i of the matrices (16), (18),(19) verify the condition $|\lambda_i| \leq 1$.

We can then state the following

Theorem 3.1 *The stability condition for the finite differences systems, defined in §2 and §3, modelling a linear rod clamped or hinged in both end points or clamped in the first end point and hinged or free to move in the last one is given by:*

$$\Delta t \leq \left[\frac{LT_0}{M\Delta x^2} + \frac{4L^2C}{M\Delta x^4}\right]^{-\frac{1}{2}}.$$

As a conclusion of this section we remember that in [4] it has been remarked that the difference system (8) may be seen as a second order finite differences scheme for the numerical integration of the partial differential equation:

$$\frac{\partial^2 y}{\partial t^2} = \frac{LT_0}{M}\frac{\partial^2 y}{\partial x^2} - \frac{L^2 C}{M}\frac{\partial^4 y}{\partial x^4} - g - \sigma\frac{\partial y}{\partial t} \quad 0 \le x \le L \tag{21}$$

(where $\sigma = \frac{\rho}{m}$) which models the motion of a linear rod in the continuous case. For this model the natural boundary constrains, for a rod hinged (or clamped) at the end points, are

$$\begin{cases} y(0,t) & = y(L,t) = 0 \\ \frac{\partial^2 y(0,t)}{\partial x^2} & = \frac{\partial^2 y(L,t)}{\partial x^2} = 0 \quad \left(\text{or} \quad \frac{\partial y(0,t)}{\partial x} = \frac{\partial y(L,t)}{\partial x} = 0\right) \end{cases}, \tag{22}$$

while for a rod clamped in $x = 0$ and free to move in $x = L$, are

$$y(0,t) = 0, \quad \frac{\partial y(0,t)}{\partial x} = 0, \quad \frac{\partial^2 y(L,t)}{\partial x^2} = 0, \quad \frac{\partial^3 y(L,t)}{\partial x^3} = 0. \tag{23}$$

It is easy to prove that our choic ofe the forces acting on the end points, in order to simulate the effect of the constraints at the end points, is equivalent to approximate equations (22),(23) by using second order differences. More precisely for $x = L$ we have from (22)

$$y_{n,k} = 0 \quad \text{and} \quad y_{n-1,k} - 2y_{n,k} + y_{n+1,k} = 0 \quad (\text{or} \quad y_{n+1,k} - y_{n-1,k} = 0)$$

from which

$$y_{n,k} = 0 \quad \text{and} \quad y_{n-1,k} = -y_{n+1,k} \quad (\text{or } y_{n-1,k} = y_{n+1,k}),$$

while from the last two equations of (23):

$$y_{n-1,k} - 2y_{n,k} + y_{n+1,k} = 0 \quad \text{and} \quad y_{n+2,k} - 2y_{n+1,k} + 2y_{n-1,k} - y_{n-2,k} = 0$$

from which

$$y_{n+1,k} - y_{n,k} = y_{n,k} - y_{n-1,k} \quad \text{and} \quad y_{n+21,k} - y_{n+1,k} = 2(y_{n,k} - y_{n-1,k}) - (y_{n-1,k} - y_{n-2,k}).$$

We remark again that for P_n (with mass $\frac{m}{2}$) the term due to the approximation of the fourth derivative must be halved.

In the non linear case the partial differential equation corresponding to the finite difference scheme (9) is (see [4]):

$$\frac{\partial^2 y}{\partial t^2} = \frac{LT_0}{M}\frac{\partial^2 y}{\partial x^2} - \frac{L^2 C}{M}\frac{\partial y}{\partial x}\left(\frac{1}{1+y_x^2}\frac{\partial}{\partial x}\left(\frac{y_{xx}}{1+y_x^2}\right)\right) - g - \sigma\frac{\partial y}{\partial t} \quad 0 \le x \le L \tag{24}$$

while the boundary conditions are still (22) and (23). Regarding the stability conditions we remember that a sufficient condition has been given in [4]. We think that such a condition still holds also introducing the boundary conditions as done in the linear case.

4 Numerical results and conclusions

In order to validate the previous discrete models a harmonic steel rod has been considered, with circular normal section, with physical parameters $E = 206000 N/mm^2$ (Young modulus), density $\delta = 7.86 \cdot 10^{-6} Kg/mm^3$, length L, diameter d. As external forces gravity and viscosity, with $\rho = 0.075 Kg/s$, have been considered . Relations given in §2 and §3 to compute the largest displacement f_m at equilibrium starting from a horizontal initial position have been used and the results obtained have been compared with the ones given by the classical formulas of continuum mechanics. We report only the particular case $L = 1500mm$ and $d = 10mm$:

$$\begin{aligned} f_m &= \frac{PL^3}{384EI} \simeq 0.78mm, \text{ clamped - clamped rod} \\ f_m &= \frac{5PL^3}{384EI} \simeq 3.927mm, \text{ hinged - hinged rod} \\ f_m &= \frac{PL^3}{8EI} \simeq 37.60mm, \text{ clamped - free rod} \end{aligned} \tag{25}$$

where I is the moment of inertia and P is the weight of the rod.

In all cases our numerical results agree to the second decimal digit, with the ones given by (25) when we use 81 masses. In figure 2 are shown the computed equilibrium configurations in the four different cases of boundary conditions, in order to emphasize that the different constraints at the end points (clamped-clamped, hinged-hinged, clamped-hinged and clamped-free) are well simulated by our discrete scheme.

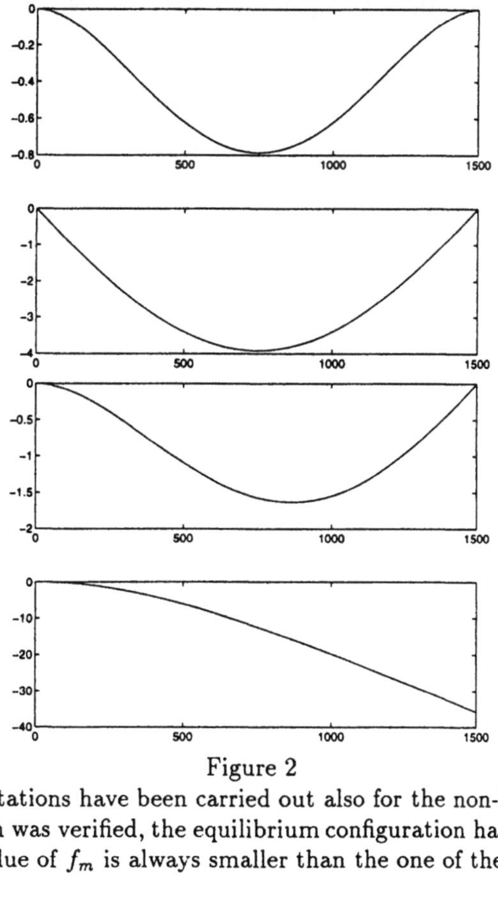

Figure 2

Numerical computations have been carried out also for the non-linear case and, when the stability condition was verified, the equilibrium configuration has always been reached and the computed value of f_m is always smaller than the one of the corresponding linear case.

We remark that, even if the numerical results are given for $\rho = 0.075 Kg/s$, we have considered also different values of ρ and, according to the stability condition (20) (which is a weak stability condition for $\rho = 0$ [4]), the computation time grows and the accuracy of the results decays when ρ decreases. In order to improve our method, regarding the computation time and the freedom of choosing ρ arbitrarily small, we are studying models which lead to unconditionally stable difference schemes.

5 Acknowledgment

We thanks our student Frattini Marco who has carried out the numerical computations.

References

[1] D. Greenspan, Arithmetic applied mathematics, Pergamon, New York, 1980.

[2] M. Frontini and L. Gotusso, Appl. Math. Modelling, 14, 489 (1990).

[3] L. Gotusso and A. Veneziani, WSSIAA,4, 295 (1995).

[4] L. Gotusso and A. Veneziani, Comp. Math. Modelling, 24, 99 (1996).

[5] L. Gotusso, G. Prouse and A. Veneziani in: Proceedings Fifth Int. Coll. on Diff. Eq., D. Bainov and A. Dishliev, (Eds.),SCT Pub, Singapore (1995) pp. 91-100.

[6] L. Gotusso, G. Prouse and A. Veneziani in: Proceedings Sixth Int. Coll. on Diff. Eq.,(1996) in press.

[7] D. Bini, M. Capovani and O. Menchi, Metodi numerici per l'algebra lineare, Zanichelli, Bologna, 1988.

7th Int. Coll. on Differential Equations, pp. 101-108
D. Bainov (Ed)

ON LOCAL AND GLOBAL SOLVABILITY OF NONLINEAR MIXED PROBLEMS MODELLING VIBRATIONS OF A STRING

C. L. FROTA[1] and N. A. LAR'KIN[2]

[1]Departamento de Matemática - Universidade Estadual de Maringá

Av. Colombo, 5790 - CEP: 87020-900 Maringá - Pr. - Brasil.

[2]The Institute of Theoretical and Applied Mechanics

Novosibirsk - 90, 630090, Russia.

ABSTRACT: We study the local solvability of the boundary value problem for the non-homogeneous Kirchhoff-Carrier equation and the global solvability of the mixed problem for the damped Carrier equation.

1. INTRODUCTION

For an arbitrary dimension n, the mixed problem associated to the Kirchhoff model (cf. Kirchhoff [7], chapter 9, § 7) is given by

$$\frac{\partial^2 u}{\partial t^2} - M(||u||^2)\Delta u = f \quad \text{in } Q = \Omega \times (0,T); \tag{1}$$

$$u = 0 \quad \text{on } \Sigma = \Gamma \times (0,T); \tag{2}$$

$$u(x,0) = u_0(x) \quad , \quad \frac{\partial u}{\partial t}(x,0) = u_1(x) \quad \text{on } \Omega; \tag{3}$$

and the mixed problem associated to the Carrier model (Carrier [3]) is

$$\frac{\partial^2 u}{\partial t^2} - M(|u|^2)\Delta u = f \quad \text{in } Q = \Omega \times (0,T); \tag{4}$$

$$u = 0 \quad \text{on } \Sigma = \Gamma \times (0,T); \tag{5}$$

$$u(x,0) = u_0(x) \quad , \quad \frac{\partial u}{\partial t}(x,0) = u_1(x) \quad \text{on } \Omega. \tag{6}$$

Here, Ω is a bounded open set of $\mathbf{R}^n$, with a smooth boundary Γ; $M : [0,\infty) \to \mathbf{R}$ is a positive real function, Δ is the Laplace operator and

$$||u||^2 = \sum_{i=1}^{n} \int_\Omega (\frac{\partial u}{\partial x_i})^2 dx \quad , \quad |u|^2 = \int_\Omega u^2 dx. \tag{7}$$

[2]Supported by CNPq-Brasil as a Visiting Professor at the State University of Maringá.

The mixed problem (1) – (3) has been studied by several authors. Some of them has called (1) by Kirchhoff-Carrier's equation, see for example Medeiros & Milla Miranda [8], Frota [6], Brito [2], Newman [9], D'Ancona & Spagnolo [4] and the references therein. Actually, the models of Kirchhoff and Carrier are distinct, and, while we have a large number of publications on Kirchhoff's equation, as far as we know, there is only one (Ebihara [5]) that in some sense involves Carrier's equation. In Ebihara [5], the author has shown local existence of classical solutions to the mixed problem for the following equation

$$\frac{\partial^2 u}{\partial t^2} - M\left(|u|, ||u||, |\Delta u|\right) \Delta u = f, \quad \text{on } Q = \Omega \times (0,T).$$

In the first part of the present paper we consider a general model when function M depends on x, t and on all of the norms $|u|^2, ||u||^2, \left|\frac{\partial u}{\partial t}\right|^2$. We prove the local existence of a unique solution to the problem

$$\rho(x,t)\frac{\partial^2 u}{\partial t^2} - M(x,t,|u|^2,||u||^2,|\frac{\partial u}{\partial t}|^2)\Delta u = f \text{ in } Q = \Omega \times (0,T); \tag{8}$$

$$u = 0 \quad \text{on } \Sigma = \Gamma \times (0,T); \tag{9}$$

$$u(x,0) = u_0(x) \quad , \quad \frac{\partial u}{\partial t}(x,0) = u_1(x) \quad \text{on } \Omega. \tag{10}$$

The dependence of M of $\left|\frac{\partial u}{\partial t}\right|^2$ seemed to be nonphysical. Nevertheless, we have found published papers devoted to this model, see Aliyev [1]. Therefore, we allowed the function M to depend also on $\left|\frac{\partial u}{\partial t}\right|^2$.

In the second part of our paper, we consider the mixed problem for the damped Carrier equation

$$\frac{\partial^2 u}{\partial t^2} - M(|\, u(t)\,|^2)\Delta u + \alpha \,|\, \frac{\partial u}{\partial t}\,|^\rho \,\frac{\partial u}{\partial t} = f(x,t). \tag{11}$$

The Cauchy problem for the Kirchhoff model

$$\frac{\partial^2 u}{\partial t^2} - (1 + \int_\Omega |\, \nabla u(t)\,|^2 \,dx)\Delta u = F(u, u_t, \nabla u) \tag{12}$$

was considered by D'Ancona & Spagnolo [4]. Under smallness conditions for the initial data, they proved global solvability of the Cauchy problem. They also mentioned that existence of global solutions to (12) without restrictions on the size of the data is an open problem.

Here, we prove existence and uniqueness of global solutions to the mixed problem for (11) without smallness conditions for the initial data and the function f.

Remark 1: Our results can be easily extended to equations of the form

$$\frac{\partial^2 u}{\partial t^2} - M(x,t,|\, u(t)\,|^2)\Delta u + F(x,t,u,u_t) = f(x,t). \tag{13}$$

2. LOCAL SOLVABILITY

Let Ω be a bounded open set of $\mathbf{R}^n$ with a smooth boundary Γ, $T > 0$. We assume that $\rho : \overline{\Omega} \times [0,T] \to \mathbf{R}$ and $M : \overline{\Omega} \times [0,T] \times [0,\infty)^3 \to \mathbf{R}$ satisfy the conditions:

$$\rho \in C^1\left(\overline{\Omega} \times [0,T]\right) \text{ with } 0 < \rho_0 \le \rho(x,t); \tag{14}$$

$$M \in C^1\left(\overline{\Omega} \times [0,T] \times [0,\infty)^3\right) \text{ with } 0 < m_0 \le M(x,t,\lambda,\xi,\theta); \tag{15}$$

$$M(x,t,\lambda,\xi,\theta) + \left|\frac{\partial M}{\partial t}(x,t,\lambda,\xi,\theta)\right| \le k_1\left(1 + \lambda^\alpha + \xi^\beta + \theta^\gamma\right); \tag{16}$$

$$\left|\frac{\partial M}{\partial \lambda}(x,t,\lambda,\xi,\theta)\right| \le k_2\left(1 + \lambda^{\alpha-1} + \xi^\beta + \theta^\gamma\right); \tag{17}$$

$$\left|\frac{\partial M}{\partial \xi}(x,t,\lambda,\xi,\theta)\right| \le k_3\left(1 + \lambda^{\alpha} + \xi^{\beta-1} + \theta^\gamma\right); \tag{18}$$

$$\left|\frac{\partial M}{\partial \theta}(x,t,\lambda,\xi,\theta)\right| \le k_4\left(1 + \lambda^{\alpha} + \xi^{\beta} + \theta^{\gamma-1}\right), \tag{19}$$

where $\rho_0, m_0, k_1, ..., k_4$ are positive constants, and α, β, γ are real numbers bigger than one. We state the following local existence theorem:

Theorem 1. *Under the assumptions* (14) – (19), *given* $u_0 \in H_0^1(\Omega) \cap H^2(\Omega)$, $u_1 \in H_0^1(\Omega)$ *and* $f \in L^4(0,T; H_0^1(\Omega))$, *there exist a positive number* $T_0 \le T$ *and a unique function* u *such that*

$$u \in L^\infty(0,T_0; H_0^1(\Omega) \cap H^2(\Omega)), \quad u' \in L^\infty(0,T_0; H_0^1(\Omega)), \quad u'' \in L^\infty(0,T_0; L^2(\Omega))$$

which is a solution to the following problem:

$$\rho(t)u''(t) - M(t, |u(t)|^2, \|u(t)\|^2, |u'(t)|^2)\Delta u(t) = f(t) \; a.e. \text{ on } [0,T_0]; \tag{20}$$

$$u = 0 \quad \text{on } \Sigma = \Gamma \times (0,T_0); \tag{21}$$

$$u(x,0) = u_0(x) \quad , \quad u'(x,0) = u_1(x) \quad \text{on } \Omega. \tag{22}$$

Remark 2: As an application of this theorem, let us consider the following function

$$M(x,t,\lambda,\xi,\theta) = \psi_0(x,t) + \psi_1(x,t)\,\lambda + \psi_2(x,t)\,\xi + \psi_3(x,t)\,\theta. \tag{23}$$

i) If $\psi_0(x,t) = k_0$, $\psi_2(x,t) = k_1$ and $\psi_1(x,t) = \psi_3(x,t) = 0$, then (20) is Kirchhoff's equation.
ii) If $\psi_0(x,t) = c_0$, $\psi_1(x,t) = c_1$ and $\psi_2(x,t) = \psi_3(x,t) = 0$, then (20) is Carrier's equation.

Proof: Let $\omega_{j_{j\in\mathbf{N}}}$ be eigenfunctions of the problem

$$\begin{cases} -\Delta\omega_j = \lambda_j\omega_j, & j = 1,2,3,\ldots . \\ \omega_{j|_\Gamma} = 0. \end{cases} \tag{24}$$

We denote the inner products in $L^2(\Omega)$ and $H_0^1(\Omega)$ as follows

$$(f,g)=\int_\Omega f(x)\,g(x)\,dx \quad , \quad ((\varphi,\psi))=\sum_{i=1}^{n}\int_\Omega \frac{\partial\varphi}{\partial x_i}(x)\,\frac{\partial\psi}{\partial x_i}(x)\,dx. \tag{25}$$

Our approach is to seek a local solution u of (20) – (22) as a limit of a suitable sequence $(u_m)_{m\in\mathbf{N}}$. Since ρ and M satisfy (14) and (15), for each $m\in\mathbf{N}$ there exist a number $T_m>0$ and a function $u_m:\Omega\times[0,T_m]\to\mathbf{R}$ of the form

$$u_m(x,t)=\sum_{j=1}^{m} g_{jm}(t)\ \omega_j(x)$$

such that for all $1\le j\le m$

$$(\rho(t)u_m''(t),\omega_j)-(M(t,\mid u_m(t)\mid^2,\|u_m(t)\|^2,\mid u_m'(t)\mid^2)\Delta u_m(t),\omega_j)=(f(t),\omega_j);$$

$$u_m(0)=\sum_{j=1}^{m}\ (u_0,\omega_j)\ \omega_j \ ,\quad u_m'(0)=\sum_{j=1}^{m}\ (u_1,\omega_j)\ \omega_j.$$

Let us denote

$$\Phi_{t,u}=\left(t,|u_m(t)|^2,\|u_m(t)\|^2,|u_m'(t)|^2\right),$$

then, we can rewrite the approximate equation in the form

$$(\rho(t)\,u_m''(t),v)-(M(\Phi_{t,u})\,\Delta u_m(t),v)=(f(t),v),\ \ \forall\ v\in V_m, \tag{26}$$

where V_m is the m-dimensional subspace spanned by $\{\omega_1,\omega_2,...,\omega_m\}$.
Estimates: We put $v=u_m''(t)$ in (26). Then

$$(\rho(t)u_m''(t),u_m''(t))\le\left[|M(\Phi_{t,u})\,\Delta u_m(t)|+|f(t)|\right]|u_m''(t)|.$$

From the above inequality and (14), we get

$$|u_m''(t)|\le\rho_0^{-1}\left[|M(\Phi_{t,u})\,\Delta u_m(t)|+|f(t)|\right]. \tag{27}$$

Taking $v=-2\Delta u_m'(t)$ in (26) we obtain

$$\frac{d}{dt}\left[\sum_{i=1}^{n}\left(\rho(t),\left(\frac{\partial u_m'}{\partial x_i}(t)\right)^2\right)+\left(M(\Phi_{t,u}),(\Delta u_m(t))^2\right)\right]=$$

$$\sum_{i=1}^{n}\left(\rho'(t),\left(\frac{\partial u_m'}{\partial x_i}(t)\right)^2\right)-2\sum_{i=1}^{n}\left(\frac{\partial\rho}{\partial x_i}(t),u_m''(t)\,\frac{\partial u_m'}{\partial x_i}(t)\right)$$

$$+\left(M'(\Phi_{t,u}),(\Delta u_m(t))^2\right)+2\,(u_m(t),u_m'(t))\left(\frac{\partial M}{\partial\lambda}(\Phi_{t,u}),(\Delta u_m(t))^2\right)$$

$$+2\,((\ u_m(t),u_m'(t)\ ))\left(\frac{\partial M}{\partial\xi}(\Phi_{t,u}),(\Delta u_m(t))^2\right)$$

$$+2\,(u_m'(t),u_m''(t))\left(\frac{\partial M}{\partial\theta}(\Phi_{t,u}),(\Delta u_m(t))^2\right)-2\,(f(t),\Delta u_m'(t)).$$

Taking into account (14) – (19) and the following inequalities

$$|u_m(t)| \leq C\,|\Delta u_m(t)|\,,\ |u'_m(t)| \leq C\,\|u'_m(t)\|\,,\ \|u_m(t)\| \leq C\,|\Delta u_m(t)|\,,$$

we can find $\kappa \in \mathbf{N}$ ($\kappa > 4$) such that

$$\frac{d}{dt}\left[\sum_{i=1}^{n}\left(\rho(t)\,,\left(\frac{\partial u'_m}{\partial x_i}(t)\right)^2\right) + \left(M(\Phi_{t,u})\,,(\Delta u_m(t))^2\right)\right] \leq$$
$$C_1\left(|f(t)|^2 + \|f(t)\|^2 + |f(t)|^4\right) + C_2\left(\|u'_m(t)\|^{2\mathrm{K}} + |\Delta u_m(t)|^{2\mathrm{K}}\right).$$

Integrating this inequality, we have

$$\|u'_m(t)\|^2 + |\Delta u_m(t)|^2 \leq C_3\left[1 + \int_0^t \left(\|u'_m(\tau)\|^{2\mathrm{K}} + |\Delta u_m(\tau)|^{2\mathrm{K}}\right)d\tau\right].$$

Consequently, there exist T_0, $0 < T_0 < T$ and a constant C_4 such that

$$\|u'_m(t)\|^2 + |\Delta u_m(t)|^2 \leq C_4,\ \ \forall\ t \in [0, T_0]\,. \tag{28}$$

From (27) and (28), choosing a suitable subsequence $(u_{m_\nu})_{\nu\in\mathbf{N}}$ of $(u_m)_{m\in\mathbf{N}}$, we have a function $u : \Omega \times [0, T_0] \to \mathbf{R}$ such that

$$\begin{aligned}
&u_{m_\nu} \overset{*}{\rightharpoonup} u \ \text{ in } \ L^\infty(0, T_0; H_0^1(\Omega) \cap H^2(\Omega)),\\
&u'_{m_\nu} \overset{*}{\rightharpoonup} u' \ \text{ in } \ L^\infty(0, T_0; H_0^1(\Omega)),\\
&u''_{m_\nu} \overset{*}{\rightharpoonup} u'' \ \text{ in } \ L^\infty(0, T_0; L^2(\Omega));
\end{aligned}$$

and u satisfies (20) – (22). Uniqueness is proved by standard arguments, see Frota [6].

3. GLOBAL SOLVABILITY

Theorem 2. *Let T be a finite positive number, $\rho > 1$ and the function $M : [0,\infty) \to \mathbf{R}$ satisfies the conditions*

$$M \in C^1([0,\infty)) \ \text{ with } \ 0 < m_0 \leq M(\lambda),\ \ \forall \lambda \geq 0; \tag{29}$$

$$\frac{|\,M'(\lambda)\lambda^{\frac{1}{2}}\,|}{M(\lambda)} \leq k_1,\quad \forall\,\lambda \geq 0; \tag{30}$$

then for each $u_0 \in H_0^1(\Omega) \cap H^2(\Omega), u_1 \in H_0^1(\Omega) \cap L^{2\rho+2}(\Omega)$ and $f \in L^2(0,T;L^2(\Omega))$ with $f' \in L^2(0,T;L^2(\Omega))$, there exists a unique function u in the class

$$u \in L^\infty(0,T;H_0^1(\Omega) \cap H^2(\Omega)),\ \ u' \in L^\infty(0,T;H_0^1(\Omega)),\ \ u'' \in L^\infty(0,T;L^2(\Omega))$$

such that

$$u'' - M(|\,u\,|^2)\Delta u + |\,u'\,|^\rho\,u' = f \quad \text{a.e. in } Q = \Omega \times (0,T); \tag{31}$$

$$u = 0 \quad \text{in } \Sigma = \Gamma \times (0,T); \tag{32}$$

$$u(0) = u_0\,, \qquad u'(0) = u_1\,. \tag{33}$$

Proof: Let $\{\omega_j\}_{j\in\mathbf{N}}$ be defined by (24). For each $m \in \mathbf{N}$ we construct the function

$$u_m(x,t) = \sum_{j=1}^{m} g_{jm}(t)\omega_j(x), \quad x \in \Omega \text{ and } t \in [0, T_m]$$

which is a solution to the following problem

$$\begin{cases} (u''_m(t), \omega_j) - M(|\, u_m(t)\,|^2)(\Delta u_m(t), \omega_j) + (|\, u'_m(t)\,|^\rho\, u'_m(t), \omega_j), \;\; 1 \le j \le m; \\ u_m(0) = u_{0m} = \sum_{j=1}^{m} (u_0, \omega_j)\, \omega_j\,; \\ u'_m(0) = u_{1m} = \sum_{j=1}^{m} (u_1, \omega_j)\, \omega_j. \end{cases}$$

Taking into account (29), we can rewrite this problem in the form

$$\frac{(u''_m(t), v)}{M(|\, u_m(t)\,|^2)} - (\Delta u_m(t), v) + \frac{(|\, u'_m(t)\,|^\rho\, u'_m(t), v)}{M(|\, u_m(t)\,|^2)} = \frac{(f(t), v)}{M(|\, u_m(t)\,|^2)}, \quad \forall v \in V_m. \quad (34)$$

Estimate 1. Taking $v = 2u'_m(t)$ in (34), we obtain

$$\frac{d}{dt}\left[\frac{1}{M(|\, u_m(t)\,|^2)}\,|\, u'_m(t)\,|^2 + \|u_m(t)\|^2\right] + \frac{2}{M(|\, u_m(t)\,|^2)}\left[\,|\, u'_m(t)\,|^{\rho+2}_{L^{\rho+2}(\Omega)}\right.$$

$$\left. - \frac{|\, M'(|\, u_m(t)\,|^2)\,|\, u_m(t)\,||}{M(|\, u_m(t)\,|^2)}\,|\, u'_m(t)\,|^3\right] \le \frac{2}{M(|\, u_m(t)\,|^2)}\,|\, f(t)\,||\, u'_m(t)\,| \qquad (35)$$

Taking into account (30) and employing the Young inequality, we get

$$\frac{d}{dt}\left[\frac{1}{M(|\, u_m(t)\,|^2)}\,|\, u'_m(t)\,|^2 + \|u_m(t)\|^2\right] + C_1\frac{1}{M(|\, u_m(t)\,|^2)}\,|\, u'_m(t)\,|^{\rho+2}_{L^{\rho+2}(\Omega)} \le$$

$$C_2 \;+\; C_3\,|\, f(t)\,|^{\frac{\rho+2}{\rho+1}},$$

where C_1, C_2 and C_3 are positive constants which do not depend on t and m. Integrating over $(0,t)$, we obtain

$$\frac{|\, u'_m(t)\,|^2}{M(|\, u_m(t)\,|^2)} + \|u_m(t)\|^2 + C_1\int_0^t \frac{|\, u'_m(\tau)\,|^{\rho+2}_{L^{\rho+2}(\Omega)}}{M(|\, u_m(\tau)\,|^2)}\,d\tau \le C_4. \qquad (36)$$

We get from here

$$\|u_m(t)\|^2 \le C_4, \qquad \forall t \in [0, T_m], \qquad (37)$$

and, since $u_m(t) \in H_0^1(\Omega)$, $\forall t \in [0, T_m]$, Poincaré's inequality gives

$$|\, u_m(t)\,|^2 + \|u_m(t)\|^2 \le C_5, \quad \forall t \in [0, T_m].$$

This allows to extend solutions u_m to the whole interval $[0, T]$. Moreover, since M is C^1, we obtain

$$0 < m_0 \le M(|\, u_m(t)\,|^2) < C_6, \quad \forall t \in [0, T]; \qquad (38)$$

Then (36) gives

$$| u_m'(t) |^2 + \|u_m(t)\|^2 + \int_0^T \|u_m'(t)\|_{L^{\rho+2}(\Omega)}^{\rho+2} dt \leq C_7, \quad \forall t \in [0,T]. \tag{39}$$

Estimate 2. From equation (34), we have

$$| u_m''(0) | \leq C_8 \left(| \Delta u_{0m} | + \|u_{1m}\|_{L^{2\rho+2}(\Omega)}^{\rho+1} + | f(0) | \right)$$

which implies

$$| u_m''(0) |^2 \leq C_9. \tag{40}$$

Differentiating (34) with respect to t and taking $v = 2u_m''(t)$, we obtain

$$\frac{d}{dt}\left[\frac{| u_m''(t) |^2}{M(| u_m(t) |^2)} + \|u_m'(t)\|^2\right] + \frac{2(\rho+1)}{M(| u_m(t) |^2)}(| u_m'(t) |^\rho, (u_m''(t))^2) =$$
$$\frac{4M'(| u_m(t) |^2)}{(M(| u_m(t) |^2))^2}(u_m'(t), u_m(t))\,(| u_m'(t) |^\rho\, u_m'(t), u_m''(t))$$
$$+\frac{2M'(| u_m(t) |^2)}{(M(| u_m(t) |^2))^2}(u_m'(t), u_m(t))\; | u_m''(t) |^2 + \frac{2}{M(| u_m(t) |^2)}(f'(t), u_m''(t))$$
$$-\frac{4M'(| u_m(t) |^2)}{(M(| u_m(t) |^2))^2}(u_m'(t), u_m(t))\,(f(t), u_m''(t)).$$

This relation and (38) – (40) imply

$$| u_m''(t) |^2 + \|u_m'(t)\|^2 + C_{10}\int_0^t (| u_m'(\tau) |^\rho, (u_m''(\tau))^2) d\tau \leq C_{11}(1 + \int_0^t | u_m''(\tau) |^2 d\tau),$$

whence

$$| u_m''(t) |^2 + \|u_m'(t)\|^2 \leq C_{12}. \tag{41}$$

Estimate 3. Taking $v = -2\Delta u_m'(t)$ in (34), we obtain

$$\frac{d}{dt}\left[\frac{\|u_m'(t)\|^2}{M(| u_m(t) |^2)} + | \Delta u_m(t) |^2\right] + \frac{2(\rho+1)}{M(| u_m(t) |^2)}\sum_{i=1}^n \int_\Omega | u_m' |^\rho \left(\frac{\partial u_m'}{\partial x_i}\right)^2 dx =$$
$$-2\frac{M'(| u_m(t) |^2)}{(M(| u_m(t) |^2))^2}(u_m'(t), u_m(t))\, \|u_m'(t)\|^2 + 2\,(f(t), -\Delta u_m'(t)).$$

Integration over $(0,t)$ and some calculations give

$$| \Delta u_m(t) |^2 \leq C_{13}. \tag{42}$$

Estimates (39), (41) and (42) allow us to pass to the limit as $m \to \infty$ and get a solution of (31) – (33). One can prove uniqueness in usual way.

REFERENCES

[1] ALIYEV, F. A., On a quasilinear hyperbolic equation. *Soviet Math. Dokl.*, 300, 6, (1988), pp. 1311-1315.

[2] BRITO, E. H., The damped elastic stretched string equation generalized: existence, uniqueness, regularity and stability. *Applicable Anal. 13*, (1982), pp. 219-233.

[3] CARRIER, G. F., On the non-linear vibration problem of the elastic string. *Quart. Appl. Math. 3*, (1945), pp. 157-165.

[4] D'ANCONA, P.,& SPAGNOLO, S., Nonlinear Pertubations of the Kirchhoff Equation. *Comm. on Pure and Applied Math., Vol. XLVII*, (1994), pp. 1005-1029.

[5] EBIHARA, Y., On the existence of local smooth solutions for some degenerate quasilinear hyperbolic equations. *An. Acad. brasil. Ciênc. 57*, (2), (1986), pp. 145-152.

[6] FROTA, C. L., Non local solutions of a nonlinear hyperbolic partial diferential equation. *Portugaliae Mathematica 51*, (1994), pp. 455-473.

[7] KIRCHHOFF, G., Vorlesungen Über Mechanik. *Tauber Leipzig*, (1883).

[8] MEDEIROS, L. A. & MILLA MIRANDA, M., On a nonlinear wave equation with damping. *Revista Matemática de la Universidad Complutense de Madrid 3*, (2 y 3) (1990), pp. 213-231.

[9] NEWMAN, W. G., Global solutions of a nonlinear string equation. *Journal of Mathematical Analysis and Applications 192*, (1995), pp. 689-704.

7th Int. Coll. on Differential Equations, pp. 109-113
D. Bainov (Ed)

Poincaré-Bendixson Type Theorems for n Dimensional Spaces

Takao Fujimoto and Yoshihisa Fujimoto*

Department of Economics, University of Okayama, Okayama, 700, Japan
Department of Mathematics, Toyo University, Kawagoe-shi, 350, Japan

Abstract

The n-dimensional autonomous systems are discussed under some appropriate geometric conditions. It is shown that a Poincaré-Bendixson type theorem holds. There exists a solution whose orbit stays on the boundary of a certain set for the above systems.

Keywords: Autonomous system, Poincaré-Benedixson theorem, Periodic solution

1 Introduction

A well known theorem on the existence of periodic solutions due to Poincaré [8] and Bendixson [1] was established for the spaces of dimension two. Various applications for the systems on the plane thus include a set of conditions which are proved to be sufficient to satisfy the prerequisites for Poincaré-Bendixson theorem. The same has been true for the nonlinear second order differential equations.

Rauch [9] considered a third order autonomous equation which can be decomposed into a linear and a nonlinear part of special type. This result was extended by Williamson [12] into n-th order equations using the 'torus principle' (Pliss [7]). Cronin [3] took up an n-dimensional nonautonomous system whose nonautonomous part is a function of a fixed period, and showed the existence of a solution of the same fixed period under a set of assumptions. As she noted, these assumptions remind us of a popular variant of Poincaré-Bendixson theorem: if a given autonomous system on the plane has an unstable critical point, and if there is a simple closed curve C containing this unstable critical point but no other critical point in its interior or on

*supported by the Special Reserch Funds of Toyo University.

C, and if every trajectory which intersects C crosses C inwardly, then the system has a periodic solution.

In this note, we consider n-dimensional autonomous systems which share the above geometric characteristic, and show that there exists a solution whose orbit stays on the boundary of a certain set. Some comments are made concerning the case of dimension less than four.

2 Assumptions and The Main Theorem

Let us consider the autonomous system

$$\dot{x} = f(x) \tag{1}$$

where x is an n-vector in the n-dimensional Euclidean space R^n. We make the following assumptions.

Assumption 1. The system (1) has a unique solution when the initial vector $x(0)$ is given in a domain D which is homeomorphic to an open ball.

Assumption 2. Each solution starting in D is continuous (most probably differentiable in many applications) with respect to x and the initial point, and each orbit starting from a point on ∂D remains on the boundary or leaves the boundary inwardly, and is defined for all $t \geq 0$, while the negative semi-orbit starting from a point on ∂D never enters into the interior of D.

Assumption 3. There exists an open set S in D which is also homeomorphic to an open ball and each orbit starting from a point on ∂S(the boundary of S) either remains on the boundary or leaves the boundary outwardly.

Assumption 4. There exists no critical point on the set $D - S$.

Denote by $x(t, x_0)$ the solution with the initial vector x_0 in D, and fix a positive scalar T. And define a mapping F_T on D from x_0 to $x(T, x_0)$. This mapping F_T is a homeomorphism from D to $F_T(D)$, and satisfies

$$D \supset F_T(D) \supset F_T(S) \supset S.$$

We define $F_T^2(D) \equiv F_T(F_T(D))$, and similarly the symbol $F_T^i(D)$ for the i-th iterate of the transformation F_T. And we put $F_T^0(D) \equiv D$.

Now we have

Main Theorem. *Given the above assumptions 1-4, there exists a solution which remains on the boundary of a certain open set in D.*

proof. Let $\overline{X}$ denote the closure of a set X, and consider a sequence of closed sets $\overline{D} \supset \overline{F_T(D)} \supset \overline{F_T^2(D)} \supset \cdots \supset \overline{F_T^i(D)} \supset \cdots$, each of which contains the open set S. Put $L \equiv \bigcap_{i=1}^{\infty} \overline{F_T^i(D)}$. The set L is closed and contains S. Let L° be the interior of L.

We now prove that each orbit starting from a point ∂L° remains there. First, we pick up an arbitrary point x on ∂L°, and show that $F_T(x) \in \partial L^\circ$. If $F_T(x) \in L^\circ$, then there is a neighbourhood of x which is mapped into L° by F_T and contains an exterior point of $F_T^i(D)$ for some i (i may be 0) because of Assumption 2. (Fig.1). This contradicts the definition of L. On the other hand, if $F_T(x) \notin \overline{L^\circ}$, some points of L° are mapped into the outside of $\overline{L^\circ}$, contradicting the monotonicity of the sequence $\{F_T^i(D)\}_{i=1,2,\cdots}$.

Next we prove that even if we change the designated scalar T to another T', we end up with the same set L. Suppose that these sets are distinct, and without loss of generality suppose also that $T' < T$ and a point x on ∂L° is in the interior of L', L'°, where L' is the 'limit' set corresponding to T'. (Note that the word 'limit' set here is used in the set-theoretical meaning.) Then, within L'° there exists a neighbourhood of x which contains a point P on an orbit which starts from a boundary point Q of D such that $F_T^i(Q) = P$ for some i. Now by Assumption2, we can find a point Q' on this orbit which is either an exterior point of D or on the boundary of D such that $F_{T'}^j(Q') = P$ for some j. This implies L'° cannot be a 'limit' set of monotone shrinking sequence . A contradiction.

Thus we have shown that each orbit starting from a point on ∂L° remains there. This completes the proof. ∎

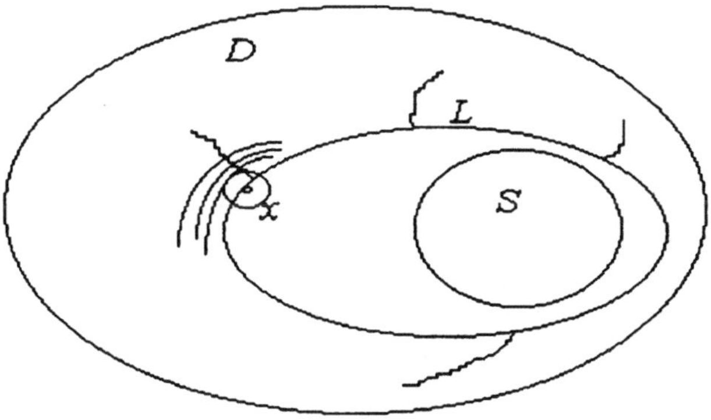

Fig. 1

3 Dimensions 2 and 3

In the case of dimension two, it has been hoped to show by using our result(So far, unsuccessful).

Theorem. *When the dimension is two, given Assumptions* 1-4, *the system (1) has a periodic solution.*

If one can show in the three dimensional case that the interior of the limit set L° is homeomorphic to an open ball, thus ∂L° is homeomorphic to the sphere S^2, then we can apply Poincaré-Bendixson theorem (or our own result) to the trajectories on the two-dimensional sphere.

4 Remarks

(1) The trouble with our result is that the boundary set ∂L° need not be homeomorphic to a hypersphere when the dimension n is greater than two. Is the interior of the limit set connected? Assumption 4 should have played some role. (Cf. Yoneyama's example in Lefschetz [5].)

(2) The fact that a solution starting from a point of the boundary ∂L° remains there has nothing to do with periodicity. (In economics, however, this property, rather than the exact periodicity, can be used in business cycle theory.)

(3) When we impose more conditions on domains D and f(e.g., convexity or star-convexity etc. and the preservation of these properties by f), the sequence $\{F_T^i(D)\}_{i=1,2,\cdots}$ may converge to L° which is homeomorphic to an open ball. (Cf. Tarski [11].)

(4) In a dual way, with suitable modifications in Assumptions 2 and 3, we can conceive of the 'limit' set of expanding sequence $\{F_T^i(D)\}_{i=1,2,\cdots}$. If two 'limit' sets coincide, the boundary ∂L° is 'stable'.

(5) We avoid the use of limit points of a semi-orbit, and the notion of limit sets in the traditional sense. (For the conventional proof, see Coddington and Levinson [2] or Lefschetz [6].) More importantly, our method, if sound enough, is equally useful to prove a proposition similar to our Main Theorem for the system of nonlinear <u>difference</u> equations, possibly even in Banach spaces. The first motivation of the authors was to prove the existence of a <u>quasi</u>-periodic solution for <u>discrete</u> versions of Goodwin growth cycle model. (See Goodwin [4] and Semmler(ed.) [10].)

References

[1] I. Bendixson. Sur les courbes définies par des équations différentielles. Acta Math.,Vol. 24 (1901) 1-88

[2] E. Coddinton & N. Levinson. *Theory of Ordinary Differential Equations.* McGraw-Hill, 1955.

[3] Jane Cronin. *Fixed Points and Topological Degree in Nonlinear Analysis.* Mathematical Surveys No. 11, Amer. Math. Soc., 1964.

[4] R.M. Goodwin. A growth cycle. in C.H.Feinstein(ed.), *Socialism, Capitalism, and Economic Growth*, Cambridge Univ. Press, 1967.

[5] S. Lefschetz. *Introduction to Topology.* Princeton Univ. Press, 1949.

[6] ______ *Differential Equations: Geometrical Theory, 2nd ed.* Interscience, 1963.

[7] V.A. Pliss. *Nonlinear Problems of the Theory of Oscillations.* Academic Press, 1966.

[8] H. Poincaré. Sur les courbes définies par une équation différentielle. *Œubres*, Vol. 1, Gauthier-Villars, 1892.

[9] L.L. Rauch. Oscillations of a third order nonlinear autonomous system. *Contributions to the Theory of Nonlinear Oscillations*, Ann. of Math. Studies. No. 20, Princeton Univ. Press, 1950, 35-88.

[10] W. Semmler(ed.) *Competition, Instability, and Nonlinear Cycles.* Lecture Notes in Economics and Mathematical Systems 275, Springer, 1986.

[11] A. Tarski. A lattice-theoretical fixpoint theorem and its applications. Pacific J. Math., Vol. 5 (1955) 285-309.

[12] D. Williamson. Periodic motion in a class of nth-order autonomous differential equations. J. Math. Appl., Vol. 53 (1976) 669-679.

7th Int. Coll. on Differential Equations, pp. 115-122
D. Bainov (Ed)

THE EFFECT OF HETEROGENEITY IN SHARING RATES AND CHOICE OF SHOOTING GALLERY ON THE SPREAD OF HIV AMONGST SHARING INJECTING DRUG USERS

David Greenhalgh

Department of Statistics and Modelling Science, University of Strathclyde, Livingstone Tower, 26 Richmond Street, Glasgow G1 1XH, U.K.

ABSTRACT

HIV (Human Immunodeficiency Virus) and AIDS (Acquired Immune Deficiency Syndrome) spreads amongst drug injectors by sharing syringes. "Shooting galleries" in which addicts rent injection equipment are common in situations where possession of injecting equipment is illegal. Data on the sharing and injecting habits of drug users clearly exhibits heterogeneity in equipment sharing rates. In this paper we examine the implications of this heterogeneity on the spread of HIV and AIDS amongst injecting drug users. We assume that drug users are divided into groups which share injection equipment at different rates, that they have different preferences for different shooting galleries and different probabilities of cleaning their needles before use. These probabilities may also depend on the shooting gallery. We describe two models, first where the sizes of the populations of the different types of drug users are held constant and a more realistic second model where recruitment of new drug users and the attrition of the numbers of drug users due to HIV and AIDS are taken into account. Results of an equilibrium and stability analysis are presented for both models.

For the first model we find that there is a unique threshold parameter R_o which determines the behaviour of the system. There is always a disease-free equilibrium and if $R_o \leq 1$ then this is unique and globally stable so that if $R_o \leq 1$ the disease will die out. If $R_o > 1$ then there is in addition a unique endemic equilibrium. If HIV is introduced into the population of addicts then the fraction of addicts and syringes that are infected will tend to their unique endemic equilibrium values. These are global results. Some less complete results are presented for the second model.

KEYWORDS:- HIV, AIDS, Drug Users, Equilibrium and Stability Analysis.

1. INTRODUCTION

This paper considers a mathematical model for the spread of HIV and AIDS amongst injecting intravenous drug addicts (IVDUs) sharing injecting equipment in shooting galleries. A shooting gallery is defined as any place where addicts meet to share injection equipment. To give an estimate of the sort of numbers which we are talking about there are an estimated 9,400 injecting drug users in Glasgow, U.K. a city with a population size of roughly two million and an estimated 2,300 injecting drug users in New Haven, Connecticut, U.S.A. [1]. The model is based on a simple model for the spread of HIV and AIDS amongst injecting drug-users due to Kaplan [2]. This model assumes a constant size population of n sharing injecting intravenous drug addicts who inject drugs in shooting galleries and that there are m shooting galleries, equivalently m 'kits' of injection equipment in circulation.

However this simple model has quite a few unrealistic assumptions. We shall concentrate on making two of these more realistic. The first is the assumption that all addicts share injection equipment at the same rate whereas in practice different addicts share at different rates. This has been verified by a wide range of studies from different parts of the world, including data from attendors at U.K. needle exchanges [3], injecting intravenous drug addicts in Sydney, Australia [4] and Glasgow [5]. The second unrealistic assumption is that all addicts are equally likely to visit all shooting galleries. If we are talking about a large city such as Glasgow then addicts will be more likely to visit some shooting galleries rather than others. The differences in preferences could be due to geographical location, social preference or other factors. We shall modify Kaplan's basic model to take these factors into consideration.

Our model considers n addicts divided into p groups, which have sizes $n_1, n_2, \ldots n_p$ with $n_1 + n_2 + \ldots + n_p = n$. The addicts in the different groups visit shooting galleries at different rates with addicts in group i visiting at rate λ_i. In addition there are in total m needles, which are divided into q shooting galleries containing $m_1, m_2, \ldots m_q$ needles where $\sum_{j=1}^{q} m_j = m$. When an addict in group i visits a shooting gallery he or she chooses shooting gallery j on any given visit with probability p_{ij} where $\sum_{j=1}^{q} p_{ij} = 1$. Different addicts choose different shooting galleries independently and when an addict visits shooting gallery j he or she chooses a needle from that shooting gallery at random.

We shall consider two models. The first model assumes that the sizes of the different groups of addicts remain constant, so that all addicts (whether susceptible or infected) who cease sharing injection equipment are immediately replaced by new susceptible addicts. This assumption is perfectly reasonable for susceptible addicts as in the absence of HIV and AIDS the group sizes are in equilibrium so that a susceptible addict who ceases to share injection equipment will immediately be replaced by a new susceptible addict. However it is less reasonable for infected addicts as one of the reasons why infected addicts cease sharing injection equipment is that they develop HIV/AIDS. Thus we would expect HIV/AIDS to reduce the group sizes. However this assumption does make the model a lot easier to analyse mathematically. Hence in our first model, Model A, we assume that the group sizes of the different types of addicts remain constant, whilst the second more realistic model, Model B, takes account of recruitment of addicts into the different groups, death of infected addicts and cessation of sharing.

2. MODEL A.

Model A makes the following assumptions:

1 All sharing of drug injection equipment occurs in shooting galleries. Here a shooting gallery is simply defined to be a place where addicts meet to sequentially rent

injection equipment. We assume that sharing injection equipment is the only way addicts can become infected.

2 Each addict in group i visits shooting galleries according to a Poisson process of rate λ_i.

3 In practice addicts receive counselling to clean their needles before sharing. Almost all addicts do clean their needles before use using some method [5], but only rinsing with bleach, alcohol or boiling water are thought to be effective. Goldberg et al. [5] indicate that 44% of their sample of drug users clean their needle before use using one of these methods. Kaplan and O'Keefe, [1], show that 84% of New Haven needle exchange first time attendees self-report cleaning with bleach before use. It is unclear whether these self-reported syringe cleaning rates are accurate as there may be social pressure to give the 'correct' answer.

Goldberg et al. [5] demonstrate clearly that the probability that an addict cleans his or her syringe before use depends on the rate at which he or she shares injection equipment. More frequent users are less likely to clean their equipment before use. It also seems reasonable to assume that the probability that the addict cleans his or her syringe before use depends on the shooting gallery. An addict is more likely to clean his or her equipment before use in a friend's house than shooting up in a backstreet or park where cleaning facilities are less likely to be available. Hence we assume that $\xi = \xi_{ij}$: the probability that an addict cleans his or her syringe before use depends on both i, the type of addict, and j, the shooting gallery.

4 Following Kaplan [2] we assume that injection equipment becomes infectious if used by an infected addict.

5 Also following Kaplan [2] we suppose that when an infectious syringe is used once by a susceptible addict the act of injection will 'flush' the equipment (or rid the equipment of infectious blood) with probability θ. The idea behind flushing is that a syringe contains a relatively small amount of infectious blood compared with the amount of blood injected into the syringe from a susceptible addict, hence it may be possible for no infectious blood to be left in the syringe after the addict has injected. In his original paper Kaplan took $\theta = 0.25$. Allard [6] casts doubt on whether flushing is a biologically realistic phenomenon. Later simulation models of the spread of HIV and AIDS amongst injecting drug users due to Peterson et al. [7] and Blower et al. [8] do not incorporate flushing. In the presence of uncertainty we have retained Kaplan's original flushing assumption in our model.

6 We assume that any uninfected addict using an infectious syringe once is considered 'exposed' to HIV and becomes infected with probability α.

7 As previously discussed we follow Kaplan [2] and assume that the size of group i remains constant at n_i. Thus infected addicts who cease sharing injection equipment for whatever reason are immediately replaced by a susceptible addict of the same type. Thus the 'birth' rate of new susceptible addicts and the 'death' rate of HIV infected addicts are both assumed to have a common value μ. Of course

here by 'death' of an infected addict we mean only that the addict ceases sharing injection equipment.

With these assumptions the arrival rate of type i addicts at a given needle in shooting gallery j is

$$\Lambda_{ij} = \lambda_i p_{ij} \frac{n_i}{m_j} \qquad i = 1, 2, \ldots p \quad j = 1, 2, \ldots q.$$

We denote the fraction of addicts in group i who are infected at time t by $\pi_i(t)$ and the fraction of needles in shooting gallery j that are infected at time t by $\beta_j(t)$. Then the differential equations which describe the spread of the disease can be derived as

$$\frac{d\pi_i}{dt} = (1-\pi_i)\left(\sum_{j=1}^{q} \lambda_i p_{ij}(1-\xi_{ij})\alpha\beta_j\right) - \mu\pi_i \tag{1}$$

and

$$\frac{d\beta_j}{dt} = \sum_{i=1}^{p} \Lambda_{ij}\pi_i - \sum_{i=1}^{p} \Lambda_{ij}\beta_j\{1 - (1-\pi_i)(1-\xi_{ij})(1-\theta)\}$$

with suitable initial conditions $0 \leq \pi_i(0),\ \beta_j(0) \leq 1$ for $i = 1, 2, \ldots p$ and $j = 1, 2, \ldots q$. These equations are fully derived in Greenhalgh [9].

3. THE BASIC REPRODUCTION NUMBER

A key parameter in the epidemiology of infectious diseases is the basic reproduction number R_o. This is usually defined as the expected number of secondary cases produced by a single infectious case of disease entering a population where disease is absent. We expect an epidemic to occur if and only if R_o exceeds one. Applying this idea to our model consider a single infected type j syringe entering a population consisting of m uninfected syringes and n susceptible addicts. This syringe is visited by type i addicts at rate Λ_{ij}. Each of these addicts cleans the syringe before use with probability ξ_{ij} and independently flushes the syringe with probability θ. Thus the probability that the syringe is either flushed or cleaned is $1 - (1-\theta)(1-\xi_{sj})$. The total rate at which the syringe is rendered uninfectious is $\sum_{s=1}^{p} \Lambda_{sj}(1-(1-\theta)(1-\xi_{sj}))$. Hence the total time that the syringe remains infected is

$$\frac{1}{\sum_{s=1}^{p} \Lambda_{sj}(1-(1-\theta)(1-\xi_{sj}))}.$$

During this time type i addicts visit the syringe at rate Λ_{ij}. A fraction $(1-\xi_{ij})\alpha$ of these addicts both do not clean the syringe before use and are infected. Hence during this time a total of

$$\frac{\Lambda_{ij}(1-\xi_{ij})\alpha}{\sum_{s=1}^{p} \Lambda_{sj}(1-(1-\theta)(1-\xi_{sj}))}$$

type i addicts are infected by the syringe. Each infected type i addict continues to share injecting equipment for time $(1/\mu)$ during which time he or she infects $(1/\mu)\lambda_i p_{ik}$

type k syringes. Hence the number of secondary type k syringes left infected by a single type j syringe is

$$Q_{jk} = \sum_{i=1}^{p} \frac{\Lambda_{ij}(1-\xi_{ij})}{\sum_{i=1}^{p} \Lambda_{sj}(1-(1-\theta)(1-\xi_{sj}))} \frac{\alpha}{\mu} \lambda_i p_{ik}.$$

The basic reproduction number is the largest eigenvalue of this $q \times q$ matrix $\mathbf{Q}$. To see why this is the case, initially there is just one infected type j syringe which can be represented by the vector $\mathbf{e}_j = (0,0,\ldots,1,0,\ldots,0)$ with a '1' in the j'th position. At the second generation there are $Q_{jk} = (\mathbf{Q}^T\mathbf{e}_j)_k$ infected type k syringes so the vector of infected types of syringes is $\mathbf{Q}^T\mathbf{e}_j$. Each of these infected syringes then goes on to infect a third generation of infected syringes. Assuming that both n, the total number of addicts, and m, the total number of syringes, are very large, in the limit each of the second generation of infected syringes infects a third generation of infected syringes independently of the other infected syringes. Thus if n and m are both very large then the total number of type k syringes infected in the third generation is given approximately by $(\mathbf{Q}^{T^2}\mathbf{e}_j)_k$. Similarly the number of type k syringes infected at the fourth generation is approximately $(\mathbf{Q}^{T^3})\mathbf{e}_j$ and at the N'th generation approximately $(\mathbf{Q}^T)^{N-1}\mathbf{e}_j$.

But provided that α, θ and μ are all positive and for each $i = 1,2,\ldots p$ and $j = 1,2,\ldots q$ p_{ij} is strictly positive then Q_{jk} is a matrix with each entry strictly positive and hence has a unique positive eigenvector $\mathbf{e} > \mathbf{0}$, with eigenvalue $\rho(\mathbf{Q}) = \rho(\mathbf{Q}^T)$ [10]. (Recall that the spectral radius $\rho(\mathbf{Q})$ of a matrix $\mathbf{Q}$ is defined to be the modulus of the largest eigenvalue of $\mathbf{Q}$.) Moreover $(\mathbf{Q}^T)^{N-1}\mathbf{e}_j \approx \rho(\mathbf{Q}^T)^{N-1}\mathbf{e}_j$ if N is large. Hence if $R_o = \rho(\mathbf{Q}) > 1$ then we expect the disease to take off and if $R_o < 1$ then we expect the disease to die out.

Similarly consider a single infected type i addict entering a population of n susceptible addicts and m uninfected syringes, where n and m are large. Then the number of second generation type ℓ addicts infected by this single addict is

$$\hat{\mathbf{Q}}_{i\ell} = \sum_{j=1}^{q} \frac{\alpha}{\mu} \lambda_i p_{ij} \frac{\Lambda_{\ell j}(1-\xi_{\ell j})}{\sum_{s=1}^{p} \Lambda_{sj}(1-(1-\theta)(1-\xi_{sj}))}.$$

So, arguing in a similar fashion to before, we also expect R_o to be the largest eigenvalue of this $p \times p$ matrix $\hat{\mathbf{Q}}$. In fact we can show that $\mathbf{Q}$ and $\hat{\mathbf{Q}}$ have the same spectral radius [9]. We expect that the disease takes off if $R_o > 1$ and dies out if $R_o < 1$. In the next Section we shall formally state a result which proves this.

4. RESULTS FOR MODEL A

We again assume that α, θ, μ and p_{ij} are all strictly positive and $1 > \xi_{ij}$ for $i = 1,2,\ldots p$ and $j = 1,2,\ldots q$. This avoids having groups of addicts or shooting galleries that the disease cannot reach. Our main result is the following theorem:

Theorem 1

(a) If $R_o \leq 1$ then the system of equations (1) has a unique equilibrium solution where the disease has died out in each group of addicts and each shooting gallery. Whatever the initial state the disease will eventually die out.

(b) If $R_o > 1$ then there is still the equilibrium where the disease has died out. However there is also a unique endemic equilibrium. Provided that initially disease is present in at least one group of addicts or at least one shooting gallery then both $\pi_i(t)$, the fraction of type i addicts who are infected at time t and $\beta_j(t)$, the fraction of needles in shooting gallery j that are infected at time t tend to their unique endemic equilibrium values as $t \to \infty$.

Proof

See [9].

This completes our analysis of Model A where we assumed that the sizes of the different groups of addicts remained constant. In the next Section we shall describe the more realistic Model B where recruitment of new drug users into the population and the different reasons why drug users cease sharing injection equipment are taken into consideration.

5. MODEL WITH VARYING POPULATION SIZE (MODEL B).

We suppose that in the absence of HIV/AIDS drug users cease sharing injection equipment at rate μ. Caulkins and Kaplan [11] estimate that before the advent of HIV an average drug-user injected for roughly eight years in the U.S.A. which corresponds to a value of μ of 0.125/year. In addition infected drug users cease sharing injection equipment because they develop HIV/AIDS at rate σ. We take $(1/\sigma)$ to be the average AIDS incubation period so $\sigma = 0.102$ /year [12]. Caulkins and Kaplan [11] discuss recruitment of new drug addicts into the drug-using population. They suggest a recruitment rate proportional to N^ν where N is the population size.

In the model we assume that new addicts are recruited into the i'th group at rate $c_i n^\nu$ $(0 \leq \nu < 1)$. Then we find that there is a unique disease-free equilibrium $\mathbf{n}_o^* = (n_{01}^*, n_{02}^*, \dots n_{op}^*)$ where $n_{oi}^* = \left(c_i / \sum_{\ell=1}^{p} c_\ell\right) \left(\sum_{\ell=1}^{p} c_\ell/\mu\right)^{\frac{1}{1-\nu}}$ for $i = 1, 2, \dots p$. In the previous model we assumed that n, the total number of addicts, and m, the total number of syringes were constant, but here the total number of addicts varies with time. It is reasonable to assume that the total number of syringes m also varies with time in such a way that $\gamma = n/m$, the gallery ratio, or number of needles per addict, also remains constant. We also assume that the proportion of the total number of needles which are in shooting gallery j remains constant over time. Then the differential equations which describe the progress of the disease are

$$\frac{dn_i}{dt} = c_i n^\nu - \mu n_i - \sigma \pi_i n_i,$$

$$\frac{d\beta_j}{dt} = \sum_{i=1}^p \Lambda_{ij}\pi_i - \sum_{i=1}^p \Lambda_{ij}\beta_j\{1-(1-\pi_i)(1-\theta)(1-\xi_{ij})\} - \frac{\beta_j}{n}\frac{dn}{dt},$$

and
$$\frac{d\pi_i}{dt} = (1-\pi_i)\left(\sum_{j=1}^q \lambda_i p_{ij}(1-\xi_{ij})\alpha\beta_j\right) - (\mu+\sigma)\pi_i - \frac{\pi_i}{n_i}\frac{dn_i}{dt},$$

for $i = 1, 2, \dots p$ and $j = 1, 2, \dots q$.

6. RESULTS FOR MODEL B

Again we can write down an expression for the basic reproduction number $R_o = R_o(\mathbf{n}_o^*)$. This is the same expression as for Model A (the common spectral radius of the matrices $\mathbf{Q}$ and $\hat{\mathbf{Q}}$) with n_i replaced by n_{oi}^* for each i and μ replaced by the new cessation of sharing rate of infected addicts in Model B, which is $(\mu+\sigma)$. We have the following Theorem for Model B.

Theorem 2

For the model with recruitment:

A. There is a unique disease-free equilibrium solution $\hat{\pi}_{oi} = 0, \hat{\beta}_{oj} = 0$ and $\hat{n}_{oi} = n_{oi}^*$ $i = 1, 2, \dots p$, $j = 1, 2, \dots q$. This equilibrium is locally stable if $R_o(\mathbf{n}_o^*) < 1$.

B. If $R_o(\mathbf{n}_o^*) > 1$ then the above equilibrium exists, is the unique disease-free equilibrium and is unstable.

The proof of this Theorem and further results are discussed in [9].

However the behaviour of the model with recruitment and death of drug users can be substantially more complicated than the behaviour of the model with constant population sizes. In particular the endemic equilibrium may be locally unstable if γ is small and ν is near one, even for the case $p = q = 1$, where there is just one group of addicts and one shooting gallery. In this case we expect that limit cycles arise from the stable endemic equilibrium by Hopf bifurcation.

In the case where $p = q = 1$, if we linearise about the endemic equilibrium the characteristic equation is

$$\omega^3 + a_1\omega^2 + a_2\omega + a_3 = 0$$

where $a_1 > 0$, $a_3 > 0$. Then by the Routh-Hurwitz conditions [13] a necessary and sufficient condition for local stability is that $\Delta = a_1 a_2 - a_3 > 0$. But we can show that if $\nu \approx 1$ and γ is small then $a_2 < 0$. Hence as ν and γ change along a curve parameterised as $\theta \in [\theta_a, \theta_b]$ where $\theta_o \in [\theta_a, \theta_b]$, then the characteristic equation has a pair of complex conjugate roots $\eta(\theta) \pm i\xi(\theta)$ which cross the real axis at $\theta = \theta_o$. If in addition the transversality condition $\eta'(\theta_o) \neq 0$, equivalent to $\Delta'(\theta_o) \neq 0$ is satisfied then the crossing is transversal and there is a Hopf bifurcation at $\theta = \theta_o$. Moreover Hopf bifurcation can occur for realistic parameter values. Further details are given in [9].

7. SUMMARY AND CONCLUSIONS

In this paper we have described two mathematical models for the spread of HIV and AIDS amongst a population of injecting intravenous drug-users. Both models incorporated heterogeneity amongst drug-users corresponding to the rate at which they visit shooting galleries. For Model A the different group sizes were held constant whereas in Model B the recruitment and loss of sharing injecting drug addicts into the population was taken into consideration.

For Model A we found that there was a unique threshold value R_o which determined the behaviour of the disease. For $R_o \leq 1$ we expect that the disease dies out whereas for $R_o > 1$ we expect that the disease tends to a unique endemic equilibrium. For Model B we found that there was a corresponding threshold value $R_o(\mathbf{n}_o^*)$. For $R_o(\mathbf{n}_o^*) < 1$ the unique disease-free equilibrium was locally asymptotically stable whereas for $R_o(\mathbf{n}_o^*) > 1$ the disease-free equilibrium was unstable. We found that for Model B limit cycle behaviour was possible, even for the case $p = q = 1$.

REFERENCES

[1] E.H. Kaplan and E. O'Keefe, Interfaces, 23, 7-26 (1993).

[2] E.H. Kaplan, Reviews of Infectious Diseases, 11, 289-298 (1989).

[3] G.V. Stimson, L.J. Alldritt, K.A. Dolan, M.C. Donoghoe and R.A. Lart, Injecting Equipment Exchange Schemes. Final report. Monitoring Research Group, Sociology Department, Goldsmith's College, New Cross, London, England. (1988)

[4] J. Wolk, K. Wodak, A. Morlet, J.J. Guinan and J. Gold, Med. J. Australia, 152, 453-458 (1988).

[5] D.J. Goldberg, M. Frischer, G. Hay, S.T. Green, A. Taylor, and N. McKeganey, Unpublished manuscript. (1997)

[6] R. Allard, J. AIDS, 3, 1010-1016 (1990).

[7] D. Peterson, K. Willard, M. Altmann, L. Gatewood, and G. Davidson, J. AIDS, 3, 1086-1095, (1990).

[8] S.M. Blower, D. Hartel, H. Dowlatabi, R.M. Anderson and R.M. May, Phil. Trans. Roy. Soc. Lond. Series B., 321, 171-187 (1991).

[9] D. Greenhalgh, Math. Biosci., 136, 141-186, (1996).

[10] R.S. Varga, Matrix Iterative Analysis. Prentice-Hall, New Jersey (1962).

[11] J.P. Caulkins and E.H. Kaplan, Interfaces, 21, 50-63 (1991).

[12] I.M. Longini, W.S. Clark, R.H. Byers, J.W. Ward, W.W. Darrow, G.F. Lemp and H.W. Hethcote, Statistics in Medicine, 8, 831-843 (1989).

[13] R.M. May, Stability and Complexity in Model Ecosystems. (Second Edition), Princeton University Press, Princeton.

7th Int. Coll. on Differential Equations, pp. 123-130
D. Bainov and V. Covachev (Eds)

Nontrivial Solutions for a Fourth Order O.D.E. with Singular Boundary Conditions.

M. R. GROSSINHO* and T. F. MA**

* CMAF, Universidade de Lisboa, Av. Prof. Gama Pinto, 2, 1699 Lisboa Codex, Portugal.
E-mail: mrg@ptmat.lmc.fc.ul.pt.
** Universidade Estadual de Maringá, Av. Colombo 3690, 7020 Maringá-PR, Brazil.
E-mail: matofu@uem.br.

Abstract. We prove existence results for a fourth order equation with singular nonlinear boundary conditions. We use a direct variational formulation and also a dual variational procedure, in the spirit of the dual action principle of Clarke, when the nonlinerities are monotonous.
Key words. Nonlinear value problem, singular boundary conditions, variational methods, duality.

1 Introduction

In this paper, we are concerned with the existence of solutions to the nonlinear fourth order problem:

$$u^{(iv)} + g(x, u) = 0 \tag{1.1}$$

$$u''(0) = u''(1) = 0 \tag{1.2}$$

$$u'''(0) = -f(u(0)) \tag{1.3}$$

$$u'''(1) = f(u(1)) \tag{1.4}$$

where $g : (0,1) \times I\!R \longrightarrow \mathbf{R}$ is a Carathéodory function and $f : I \subset I\!R \longrightarrow \mathbf{R}$ is a real continuous function not defined on the whole $I\!R$. This fact raises a singular situation with respect to the boundary conditions. More precisely, f is defined in $I = (a, b)$ where

$$-\infty \leq a < 0 < b \leq +\infty.$$

We consider the cases when a, $b \in I\!R$ and $a \in I\!R$, $b = +\infty$. The case $a = -\infty$, $b \in I\!R$ is analagous to the last one. The function f is also supposed to satisfy

$$f(s) = 0 \quad \text{iff} \quad s = 0 \tag{1.5}$$

In Section 1, for a suitable variational setting, we obtain solutions of the above problem as critical points of an associated functional. In section 2, we assume that f and g are monotonous and some existence results are derived by using a dual variational procedure, in the spirit of the dual action principle of Clarke [1].

The results presented here complement those of [2], where the authors were concerned with the existence of symmetric solutions and also those of [3] where the domain of f is $I\!R$, which does not raise a singular problem.

2 Singular problem via variational formulation

Let $\|\cdot\|_p$ denote the L^p-norm and consider the Sobolev space $E = H^2(0,1)$ with the norm

$$\|u\|_E^2 = \|u''\|_2^2 + \|u\|_2^2.$$

Since f is not defined on the whole $\mathbb{R}$, we shall work on an open subset Ω of E.

More precisely, let

$$\Omega := \left\{u \in H^2(0,1); \quad a < u(0),\, u(1) < b\right\},$$

where $a, b \in \mathbb{R}$ or $a \in \mathbb{R}$ and $b = +\infty$, and consider the functional $J : \Omega \subset E \longrightarrow \mathbb{R}$ defined by

$$J(u) := \frac{1}{2}\int_0^1 |u''(x)|^2 dx + \int_0^1 G(x, u(x))dx + F(u(0)) + F(u(1))$$

where $G(x,u) = \int_0^u g(x,t)dt$ and $F(s) = \int_0^s f(t)dt$ are primitives of g and f, respectively. Then J is well defined. Moreover J is of class C^1 and weakly lower semicontinuous in Ω and, for all $\phi \in E$,

$$J'(u)\phi = \int_0^1 u''(x)\phi''(x)dx + \int_0^1 g(x, u(x))\phi(x)dx + f(u(0))\phi(0) + f(u(1))\phi(1).$$

Of course, $u \in \Omega$ is a critical point of J if and only if it is a solution of the problem (1.1)–(1.4) that belongs to $W^{4,1}(0,1)$.

Remark 1.1: When applying variational methods, it is required, sometimes, that J satisfies the Palais-Smale condition (or (PS) condition) in $\Omega \subset E$, that is, if (u_n) is a sequence in $\Omega \subset H^2(0,1)$ s.t. $J(u_n)$ is bounded and $J'(u_n) \longrightarrow 0$ in the dual $(H^2(0,1))^*$ then (u_n) has a strongly convergent subsequence.

We observe that, to prove that such a sequence (u_n) has a convergent subsequence is equivalent to prove that (u_n) is bounded since the fact that g is L^1-Carathéodory and f is continuous implies that J' is a compact perturbation of the identity (see e.g. [4] appendix B).

For the reader's convenience, we recall some spectral properties of the linear eigenvalue problem associated to problem (1.1)-(1.4) that can be found in [5] or [6]. Consider the problem

$$(L) \quad \begin{cases} u^{(iv)} = \lambda u \\ u''(0) = u''(1) = 0 \\ u'''(0) = u'''(1) = 0 \end{cases}$$

This eigenvalue problem possesses a sequence of eigenvalues (λ_k) , $k \geq -1$, such that $\lambda_{-1} = \lambda_0 = 0$ and $0 < \lambda_1 < \lambda_2 \leq \lambda_3 \leq ...$

The corresponding eigenfunctions $\{1, x - \frac{1}{2}, \phi_1, \phi_2, ...\}$ provide an orthogonal basis for both $L^2(0,1)$ and $H^2(0,1)$. Then, $H^2(0,1) = V \oplus W$, with $V = \text{Span}\{1, x\}$ and $W = V^\perp \cap H^2(0,1)$. As notation, we put for $u \in H^2(0,1)$, $u = v + w$, with $v \in V$ and $w \in W$. By the variational characterization of the eigenvalue λ_1 , it is clear that for all $w \in W$

$$\int w''^2 \geq \lambda_1 \int w^2.$$

Remark 1.2: By the above inequality, the norm $\|w''\|_2$ is equivalent to $\|\cdot\|$ in W.

Theorem 1 *Assume that there exists $L \in L^1(0,1)$ and $\varepsilon \geq 0$ such that*

$$G(x,u) \geq -\varepsilon|u|^2 + L(x) \quad \textit{for a.e. } x \in (0,1) \tag{G1}$$

and suppose that

$$F(s) \longrightarrow +\infty \quad \text{if } s \to a \text{ or } s \to b. \tag{F1}$$

If $b = +\infty$, suppose moreover that, for some $C > 0$,

$$F(s) > C|s|^2 \text{ if } s > 0. \tag{F2}$$

Then, if ε is small enough, problem (1.1)–(1.4) has a solution that minimizes the functional J in Ω.

Proof. First, observe that

$$\|u\|_*^2 = \|u''\|_2^2 + |u(0)|^2 + |u(1)|^2$$

provides a norm in $E = H^2(0,1)$ equivalent to $\|\cdot\|_E$. In fact, it is clear that $\|\cdot\|_*$ is a seminorm of E . Let $u \in E$ be such that $\|u\|_* = 0$. Then $\|u''\|_2 = u(0) = u(1) = 0$ which shows that u satisfies $\|u\|_{H^2 \cap H_0^1} = 0$. Thus $u \equiv 0$.

It is clear from the assumptions on f that for some $C_1 > 0$,

$$F(s) > \frac{1}{2}C|s|^2 + \frac{1}{2}F(s) - C_1 \quad \text{for all } s \in (a,b)$$

Then

$$\begin{aligned} J(u) &\geq \frac{1}{2}\|u''\|_2^2 - \varepsilon\|u\|_2^2 - \|L\|_1 + \frac{C}{2}(|u(0)|^2 + |u(1)|^2) + \frac{1}{2}\{F(u(0)) + F(u(1))\} - C_1 \\ &\geq \min\{\frac{1}{2}, \frac{C}{2}\}\|u\|_*^2 - \varepsilon\|u\|_2^2 + \frac{1}{2}\{F(u(0)) + F(u(1))\} - C_2 \end{aligned}$$

for some $C_2 > 0$. If ε is small enough, we can write for some $C_3 > 0$,

$$J(u) \geq C_3\|u\|_*^2 + \frac{1}{2}\{F(u(0)) + F(u(1))\} - C_2$$

So,

$$J(u) \longrightarrow +\infty \quad \text{if} \quad \|u\|_* \to +\infty$$

and by (F1),

$$J(u) \longrightarrow +\infty \quad \text{if} \quad u \to \partial\Omega.$$

Therefore, since J is weakly lower semicontinuous, it has a minimum in Ω (see [7]). □

Next, we remark that if $b = +\infty$ but the requirement (F2) is dropped, then we have still another minimizing result if we assume that

$$G(x,u) \geq L(x) \quad \text{for some } L \in L^1(0,1). \tag{G2}$$

Theorem 2 *Let $b = +\infty$ and assume that (F1) and (G2) hold. Then problem (1.1)–(1.4) has a solution.*

Proof. Let us prove that the functional J is coercive. From (1.5), F cannot change sign which together with (F1) implies $F \geq 0$. For $u \in E$, we put $u = v + w$ with $v \in V$ and $w \in W$, According to the notations introduced above. From $(G2)$ and $(F1)$ we easily observe that

$$J(u) \geq \frac{1}{2}\|w''\|_2^2 - \|L\|_1 - 2C \tag{2.1),}$$

which shows that J is bounded below. Then, if the (PS) condition holds in Ω, J is coercive (see [8]) and the thesis follows. Let $u_n = v_n + w_n$ be a sequence in Ω such that

$$J(u_n) \text{ is bounded and} J'(u_n) \longrightarrow 0 \quad \text{in the dual } E^*$$

From (2.1) it is clear that $\|w_n''\|_E \leq M$, for some $M \in \mathbf{R}^+$, independently of n. Suppose, by contradiction, that $\|v_n\|_E \longrightarrow +\infty$. Then, for a subsequence, as v_n are linear functions, that is, $v_n = a_n x + b_n$ with $a_n, b_n \in \mathbb{R}$, either $|v_n(0)| \to +\infty$ or $|v_n(1)| \to +\infty$. But

$$|u_n(x)| \geq |v_n(x)| - |w_n(x)| \geq |a_n x + b_n| - \alpha M \longrightarrow +\infty.$$

and therefore either $|u_n(0)| \to +\infty$ or $|u_n(1)| \to +\infty$., which cannot happen since $u_n \in \Omega$. Thus (u_n) is bounded and, by Remark 1.1, (PS) condition holds, that is, (u_n) has a convergent subsequence (in $\overline{\Omega}$). Moreover, since $J(u_n)$ is bounded, by condition (F_1), the sublimit of (u_n) belongs to Ω. Then (PS) condition holds in Ω and J is coercive. Therefore J has a minimum in Ω and the result follows. □

Remark 1.3: We observe that $u \equiv k$ is a solution of (1.1)–(1.4) and, by (1.5), it is the only constant solution of (1.1)–(1.4). If, however, we add the hypothesis

$$g(x,0) \neq 0 \quad \text{for} \quad x \in \Omega \subset (0,1), \ |\Omega| \neq 0$$

then the solutions cannot be trivial.

3 Singular problems via dual variational procedure

Let $-\infty < a < 0 < b < +\infty$ and suppose that the functions f and g satisfy the following hypotheses (H):

$(H)(i)$ The function $f : (a,b) \to \mathbb{R}$ is continuous, strictly monotonous, onto and $f(0) = 0$.

(ii) The function $g : [0,1] \times \mathbb{R} \to \mathbb{R}$ is continuous, strictly monotonous in the second variable and onto.

This section parallels, in some way, Section 3 of [2] where the authors studied problem (1.1)–(1.4) in the framework of symmetric solutions, assuming symmetric properties on g. Here, we do not consider any symmetry. For the sake of clearness, we develope the general setting of the problem. As for the existence results, we sketch briefly the proofs and refer [2] where, under adequate modifications, the reader can find details.

We recall some basic facts. If we put

$$F(t) = \int_0^t f(s)ds \qquad \text{and} \qquad G(x,u) = \int_0^1 g(x,t)dt$$

we have that F and G are strictly convex functions and if we consider the respective Fenchel-Legendre transforms, F^* and $G^*(x,\cdot)$, it is known that they satisfy the following relations:

$$\forall x \in (0,1) \quad G^*(x,v) = vu - G(x,u) \quad \text{if} \quad v = g(x,u)$$

$$F^*(s) = st - F(t) \quad \text{if} \quad s = f(t).$$

We shall denote by $g^*(x,\cdot)$ and f^* the inverse functions of $g(x,\cdot)$ and f, respectively. Then

$$F^*(s) = \int_0^s f^*(t)dt \quad \textit{and} \quad G^*(x,v) = \int_0^v g^*(x,t)dt + G^*(x,0)$$

If $F_1 \leq F_2$ then $F_1^* \geq F_2^*$. In an analogous way, if $G_1(x,.) \leq G_2(x,.)$ then $G_1^*(x,.) \geq G_2^*(x,.)$.

Let us consider the associated linear problem

$$(\mathrm{W}) \begin{cases} w^{(iv)} = v \in L^p(0,1) & p \geq 1 \\ w''(0) = w''(1) = 0 \\ w'''(0) = \alpha \in \mathbb{R} \\ w'''(1) = \beta \in \mathbb{R} \end{cases}$$

Lemma 3 *Given $v \in L^1(0,1)$ and $\alpha, \beta \in \mathbb{R}$, problem (W) has a solution w if and only if*

$$\alpha = \int_0^1 (x-1)v(x)dx \text{ and } \beta = \int_0^1 xv(x)dx \tag{3.1}$$

Moreover, if the solution w satisfies $w(0) = w(1) = 0$, then it is unique.

Proof. The necessary condition is trivial. As for the sufficient condition, it is easy to see that the function

$$w(x) = \frac{1}{6}\int_0^x (x-t)^3 v(t)dt + \frac{x^3}{6}\int_0^1 (t-1)v(t)dt + x\int_0^1 (t^3 - 3t^2 + 2t)v(t)dt$$

is a solution of problem (W) provided (3.1) holds.

Let w_1, w_2 be two solutions of (W) such that, for $x \in \{0,1\}$, $w_1(x) = w_2(x) = 0$. Hence, $u(x) = w_1(x) - w_2(x)$ satisfies

$$\begin{cases} u^{(iv)} = 0 \\ u(0) = u(1) = 0 \\ u''(0) = u''(1) = 0 \end{cases}$$

which implies that $u \equiv 0$. Then $w_1 = w_2$.□

By the above lemma, we can define the operator $K : L^p(0,1) \longrightarrow C^0([0,1])$

$$Kv(x) = \frac{1}{6}\int_0^x (x-t)^3 v(t)dt + \frac{x^3}{6}\int_0^1 (t-1)v(t)dt + x\int_0^1 (t^3 - 3t^2 + 2t)v(t)dt,$$

that associates to every $v \in L^p(0,1)$ the unique solution w of (W) such that $w(0) = w(1) = 0$.

Lemma 4 *The above defined linear operator $K : L^p(0,1) \longrightarrow C^0([0,1])$ satisfies the following conditions*

(i) There exists $k > 0$ such that $|Kv(x)| \leq k\|v\|_p$ for all $v \in L^p(0,1)$ and $x \in [0,1]$.

(ii) $0 \leq \int_0^1 (Kv)vdx \leq k\|v\|_p^2$.

(iii) K is completely continuous.

Proof (i) follows directly from the definition of definition of K. As for (ii), we observe that $\int_0^1 (Kv)v = \int_0^1 (Kv)''^2$. Then, we obtain the conclusion from (i). As for (iii), let v_n be a sequence that converges weakly to some v in $L^p(0,1)$. Since (v_n) is bounded, by (i), (Kv_n) is uniformly bounded in $C^0([0,1])$. Then, easy computations show that (Kv_n) is also equicontinuous. Thus, by Arzelá-Ascoli theorem, and passing to a subsequence if necessary, there is $z \in C^0([0,1])$ such that $(Kv_n) \to z$ uniformly. On the other hand, since $v_n \to v$ weakly in $L^p(0,1)$, it follows easily that $Kv_n \to Kv$ pointwisely in $(0,1)$ and then $z = Kv$. □

Let J^* be the functional defined in $L^p(0,1)$ as follows

$$J^*(v) = \frac{1}{2}\int_0^1 (Kv)vdx + \int_0^1 G^*(x,v)dx + F^*(\alpha_v) + F^*(\beta_v)$$

where $\alpha = -\int_0^1 (x-1)v(x)dx$ and $\beta = -\int_0^1 xv(x)dx$. From the properties of K, G^* and F^*, it is clear that J^* is a C^1 functional, weakly lower semicontinuous and its derivative is given by

$$J^{*\prime}(v)\phi = \int_0^1 (Kv)\phi dx + \int_0^1 g^*(x,v)\phi dx + f^*(\alpha_v)\alpha_\phi + f^*(\beta_v)\beta_\phi$$

for all $\phi \in L^p(0,1)$.

Theorem 5 *Let $v \in L^p(0,1)$ be a critical point of J^*. Then, there is a linear function $h(x) = ax+b$ such that $u := h - Kv$ is a solution of*

$$(P)\begin{cases} u^{(iv)} + g(x,u) = 0 \\ u''(0) = u''(1) = 0 \\ u'''(0) = -f(u(0)) \\ u'''(1) = -f-(u(1)) \end{cases}$$

If, moreover f is odd, then u is a solution of problem (1.1)–(1.4).

Proof. Let v be a critical point of J^*. For every $\phi \in L^p(0,1)$ such that $\int_0^1 \phi(x)dx = 0$ and $\int_0^1 x\phi(x)dx = 0$, we have

$$\int_0^1 (Kv + g^*(x,v))\phi(x)dx = 0$$

Then, there is a linear function $h(x) = ax + b$ such that

$$Kv + g^*(\cdot, v) = h.$$

If we put $u := h - Kv = g^*(.,v)$, since $g^* = g^{-1}$, we obtain $v = g(x,u)$ and as $u^{(iv)} = -K(v)(iv) = -v$, we have then

$$u^{(iv)} + g(.,u) = h \ \ in \ \ (0,1).$$

From the definiton of Kv, it is clear that

$$u''(0) = u''(1) = 0 \ \text{ and } \ h(x) = (u(1) - u(0))x + u(0).$$

Now we take $\phi \in L^p(0,1)$ such that $\int_0^1 \phi dx = \alpha_\phi \neq 0$ but $\beta_\phi = -\int_0^1 xv(x)dx = 0$. Then

$$0 = \int_0^1 h(x)\phi(x)dx + f^*(\alpha_v)\alpha_\phi = (u(0) - f^*(\alpha_v))\int_0^1 \phi(x)dx$$

which implies $f^*(\alpha_v) = u(0)$ and, therefore, $f(u(0)) = \alpha_v$. Since

$$u'''(0) = (h - Kv)'''(0) = (-Kv)'''(0) = -\alpha_v$$

we conclude that

$$u'''(0) = -f(u(0)).$$

Finally, take $\phi \in L^p(0,1)$ such that $\int_0^1 \phi(x)dx = 0$ but $\beta_\phi \int_0^1 x\phi(x)dx \neq 0$. In this case $\alpha_\phi = \beta_\phi \neq 0$ and

$$0 = \int_0^1 h(x)\phi(x)dx + f^*(\alpha_v)\alpha_\phi + f^*(\beta_v)\beta_\phi = [u(1) - u(0) + f^*(\alpha_v) + f^*(\beta_v)]\beta_\phi.$$

Observing that $f^*(\alpha_v) = u(0)$, we derive that $f^*(\beta_v) = -u(1)$ and therefore $\beta_v = f(-u(1))$. Since $u'''(1) = (h - Kv)'''(1) = -\beta_v$ we conclude that $u'''(1) = -f(-u(1))$. This ends the proof. □

Theorem 6 *Let the hypotheses (H) be satisfied. If $g(x,\cdot)$ is an increasing function such that*

$$\frac{C_1}{p}|u|^p - D_1 \leq G(x,u) \leq \frac{C_2}{p}|u|^p + D_2$$

for $x \in (0,1)$, $u \in \mathbb{R}$, where C_1, C_2, D_1, D_2 are positive constants and $p > 1$, then problem (1.1)–(1.4) has a solution if f is odd.

Proof. If f is increasing, using the properties of the Fenchel-Legendre transform and the assumptions on f and g, easy computations show that the functional J^* is coercive in $L^{p'}$, which, together with the fact that it is also weakly lower semicontinuous, implies that J has a minimum. If f is decreasing, putting $\bar{f} := -f$ and considering $\bar{F}^*$, where $\bar{F}$ is a primitive of $\bar{f}$, the functional

$$J_1^*(v) = \frac{1}{2}\int_0^1 (Kv)v\,dx + \int_0^1 G^*(x,v)\,dx - \bar{F}^*(\alpha_v) - \bar{F}^*(\beta_v)$$

is well defined and the above arguments used for J^* also show that J_1^* has a critical point. We observe that arguing as in Theorem 4 we can derive that, if $v \in L^{p'}$ is a critical point of J_1^*, there is a linear function h such that $u := h - Kv$ is a solution of problem (P) with the last two conditions replaced by

$$u'''(0) = f(-u(0)) \quad \text{and} \quad u'''(1) = f(u(1))$$

Since f is odd, by Theorem 4 the critical points are solutions of problem (1.1)–(1.4). □

Theorem 7 *Let the hypotheses (H) be satisfied. If $g(x,\cdot)$ is a decreasing function such that*

$$\frac{C_1}{p}|u|^p - D_1 \le -G(x,u) \le \frac{C_2}{p}|u|^p + D_2$$

for $x \in (0,1)$, $u \in \mathbb{R}$, where C_1, C_2, D_1, D_2 are positive constants, $1 < p < 2$ and, in case $p = 2$, $C_2 < \frac{1}{2k}$ (where k is the constant given by Lemma 3 (ii)), then problem (1.1)–(1.4) has a solution if f is odd.

Proof. Define $\bar{g} := -g$ and consider $\bar{G}^*$, where $\bar{G}$ is a primitive of $\bar{g}$. Suppose that f is increasing. Then the functional

$$J_2^*(v) = \frac{1}{2}\int_0^1 (Kv)v\,dx - \int_0^1 \bar{G}^*(x,v)\,dx + F^*(\alpha_v) + F^*(\beta_v)$$

is well defined in $L^{p'}$. As in Theorem 5, we can derive that, if $v \in L^{p'}$ is a critical point of J_2^*, then there is a linear function h such that $u := Kv - h$ is a solution of problem (P) with the last two conditions replaced by

$$u'''(0) = f(-u(0)) \quad \text{and} \quad u'''(1) = f(u(1)).$$

Hence, if f is odd, u is a solution of problem (1.1)–(1.4). Arguing as in the previous theorem, we can see that $-J_2^*$ is coercive and weakly lower semicontinuous, and therefore has a critical point.

If f is decreasing, the proof is analogous, considering the functional

$$J_3^*(v) = \frac{1}{2}\int_0^1 (Kv)v\,dx - \int_0^1 \bar{G}^*(x,v)\,dx - \bar{F}^*(\alpha_v) - \bar{F}^*(\beta_v),$$

where $\bar{F} = \int_0^s \bar{f}(t)\,dt$ with $\bar{f} = -f$. □

Next result concerns the existence of nontrivial solutions of problem (1.1)–(1.4) supposing $g(x,0) \equiv 0$.

Theorem 8 *Let the hypotheses (H) be satisfied. Let $g(x,\cdot)$ be a decreasing function such that, for some $p > 2$,*

$$\frac{C_1}{p}|u|^p \le -G(x,u) \le \frac{C_2}{p}|u|^p$$

$$pG(x,u) \ge g(x,u)u - C_3,$$

for $x \in (0,1)$, $u \in \mathbb{R}$, where C_1, C_2 and C_3 are positive constants. If the function f is increasing and odd and satisfies

$$\lim_{s\to 0} \frac{f(s)}{s} = +\infty,$$

then problem (1.1)-(1.4) has a nontrivial solution.

Proof. Denote $\bar{g} = -g$ as in the previous theorem. Considering the functional J_2^* defined before and the assumptions on f and g and using the properties of the Fenchel-Legendre transforms G^* and F^*, it can be seen that $-J_2^*$ satisfies the geometrical conditions of Mountain Pass Lemma (see [9], [4], [10] and [11]) and the following *Palais–Smale condition* $(PS)_c$: given $c \in \mathbb{R}$, a functional $J \in C^1(E, \mathbb{R})$, where E is a Banach space, is said to satisfy $(PS)_c$ if, whenever $u_n \in E$ is such that $J(u_n) \longrightarrow c$ and $J'(u_n) \longrightarrow 0$, then c is a critical value of J.

Hence, by the Mountain Pass Lemma, the functional $-J_2^*$ has a critical point. If f is odd, by the considerations made in the proof of the previous theorem, we derive the existence of a solution of problem (1.1)–(1.4). □

ACKNOWLEDGEMENT: The authors thank the support of CMAF and of Praxis XXI project 2/21/MAT/125/94

References

[1] F. Clarke, *A classical variational principle for Hamiltonian trajectories*, Proc. Amer. Math. Soc., **76**, (1979), 186-188.

[2] M.R. Grossinho and T.F. Ma, *Symmetric equilibria for a beam with a nonlinear elastic foundation*, Portugaliae Math. **51, 3**, (1994), 375-393.

[3] M.R. Grossinho and T.F. Ma, *Existence results for a fourth order ODE with nonlinear boundary conditions*, Proc. of CDE'V, Univ. Rousse, Bulgaria, ed. S.Bilchev and S.Tersian,(1995), 34-41.

[4] P. Rabinowitz, *Minimax methods in critical point theory with applications to differential equations*, CBMS Reg. Conf. Series in Math. 65, AMS, Providence, (1986).

[5] E. Feireisl, *Non-zero time periodic solutions to an equation of Petrovsky type with nonlinear boundary conditions: slow oscillations of beams on elastic bearings*, Preprint, (1991).

[6] W. Weaver Jr, S. Timoshenko and D. H. Young, *Vibrations problems in engineering*, 5th ed, John Wiley and Sons, New York, (1990).

[7] D. G. de Figueiredo, *Lectures on Ekeland variational principle with applications and detours*, TATA Inst. Fund. Res., Springer Verlag, Heidelberg, (1989).

[8] L. Caklovic, S. Li and M. Willem, *A note on Palais-Smale condition and coercivity*, Diff. Int. Eqs., **3**, (1990), 799-800.

[9] A. Ambrosetti and P. Rabinowitz, *Dual variational methods in critical point theory and applications*, J. Funct. Anal., **14**, (1973), 349-381.

[10] J. Mawhin and M. Willem, *Critical point theory and Hamiltonian systems*, Appl. Math. Sc. 74, Springer Verlag, New York, (1989).

[11] H. Brezis, J.-M. Coron and L. Nirenberg, *Free vibrations for a nonlinear wave equation and a theorem of P. Rabinowitz*, Comm. Pure Appl. Math., **33**, (1980), 667-689.

7th Int. Coll. on Differential Equations, pp. 131-137
D. Bainov (Ed)

THE LIMITING PROBLEM IN A DOUBLE FREE BOUNDARY PROBLEM

YoonMee Ham
Department of Mathematics
Kyonggi University
Suwon 442-760, Korea

ABSTRACT

In [3], they showed the existence of the Hopf bifurcation in the parabolic double free boundary problem with Pushchino dynamics for a finite diffusion constant. In this paper, we consider a limiting problem which is obtained as the diffusion constant tends to infinity. We shall investigate the behaviors of the solutions for the limiting problem.

Key words: free boundary, Hopf bifurcation, Poincare-Bendixon theorem

1 Preliminaries and Approximating Solutions

The parabolic free boundary problem with Pushchino dynamics is given by (see in [3])

$$\text{(P)}\qquad \begin{cases} v_t = Dv_{xx} - (c_1+b)v + c_1H(x-s(t)) - c_1H(x-p(t)) \\ \qquad\qquad \text{for } (x,t)\in\Omega^-\cup\Omega^+, \\ v_x(0,t) = 0 = v_x(1,t) \quad \text{for } t>0, \\ v(x,0) = v_0(x) \quad \text{for } 0\le x\le 1, \\ \tau\dfrac{ds}{dt} = C(v(s(t),t)) \quad \text{for } t>0, \\ \tau\dfrac{dp}{dt} = -C(v(p(t),t)) \quad \text{for } t>0, \\ s(0) = s_0 \quad \text{for } 0<s_0<1, \\ p(0) = p_0 \quad \text{for } 0<p_0<1, \end{cases}$$

where $v(x,t)$ and $v_x(x,t)$ are assumed continuous in $\Omega = (0,1)\times(0,\infty)$. Here, D is a positive diffusion constant and, c_1+b and c_1 are positive constants. Moreover, $\Omega^- =$

$\{(x,t)\in\Omega : 0<x<s(t), p(t)<x<1\}$ and $\Omega^+ = \{(x,t)\in\Omega : s(t)<x<p(t)\}$. The velocity function $C(\cdot)$ of the free boundary $s(t)$ and $p(t)$ are represented by

$$C(v) = \frac{2v - \frac{c_1-2a}{c_1+c_2}}{\sqrt{(\frac{c_1-a}{c_1+c_2} - v)(v + \frac{a}{c_1+c_2})}}$$

where a is a positive constant and $-c_1 < b < \frac{c_1(c_2+a)}{c_1\ a}$ with positive constants c_1, c_2.

The well posedness and the Hopf bifurcation for a finite diffusion constant of this problem was shown in [3,4]. In this paper, we shall show the behavior of the solutions as a constant D tends to infinity. We will construct a sequence of solutions $\{(v_m, s_m, p_m)\}_{m=1}^{\infty}$ as $D\uparrow\infty$. In order to this, we introduce the following lemma:

Lemma 1.1 *Let $Q_T = (0,1)\times(0,T)$. We have the following:*

(i) $s\in C^1([0,T])$ and $\|s'\|_{L_\infty(0,T)} < \infty$,
$p\in C^1([0,T])$ and $\|p'\|_{L_\infty(0,T)} < \infty$

(ii) For some constant M, $-M\le v\le M$ in $[0,1]\times[0,T]$.

(iii) For any $\eta\in(0,T)$, $\|v\|_{C^{1/2,1/4}([0,1]\times[\eta,T])}\le\tilde{c}$ where $\tilde{c}$ depend on η.

Moreover,

(iv) $\int\int_{Q_T} v_x^2 dx d\mu \le \frac{\alpha}{D}$ for some constant α.

(v) For any $\eta\in(0,T)$, there exists a constant β which does not depend on η such that

$$\int_\eta^T\int_0^1 v_t^2 dx d\mu \le \frac{\beta}{\eta}\int^1 v_x^2(t)dx \le \frac{\beta}{Dt}$$

for almost everywhere $t\in(0,T)$.

Proof: The proofs of (i), (ii) and (iii) are refer to [1, 4]. If we multiply the first equation of (1) to v and integrates on $(0,1)\times(0,T)$, and we use the integration by parts and the boundary conditions then (iv) is obtained. For (v), we let $\eta\in(0,T)$ and $\mu\in(0,t)$. Multipling μv_t and integrating on Q_t of (1), then

$$\begin{aligned}\int\int_{Q_t}\mu v_t^2\,dx d\mu &= -Dt\int_0^1 v_x^2(x,t)dx + D\int\int_{Q_t} v_x^2 dx dt + \int\int_{Q_t} -(c_1+b)\mu v v_t dx d\mu \\ &+ \int\int_{Q_t}(\mu c_1(H(x-s(\mu)) - H(x-p(\mu)))v_t)dx d\mu.\end{aligned}$$

By (ii), we obtain

$$\int\int_{Q_t} \mu v_t^2 dx d\mu + Dt \int_0^1 v_x^2(x,t)dx \le C_1 + D \int\int_{Q_t} v_x^2 dx dt$$

for some constant C_1. Therefore, for a constant C, the following inequality is obtained

$$(1) \qquad \int\int_{Q_t} \mu v_t^2 dx d\mu + Dt \int_0^1 v_x^2(x,t)dx \le C.$$

Let $\eta \in (0,T)$, $t \in (0,T)$ and $\mu \in (\eta, T)$. From 1, it is easily obtained

$$\int_0^1 v_x^2(x,t)dx \le \frac{C}{Dt}.$$

We divide the equation 1 by t, then we have

$$\begin{aligned} \frac{C}{t} &\ge \frac{1}{t}\int_0^t \int_0^1 \mu v_t^2(x,\mu) dx d\mu + D\int_0^1 v_x^2 dx \\ &= \int_0^1 t v_t^2(x,t)dx + D\int_0^1 v_x^2 dx. \end{aligned}$$

Again, divide by t and integrate on (η, T), then

$$\begin{aligned} \int_\eta^T \int_0^1 v_t^2 dx dt \le \left| \int_\eta^T \int_0^1 v_t^2 dx dt \right| &\le \left| \tfrac{1}{\eta} - \tfrac{1}{T} - D\log(\tfrac{T}{\eta}) \int_0^1 v_x^2 dx \right| \\ &\le \frac{\beta}{\eta}\int_0^1 v_x^2 dx \end{aligned}$$

for some positive constant β. □

From (i) and (ii) in the above lemma 1.1, we apply the Azela-Ascoli theorem and thus we obtain the following theorem;

Theorem 1.2 *There exists a sequence D_m such that $D_m \to \infty$ as $m \to \infty$ and a corresponding sequence of solutions (v_m, s_m, p_m) and (ξ, ϕ, ψ) satisfying*

(a) $(\xi, \phi, \psi) \in (C([0,1] \times (0,T)) \cap L_2(0,t : H^1(0,1))) \times (C^{0,1}([0,T])^2$.

(b) $v_m \to \xi$, *uniformly in* $[0,1] \times [\eta, T]$ *for all* $\eta \in (0,T)$.

(c) $v_m \to \xi$, *in* $L_2(Q_T)$ *and almost everywhere in* Q_T.

Moreover, we obtain

(d) $v_{mx} \to \xi_x$, *weakly in* $L_2(Q_T)$.

(e) $v_{mt} \to \xi_t$, *weakly in* $L_2((0,1) \times (\eta, T))$ *for all* $\eta \in (0,T)$

(f) $s_m \to \phi$ *and* $p_m \to \psi$, *uniformly in* $[0,T]$ *and weakly in* $H^1(0,T)$.

In the next two theorems, we shall find a problem which has a solution (ξ, ϕ, ψ). We call this system a limiting problem.

Theorem 1.3 *The limiting function ξ is the only function of t.*

Proof: By lemma 1.1, we have

$$\frac{\beta}{\eta}\int_0^1 {v_{mx}}^2 dx \le \frac{\beta}{tD_m}$$

and $v_{mx} \to \xi_x$, weakly in $L_2(Q_T)$. Thus,

$$\begin{aligned} \int\int_{Q_T} {\xi_x}^2 &\le \liminf \int\int {v_{mx}}^2 \\ &\le \liminf \frac{\alpha}{D_m} \\ &= 0. \end{aligned}$$

Therefore, we have $\xi_x = 0$ a.e. and thus ξ is a function of t. □

We now show the limiting problem of solutions.

Theorem 1.4 *If we have a sequence of solutions (v_m, s_m, p_m) of (P) which converge to (ξ, ϕ, ψ) as $m \to \infty$ then (ξ, ϕ, ψ) satisfy the following differential system*

$$(2) \qquad \begin{cases} \xi'(t) = -(c_1 + b)\xi + c_1(1 - \phi(t)) - c_1(1 - \psi(t)) & \text{for } t \in (0, T] \\ \tau\phi'(t) = C(\xi(t), t) & \text{for } t \in (0, T) \\ \tau\psi'(t) = -C(\xi(t), t) & \text{for } t \in (0, T) \\ \xi(0) = \int_0^1 v_0 \\ \phi(0) = s_0, \ \psi(0) = p_0. \end{cases}$$

Proof: Since $s_m(t)$ satisfies that

$$s_m(t) - s_0 = \frac{1}{\tau}\int_0^t C(v_m(s_m(\nu), \nu))d\nu,$$

the limit ϕ satisfy that

$$\phi(t) - s_0 = \frac{1}{\tau}\int_0^t C(\xi(\nu), \nu))d\nu.$$

Therefore

$$\begin{cases} \tau\phi' = C(\xi(t), t), & t \in (0, T) \\ \phi(0) = s_0 \end{cases}$$

Similarily, $\psi(t)$ can be obtained. We now show about the limit of ξ'. Since $v_m \to \xi$, uniformly in $[0,1] \times [\eta, T]$ for all $\eta \in (0,T)$, $\int_0^1 v_m(x,t)dx \to \xi$, uniformly in $[\eta, T]$ for all $\eta \in (0,T)$. Let η and t in $(0,T)$ be such that $\eta < t$ and integrate the first equation in (2) on $(0,1) \times (\eta, t)$, then we obtain

$$\begin{aligned}\int_0^1 v_m(x,t)\,dx &= \textstyle\int_0^1 v_m(x,\eta)dx - \int_\eta^t \int_0^1 (c_1+b)v_m dx d\mu \\ &\quad + \textstyle\int_\eta^1 \int_0^1 \Big(c_1 H(x-s_m) - c_1 H(x-p_{(m)})\Big) dx d\mu.\end{aligned}$$

We deduce that

$$\begin{aligned}\int_0^1 v_m(x,t)\,dx &= \textstyle\int_0^1 v_m(x,\eta)dx - \int_0^t \int_0^1 (c_1+b)v_m dx d\mu \\ &\quad \textstyle\int_0^t \int_0^1 \Big(c_1 H(x-s_m) - c_1 H(x-p_m)\Big) dx d\mu.\end{aligned}$$

In order to show that $\int_0^1 v_m(x,t)dx$ converges to ξ, calculate the difference between them

$$\begin{aligned}&\Big|\int_0^t \int_0^1 \Big(-(c_1+b)v_m + c_1 H(x-s_m) - c_1 H(x-p_m)\Big) dx d\mu \\ &\quad - \int_0^t \int_0^1 \Big(-(c_1+b)\xi + c_1 H(x-\phi) - c_1 H(x-\psi)\Big) dx d\mu\Big| \\ &\le \int_0^t \int_0^1 (c_1+b)\,|v_m - \xi|\,dx d\mu \\ &\quad + c_1 \int_0^t \int_0^1 |H(x-s_m) - H(x-\phi)|\,dx d\mu \\ &\quad + c_1 \int_0^t \int_0^1 |H(x-p_m) - H(x-\psi)|\,dx d\mu \\ &\le (c_1+b)\sqrt{T}\;\|v_m - \xi\|_{L_2(Q_T)} + c_1 T\,\|s_m - \phi\|_{L_\infty(0,T)} \\ &\quad + c_1 T\,\|s_m - \phi\|_{L_\infty(0,T)} \\ &\to 0 \quad \text{as} \quad m \to \infty.\end{aligned}$$

Thus, we obtain that

$$\begin{aligned}\xi(t) &= \textstyle\int_0^1 v_m(x,\eta)dx - \displaystyle\int_0^t \int_0^1 (c_1+b)\xi dx d\mu \\ &\quad \int_0^t \int_0^1 \Big(c_1 H(x-\phi) - c_1 H(x-\psi)\Big) dx d\mu.\end{aligned}$$

Hence we prove the theorem. □

The limiting problem of (1) has at most one solution so that in fact $v \to \xi$, $s \to \phi$ and $p \to \psi$ as $D \uparrow \infty$ (see [1, 2])

2 The stability for the limiting problem

In this section, we shall examine the stability of solutions for the problem (2). The stationary solutions (ξ^*, ϕ^*, ψ^*) of (2) is a solutions of the following problem

$$(3) \qquad \begin{cases} 0 = -(c_1 + b)\xi - c_1\phi + c_1\psi \\ 0 = \frac{1}{\tau}C(\xi) \\ 0 = -\frac{1}{\tau}C(\xi) \end{cases}$$

The equation $C(\xi) = 0$ has a solution $\xi = \xi^* = \frac{c_1 - 2a}{2(c_1 + c_2)}$. From the first equation in (2), ϕ^* and ψ^* satisfy that

$$-c_1\phi^* + c_1\psi^* = (c_1 + b)\xi^*.$$

We finally obtain the following theorem:

Theorem 2.1 *The critical point (ξ^*, ϕ^*, ψ^*) is a stable equlibrium point of (2).*

Proof: The linearlized eigenvalue problem at (ξ^*, ϕ^*, ψ^*) of (3) is

$$\begin{cases} -(c_1 + b)\xi - c_1\phi + c_1\psi = \lambda\xi \\ \frac{1}{\tau}C'(\xi^*)\xi = \lambda\phi \\ -\frac{1}{\tau}C'(\xi^*)\xi = \lambda\psi. \end{cases}$$

where $C'(\xi^*) = \dfrac{c_1}{2(c_1 + c_2)} > 0$. The eigenvalues are given by

$$(c_1 + b + \lambda)\lambda^2 + c_1\frac{C'(\xi)}{\tau}\lambda = 0$$

and thus, the eigenvalues are

$$\lambda = 0,\ \lambda = \frac{-(c_1 + b) \pm \sqrt{(c_1 + b)^2 - 8c_1\frac{C'(\xi^*)}{\tau}}}{2}.$$

Therefore, the eigenvalues have negative real parts thus, the equlibrium solutions are locally stable. □

References

[1] A. Fasano and M. Primicerio , General Free-Boundary Problem for the Heat Equation I. J. of Mathematical Analysis and Applications, 57, 694-723 (1977).

[2] J. K. Hale, Ordinary differential equations, Wiley-interscience, New York (1969)

[3] Y. M. Ham and S. S. Yum, The well posedness in a parabolic double free boundary problem , J. of PDE, 8, No.3, 211-218 (1995).

[4] Y. M. Ham and B. I. Seung, The Hopf Bifurcation in a Parabolic Free Boundary Problem with Pushchino Dynamics, J. Korean Math. Soc. 32, No.2, 237-250 (1995)

7th Int. Coll. on Differential Equations, pp. 139-146
D. Bainov (Ed)

On a Liénard System with three equilibrium points

Makoto **HAYASHI**
Department of Mathematics, College of Science and Technology,
Nihon University, Funabashi 274, Japan

Keywords: Liénard system, Closed orbit, Invariant set
MOS Subject Classification: 34C05, 34C25, 34C37

Abstract

Sufficient conditions for the Liénard system with three equilibrium points to have no non-trivial closed orbits and to have exactly one closed orbit are given. In this paper the how to apply our results to the Bogdanov-Takens system with a symmetric cusp of order 2 which is famous in Bifurcation Theory is stated.

1 Introduction

This paper is a full account of the results announced by the author in a talk[H4] at the Seventh International Colloquium on Differential Equations, Plovdiv, 1996.

We consider the Liénard system of the following form.

$$\begin{cases} \dot{x} = y - F(x) \\ \dot{y} = -g(x), \end{cases} \tag{L}$$

where $(\dot{\ })=(d/dt)$, F and g are locally Lipschitz continuous on $\mathbf{R}$. We assume the condition

[C1] $F(0) = g(0) = 0$ and there exist a_1, a_2 such that $a_1 < 0 < a_2$, $g(a_1) = g(a_2) = 0$ and $g(x)(x - a_1)x(a_2 - x) > 0$ if $(x - a_1)x(a_2 - x) \neq 0$.

Then system (L) has three equilibrium points $(0,0)$, $(a_1, F(a_1))$ and $(a_2, F(a_2))$. Usually, the case where the system has only one equilibrium point $(0,0)$ is discussed. The uniqueness of the closed orbit in such case has been discussed by [Xi], [Zh] and [H1], etc. Our first purpose is to give sufficient conditions for system (L) with above mentioned three equilibrium points to have no non-trivial closed orbits and to have exactly one closed orbit. Our main results will be stated in Section 2 as Theorem 2.1 and Theorem 2.2. They here aren't proved from the reason that we want to give the how to apply our method to the following Bogdanov-Takens system.

$$\begin{cases} \dot{x} = y - \frac{1}{3}(x^3 - 3\mu_2 x) \\ \dot{y} = x^3 + \mu_1 x, \end{cases} \tag{BT}$$

where μ_1 and μ_2 are parameters. System (BT) is an important example in Bifurcation Theory. Our results for system (BT) will be stated in Section 3 as Theorem 3.1 and

Theorem 3.2. They will be proved in Section 4. In [CRW], it was proved that system (BT) has at most one closed orbit by using [Zh]. Our results improve that of [CRW]. In fact, sufficient conditions for system (BT) to have no non-trivial closed orbits and to have exactly one closed orbit are given. We give an exsistence region of the closed orbit, too. The results are derived by constructing some invariant set of system (BT), combining [H1] and Poincaré's index theorem. In Section 5 the fact that a unique non-trivial closed orbit of system (BT) is not algebraic will be given by [H2].

2 Main Results

To state our results, we use the following condition for system (L).

[C2] $\exists\alpha_1 < 0$ and $\exists\alpha_2 > 0$ such that $F(\alpha_1) = F(\alpha_2) = 0$, $xF(x) < 0$ for $\alpha_1 < x < 0$ and $0 < x < \alpha_2$.

We also consider the following conditions for system (L).

[C3$_a$] $\exists x_1 \in [\alpha_1, 0)$ such that $\frac{1}{2}F^2(x_1) + G(x_1) \geq G(\beta_2)$,

[C3$_b$] $\exists x_2 \in (0, \alpha_2]$ such that $\frac{1}{2}F^2(x_2) + G(x_2) \geq G(\beta_1)$,

where $\beta_1 = \max(\alpha_1, a_1)$, $\beta_2 = \min(\alpha_2, a_2)$ and $G(x) = \int_0^x g(\xi)d\xi$. In what follows, let c_1 be a point in the interval $(\alpha_1, 0)$ where $F(x)$ attains its maximum in the interval. Similarly assume that $F(x)$ attains its minimum in $(0, \alpha_2)$ at c_2. Our first main result is stated as follows.

Theorem 2.1 *If the Liénard system* (L) *satisfies the conditions* [C1], [C2] *and besides the condition*

[C4] $(a_2 \leq \alpha_2$ *and* [C3$_a$]) *or* $(\alpha_1 \leq a_1$ *and* [C3$_b$]),

then the system has no non-trivial closed orbits.

We need the following notation to state an existence theorem of a closed orbit of system (L). For $i = 1, 2$ we write

$$\Phi_i = (\sqrt{2|G(a_i) - G(\alpha_i)|} - |F(c_i)|)^2 + 2G(\alpha_i),$$

$$\Psi_i(\epsilon) = (\sqrt{(|F(a_i)| - \epsilon)^2 + 2|G(a_i) - G(\alpha_i)|} + |F(c_i)|)^2 + 2G(\alpha_i),$$

where ϵ is a real parameter. Our second main result is the following.

Theorem 2.2 *Assume that the Liénard system* (L) *satisfies the conditions* [C1], [C2] *and*

[C5] $a_1 < \alpha_1$ *and* $\alpha_2 < a_2$,
$F(x)$ is monotone increasing for $a_1 < x < \alpha_1$ *and* $\alpha_2 < x < a_2$,

[C6] *There exist* ϵ_1, ϵ_2 *such that* $|F(a_i)| \geq \epsilon_i > 0$ $(i = 1, 2)$ *and*
$$F(c_1) - F(c_2) \leq \min(\Psi_1(\epsilon_1)^{1/2} - \Phi_2^{1/2}, \Psi_2(\epsilon_2)^{1/2} - \Phi_1^{1/2}).$$

Also assume that the condition

[C7] $(G(\alpha_1) < G(\alpha_2)$ *and* [C3$_a$]) *or* $(G(\alpha_2) < G(\alpha_1)$ *and* [C3$_b$]) *or* $G(\alpha_1) = G(\alpha_2)$

holds. Then the system has exactly one closed orbit crossing the segments $a_1 < x < \alpha_1$ *and* $\alpha_2 < x < a_2$ *on the* x*-axis.*

Remark. If F and g are odd functions and $\alpha_2 < x < a_2$, then the conditions [C3$_a$] and [C3$_b$] automatically hold. See the proof for system (BT) in Section 4.

3 Application to system (BT)

In this section our aim is to seek sufficient conditions for the parameters μ_1 and μ_2 in system (BT) in order that the system has no non-trivial closed orbits and has exactly one closed orbit. The following is our first result.

Theorem 3.1 *If the Bogdanov-Takens system* (BT) *satisfies the condition*

[C8] $\mu_1 \geq 0$ *or* $0 < -\mu_1 \leq 3\mu_2$,

then the system has no non-trivial closed orbits.

We set $F(x) = \frac{1}{3}(x^3 - 3\mu_2 x)$ and $g(x) = -x^3 - \mu_1 x$. Our second result for system (BT) is stated as follows.

Theorem 3.2 *If the Bogdanov-Takens system* (BT) *satisfies the following conditions*

[C6′] $\exists\epsilon$ *such that* $|F(\sqrt{3\mu_2}\,)| \geq \epsilon > 0$ *and* $-2F(\sqrt{\mu_2}\,) \leq \Psi_2(\epsilon)^{1/2} - \Phi_1^{1/2}$,

[C9] $0 < 3\mu_2 < -\mu_1$,

then the system has exactly one closed orbit crossing the segments $-\sqrt{-\mu_1} < x < -\sqrt{3\mu_2}$ *and* $\sqrt{3\mu_2} < x < \sqrt{-\mu_1}$.

We shall present phase portraits of two systems to which Theorem 3.2 can be applied. System (BT) in Example 3.1 below satisfies all conditions in Theorem 3.2. So we see that the system has exactly one closed orbit crossing the segments $-2 < x < -1$ and $1 < x < 2$ on the x-axis as shown in Figure 3.1. Example 3.2 is the case where system (BT) satisfies the condition [C9] but not the condition [C6′]. So we can show that the system has no non-trivial closed orbits as shown in Figure 3.2.

Example 3.1

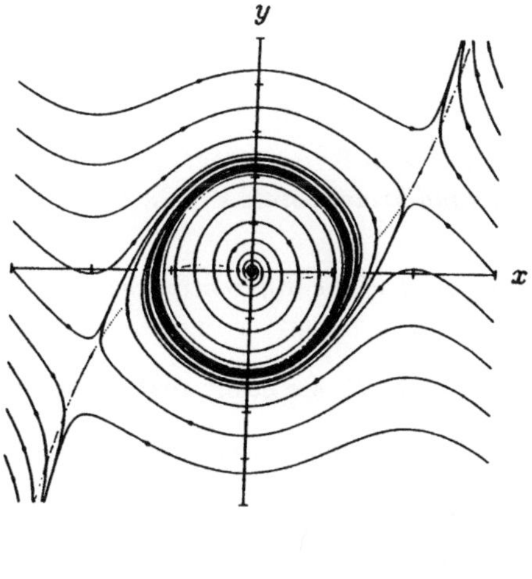

Figure 3.1. Phase portrait for system (BT) with $F(x) = \frac{1}{3}(x^3 - x)$ and $g(x) = -x^3 + 4x$.

Example 3.2

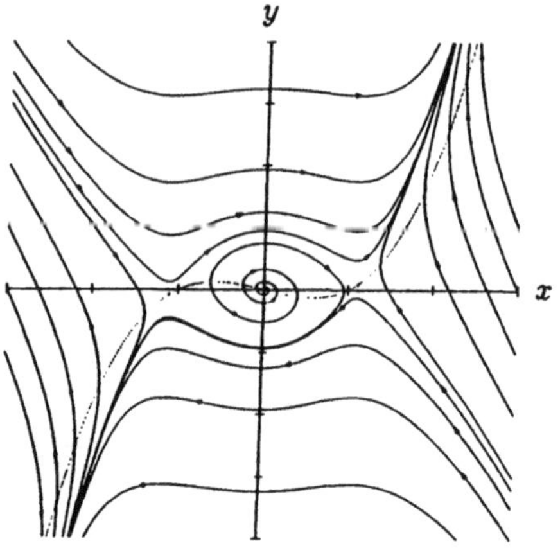

Figure 3.2. Phase portrait for system (BT) with
$F(x) = \frac{1}{3}(x^3 - x)$ and $g(x) = -x^3 + 1.21x$.

4 Proofs

To prove Theorem 3.1 and Theorem 3.2, we shall prepare some lemmas. We use here Poincaré's concept of the index of a closed curve or of a point in $\mathbf{R}^2$ with respect to a vector field in $\mathbf{R}^2$. We denote the index of a curve C and a point P by *ind*(C) and *ind*(P), respectively. As for this concept we have the following famous lemma.

Lemma 4.1 (Poincaré) *Let* $\vec{f}(\mathrm{P})$ *be a vector field in* $\mathbf{R}^2$ *all whose equilibrium points are isolated. Then the following statements* (1), ..., (4) *hold.*
(1) *If* P *is an equilibrium point other than a saddle, then* $ind(\mathrm{P}) = 1$.
(2) *If* P *is a saddle equilibrium point, then* $ind(\mathrm{P}) = -1$.
(3) *If* C *is a closed Jordan curve in* $\mathbf{R}^2$ *such that* $\vec{f}(\mathrm{P}) \neq \vec{0}$ *for all* $\mathrm{P} \in \mathrm{C}$ *and there are a finite number of equilibrium points* $\mathrm{P}_1, \ldots, \mathrm{P}_n$ *inside the curve* C*, then* $ind(C) = \sum_i ind(\mathrm{P}_i)$.
(4) *If* C *is a closed solution orbit of the differential equation* $\vec{x}'(t) = \vec{f}(\vec{x}(t))$*, then* $ind(C) = 1$.

For the proof see [J-S] or [Le]. We see easily that the following lemma holds.

Lemma 4.2 *If* $\mu_1 \geq 0$*, system* (BT) *has no non-trivial closed orbits.*

Proof. If $\mu_1 \geq 0$, system (BT) has a unique equilibrium point at the origin, and it is a saddle point. Thus if the system has a closed orbit, it contradicts Lemma 4.1. □

As for the indices of the three equilibrium points of system (BT), the following holds.

Lemma 4.3 *If* $\mu_1 < 0$*, system* (BT) *has three equilibrium points* $\mathrm{P}_0 = (0,0)$, $\mathrm{P}_1 = (-\sqrt{-\mu_1}, F(-\sqrt{-\mu_1}))$ *and* $\mathrm{P}_2 = (\sqrt{-\mu_1}, \mathrm{F}(\sqrt{-\mu_1}))$. *Also* $ind(\mathrm{O}) = +1$, $ind(\mathrm{P}_1) = ind(\mathrm{P}_2) = -1$.

From Lemma 4.1 and Lemma 4.3 we obtain the following lemma.

Lemma 4.4 *If system* (BT) *has a closed orbit C, its inside $\mathring{C}$ contains the equilibrium point* P_0, *but neither of the two equilibrium points* P_1 *and* P_2.

Proof of Theorem 3.1. From Lemma 4.2 we assume the condition $0 < -\mu_1 \leq 3\mu_2$. By Lemma 3 in [H1] we see that no non-trivial closed orbit of system (BT) exists in the strip $-\sqrt{-\mu_1} < x < \sqrt{-\mu_1}$. Therefore, if system (BT) has a non-trivial closed orbit C, it must meet the outside of the strip $-\sqrt{-\mu_1} < x < \sqrt{-\mu_1}$. From this fact amd Lemma 4.3 we see that C must meet at least one of the two vertical lines $x = -\sqrt{-\mu_1}$ and $x = \sqrt{-\mu_1}$. Let us express C as $C = \{(\phi(t), \psi(t)) \mid 0 \leq t \leq T\}$, where the pair $(\phi(t), \psi(t))$ is a solution of system (BT) and T is the smallest period of the solution. The line $x = -\sqrt{-\mu_1}$ consists of two open half lines $l^+ = \{(-\sqrt{-\mu_1}, y) \mid y > F(-\sqrt{-\mu_1})\}$, $l^- = \{(-\sqrt{-\mu_1}, y) \mid y < F(-\sqrt{-\mu_1})\}$ and the equilibrium point P_1. C does not meet P_1. If C meets l^+ at a point $(\phi(t_0), \psi(t_0))$ with $0 \leq t_0 < T$, then $\phi(t_0) = -\sqrt{-\mu_1}$, $\psi(t_0) > F(-\sqrt{-\mu_1})$ and $\phi'(t_0) = \psi(t_0) - F(\phi(t_0)) = \psi(t_0) - F(-\sqrt{-\mu_1}) > 0$. Hence the point $(\phi(t), \psi(t))$ crosses the line l^+ from left to right at $t = t_0$. From this fact it follows that the point $(\phi(t), \psi(t))$ cannot cross l^+ more than once in the time interval $0 \leq t < T$ without crossing the lower half line l^-. Similarly we see that $(\phi(t), \psi(t))$ cannot cross l^- more than once without crossing l^+. From these facts we conclude that, if C meets the line $x = -\sqrt{-\mu_1}$, it crosses the two half lines l^+ and l^- once each in the time interval $0 \leq t < T$. It follows from this fact that, if C meets the line $x = -\sqrt{-\mu_1}$, the equilibrium point P_1 is in the inside $\mathring{C}$ of C. This contradicts Lemma 4.4. Similarly, if C meets the line $x = \sqrt{-\mu_1}$, then we see that $\mathring{C}$ contains the point P_2, which again contradicts Lemma 4.4. □

Now we shall prove Theorem 3.2. The following lemma is a key to prove Theorem 3.2.

Lemma 4.5 *If system* (BT) *satisfies the condition* [C9] *and has a non-trivial closed orbit, then it must cross the segments $-\sqrt{-\mu_1} < x < -\sqrt{3\mu_2}$ and $\sqrt{3\mu_2} < x < \sqrt{-\mu_1}$ on the x-axis.*

To prove Lemma 4.5, we construct a closed plane curve Υ by the same method as in [H1]. We consider a closed plane curve defined by the equation

$$\tfrac{1}{2}y^2 + G(x) = G(-\sqrt{3\mu_2}).$$

The function $y = \varphi^-(x) = -\sqrt{2(G(-\sqrt{3\mu_2}) - G(x))}$ expresses the part in the lower half plane $y \leq 0$ of the above curve. From $G(-\sqrt{3\mu_2}) = G(\sqrt{3\mu_2})$, the graph of $y = \varphi^-(x)$ intersects with the gragh of $y = F(x)$ in the range $0 < x \leq \sqrt{3\mu_2}$. So letting

$$x_2 = \sup\left\{x \in (0, \sqrt{3\mu_2}\,] \mid \varphi^-(x) - F(x) = 0\right\},$$

we have $0 < x_2 \leq \sqrt{3\mu_2}$ and $\varphi^-(x_2) = F(x_2)$. Let Υ^- be the union of the graph of the equation $y = F(x)$ for $x_2 \leq x \leq \sqrt{3\mu_2}$ (the curve BC in Figure 4.1) and that of the equation $y = \varphi^-(x)$ for $-\sqrt{3\mu_2} \leq x < x_2$ (the curve AB in Figure 4.1).

Similary we consider the function $y = \varphi^+(x) = \sqrt{2(G(-\sqrt{3\mu_2}) - G(x))}$ expresses the part in the upper half plane $y \geq 0$. Letting

$$x_1 = \inf\left\{x \in [-\sqrt{3\mu_2}, 0) \mid F(x) - \varphi^+(x) = 0\right\},$$

we have $-\sqrt{3\mu_2} \leq x_1 < 0$ and $F(x_1) = \varphi^+(x_1)$. Let Υ^+ be the union of the graph of the equation $y = \varphi^+(x)$ for $x_1 \leq x \leq \sqrt{3\mu_2}$ and that of the equation $y = F(x)$ for $-\sqrt{3\mu_2} \leq x < x_1$. Υ^+ is the curve ADC in Figure 4.1. Let $\Upsilon = \Upsilon^+ \cup \Upsilon^-$ and Ω be the domain surrounded by Υ. Υ is a closed curve that surrounds the origin.

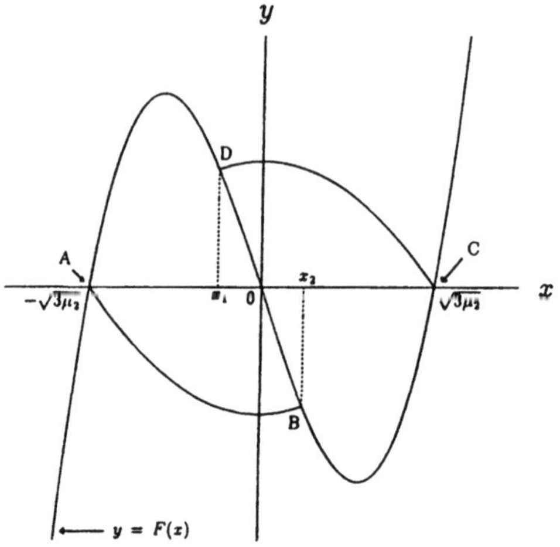

Figure 4.1

Proof of Lemma 4.5. System(BT) has no closed orbits in the domain Ω by Lemma 3 in [H1] since Ω is included in the strip $-\sqrt{3\mu_2} < x < \sqrt{3\mu_2}$. On the other hand, by Lemma 1 in [H1], the set Ω^c is an invariant set of the system. Thus if the system has a closed orbit, it exists outside the domain Ω. But, by the same discussion as in Theorem 3.1, we see that the closed orbit must be in the strip $-\sqrt{-\mu_1} < x < \sqrt{-\mu_1}$ and contain the origin in its inside. Therefore the closed orbit must cross the segments $-\sqrt{-\mu_1} < x < -\sqrt{3\mu_2}$ and $\sqrt{3\mu_2} < x < \sqrt{-\mu_1}$ on the x-axis. □

Next we define a new domain including the domain Ω. We construct the closed curve $\Gamma = \Gamma^+ \cup \Gamma^-$ by the following method, where the curves Γ^+ and Γ^- express the part in the half plane $y \geq 0$ and $y < 0$, respectively. First we construct the curve $\Gamma^+ = \cup_{i=1}^{6} \Gamma_i^+$. We define three functions $V_1(x,y)$, $V_2(x,y)$ and $V_3(x,y)$ by

$$\begin{cases} V_1(x,y) = \frac{1}{2}y^2 + G(x) & \text{for } x \in [-\sqrt{3\mu_2}, -\sqrt{-\mu_1}\,] \cup [\sqrt{3\mu_2}, \sqrt{-\mu_1}\,], \\ V_2(x,y) = \frac{1}{2}(y - F(-\sqrt{\mu_2}\,))^2 + G(x) & \text{for } x \in [-\sqrt{3\mu_2}, 0], \\ V_3(x,y) = \frac{1}{2}(y - F(\sqrt{\mu_2}\,))^2 + G(x) & \text{for } x \in [0, \sqrt{3\mu_2}\,]. \end{cases}$$

Let E in Figure 4.2 be the point $(\sqrt{-\mu_1}, F(\sqrt{-\mu_1}) - \epsilon)$, where ϵ be a positive constant satisfying [C6']. We set

$$\Gamma_1^+ = \{(\sqrt{-\mu_1}, y) \mid 0 \leq y \leq F(\sqrt{-\mu_1}\,) - \epsilon\}.$$

Γ_1^+ is the segment EQ. Γ^+ is the arc EF in Figure 4.2 which is defined by the equation

$$V_1(x,y) = \frac{1}{2}(F(\sqrt{-\mu_1}\,) - \epsilon)^2 + G(\sqrt{-\mu_1}\,)\ (x \in [\sqrt{3\mu_2}, \sqrt{-\mu_1}\,]).$$

The ordinate of F is given by $\eta_F = ((F(\sqrt{-\mu_1}\,) - \epsilon)^2 + 2(G(\sqrt{-\mu_1}\,) - G(\sqrt{3\mu_2}\,)))^{1/2}$. Γ_3^+ is the arc FG in Figure 4.2 which is defined by the equation

$$V_3(x,y) = \frac{1}{2}(\eta_F - F(\sqrt{\mu_2}\,))^2 + G(\sqrt{3\mu_2}\,).$$

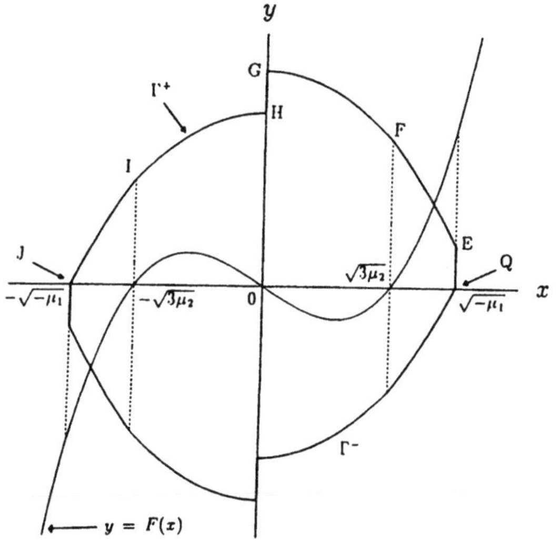

Figure 4.2

The ordinate of G is given by $\eta_G = F(\sqrt{\mu_2}\,)+\Psi_2^{1/2}(\epsilon)$. Let J denote the point $(-\sqrt{-\mu_1}, 0)$. Γ_4^+ is the arc IJ in Figure 4.2 which is defined by the equation

$$V_1(x,y) = G(-\sqrt{-\mu_1}\,).$$

The ordinate of I is given by $\eta_I = (2(G(-\sqrt{-\mu_1}\,) - G(-\sqrt{3\mu_2}\,)))^{1/2}$. Γ_5^+ is the arc HI in Figure 4.2 which is defined by the equation

$$V_2(x,y) = \frac{1}{2}(\eta_I + F(-\sqrt{\mu_2}\,))^2 + G(-\sqrt{3\mu_2}\,).$$

The ordinate of H is given by $\eta_H = F(-\sqrt{\mu_2}\,) + \Phi_1^{1/2}$. Note that $\eta_H \le \eta_G$ by [C6′]. We set

$$\Gamma_6^+ = \{(0,y) \mid \eta_H \le y \le \eta_G\}.$$

Γ_6^+ is the segment GH in Figure 4.2. Thus Γ^+ is the curve defined by the above six arcs.

By a similary way we can construct the curve Γ^- defined by the six arcs as shown in Figure 4.2. Γ is a closed curve that surrounds the origin and includes the domain Ω. Let $\tilde{\Omega}$ be the domain surrounded by Γ. As for the domain $\overline{\Omega} = \tilde{\Omega} - \Omega$ the following holds.

Lemma 4.6 *Assume that the conditions* [C6′] *and* [C9] *hold. A solution orbit of the system starting from a point in* $\overline{\Omega}$ *stays in* $\overline{\Omega}$.

Proof. By investigating the direction of the field of velocity vectors defined by system (BT) on Γ, we see that a solution orbit starting from a point in $\overline{\Omega}$ cannot go outside $\overline{\Omega}$ through Γ. On the other hand, since Ω^c is an invariant set of the system, a solution orbit of the system starting from a point in $\overline{\Omega}$ must stay in $\overline{\Omega}$. □

Using Lemma 4.5 and Lemma 4.6, we can now easily prove Theorem 3.2.

Proof of Theorem 3.2. By Lemma 4.5 and Lemma 4.6, a solution orbit starting from a point in $\overline{\Omega}$ must stay in $\overline{\Omega}$. Thus from the well-known Poincaré–Bendixson theorem, we see that the system has at least one closed orbit in $\overline{\Omega}$. On the other hand, by the same method as [Hi], we can show that the system has at most one closed orbit in $\overline{\Omega}$. Hence system (BT) has exactly one closed orbit in $\overline{\Omega}$. □

5 Appendix

In [H2], we gave a necessary and sufficient condition under which the polynomial Liénard system has an algebraic invariant curve. The result in [H2] is stated as follows.

Theorem 5.1 *If the polynomial Liénard system* (L) *satisfies the conditions* $F \neq const$ *and* $\deg F \geq \deg g$, *then the system has no non-trivial algebraic closed orbits.*

Thus the following result holds.

Corollary 5.2 *The unique non-trivial closed orbit of system* (BT) *is not algebraic.*

Remark. Our methods can be applied to the following Bogdanov-Takens system with a symmetric cusp of order 3, too.

$$\begin{cases} \dot{x} = y \\ \dot{y} = -(x^4 - \mu_3 x^2 - \mu_2)y + x^3 + \mu_1 x, \end{cases}$$

where μ_1, μ_2 and μ_3 are parameters. See [H3].

References

[CRW] Li. Chengzhi, C. Rousseau and X. Wang. : *A simple proof for the unicity of the limit cycle in the Bogdanov-Takens system.* Canad. Math. Bull. **33(1)** (1990), 84–92.

[H1] M. Hayashi. : *On the uniqueness of the closed orbit of the Liénard system.* To appear in Math. Japonica.

[H2] ______. : *On polynomial Liénard systems which have invariant algebraic curves.* To appear in Funkcialaj. Ekvacioj.

[H3] ______. : *On the closed orbit of the Bogdanov-Takens system with a symmetric cusp of order* 3. (Preprint).

[H4] ______. : *On a Liénard system with three equilibrium points.* Colloquium talk, Seventh Internatinal Colloquium on Differential Equations, Plovdiv, 1996.

[Hi] M. Hirano. : *On the Uniqueness of the Periodic Orbit of Liénard Equation.* Math. Japonica. **36** (1991), 1193–1199.

[J–S] D. W. Jordan and P. Smith. : *Nonlinear Ordinary Differential Equations.* 2nd Ed. Oxford University Press Inc., 1987.

[Le] S. Lefschetz. : *Differential Equations : Geometric Theory.* 2nd Ed. Dover Publications Inc., 1977.

[Xi] Z. Xianwu. : *On the Uniqueness of limit cycle of Liénard system.* Scientia Scinica (Series A), **25(6)** (1982), 583–592.

[Zh] Z. Zhifen. : *Proof of the uniqueness theorem of limit cycles of generalized Liénard equations.* Appl. Anal. **23** (1986), 63–76.

7th Int. Coll. on Differential Equations, pp. 147-154
D. Bainov (Ed)

RE-EXAMINATION ON VALIDITY OF ANALYTIC SOLUTION OF BESSEL'S EQUATION TO ESTIMATE THE MAGNITUDE OF FLUCTUATION OF STEM WOOD TEMPERATURE IN STANDING *Pinus densiflora* TREES

Yoshihiro HAYASHIDA

Cosmo Land Planning, 103 Maruei mansion, Higashi-chyo 3-5-8, Koonosu, Saitama 365, Japan

Abstract

Woody-tissue temperature was measured at the depth of 0, 5 and 10 cm beneath the bark along the radial direction in two seasons at various heights of nearly 70-year-old *Pinus densiflora* trees. In summer and autumn, each measurement lasted during ten successive days. If we consider tree stem as a part of infinite cylindrical system at any given radius, the inner temperature field is a function of surface temperature amplitude, thermal diffusivity of wood and angular velocity of applied surface temperature oscillation. Under natural conditions, it is hard to expect to get such an idealized one. However, if we choose carefully circumference conditions, woody-tissue temperature field reveals exactly the same result from which the Bessel's equation shows us.

Key words: Kelvin' function (ber$x$+i bei$x$), sine wave (periodic solution), thermal diffusivity, cooling or warming effect of ascending sap flow

1 INTRODUCTION

Previous study about temperature field of pine tree stem conducted by Herrington [1] concluded that strong heat resistance layer exists between bark and xylem. Its resistance to prevent thermal conductance was estimated to be nearly 20%, namely wood temperature in any radial position except surface showed 20% lower than that of analytic solution of Bessel's equation. He measured temperatures of stem at 1.4 m high. He assumed that cooling effect of sap flow that decreases temperature amplitude would disappear over the stem position 1.0 m high. However, sap flow velocity in pine tree [2,3] ranged from 15 to 25 cm an hour in the daytime in summer. These values correspond to those of water columns equal to its stem diameters, real sap flow velocities in tracheid vessel reach at least double figures of above mentioned velocities. In his case, cooling effect was visible at the height of 0.86 m. This fact suggests that negligible cooling effect by ascending sap flow is still disputable even at the height of DBH in the case of Herrington [1].

Over a century ago, the cooling effect of sap flow at stem base was recognized [3]. Herrington [1] summarized previous studies regarding the effect of sap flow in details, and guessed nearly 0.5 m downward from effected position would be free from this effect. Our previous study [4] also showed that cooling effect was clearly noticeable over 1.3 m above ground in pine tree, though it was larger than that of Herrington's tree. In that experiment, temperature of wood only 10 cm deep from the bark was measured. Since sap flow ascends mainly peripheral region of stem in conifer, we further measured temperatures of 0, 5, and 10 cm deep from the surface at various heights of the same pine tree plus another one.

This study reports that under certain conditions that exclude the cooling effect of sap flow reveal no special heat resistance layer exists under the bark.

2 DATA COLLECTION AND ANALYSIS

Temperature data were collected at the Experimental Station at Tanashi [6,7] in 1981. In summer experiment(Tree 1 [5]), owing to strong cooling effect of stem base, only one stem position 5.0 m high (diameter=31 cm) was selected. In fall experiment, two trees including Tree 1 were selected. Tree 1 was measured at the height of 0.7 m (diameter=43 cm). Tree 2 was measured at the height of 0.7 m (diameter=37 cm) and 5.0 m (diameter=28 cm). Temperature record system was the same as that described by Matsumoto et al. [8] and Negisi et al. [5]. Air temperature (1.3 m above ground) and soil temperature (-0.3 and -1.0 m) were also recorded. Wood temperatures at northern side of the stem were recorded in every case to prevent direct sun light. During these experiments, respiration chambers were attached to the stem, small chambers at summer (100 cm^2) and large chambers at fall (1000 $\sim$ 1600 cm^2). The chamber covered temperature measuring hole. Chambers were covered with aluminum foil to shieled radiation and air was circulated by pump to measure CO_2 concentration. This respiration measurement aimed to detect daytime depression of respiration affected by sap flow [2,6,7] and/or xylem temperature. During fall experiment, Tree 2 was completely defoliated using chain saw to stop the sap flow. Tree 2 was felled finally and measured dry weight, water content, leaf mass, etc. Each temperature was recorded automatically at the interval about 30 min. and all the data were punched out on paper tape. Later, the data were processed using the main frame of the University of Tokyo and stored in magnetic tape.

Temperature data in each depth were regressed as a sine wave in each day using the linear regression package of *Mathematica* [8]. Surface temperature was expressed as $a + b\sin x$, inner temperature was expressed as $a' + b'\sin(x - \phi)$, where a or a' denotes daily mean xylem temperature, b or b' denotes temperature amplitude of each position, and ϕ denotes phase lag of sine wave from the surface. Soil temperature recorded annually was also expressed as sine wave to estimate soil thermal diffusivity.

2.1 Analytic Solution of Stem

Circular infinite cylindrical system, where temperature gradient occurs in the radial direction only and heat production by respiration can be negligible, is treated as one dimensional. In this case, appropriate form of the heat equation is

$$\frac{\partial T}{\partial t} = \alpha\left(\frac{\partial^2 T}{\partial r^2} + \frac{1}{r}\frac{\partial T}{\partial r}\right) \tag{1}$$

where T = temperature, °C

t = time, sec.

α = thermal diffusivity, m^2/s ($\alpha > 0$)

r = radius, m ($r > 0$).

Equation (1) assumes that the thermal properties of the stem cylinder are isotropic, homogeneous and invariant with temperature. Considering the solution of equation (1) with the condition that has a time-periodic boundary conditions proportional to a sine wave (time-factor $e^{i\omega t}$; period $= 2\pi/\omega$) [10,11], we can write in the form:

$$\frac{d^2T}{dr^2} + \frac{1}{r}\frac{dT}{dr} - i\frac{\omega}{\alpha}T = 0 \tag{2}$$

The general solution of (2) is

$$T = C_1 I_0(kr\sqrt{i}) + C_2 K_0(kr\sqrt{i}), \tag{3}$$

where

$$k^2 = \frac{\omega}{\alpha}, \tag{4}$$

and where C_1 and C_2 are arbitrary constants. Even when $r = 0$, the center temperature is finite value, hence C_2 must be zero. Accordingly, equation (3) is written as follows [10],

$$T = C_1 I_0(kr\sqrt{i}) = C_1 I_0\left(\frac{1+i}{\sqrt{2}}kr\right) = C_1\{ber(kr) + i\,bei(kr)\} \tag{5}$$

If the initial temperature is zero and the surface temperature ($r = R$) is $T_R = A\sin(\omega t)$, then the inner temperature field ($T_r; r = r_a$) becomes as follows [1,11],

$$T_r = A\left(\frac{ber^2(kr) + bei^2(kr)}{ber^2(kR) + bei^2(kR)}\right)^{1/2} \sin(\omega t - \phi) \tag{6}$$

and

$$\phi = \tan^{-1}\left(\frac{bei(kR)ber(kr) - ber(kR)bei(kr)}{ber(kR)ber(kr) + bei(kR)bei(kr)}\right) \tag{7}$$

Equation (7) denotes argument difference between $I_0(kR\sqrt{i})$ and $I_0(kr\sqrt{i})$ on the Gaussian plane.

2.2 Analytic Solution of Soil

Herrington's summary [1] of previous study about the cooling or warming effect by sap flow at base of stem suggests that it mainly depends on tree species (diffuse porous or ring porous, deciduous or evergreen) tree sizes and site conditions. Warming effect of stem base is recognized in midwinter in spruce [1] and pine [5], but not deciduous tree [5]. For example, soil daily mean temperature 0.3 m deep at Tanashi is well correlated with the daily mean temperature 10 cm deep at the height of 0.3 m of Tree 1's stem. However, another two tree species standing near the tree 1 do not show the same course in the same depth and the same height [5]. Since the soil temperature in each depth must be same in each season, these differences are caused from each species character and its sample tree size. It is useful to estimate the soil temperatures at various depth to compare these points in each other. To clarify the soil temperature field in various season, I applied essentially the same method used in the previous sub-section 2.1.

The equation that has been used in this case is [12,13],

$$\frac{\partial T}{\partial t} = \alpha\frac{\partial^2 T}{\partial x^2} \tag{8}$$

where α = thermal diffusivity of soil, m^2/s ($\alpha > 0$)

x = soil depth, m.

The solution of (8) where surface temperature is given by $T = A\sin(\omega t)$ is well known [12,13],

$$T(x,t) = A_0 + Ae^{-\sqrt{\frac{\omega}{2\alpha}}x}\sin\left(\omega t - \sqrt{\frac{\omega}{2\alpha}}x\right) \tag{9}$$

where $T = A_0$ when $x = \infty$, and A is temperature amplitude when $x = 0$. The thermal diffusivity of soil is

$$\alpha = \frac{\omega}{2\left(\frac{\ln(aA)}{x}\right)^2} \tag{10}$$

where a is an amplitude of temperature in the soil depth from origin. In this study, soil surface temperature was not measured, the origin of $x = 0$ was substituted to appropriate depth near the soil surface. For example, in the case in which the annual course of soil temperature was measured in the depth of 0.3 and 1.0 m, surface temperature was assumed at the temperature of 0.3 m of depth, then temperature amplitude in the depth 1.0 m corresponds to the depth of 0.7 m in the soil.

3 RESULT AND DISCUSSION

The thermal diffusivity of resinosa pine wood was calculated by Herrington [1] using experimental formula and stem wood properties (water content; 78%, specific gravity; 0.40). Average volumetric heat capacity was 2030.7KJ/m^3, and average thermal conductivity was 0.29W/m · K, then thermal diffusivity is calculated as 1.43×10^{-7} m^2/s, which is very similar value of thermal diffusivity of water at 20°C. According to Maku [14], experimetally measured thermal diffusivity of pine wood (*Pinus densiflora*) in the condition of very low water content (17.3%) and high specific gravity of 0.72 is 1.49×10^{-7} m^2/s at 100°C. In this study, the same thermal diffusivity of resinosa pine was used, because stem wood water content and estimated specific gravity of Tree 2 were similar to those from Herrington's red pine data. Calculated thermal diffusivity of stem center by experimental formula is up to 15% higher than that of peripheral part, but Herrington and I used the same value of thermal diffusivity of sapwood considering that sap wood plays main role of heat conduction from the surface to inner part.

Although summer and fall experiments covered ten successive days, seven or five days were suitable for applying sine wave to the temperature oscillation of stems because of the shift of daily mean temperature between days. This method requires that the stem should be in a nearly steady periodic state for at least one day before any analysis is attempted [1]. Results of summer experiment show that average response of wood temperature is in excess of 20% higher than that of analytic solution. When surface temperature amplitude is nearly 2°C, inner temperature amplitude at the depth of 10 cm is expected to be reduced to 0.8 °C, then the amplitude difference of 20 ~ 25 % corresponds to only 0.2 °C. Hence it is concluded that the notice of cooling effect of sap flow in xylem, if any, would be impossible in the case of Herrington's resinosa pine, which has the 28.4 cm diameter.

While the phase shift at the depth of 10 cm is expected to be delayed 6.4 hours, measured average phase shift was nearly 5.5 hours. Since the phase shift in 5 cm deep was well agreed with that of calculated time (3.2 hours), the difference was due to the early shift at the depth of 10 cm or more. This height of stem is considered to be free from the effect of sap flow [5,6,7], and it is not conceivable that the center of matured coniferous stem is sensible to the heat production by respiration [15]. We must therefore conclude that the conduction of heat from the south side of stem produces the enhanced stem temperature rise near the center [5,16,17]. Indeed, when we measured stem surface temperatures at four directions, the south surface temperature was 1 ~ 1.5°C higher than that of the north side according to the solar radiation. This duration of tendency maintained about twelve hours or more in the occasion of late August in 1981.

In fall experiment, daily mean air temperature was nearly 14 ~ 16°C, which was about 10°C lower than that of summer, and corresponds to late summer temperature of the Connecticut forest [1]. This drop of temperature lowers the ascent of sap. Morikawa [17] studied seasonal variation in transpiration of *Chamaecyparis obtusa* at Tanashi. The relation between daily transpiration (T_r) and mean air temperature (T°C) on clear days, is $Tr = 0.127T - 0.289$, where T_r denotes gH_2O/gd.w./day. This equation was obtained by cut leaf method, and he later confirmed this relation using heat pulse method in intact trees [3], especially in dominant ones. The equation is not simply applied for pine tree, but it gives a certain criterion. Rough estimate of ten degree drop of temperature suggests that transpiration is nearly half of that of summer.

During fall experiment, the stem mean temperatures in 10 cm deep of Tree 1 were higher than those of Tree 2. This is due to warming effect of sap flow in this season [5]. Before defoliation, the temperature difference was small in both trees, but the final experiment day, the four days later from defoliation, that difference increases up to three degrees. Daily mean stem temperature differences in radial direction at 0.7 m high was less than 0.5°C in both trees. However, mean xylem temperature is higher in Tree 1 and lower in Tree 2 than that of air [7]. Maximum difference between xylem mean temperature and mean air is nearly 1.5 °C in both trees in each day.

The results of fall experiment show that measured temperature field agrees with an error of not over ±5% to the expected gradient by analytic solution in both trees, though analysis is made only for five days. Even at the 5.0 m high of Tree 2, the excess of response was not found. This is probably due to the large respiration chamber, which surrounded measured stem surface including south side to shield solar radiation. The daily mean xylem temperature in each radial position of this height is less than 0.3°C, and maximum difference between xylem mean temperature and mean air is less than 1°C.

Vertical temperature field of soil in this period affects to decrease stem temperature difference in longitudinal direction near the ground. Average temperatures 0.3 m and 1.0 m deep are nearly the same in 17±0.5°C. This is due to the soil thermal diffusivity and annual course of air temperature. The soil thermal diffusivity estimated from equation (10) is nearly 1.8×10^{-7} m^2/s. We thus obtain from (9) the equation

$$T(x,t) = A_0 + 9.3e^{-0.73x} \sin(\omega t - 0.73x) \qquad (11)$$

which expresses every one meter deep in the soil, soil temperature amplitude oscillation diminishes nearly 52%, and its wave delays nearly 42 days. An example of A_0 in one meter deep in the stand was 13.8°C, and if this temperature is assumed to be an approximate value of A_0, then the soil temperature in 2.0 m deep is expressed as $13.8+2.2\sin(\omega t-1.46)$, which has yearly maximum temperature 16.0°C from late Octobe to early November. Thus, the October experiment in this site gives fairly ideal condition to regard the stem near ground as an infinite cylindrical system. Oota et al. [19] showed that no stem temperature drop was occurred when the water that has the same temperature of the surrounding is conducted through stem section. Under natural circumference, it is hard to prevent the effect of sap flow. Accordingly, we must carefully select the occasion that stem has nearly the same stable temperature in longitudinal direction at least one meter or more and nearly the same daily mean temperature considering sap flow velocity.

ACKNOWLEDGEMENT

I wish to express my great thanks to professor Ken'itiroo Negisi for his guidance and encouragement. Thanks are also due to members of Tokyo University Forest Experimental Station, Tanashi, Tokyo for their aid in measurements. This study was supported in part by a fund from the Ministry of Education (56480045). I ask all those who have helped me with information and guidance, or who have otherwise facilitated my research, to accept this acknowledgement of my gratitude.

REFERENCES

1. L.P.Herrington, Yale Univ. Sch. For., 73, 1-79, (1969).
2. K.Negisi, J. Jpn. For. Soc., 61, 88-93, (1979).
3. Y.Morikawa, Bull. Tokyo Univ. Forests, 66, 251-297, (1974).
4. Th.Hartig, Allg. Forst - u. Jagdztg., 49, 145-152, (1874).
5. K.Negisi, Y.Hayashida, and K.Yagi, Bull. Tokyo Univ. Forests, 72, 1-14, (1982).
6. Y.Hayashida et al., Trans. 93rd. Mtg. Jpn. For. Soc., 237-238, (1982a).
7. Y.Hayashida et al., Trans. 93rd. Mtg. Jpn. For. Soc., 239-240, (1982b).
8. Y.Matsumoto et al., J. Jpn. For. Soc., 64, 149-154, (1982).
9. S.Wolfram, *Mathematica*, Addison-Wesley, Redwood City, CA (1992).
10. F.Bowman, Introduction to Bessel Function, Longmans, Green and Co. Ltd. (1938).
11. H.S.Carslaw and J.C.Jaeger, Conduction of heat in solids, 2nd ed., Oxford Univ. Press, Oxford, pp. 193-201, (1959).

12. M.Shoji, Heat transfer textbook. Univ. Tokyo Adv. Mech. Engineering 6., Univ. Tokyo Press, Tokyo, 63, (1995).
13. T.Yawata, Physics of soil, Univ. Tokyo Press, Tokyo, pp. 124-136, (1975).
14. Y.Maku, Wood Research, Kyoto Univ., 6, 43-50, (1951).
15. T.Higuchi, K.Fukazawa, and M.Shimada, Res. Bull. College. Exp. Forests Hokkaido Univ., 25, 167-195, (1967).
16. R.W.Derby and D.M.Gates, Amer. J. Bot., 53, 580-587, (1966).
17. G.Mitscherlich, in: Wald, Wachstum und Umwelt. 2. Bd: Waldklima und Wasserhaushalt., Frankfurt (Main): Säuerlander, (1971) pp. 92-161.
18. Y.Morikawa, J. Jpn. For. Soc., 52, 259-262, (1970).
19. T.Oota et al., J. Jpn. Soc. Revegetation Tech., 17, 18-24, (1992).
20. N.Seki (Ed.), Engineering of heat transfer, Motikita Pub. Co., Tokyo, (1988), p. 242.
21. The Tokyo Univ. For., (ed.) The Tokyo University Forest, since 1894, p. 36, (1994).
22. K.Kanmera,M.Hashimoto, and T.Matsuda, Geological features in Japan, Iwanami Syoten, Tokyo, pp. 163-244, (1992).
23. The Tokyo Univ. For., (ed.) Bull. Tokyo Univ. For. Report., 26, 1-83, (1981).
24. K.Watanabe, Trans. 61st. Mtg. Geodetic Soc. Jpn., 81-82, (1984).
25. K.Bjor, Meddr. norske SkogforsVes., 108, 429-526, (1971).
26. W. Larcher, Physiological plant ecology, Springer Verlag, Berlin Heiderberg New York, (1975), p. 193.

APPENDIX

The soil thermal diffusivity at Tanashi shows characteristics of its loam soil. Its value with no water is reported as $2.6 \times 10^{-7} m^2/s$ [20]. It is well known that increase of soil water content decreases thermal diffusivity [13]. During summer experiment, soil water content 0.5 m deep was nearly 50 weight percent except the next day of rain. The specific heat of loam is 2.8 KJ/Kg K and the density is 1230 Kg/m^3 [20]. Increase of water content gradually decreases the loam soil thermal diffusivity. Annual rain fall in this site reaches nearly 1400 mm, and the annual mean temperature is 13.7°C [21].

Except upper layer of black soil to 0.5 m deep, loam soil reaches nearly 10 m deep [21]. This loam soil, so called 'Kantou Loam' [22], is tephra-fall layers from Mt. Fuji nearly 10 ~ 30 thousand years or more ago. By the effect of westerlies, the area having this layers is mainly restricted in Kantou Plain. Topography also affects the distribution characteristic. For example, the nearest University Forest at Chichibu, about 50 km NW of Tanashi, has no loam layer especially the foot of mountain. Soil thermal diffusivity at nursery of Chichibu is estimated to be nearly $1.7 \times 10^{-6} m^2/s$ [23]. $7.5 \times 10^{-7} m^2/s$, $2.0 \times 10^{-6} m^2/s$ and $7.6 \times 10^{-7} m^2/s$ are examples of thermal diffusivity of another Experimental Forest' nursery soil in Chiba, Aichi, and Hokkaidou, respectively [23]. Watanabe estimated soil thermal diffusivity to be $5.9 \times 10^{-7} m^2/s$ at Yamazaki in Kyoto [24]. Likewise, $1.0 \times 10^{-6} m^2/s$ and $1.1 \times 10^{-6} m^2/s$ are examples from european soil in Norway [25] and Germany [26], respectively. These estimates were calculated using (10).

Our study on xylem temperatures of three coniferous species at Tanashi [5] showed strong cooling effect of sap flow in growing season. Its effect was noticed even at the height of 5.0 m in *Metasequoia glyptostroboides*. This is partly due to the low soil thermal diffusivity of loam soil. In growing season, it causes very strong thermal gradint in longtudinal direction of tree root system compared with air temperature. Only in fall, this thermal gradient disappears. Experiment in another site will be quite different from our result, as in the case of Herrington's pine.

7th Int. Coll. on Differential Equations, pp. 155-163
D. Bainov (Ed)

Some modernizations The Gronwall-Bellman Lemma and its application to differential equations with delay

Danuta Jama
Aleksander Michał Nawrat

Silesian Technical University
Institute of Mathematics, Dept. of Differential Equations
Kaszubska Street 23, 44-100 Gliwice
e-mail: olek@zeus.polsl.gliwice.pl

Abstract

In this paper we will fix our attention on research differential equations with delay. In this purpose we will prove some theorems and lemma, which be some modification well known from ordinary differential equations Gronwall-Bellman Lemma. That will be also proof some integral inequalities, which may found practical application. From the practical regard we will apply Lesbeuqe measure, which will much simplificate after-mentioned proofs. We also will use classical Gronwall-Bellman Lemma, which is starting point to this paper.
The main result this paper are theorems, which may found application to research of stability differential equations with delay.

Mathematics Subject Classification: 34K20
Key words and phrases: differential equations with delay, Gronwall-Bellman stability.

1. Introduction

From historical ground some differential equations with delay we can even found in papers of L. Euler, but systematic research about these equations began in XX century. In this time applied parts of mathematics like control theory have used very often this equations. In years 1950-1970 this equation began to be more important in the different research areas like applied physics, some technical aspects or even in economy and biology.

On begin of this paper we want to give some definition, which will characterize the differential equation with delay.

Definition 1 *Differential equations with deviation these are such a differential equations in which the unknown function appears by different value of arguments.*

For example:

$$\dot{x}(t) = f(t, x(t), x[t - \tau(t)]), \tag{1}$$

$$\dot{x}(t) = f(t, x(t), x(t-\tau_1), x(t-\tau_2)), \tag{2}$$

$$\ddot{x}(t) = f(t, x(t), \dot{x}(t), x[t-\tau(t)], \dot{x}[t-\tau(t)]), \tag{3}$$

Definition 2 *Differential equation with deviation we will call differential equation with delay, if the highest grade derivate from unknown function appears by the same argument value and this argument is not higher then the other arguments from unknown function and is derivates which comes to the equation.*

For example that can be equations like 1, 2, 3 but with added condition on the delay function like this:

$$\tau(t) \geq 0,\ \tau_1 > 0 \wedge \tau_2 > 0,\ t \geq 0.$$

For a simple ordinary differential equation with delay of the form:

$$\dot{x}(t) = f(t, x(t), x[t-\tau_1(t)], \dots, x[t-\tau_m(t)]), \tag{4}$$

primary initial-value problem is based on determining continuity solution $x(t)$ of equation (4) for $t \geq t_0$, with the condition, that $x(t) = \varphi(t)$ on initial set E_{t_0}, set from point t_0 and from value $t - \tau_i$, $(i = 1, \dots, m)$, which are smaller than t_0 for $t \geq t_0$, where $\varphi(t)$ is given continuity function called initial function.
The solution $x_\varphi(t)$ is said to be solution of equation (4) with defined initial function $\varphi(t)$. Now we will introduce some basic definitions concerning ordinary differential equations with delay of the form (4).

Definition 3 *The solution $x_\varphi(t)$ of equation (4) is said to be stable, if for every $\varepsilon > 0$ exists such a $\delta(\varepsilon) > 0$, that from inequality $||\varphi(t) - \psi(t)|| \leq \delta(\varepsilon)$ on the initial set follows this inequality $||x_\varphi(t) - x_\psi(t)|| \leq \varepsilon$ for $t \geq t_0$, where $\varphi(t)$ is arbitrary continuous initial function.*

Definition 4 *The solution x_φ of equation (4) is said to be asymptotically stable, if:*
(1) it is stable,
(2) for each $t_0 \geq 0$ there is a $\delta_1 > 0$ such that every solution $x_\psi(t)$ which satisfies $||\varphi(t) - \psi(t)|| < \delta_1$ will also satisfy

$$\lim_{t\to\infty} [x_\varphi(t) - x_\psi(t)] = 0,$$

where $\varphi(t)$ is arbitrary continuous initial function.

Notice 1 *For change of variables: $y(t) = x(t) - x_\varphi(t)$ we can convert research of stability optional solution x_φ equation (4) on research of stability trivial solution $y(t) \equiv 0$.*

From this reason we will research stability of trivial solution.

First we will use the classical Gronwall-Bellman Lemma which was the starting point to this paper.

Lemma 1 *(Gronwall-Bellman [2]) Let be given some constant $C \in \Re$, and some continuous and definite functions on interval I, u, v, and $v(t) \geq 0$ for $t \in I$. If for a certain $t_0 \in I$ and for any $t \in I$, $t \geq t_0$, fulfilled is condition:*

$$u(t) \leq C + \int_{t_0}^{t} u(s)v(s)ds,$$

then for any $t \in I$ fulfilled is this:

$$u(t) \leq C \exp\left\{ \int_{t_0}^{t} v(s)ds \right\}.$$

Proof of this Lemma can be found in works [1] and [2].
Then we will prove some modernizations of this Lemma, Lemma 2 and two Theorems, Theorem 1 and 2, which are some modifications from well known theorems from ordinary differential equations. This Theorems can have applications in research stability in the sense of first Lapunov method also well known from ordinary differential aquations.

2. Main result

Now we will prove some modernization's well known Gronwall-Bellman Lemma.

Lemma 2 *Let be given some constans C, $\Re$ and some continuous and define functions on [A,B], u, v, and $v(t) \geq 0$ for $t \in [A, B]$. Besides let $u(t)$ be non-negative function aqual zero outside $[A, B]$. Let for a measurable set $E \subset [A, B]$ the following condition will be satisfied:*

$$m\gamma(E) \leq Nm(E),$$

where $\gamma(E)$ is set of this $t \in [A, B]$, for which $t - \tau(t) \in E$ and $m(E)$ Lesbeque measure from set E.

If for a certain $t_0 \in [A, B]$ and for any $t \in [A, B]$, $t > t_0$ fulfilled is condition:

$$u(t) \leq C + \int_{t_0}^{t} u(s - \tau(s))v(s)ds,$$

then for any $t \in [A, B]$ fulfilled is this:

$$u(t) \leq C \exp\left\{ N \int_{t_0}^{t} v(s)ds \right\}.$$

Proof On the beginning we will concentrate that under above conditions is hold following inequality:

$$\int_{A}^{B} u(s - \tau(s))ds \leq N \int_{A}^{B} u(s)ds.$$

Sufficiency
We build upper Lesbeque grand total for both integral:

$$S = \sum_k y_{k+1} m e_k,$$

$$S' = \sum_k y_{k+1} m e'_k,$$

where

$$e_k \ - \ \text{set this } t \in [A, B], \text{ for which } y_k \le u(t - \tau(t)) < y_{k+1},$$

$$e'_k \ - \ \text{set this } t \in [A, B], \text{ for which } y_k \le u(t) < y_{k+1}.$$

It is easy to prove that:

$$e_k \subset \gamma(e'_k).$$

In that case:

$$m e_k \le m\gamma(e'_k) \le N m e'_k,$$

and

$$S \le N S'.$$

Limiting transition finish proof of sufficency.

Necessity:

Let for some $E \subset [A, B]$:

$$m\gamma(E) > N m E.$$

We define $u(t)$ by:

$$f(x) = \begin{cases} 1 & t \in E \\ 0 & t \notin E \end{cases}$$

Therefore:

$$\int_A^B u(s - \tau(s)) ds = m\gamma(E) > N m E = N \int_A^B u(s) ds.$$

So our inequality is proved.

This earlier proof Lemma 2 we will use in the proof of the next Theorem.

Theorem 1 *Let function $u(t)$, continuous and positive for any t, $\tilde{t} \in [A, B]$ satisfy an inequality:*

$$u(t) \le u(\tilde{t}) + \int_t^{\tilde{t}} u(s - \tau(s)) v(s) ds, \tag{5}$$

where $v(t)$ is continuous function in interval $[A, B]$ and $v(t) \le 0$ fot this $t \in [A, B]$. Let for a measurable set $E \subset [A, B]$ the following condition is be satisfied:

$$m\gamma(E) \le N m(E)$$

where $\gamma(E)$ is set of this $t \in [A, B]$, for which $t - \tau(t) \in E$ and mE Lesbegue measure from set E.

Then for this $A \le t_0 \le t \le B$ is fulfilled this estimation:

$$u(t_0) \exp\left\{-N \int_{t_0}^{t} v(s) ds\right\} \le u(t) \le u(t_0) \exp\left\{N \int_{t_0}^{t} v(s) ds\right\}. \tag{6}$$

Proof From inequality (5) for $t \geq \tilde{t}$, we have:

$$u(t) \leq u(\tilde{t}) + \int_{t}^{\tilde{t}} u(s - \tau(s))v(s)ds.$$

Hence on the ground from Lemma 2 we obtain:

$$u(t) \leq u(\tilde{t}) + \exp\left\{\int_{t}^{\tilde{t}} v(s)ds\right\}. \tag{7}$$

Similiar, from inequality (5) for $t \leq \tilde{t}$, we have:

$$u(t) \leq u(\tilde{t}) + \int_{t}^{\tilde{t}} u(s - \tau(s))v(s)ds \leq u(\tilde{t}) + N \int_{t} \tilde{t} u(s)v(s)ds = u(\tilde{t}) + \int_{t} \tilde{t} u(s)Nv(s)ds \leq$$
$$u(\tilde{t}) \exp\left\{\int_{t}^{\tilde{t}} Nv(s)ds\right\} = u(\tilde{t}) \exp\left\{N \int_{t}^{\tilde{t}} v(s)ds\right\}.$$

Hence

$$u(t) \leq u(\tilde{t}) \exp\left\{N \int_{t}^{\tilde{t}} v(s)ds\right\},$$

we obtain

$$u(t) \exp\left\{-N \int_{t}^{\tilde{t}} v(s)ds\right\} \leq u(\tilde{t}).$$

Now we change $t \to \tilde{t}$ and $\tilde{t} \to t$ and hence

$$u(\tilde{t}) \exp\left\{-N \int_{\tilde{t}}^{t} v(s)ds\right\} \leq u(t). \tag{8}$$

Substituting $\tilde{t} = t_0$ in inequalities (7) and 8 we obtain estimation (6). The proof of the Theorem 1 is complete.
To prove the next Theorem we will need the following Definition:

Definition 5 *Let function $f(t) : [t_0, \infty) \to C$, then number defined by:*

$$\kappa[f] = \overline{\lim_{t \to \infty}} \frac{1}{t} \cdot \ln |f(t)|,$$

we will call as index from expotetial increase or characteristic exponent.

Theorem 2 *Consider linear homogeneous system:*

$$\dot{x}(t) = A(t) \cdot x(t - \tau(t)), \tag{9}$$

where $A(t)$ is continuous square matrix grade n in interval $[A, \infty)$.
Let for a measurable set $E \subset [A, B]$ the following condition will be satisfied:

$$m\gamma(E) \leq Nm(E)$$

where $\gamma(E)$ is set of this $t \in [A, B]$, for which $t - \tau(t) \in E$ and mE Lesbeque measure from set E.
If matrix $A(t)$ is bounded, that means if

$$||A(t)|| \leq C \leq \infty,$$

then every non-vanishing real or complex solution $x = x9t)$, $(A \leq t_0 \leq t \leq \infty)$ systems (9) have finite characteristic exponent.

Proof Let $x = col[x_1, \ldots, x_n]$ be non-vanishing solution linear homogeneous systems (9), and t_0 and t be points from interval (A, ∞). From equation (9) we have:

$$x(t) = x(t_0) + \int_{t_0}^{t} A(s)x(s - \tau(s))ds,$$

hence

$$||x(t) \leq ||x(t_0)|| + \int_{t_0}^{t} ||A(s)|| \cdot ||x(s - \tau(s))|| \cdot |ds|.$$

Hence on the ground from Theorem 1 we obtain

$$||x(t_0)|| \exp\left\{-N \int_{t_0}^{t} ||A(s)|| \cdot ds\right\} \leq ||x(t)|| \leq ||x(t_0)|| \cdot \exp\left\{N \int_{t_0}^{t} ||A(s)|| \cdot ds\right\}.$$

Taking this fact into consideration:

$$\kappa\left[\frac{||x(t)||}{||x(t_0)||}\right] = \kappa[x(t)],$$

we obtain

$$\kappa\left[\exp\left\{-N \int_{t_0}^{t} ||A(s)|| \cdot ds\right\}\right] \leq \kappa[x(t)] \leq \kappa\left[\exp\left\{N \int_{t_0}^{t} ||A(s)|| \cdot ds\right\}\right],$$

or

$$-\underline{A} \leq \kappa[x(t)] \leq \overline{A},$$

where

$$\underline{A} = \underline{\lim}_{t \to \infty} \frac{N}{t} \int_{t_0}^{t} ||A(s)|| \cdot ds,$$

$$\overline{A} = \overline{\lim}_{t \to \infty} \frac{N}{t} \int_{t_0}^{t} ||A(s)|| \cdot ds,$$

Since matrix $A(t)$ is bounded from assumption, that all characteristic exponent non-vanishing solution $x(t)$ linear homogeneous system (9) lie in interval $[-\underline{A}, \overline{A}]$ include in interval $[-C, C]$. This completes the proof of Theorem 2.

Theorem 3 *Fundamental solution system of homogeneous linear differential equation*

$$\dot{x}(t) = A(t) \cdot x(t - \tau(t)), \tag{10}$$

is normal if and only if is incomprassible.

Proof

Necessity:

We assume that the system is incompressible and we will show that it is normal.
Hip. Assume that it is othervise, that means that exists fundamental solution system:

$$Y(t) = \{y^1(t), \ldots, y^n(t)\},$$

such that:

$$\Sigma_Y \leq \Sigma_X, \tag{11}$$

where

$$\Sigma_Y = \sum_{s=1}^{m} \nu_s \cdot \alpha_s, \ \Sigma_X = \sum_{x=1}^{m} \eta_s \cdot \alpha_s, \ \sum_{s=1}^{m} \nu_s = \sum_{s=1}^{m} \eta_s = n,$$

but α_s that are characteristic exponents from non-vanishing solutions from differential equation (10) and ν_s, η_s, $(s = 1, \ldots, m)$, that are numbers of linear indepedent solutions with characteristic exponent α_s included correspondingly in fundamental systems $Y(t)$ and $X(t)$. We will dispose the solutions from $Y(t)$ and $X(t)$ in the sequence of increase their characteristic exponents.

Let $N'_s = \sum_{k=1}^{s} \eta_k$ $(s = 1, \ldots, m)$ be a maximal number of linear indepedent solutions which have characteristic exponent α_s. We will follow the new symbol

$$N'_s = \sum_{k=1}^{s} \nu_s, \ (s = 1, \ldots, m).$$

Theorem A [4]

The number N_s is equal to dimension of the linear subspace M_s, that means:

$$N_s = \dim M_s, \ (s = 1, \ldots, m).$$

Theorem B [4]

Let be the fundamental system $X(t)$ incompressible, η_s $(s = 1, \ldots, m)$ be tha number of solutions with characteristic exponent α_s, and N_s be maximal number of linear indepedent solutions with characteristic exponent α_s, that is hold this equality

$$\eta_1 + \eta_2 + \ldots + \eta_s = N_s, \ (s = 1, \ldots, m).$$

The proofs of these Theorems can be found in every course book for Ordinary Differential Equations.

On the ground of Theorem A and B we obtain:

$$N'_1 \leq N_1, N'_2 \leq N_2, \ldots, N'_{m-1} \leq N_{m-1}, N'_m = N_m = n,$$

where

$$\nu_s = N'_s - N'_{s-1}, \eta_s = N_s - N_{s-1}, \ (s = 1, \ldots, m; N'_0 = N_0 = 0).$$

Hence, we obtain

$$\Sigma_Y = \sum_{s=1}^{m} \nu_s \cdot \alpha_s = \sum_{s=1}^{m} (N'_s - N'_{s-1}) \cdot \alpha_s = N'_m \alpha_m - \sum_{s=1}^{m-1} N'_s(\alpha_{s+1} - \alpha_s) \geq$$
$$N_m \alpha_m - \sum_{s=1}^{m-1} N_s \cdot (\alpha_{s+1} - \alpha_s) = \sum_{s=1}^{m} \eta_s \cdot \alpha_s = \Sigma_X$$

This is contradiction with inequality (11). On the ground of this, when the system $X(t)$ is incompressible that is normal.

Sufficency

We assume that the system $X(t)$ is normal and we will show that it is incompressible.
Hip.: Assume, that it is otherwise, that means that exists linear combination

$$y = \sum_{i=1}^{p} c_i x^{(i)}(t), \ (c_p \neq 0), \tag{12}$$

such that

$$[y] \leq \max_i \kappa[x^{(i)}(t)] = \kappa[x^{(p)}(t)], \tag{13}$$

We consider system of solution:

$$Y = \{x^{(1)}(t_0, \ldots, x^{(p-1)}(t), y(t), x^{(p+1)}(t), \ldots, x^{(n)}(t)\}$$

System Y is fundamental. Let

$$\sum_{i \neq p} a_i \cdot x^{(i)}(t) + a_p \cdot y(t) \equiv 0, \ \sum_{i=1}^{n} |a_i| \not\equiv 0, \tag{14}$$

On the ground of linear indepedence of solutions $x^{(i)}(t)$ we have $a_p \neq 0$. Substituting in equation (14) the equation (12) we obtain

$$\sum_{i \leq p} (a_i + a_p c_i) x^{(i)}(t) + a_p c_p x^{(p)}(t) + \sum_{i \leq p} a_i x^{(i)}(t) \equiv 0.$$

From this we obtain, that $a_p c_p = 0$ what is impossible, because $c_p \neq 0$ and $a_p neq 0$. That follows that our system Y is fundamental.

On the ground of (13) we obtain, that $\Sigma_Y \leq \Sigma_X$ and that is contradiction with this, that system X is normal. Then every normal system is incompressible. This completes the proof of Theorem 3.

The main results this paper are Theorems 1, 2 and 3, which may found application to research os stability differential equations with delay.

REFERENCES

[1] Richard BELLMAN and Kenneth L. COOKE, Asymptotic behavior of differential-difference equations, Memories of the American Mathematical society, 35.

[2] Richard BELLMAN and Kenneth L. COOKE, Differential-Difference Equations, New York-London (1963), Acad. Press.

[3] Л. Е. Есголц, Качественные методы в математическом анализе, Москва (1995)

[4] Х. Х. КРАСОВСКИ, Некаторые задачи теорий устойчивности движения, ФизМатГиз, (1959)

[5] А. Д. МЫШКИС, Линейные дифференцалные уравнения с запаздываюшим аргументом, Гостехниздат, (1951)

7th Int. Coll. on Differential Equations, pp. 165-172
D. Bainov (Ed)

Numerical Analysis of Pseudozeros and Pseudospectra

Shingo Kado*, Champa Lal Parihar**, Vadla Mohammad Raffee*** and Miki Tsuji**** [1]

* Toshiba Co. Ltd., Shibaura 1-1-1, Minatoku, Tokyo, 105-01, Japan
** Department of Mathematics, Holkar Science College, Indore 452 001, India
*** Computer Centre, Centre For Advanced Technology, Indore 452 013, India
**** Graduate School of Mathematics, Kyushu University 33, Fukuoka, 812-81, Japan

Keywords: Balancing, companion matrix of polynomial.

Abstract

The zeros of a polynomial p are equal to the eigenvalues of the associated companion matrix A.

The $\varepsilon-$ pseudozero set $Z_\varepsilon(p)$ is the set of zeros of all polynomials $\hat{p}$ obtained by coefficientwise perturbations of p of size $\leq \varepsilon$;

$$Z_\varepsilon(p) = \{z \in \mathbf{C}: z \in Z(\hat{p}) \text{ for some } \hat{p} \text{ with } \|\hat{p} - p\| \leq \varepsilon\}. \quad (1)$$

These sets were considered earlier by Mosier.

The ε-pseudospectrum $\Lambda_\varepsilon(A)$ is another subset of $\mathbf{C}$ defined as the set of eigenvalues of matrices $\hat{A} = A + E$ with $\|E\| \leq \varepsilon$;

$$\Lambda_\varepsilon(A) = \{z \in \mathbf{C}: z \in \Lambda(A + E) \text{ for some } E \text{ with } \|E\| \leq \varepsilon\}. \quad (2)$$

In K.-C. Toh and L. N. Trefethen [8], the stability of algorithms for polynomial zerofinding in case of simple zeros, is discussed. They show that the Matlab ROOTS algorithm of balancing the companion matrix and computing its eigenvalues, is a stable algorithm for polynomial zerofinding in case of simple zeros.

In the present paper, we consider the stability of algorithms in case of polynomials with multiple zeros and obtain the results that the algorithms via companion matrices are sometimes unstable.

First we define condition numbers for multiple zeros and multiple spectra. Next numerical analysis is performed, using the algorithms of the Jenkins-Traub method, the Durand Kerner Aberth method, the QR method which is a computational path available from SSL II and the LR method which is a standard EISPACK computational path.

[1] Research was supported by Grant-in-Aid for JSPS Research Fellow no.07 2309 from the Ministry o Education, Science, Sports and Culture of Japan, 1996.

1 Introduction.

It is pointed out by K.-C. Toh and L. N. Trefethen[8] that despite the classical nature of the subject, the relationship between these two problems has been less studied and almost numerical analysts assume that it is not a good idea to find zeros via companion matrices in principle.

However K.-C. Toh and L. N. Trefethen[8] show that $Z_{\varepsilon\|p\|}(p)$ and $\Lambda_{\varepsilon\|A\|}(A)$ are quite close to one another when A is a companion matrix of p which has been "balanced" in the usual EISPACK sense, and they conclude that finding eigenvalues of companion matrices is a stable algorithm.

In the present paper, we consider the stability of algorithms in case of polynomials with multiple zeros, which has not been on close investigation in K.-C. Toh and L. N. Trefethen[8].

2 Pseudozero and pseudospectrum sets.

We introduce some notations:

Let $\mathbf{P}$ be the set of monic polynomials of degree n with complex coefficients and, for each $p \in \mathbf{P}$, we express p as

$$p(z) = \sum_{i=0}^{n} c_i z^i, c_n = 1.$$

We put $Z(p) = \{z \in \mathbf{C}; p(z) = 0\}$.

For a $n \times n$ diagonal matrix D and its diagonal vector $d = (d_0, d_1, d_2, \cdots, d_{n-1})^\top$, the norm $\|x\|_d$ is defined over $\mathbf{C}^n$ by

$$\|x\|_d = \|Dx\|_2.$$

For $z \in \mathbf{C}$, $\tilde{z}$ denotes the vector $(1, z, z^2, \cdots, z^{n-1})^\top$.

For a diagonal matrix D, we can assign a metric on $\mathbf{P}$ by

$$\|p - \tilde{p}\|_d = \left[\sum_{i=0}^{n-1} |d_i|^2 |c_i - \tilde{c}_i|^2\right]^{\frac{1}{2}},$$

which measures the perturbations in the coefficients of p relative to the weights given by d.

The above notations are the same in K.C.Toh and L.N.Trefethen [8].

For each $p \in \mathbf{P}$, we define the ε-pseudozero set of $p(z)$ by

$$Z_\varepsilon(p; d) = \{z \in \mathbf{C} : z \in Z(\hat{p}) \text{ for some } \hat{p} \in \mathbf{P} \text{ with } \|\hat{p} - p\|_d \le \varepsilon\}. \tag{3}$$

These sets quantify the conditioning of the zerofinding problem and arise in analyzing the stability of zerofinding algorithms. For a zerofinding algorithm to be stable, the computed zeros of p should lie in a region $Z_{Cu}(p; d)$ for an appropriate choice of d, where $C = O(\|p\|_d)$ and u is the machine precision.

The set is not a matter of rounding errors and numerical stability.

For example, if the coefficients c_i are perturbed by $\varepsilon_i c_i (|\varepsilon_i| \le \varepsilon)$, the value $\|\hat{p} - p\|_d$ is smaller than $\varepsilon\|p\|_d$ and hence a zeroset of the polynomial having the perturbed coefficients is included in $Z_{\varepsilon\|p\|_d}(p; d)$.

The choice $d = \|p\|_2 p^{-1} = \|p\|_2 (c_1^{-1}, \cdots, c_n^{-1})^\top$ corresponds to coefficientwise perturbations.

2.1 Condition numbers of zeros of polynomials.

Consider a single condition number sufficing to describe what may happen to each zero.

Here we extend the result of a simple zero [8] to the case of a m multiple zero and let the condition number of a m multiple zero ξ of p be defined by

$$\kappa(\xi, p; d) = \lim_{\|\hat{p}-p\|_d \to 0} \sup_{\hat{p}} \frac{|\hat{\xi} - \xi|}{(\|\hat{p} - p\|_d)^{1/m}/(\|p\|_d)^{1/m}}$$

which is equal to

$$\kappa(\xi, p; d) = \left(\|p\|_d \frac{\|\tilde{\xi}\|_{d^{-1}}}{|p^{(m)}(\xi)|}\right)^{1/m} \tag{4}$$

as a consequence of the case of a simple zero. Thus for infinitesimal ε, ε perturbations of coefficients have $\kappa(\xi, p; d)\varepsilon^{1/m}$, i.e., $O(\varepsilon^{1/m})$ influence on the zero ξ. The component of $Z_{\varepsilon\|p\|_d}(p; d)$ around ξ is the disk of radius $\varepsilon^{1/m}\kappa(\xi, p; d)$.

3 Pseudospectra of companion matrices.

For each monic polynomial $p = \sum_{i=0}^{n} c_i z^i \in \mathbf{P}$, the companion matrix associated with p is the $n \times n$ matrix

$$A_p = \begin{pmatrix} 0 & \cdots & \cdots & -c_0 \\ 1 & \ddots & & -c_1 \\ & \ddots & \ddots & \vdots \\ & & 1 & -c_{n-1} \end{pmatrix}. \tag{5}$$

The set of eigenvalues of A_p coincides with the set of zeros of p.

Let, for $p \in \mathbf{P}$, $\varepsilon-$pseudospectrum of A_p be defined by

$$\Lambda_\varepsilon(A_p; d) = \{z \in \mathbf{C} \ : \ z \in \Lambda(\hat{A}) \text{ for some } \hat{A} \in M_n \text{ with } \|\hat{A} - A_p\|_d \le \varepsilon\} \tag{6}$$

similar to the case of polynomials.

The norm $\|\cdot\|_d$ corresponds to diagonally weighted perturbations in the entries of A_p,

$$\|\hat{A} - A_p\|_d = \|D(\hat{A} - A_p)D^{-1}\|_2. \tag{7}$$

For an eigenvalue algorithm applied to the companion matrix to be stable, the computed eigenvalues of A_p should lie in a region $\Lambda_{C_u}(A_p; d)$ for some $C = O(\|A_p\|_d)$.

Balancing the matrix A is finding a diagonal matrix T such that TAT^{-1} has the 2-norm of its ith row and ith column approximately equal for each $i = 1, \cdots, n$.

In K.C.Toh and L.N.Trefethen [8], their experiments show that one of the best choices of d corresponds to balancing, i.e. $D = T$.

3.1 Condition numbers of eigenvalues of companion matrices.

Similarly to polynomials, we define a single condition number of a companion matrix A_p sufficing to describe what may happen to each eigenvalue.

Here we extend the result of a simple eigenvalue case to a multiple eigenvalue case. Let the condition number of a m multiple eigenvalue λ of A_p be defined by

$$\kappa(\lambda, A_p; d) = \lim_{\|\hat{B}-A_p\|_d \to 0} \sup_{\hat{B}} \frac{|\hat{\lambda} - \lambda|}{\|\hat{B} - A_p\|_d^{1/m}/\|A\|_d^{1/m}}$$

and, for the coefficients $b_0, b_1, \cdots, b_{n-1}$ of $\frac{(p(z)-p(\lambda))}{(z-\lambda)} = \sum_{i=0}^{n-1} b_i z^i$, we have

$$\kappa(\lambda, A_p; d) = \left(\|A_p\|_d \frac{\|\tilde{b}(\lambda)\|_d \|\tilde{\lambda}\|_{d^{-1}}}{|p^{(m)}(\lambda)|}\right)^{1/m} \quad \text{where } \tilde{b}(\lambda) = (b_0, b_1, \cdots, b_{n-1})^{\mathsf{T}}. \tag{8}$$

Thus for infinitesimal ε, ε perturbations in the entries of A_p have $\kappa(\xi, p; d)\varepsilon^{1/m}$, i.e., $O(\varepsilon^{1/m})$ influence on the eigenvalue λ. The component of $\Lambda_{\varepsilon\|A_p\|_d}(A_p; d)$ around λ is the disk of radius $\varepsilon^{1/m}\kappa(\lambda, A_p; d)$.

4 Numerical analysis in case of multiple zeros.

In this section, we consider the stability of algorithms in case of polynomials having multiple zeros.

Jenkins and Traub in [2] proposed to test the quality of zerofinding algorithms by using polynomials with random coefficients.

We take the method to test one hundred degree-15 monic polynomials having one double zero with not random coefficients but random zeros

$$r_1 \times 10^{e_1} + ir_2 \times 10^{e_2}, \tag{9}$$

where $r_i \in [-1, 1]$ and $e_i \in [-2, 2]$ $(i = 1, 2)$, in order to confirm that the polynomial has a double zero.

The coefficients are calculated in quadruple precision in order to neglect computational error.

In our experiments, we have compared following four zerofinding methods in double precision.

1.J-T method.

The J-T method is the Jenkins-Traub program CPOLY in the IMSL library [3]. The program finds the zeros once at a time in roughly increasing order of modulus and deflates the polynomial to one of lower degree.

2.DKA method.

The DKA method is a zerofinder to use initial values by the Aberth method and then converge by the Durand-Kerner method, utilizing the relation between roots and coefficients expressed by the basic symmetry equations of roots and the Newton method of the third degree convergence.

3.QR method.

The QR method is a zerofinder via a SSL II computational path for eigenvalues of a general complex matrix (DCBLNC,DCHES2,DCHESQR) [1], based on constructing the balanced companion matrix and finding its eigenvalues by the QR method.

4.LR method.

The LR method is a standard EISPACK computational path (CBAL, CORTH, COMQR) [7], which is balancing a companion matrix and to use the modified LR method.

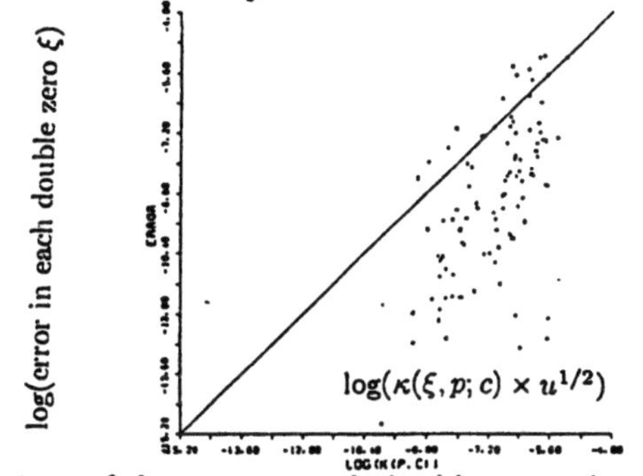

Figure 1: Comparison of the error in each double zero ξ by the J-T method with the condition number $\kappa(\xi, p; c)$, $c = \|p\|_2 p^{-1}$ of the coefficientwise perturbed zerofinding problem for 100 random polynomials.

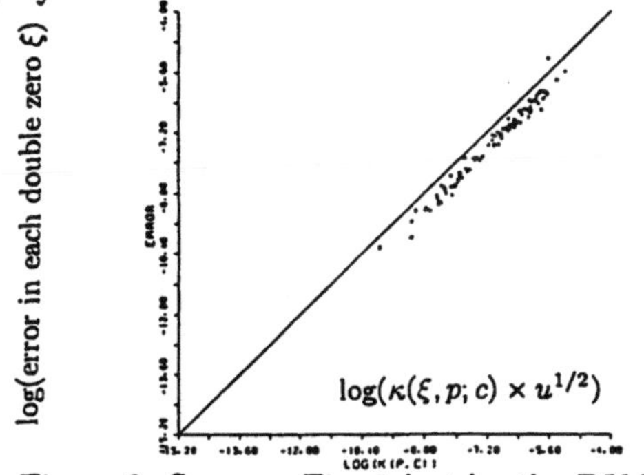

Figure 2: Same as Fig. 1 but by the DKA method.

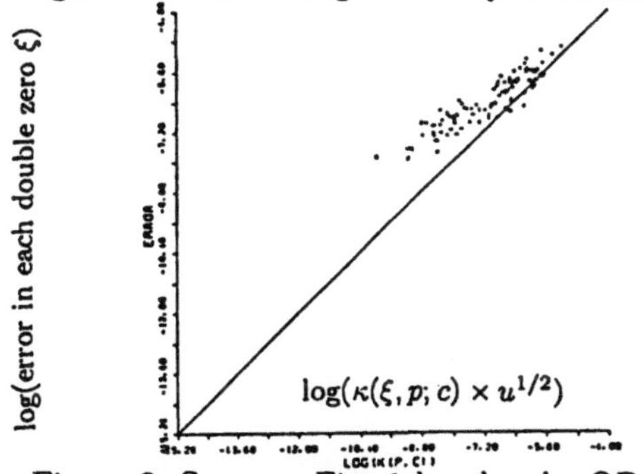

Figure 3: Same as Fig. 1 but by the QR method.

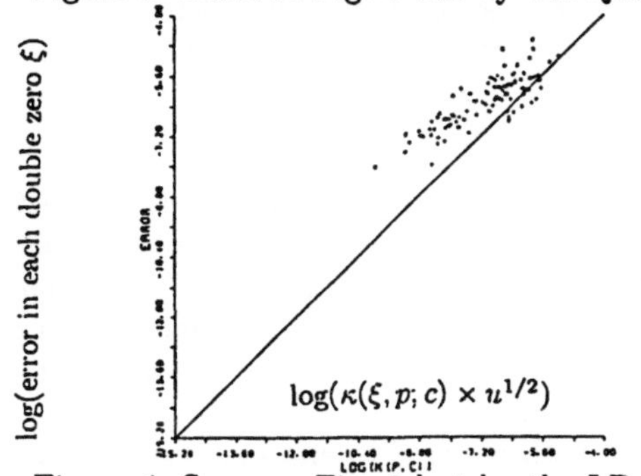

Figure 4: Same as Fig. 1 but by the LR method.

Figures 1-4 show scatter plots of the absolute error of the one double zero computed by the above four algorithms respectively versus the condition number $\kappa(\xi, p; c)$, $c = \|p\|_2 p^{-1}$ (4) of the coefficientwise perturbed zerofinding problem for the double zero of p.

Note that the scales of the errors are logarismic in Figure 1-4.

Figure 1-4 indicates that for the DKA metod, almost points are below diagonal line and match linear perturbation theory, and so the DKA method is most stable.

For the J-T method, some points appear over the diagonal line, but in many cases, the double zeros are calculated more accurately than those by the DKA method.

The QR method and the LR method have similar accuracy with each other and are somewhat unstable.

We conclude that in case of a polynomial with a double zero, the algorithms via eigenvalues of companion matrices are unstable.

5 Change of pseudozero sets and pseudospectrum sets for change from simple zeros to multiple zeros.

In this section, we study pseudozero sets and pseudospectrum sets in more detail how to change when the coefficients of the polynomials are changed from the case of simple zeros to the case of multiple zeros, using following examples.

Example 1 Comparison of the sets for the degree-20 monic polynomial p_1 with zeros equally spaced in the interval [-2.1,1.9], with the sets for the degree-20 monic polynomial $\hat{p}_1$ with zeros of p_1 excluding smallest zero and including the next smallest zero as a double zero.

Example 2 Comparison of the sets for the degree-20 monic polynomial $p_2 = z^{20} + z^{19} + z^{18} + \cdots + 1$ (Zeros are the 21st roots of unity except 1), with the sets for the degree-20 monic polynomial $\hat{p}_2$ with zeros of p_2 excluding $\exp(i2\pi/21)$ and including $\exp(i4\pi/21)$ as a double zero.

Example 3 Comparison of the sets for the degree-20 Chebyshev polynomial p_3, (Zeros are $\cos((2r+1)\pi/40)$, $r = 0, \cdots, 19$), with the sets for the degree-20 monic polynomial $\hat{p}_3$ with zeros of p_3 excluding zeros corresponding to $r = 10$, 11 and including a middle point of the 2 ones as a double zero.

Example 4 Comparison of the sets for the degree-20 monic polynomial p_4 with zeros equally spaced on a sine curve, viz, (Zeros are $(2\pi/19(k+0.5)) + i\sin(2\pi/19(k+0.5)))$, $k = -10, -9, \cdots, 9$) with the sets for the degree-20 monic polynomial $\hat{p}_4$ with zeros of p_4 excluding zeros corresponding to $k = -8, -7$ and including a middle point of the 2 ones as a double zero.

In Figure 5-8, each figure presents a graphical comparison of two pseudozero sets $Z_{\epsilon\|p\|_c}(p; c)$ $(c = \|p\|_2 p^{-1})$ of the polynomial and the corresponding pseudospectrum sets $\Lambda_{\epsilon\|A_p\|_t}(A_p; t)$ of the associated balanced companion matrix, for the above four examples. Solid lines correspond to the polynomials $p_i (i = 1, \cdots, 4)$ and dashed ones correspond to $\hat{p}_i (i = 1, \cdots, 4)$. In some cases, the dashed curves are not visible because they match the solid ones.

These figures show that the both pseudozero and pseudospectrum sets do not have a great change as zeros of polynomials are double.

6 Conclusion.

In K.-C. Toh and L. N. Trefethen [8] show finding zeros via eigenvalues of companion matrices with simple eigenvalues, is a stable algorithm.

In the present paper, we consider the stability of algorithms in case of polynomials with multiple zeros. The method used for our numerical analysis are the Jenkins-Traub method, the Durand Kerner Aberth method, the QR method and the LR method.

We compare the above four algorithms in case of double zeros and we conclude that by the J-T method, in many cases, the double zeros are calculated much accurately and it seems not to match linear perturbation theory. On the other hand, the errors in the zeros computed by the DKA metod typically match linear perturbation theory, and the DKA method is most stable, while some points by the J-T method appear over the diagonal line in Figure 1. The algorithms via eigenvalues of companion matrices are somewhat unstable in case of polynomials with double zeros.

Moreover, we show that by some exmples, both pseudozero and pseudospectrum sets do not have a great change when the coefficients of the polynomials are changed from the case of only simple zeros to the case of multiple zeros.

ACKNOWLEDGEMENT

The authors would like to express their hearty gratitude to Professor Kajiwara for his valuable suggestions.

References

[1] Fujitsu, SSL II Users guide (Science subroutine library) 1987, pp. 607.

[2] M. A. Jenkins and J. F. Traub, Principles for testing polynomial zerofinding programs, Tech. Report. Department of Computer Science, Carnegie-Mellon University, 1974.

[3] M. A. Jenkins and J. F. Traub, Numer. Math. **14** (1970), pp. 252-263.

[4] J. Kajiwara, Proceedings of the Second International Colloquium on Numerical Analysis, VSP Utrecht the Netherlands (1993), pp. 113-122.

[5] M. Mori, Fortran 77 Numerical calculation programming, The Iwanami computer science series, (1986) p.398.

[6] R. G. Mosier, Math. Comp. **47** (1986), pp. 265-273.

[7] B. T. Smith, Matrix Eigensystem Routines - EISPACK Guide, Lect. Notes Comput. Sci., vol. 6, 2nd ed., Springer, Berlin Heidelberg New York, 1976.

[8] K. - C. Toh and L. N. Trefethen, Numer. Math. **68** (1994), pp. 403-425.

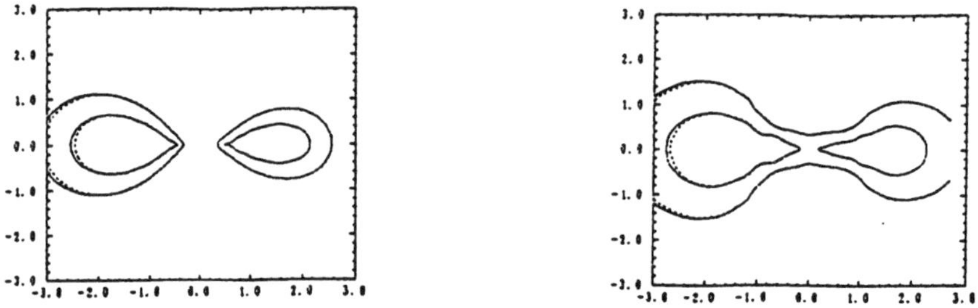

Pseudozero sets($\varepsilon = 10^{-2}, 10^{-3}$) Pseudospectrum sets($\varepsilon = 10^{-3}, 10^{-4}$)

Figure 5: For Example 1, pseudozero sets with respect to coefficientwise perturbations and pseudospectrum sets of the balanced companion matrix. Solid lines correspond to p_1 and dashed ones correspond to $\hat{p}_1$.

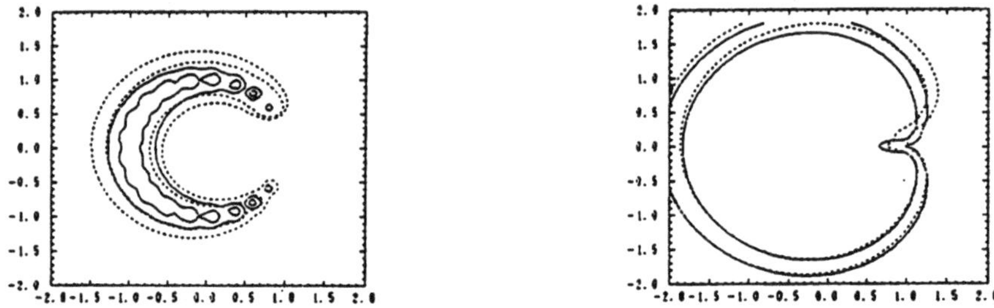

Pseudozero sets($\varepsilon = 10^{-1}, 10^{-1.2}$) Pseudospectrum sets($\varepsilon = 10^{-2}, 10^{-3}$)

Figure 6: Same as Fig. 5 but for Example 2.

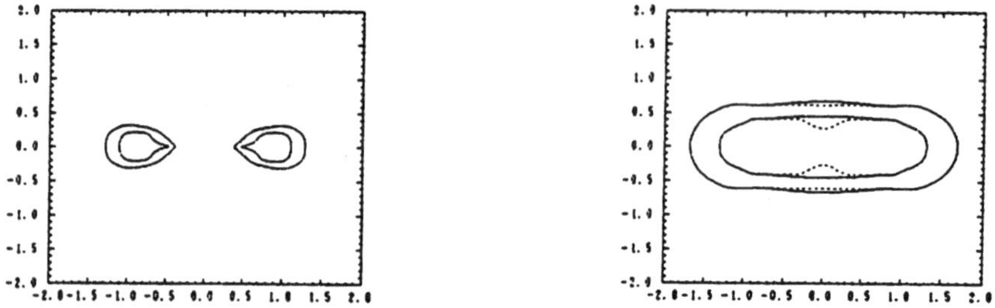

Pseudozero sets($\varepsilon = 10^{-3}, 10^{-4}$) Pseudospectrum sets($\varepsilon = 10^{-3}, 10^{-4}$)

Figure 7: Same as Fig. 5 but for Example 3.

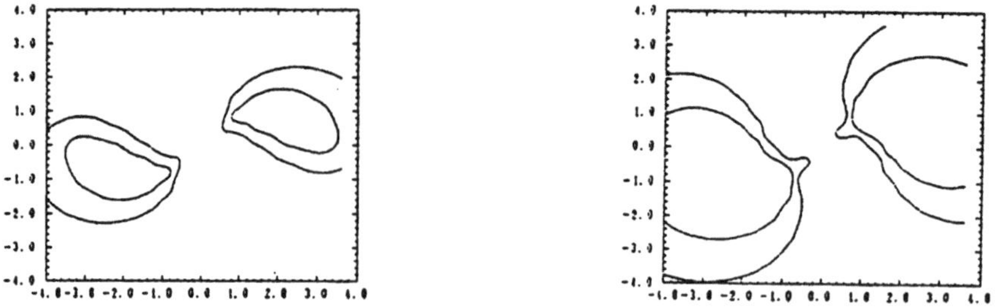

Pseudozero sets($\varepsilon = 10^{-2}, 10^{-3}$) Pseudospectrum sets($\varepsilon = 5.0 \times 10^{-5}, 1.0 \times 10^{-5}$)

Figure 8: Same as Fig. 5 but for Example 4.

7th Int. Coll. on Differential Equations, pp. 173-180
D. Bainov (Ed)

Characterization of the Holomorphy of Infinite Dimensional Domains by Oka's Principle*[1]

Graduate School of Mathematics, Kyushu University 33, Fukuoka, 812-81, Japan

Joji KAJIWARA, Xia Dong LI, Seiko OHGAI and Dong Guo ZHOU

Abstract

We characterize the condition, for infinite dimensional domains with continuous boundary to be domains of holomorphy, by validity of Oka's principle.

1 Introduction.

In one complex variable, a domain D in a complex plane $\mathbf{C}$ is a domain of holomorpy, that is, an existence domain of a holomorphic function on D by a theorem of Weierstrass. Cartan[3] stated without proof that a Cousin-I domain on $\mathbf{C}^2$ is a domain of holomorphy and Benhke-Stein[2] proved it. Oka[36] proved that a domain of holomorphy in $\mathbf{C}^n$ is Cousin-I. So, the solvability of Cousin-I problem characterizes the holomorphy of a domain in $\mathbf{C}^2$.

Concerning Cousin-II problem, Thullen[45] stated that the punctured polydisk in $\mathbf{C}^2$ is not a domain of holomorphy but is a Cousin-II domain. Oka[37] proved that a Cousin-II problem in a domain of holomorphy in $\mathbf{C}^n$ is holomorphically solvable if and only if it is solvable topologically. Sèrre called this "Oka's principle" and Thom evaluated it as the most beautiful principle in Analysis.

Let L be a complex Lie group. We denote by $\mathcal{A}_L$ the sheaf of germes of holomorphic mappings in L. We denote by ${\mathcal{E}^0}_L$ or ${\mathcal{E}^\infty}_L$ the sheaf of germes of continuous or C^∞ mappings in L. When L is the additive group $\mathbf{C}$ or the multiplicative group $\mathbf{C}-\{0\}$, the sheaf $\mathcal{A}_L$ is denoted simply by $\mathcal{O}$ or $\mathcal{O}^*$. By Kajiwara-Kazama[29], a domain D in a two dimensional Stein manifold is Stein if there exists a positive dimensional complex Lie group L such that $\mathrm{H}^1(D,\mathcal{A}_L) = 0$. By Kajiwara-Nishihara[30], a domain D in a two dimensional Stein manifold is Stein if and only if there exists a positive dimensional complex Lie group L such that Oka's principle holds for complex analytic fiber bundle with D as base space and with L as structure group, by proving that the injectivity of the canonical mapping $\mathrm{H}^1(D,\mathcal{A}_L) \to \mathrm{H}^1(D,\mathcal{E}^\infty_L)$ implies Steinness of D.

In case that the dimension is larger than 2, by Cartan[4], the domain $D := \mathbf{C}^3-(0,0,0)$ is not a domain of holomorphy but satisfies $\mathrm{H}^1(D,\mathcal{O}) = 0$. So, the vanishing of the

[1]Supported by Grant-in-Aid for Scientific Research (C) no. 08304012 from the Ministry of Education, Science, Sports and Culture of Japan, 1996.
Key Words: Cohomology Vanishing, Oka's Principle, Domain of Holomorphy.
AMS Subject Classification: 32D05, 32F15, 32L10, 32L20.

cohomology of degree 1 is not sufficient to let D be a domain of holomorphy. For an analytic coherent sheaf $\mathcal{F}$ over a Stein space and for a positive integer p, the theorem of Oka[38]-Cartan[5]-Sèrre[42] gave the cohomology vanishing $\mathrm{H}^p(D,\mathcal{F}) = 0$. Moreover, by Sèrre[43], a domain D in $\mathbf{C}^n$ is a domain of holomorphy if $\mathrm{H}^p(D,\mathcal{O}) = 0$ for $p = 1, 2, \cdots, n-1$. In other words, vanishing of a suitable cohomology with degree from 1 to $n-1$ characterizes the Seinness of n dimensional domains.

Kajiwara[27] proved that a domain D with co real 1 dimensional continuous boundary in a Stein mainfold S is Stein, if and only if there exists a positive dimensional complex Lie group L such that, for any analytic polycylinder P in S, Oka's principle holds for analytic fiber bundle with $D \bigcap P$ as base space and with L as structure group in the sense that the canonical mapping

$$\iota^\infty : H^1(D \bigcap P, \mathcal{A}_L) \to H^1(D \bigcap P, \mathcal{E}_L^\infty) \tag{1}$$

is quasi-injective, that is, the preimage of the trivial element is also trivial where $\mathcal{A}_L$ and $\mathcal{E}_L^\infty$ are, respectively, sheaves of holomorphic and C^∞ mappings into L.

The aim of the present paper is an infinite dimensionalization of the results of Kajiwara[2

In infinite dimensional case, Dineen[10] proved $\mathrm{H}^1(\Omega,\mathcal{O}) = 0$ for the structure sheaf $\mathcal{O}$ over a pseudoconvex domain Ω in a C-linear space E equipped with finite open topology. So, pseudoconvexity implies the cohomology vanishing similary to the finite dimensional case.

Kajiwara-Shon[31], however, proved, for a pseudoconvex domain Ω in the C-linear locally convex space E equipped with the finite open topology, for an analytic subset A of Ω and for any positive integer $p \leq \mathrm{codim}A-2$ the cohomology vanishing $\mathrm{H}^p(\Omega - A, \mathcal{O}) = 0$. The complement A of the open set $\Omega - A$ with respect to the open set Ω has no interior point and in case that codim $= \infty$ the cohomology vanishing of all positive degree does not implies the holomorphy of the domain. Moreover, Ohgai[35] proved, for any positive integers p and q, for a pseudoconvex domain Ω in the C-linear locally space E equipped with the finite open topology, for a q-convex C^∞ function φ on Ω and for a negative number c, the cohomology vanishing $\mathrm{H}^p(\{x \in \Omega; \varphi(x) > c\}, \mathcal{O}) = 0$. In this case, the complement $\{x \in \Omega; \varphi(x) \leq c\}$ of the open set $\{x \in \Omega; \varphi(x) < c\}$ with respect to the open set Ω may have interior points $\in \{x \in \Omega; \varphi(x) > c\}$.

So, infinite dimensional case, vanishing of cohomology of all positive degree of the structure sheaf $\mathcal{O}$ of the domain Ω does not assure the pseudoconvexity of the domain Ω.

The aim of the present paper is to infinite dimensionalize the results of Kajiwara[27] loco citato as the Main Theorem in the next section.

2 Main Theorem

Let E be a C-linear Hausdorff space and Λ be the set of finite dimensional C−linear subspaces of E. A complex valued function f on an open subset D of E is said to be *Gâtaux holomorphic* if, for any $L \in \Lambda$, the restriction of f to $D \cap F$ is holomorphic on the open subset $D \cap F$ of the finite dimensional complex space F. A complex valued Gâtaux holomorphic function f on an open subset D of E is said to be *holomorphic* if f is continuous on D. The sheaf $\mathcal{O}$ of germs of holomorphic function over E is called the *structure sheaf* of E. An open subset D of E is said to be *pseudoconvex* if, for any $F \in \Lambda$, the intersection $D \bigcap F$ is a pseudoconvex open set of the C-linear space F of finite dimension. A topology on E is said to be *finite open* if the family of open sets consists of subsets O of E such that for any $F \in \Lambda$, the intersection $O \cap F$ are open in $F \in \Lambda$.

As a **C**-linear space E with the finite open topology, we consider exclusively in the present paper the space $\Sigma\mathbf{C}$ which is a locally convex space euipped with the finite open topology and which is defined as follows:

For any integers m and n with $m < n$, we regard the complex m-space $\mathbf{C}^m$ as a subspace

$$\mathbf{C}^m = \{z = (z_1, z_2, \cdots, z_n) \in \mathbf{C}^n; z_{m+1} = z_{m+2} = \cdots = z_n = 0\} \tag{2}$$

of the superspace $\mathbf{C}^n$. Let $\pi_{m,n} : \mathbf{C}^m \to \mathbf{C}^n$ be the canonical injection. We put

$$\Sigma\mathbf{C} = \bigcup_{n \geq 1} \mathbf{C}^n, \tag{3}$$

denote, for any $n \geq 1$, by $\pi_n : \mathbf{C}^n \to \Sigma\mathbf{C}$ the canonical injection and write $\mathbf{C}^n \subset \Sigma\mathbf{C}$ too. We induce the strongest topology in $\Sigma\mathbf{C}$ so as each injection π_n is continuous. The **C**-linear space $\Sigma\mathbf{C}$ is the locally convex space with the finite open topology. Its structural sheaf is denoted by $\mathcal{O}$. It is the sheaf of germs of holomorphic functions over $\Sigma\mathbf{C}$.

Since the space $\Sigma\mathbf{C}$ is paracompact, cohomology groups over $\Sigma\mathbf{C}$ with coefficients in any sheaf of abelian groups coincides with those of Čech cohomology groups. Since we have a partition of unity by C^∞-functions of an open set in $\Sigma\mathbf{C}$, the Čech cohomology coincides with that of $\overline{\partial}$-cohomology. By Dineen[10], Colombeau-Perrot[8]-[9], Kajiwara-Shon[31] and Soraggi[44], cohomology groups of a pseudoconvex covering of a domain D in the locally convex space $\Sigma\mathbf{C}$ equipped with the finite open topology coincides with the cohomology groups of the domain D with coefficients in the structure sheaf $\mathcal{O}$.

Let Ω be a domain in E. We assume in the following theorem that it has a continuous boundary, that is, for any boundary point $z^{(0)} = (z_1^{(0)}, z_2^{(0)}, \cdots, z_n^{(0)}, \cdots)$ of Ω, there exists a neighborhood U of $z^{(0)}$ and a real valued continuous function $b(z_1, z_2, \cdots, z_{n-1}, \mathrm{Re} z_n, z_{n+1} \cdots$ in U such that

$$\Omega \cap U = \{z \in U; \mathrm{Im} z_n < b(z_1, z_2, \cdots, z_{n-1}, \mathrm{Re} z_n, z_{n+1} \cdots)\}. \tag{4}$$

"Quasi injective" in the Main Theorem means that the preimage of the neutral element is also the neutral element in the cohomology set.

Under these notations we have the following theorem:

MAIN THEOREM *Let Ω be a domain in E with continuous boundary.*

Then the following conditions are equivalent:

(1) *Ω is a domain of holomorphy over E.*

(2) *Ω is a pseudoconvex domain of holomorphy over E.*

(3) *For any complex Lie group L and an anlytic polydisc P in E, the canonical mapping*

$$\mathrm{H}^1(\Omega \cap P, \mathcal{A}_L) \to \mathrm{H}^1(\Omega \cap P), \mathcal{E}_L^0) \tag{5}$$

is injective.

(4) *For any complex Lie group L and an anlytic polydisc P in E, the canonical mapping*

$$\mathrm{H}^1(\Omega \cap P, \mathcal{A}_L) \to \mathrm{H}^1(\Omega \cap P, \mathcal{E}_L^\infty) \tag{6}$$

is injective.

(5) *For any complex Lie group L and an anlytic polydisc P in E, the canonical mapping*

$$\mathrm{H}^1(\Omega \cap P, \mathcal{A}_L) \to \mathrm{H}^1(\Omega \cap P, \mathcal{E}_L^0) \tag{7}$$

is quasi injective.

(6) *There exists a complex Lie group L such that, for any anlytic polydisc P in E, the canonical mapping*

$$\mathrm{H}^1(\Omega \cap P, \mathcal{A}_L) \rightarrow \mathrm{H}^1(\Omega \cap P, \mathcal{E}_L^\infty) \tag{8}$$

is quasi injective.

(7) *There exists a complex Lie group L such that, for any anlytic polydisc P in E, the canonical mapping*

$$\mathrm{H}^1(\Omega \cap P, \mathcal{A}_L) \rightarrow \mathrm{H}^1(\Omega \cap P, \mathcal{E}_L^0) \tag{9}$$

is quasi injective.

(8) *There exists a complex Lie group L such that, for any anlytic polydisc P in E, the canonical mapping*

$$\mathrm{H}^1(\Omega \cap P, \mathcal{A}_L) \rightarrow \mathrm{H}^1(\Omega \cap P, \mathcal{E}_L^\infty) \tag{10}$$

is quasi injective.

3 Scenario of the proof of the Main Theorem

(1) → (2) holds by the definitions. (2) → (1) holds by the affirmative solution of the Levi problem by Dineen[10]. (2) → (3) holds by Kajiwara[25]. (3) → (4), (5) → (6) and (7) → (8) follow from the canonical injection

$$\iota_\infty^0 : \mathcal{E}_L^\infty \rightarrow \mathcal{E}_L^0. \tag{11}$$

(3) → (5) and (4) → (6) holds by the definitions.

The implication (8) → (1) is the most important part of the theorem and is expained below.

In this section, we denote by S exclusively a finite dimensional Stein manifold S and by $\{D_n; n \geq 1\}$ a sequence of monotonically increasing sequence of domains D_n in S. Let $D := \cup_{n=1}^\infty D_n$ be the limit.

Let L be a complex Lie group. For each $m \leq n$, the canonical mapping $\tau_m : D_m \rightarrow D_n$ induces the canonical mapping $\pi_m^n : \mathrm{H}^1(D_m, \mathcal{A}_L) \rightarrow \mathrm{H}^1(D_n, \mathcal{A}_L)$ satisfying $\pi_n^\ell = \pi_m^\ell \pi_n^m$ for any $\ell \geq m \geq n \geq 1$. Hence the sequence $\{\mathrm{H}^1(D_n, \mathcal{A}_L); \pi_n^m\}$ of cohomology sets is an inverse system in the sense of Eilenberg-Steenrod[12]. Let $\lim_{n\rightarrow\infty} \mathrm{H}^1(D_n, \mathcal{A}_L)$ be its limit in the sense of [12]. Kajiwara[19], [22] and [26] proved the following theorem:

Theorem 1 *The canonical mapping*

$$\pi : H^1(D, \mathcal{A}_L) \rightarrow \lim_{n\rightarrow\infty} H^1(D_n, \mathcal{A}_L) \tag{12}$$

is injective.

Let

$$\tau_n^* : \mathrm{H}^1(D, \mathcal{A}_L) \rightarrow \mathrm{H}^1(D_n, \mathcal{A}_L) \tag{13}$$

be the canonical mapping. As a corollary to the above theorem, he obtained:

Theorem 2 *Let α be an element of $H^1(D, \mathcal{A}_L)$. If $\tau_n^*(\alpha)$ is a neutral element in $H^1(D_n, \mathcal{A}_L)$ for any $n \geq 1$. Then α is itsself a neutral element.*

Now we can state the scenario to prove the implication (8) $\rightarrow$ (1). By Noverraz[34] many definitions of pseudoconvexity are equivalent for our space E and it suffices to prove the pseudoconvexity of the cut $\Omega \cap \mathbf{C}^\nu$ of the domain Ω by any finite dimensional subspace $\mathbf{C}^\nu$ of E. Taking note that E is the inductive limit of $\mathbf{C}^\nu$, we prove it by the method in Kajiwara[27].

Proof of the Main Theorem Let $\mathcal{L}$ be the Lie algebra of L and $\exp : \mathcal{L} \rightarrow L$ be the exponential mapping. Let X be any element of $\mathcal{L}$.

Let $z^{(0)}$ be a boundary point of Ω. There exists a positive number ν such that $z^{(0)} \in \mathbf{C}^\nu$. We may assume that the point $z^{(0)}$ is the origin. Since Ω has a continuous boundary, there exists an open neighborhood U of $z^{(0)}$ and a real valued continuous function b on U satisfying (4) in which we may assume $n = \nu$ by the changement of the order of variables.

There exists a decreasing sequence $\{a_\mu; \mu = 1, 2, 3, \cdots\}$ of positive numbers $\{a_\mu\}$ such that

$$\{(z_1, z_2, \cdots, z_\mu) \in \mathbf{C}^\mu; |z_j| \leq 2a_j (j = 1, 2, \cdots, \mu), \mathrm{Im} z_\nu \leq b(z)\} \subset U \cap \mathbf{C}^\mu \qquad (14)$$

for $\mu = 1, 2, 3, \cdots$.

At first, we state a linguistical exposition of the proof. We take an open covering $\mathcal{U}_\nu := \{U_{\alpha\nu}, \alpha \in A\}$ of $\Omega \cap U \cap \mathbf{C}^\nu$ and consider a cocycle $\{f_{\alpha\beta}\}$ of it with coefficient $\mathcal{O}$. We shrink the neighborhood U to its relatively compact subset $P(0)$. We carry out translation

$$trans_t(z_1, z_2, \cdots, z_\nu, z_{\nu+1}, \cdots) = (z_1, z_2, \cdots, z_{\nu-1}, z_\nu + it, z_{\nu+1}, \cdots) \qquad (15)$$

for sufficiently small t which tends to 0. Then the part satisfying $\mathrm{Im} z_\nu = b(z)$ of the boundary of $\Omega \cap U \cap \mathbf{C}^\nu$ near the boundary point $z^{(0)}$ enter into the interior of $\Omega \cap U$ but other boundary poins protrude from $\Omega \cap U$. Hence we let $trans_t(\Omega \cap U)$ shrink by taking intersection $\Omega \cap U \cap P(t)$. Then $trans_t(\Omega \cap U \cap P(t) \cap \mathbf{C}^\nu)$ lies in $\Omega \cap U \cap \mathbf{C}^\nu$ compactly and the open set $\Omega \cap U \cap P(t) \cap \mathbf{C}^\nu$ increase as $t \searrow +0$. Then any holomorphic function in an open subset of the open set $trans_t(\Omega \cap U \cap P(t) \cap \mathbf{C}^\nu)$ can be extended to an ambiant open set by the pull back of the projection $\Sigma \rightarrow \mathbf{C}^\nu$. Then, the restriction of the cocycle $\{\exp(X f_{\alpha\beta})\}$ to $\Omega \cap U \cap P(t) \cap \mathbf{C}^\nu$ is a coboundary. Thus, the cocycle is a coboundary in the limit open set $\{\Omega \cap P(0) \cap \mathbf{C}^\nu\}$, obtained by the translation and shrinking. Then, the limit open set is Stein in $\mathbf{C}^\nu$. Next we state mathematical expositon below.

Let t be a positive number. For a positive number t with $t < a_\nu$, we put

$$K := \{(z_1, z_2, \cdots, z_\nu) \in \mathbf{C}^\nu; |z_j| \leq a_j - t (j = 1, 2, \cdots, \nu), \mathrm{Im} z_\nu \leq b - t\} \qquad (16)$$

which is a compact subset in $\mathbf{C}^\nu$ contained in the open set $\Omega \cap \mathbf{C}^\nu$. Then, by induction with respect to $\mu > \nu$, we can prove the existence of a decreasing sequence $\{\delta_\mu(t); \mu = \nu + 1, \nu + 2, \nu + 3, \cdots\}$ of positive numbers δ_μ with $\delta_\mu(t) < a_\mu$ such that

$$\{(z_1, z_2, \cdots, z_\mu) \in \mathbf{C}^\mu; |z_j| \leq a_j - t (j = 1, 2, \cdots, \nu),$$
$$\mathrm{Im} z_\nu \leq b(z) - t, |z_j| \leq \delta_j(t) (j > \nu)\} \subset \Omega \cap U \cap \mathbf{C}^\mu. \qquad (17)$$

for any $\mu > \nu$.

Let

$$\mathcal{U}_\nu = \{U_{\alpha\nu}; \alpha \in A\}, \qquad (18)$$

be an open covering of the intersection $\Omega \cap P(0) \cap \mathbf{C}^\nu$ and $\{f_{\alpha\beta}; \alpha, \beta \in A\}$ be any element of $\mathrm{Z}^1(\mathcal{U}_\nu, \mathcal{O})$. We consider the open covering

$$\mathcal{V}(t) := \{V_\alpha(t); \alpha \in A\}$$

$$V_\alpha(t) := \{(z_1, z_2, \cdots, z_\nu, \cdots) \in P(t); (z_1, z_2, \cdots, z_\nu) \in U_{\alpha\nu}, |z_j| < \delta_j(t)(j \geq \nu)\} \quad (19)$$

of the intersection $\Omega \cap P(t)$ of Ω and the polydisc

$$P(t) := \{((z_1, z_2, z_3, \cdots) \in E; |z_j| < a_j(j = 1, 2, \cdots, \nu), |z_j| \leq \delta_j(t)(j > \nu)\}. \quad (20)$$

Let $\rho_\nu : \Sigma \to \mathbf{C}^\nu$ be the canonical projection. Then $\{\exp(f_{\alpha\beta} o trans_t o \rho X); \alpha, \beta \in A\} \in \mathrm{Z}^1(\mathcal{V}(t), \mathcal{A}_L)$ is well defined and is an element of $\mathrm{B}^1(\mathcal{V}(t), \mathcal{E}^\infty_L)$ in the category of C^∞ by the partition of unity in the paracompact space $\Omega \cap P(t)$. Then $\{\exp(f_{\alpha\beta} \circ \rho_\nu X); \alpha, \beta \in A\} \in \mathrm{Z}^1(\mathcal{U}(t), \mathcal{A}_L)$ is a coboundary in the analytic category of the covering $\mathcal{V}_t$ of the intersection $\Omega \cap P(t)$ of the domain and the polydisc $P(t)$ in E by the assumption (8) for the open set Ω. Then by Theorem 2, the cocycle $\{\exp(f_{\alpha\beta} X); \alpha, \beta \in A\} \in \mathrm{Z}^1(\mathcal{U}_\nu, \mathcal{A}_L)$ is a coboundary. By Proposition 6 of Kajiwara[20], $\Omega \cap \mathbf{C}^\nu$ is pseudoconvex. q .e .d

References

[1] A. Andreotti et H. Grauert, *Théorèmes de finitede pour la cohomology des espaces complexs,* Bull. Soc. Math. France **90**(1962), 193-259.

[2] H. Behnke und K. Stein, *Analytische Funktionen mehrerer Veränderlichen zu vorgebenen Null-und Pol-stellenflächen,* Jber. D. M. V. **47**(1937), 177-192.

[3] H. Cartan, *Le problème de Poincaré et de Cousin pour les fonctions de plusieurs variables complexes,* C. R. Paris **198**(1934), 1284-1287.

[4] H. Cartan, *Sur les première problème de Cousin,* ibid. **207**(1938), 558-560.

[5] H. Cartan, *Idéaux de fonctions analytiques de n complexes variables,* Bull. Soc. Math. France. **78**(1950), 28-64.

[6] H. Cartan, Séminaire E.N.S., École Normare Supérieur, Paris, 1951/1952.

[7] G. Coeuré *L'équation $\bar{\partial}u = F$ en dimension infinite,* université Lille Publ. Int. **131**(1968), 6-9.

[8] J. Colombeau and B. Perri, *The $\bar{\partial}$-equation in D.F.N. spaces,* J. Math. Anal. Appl. **78**-2(1980), 466-87.

[9] J. Colombeau and B. Perri, *L'équation $\bar{\partial}$ dans les ouverts pseudo-convexes des espaces D.F.N.,* Bull. Soc. Math. France. **110**(1982), 15-26.

[10] S. Dineen, *Sheaves of holomorphic functions on infinite dimensional vector spaces,* Math. Ann. **202**(1973), 337-345.

[11] S. Dineen, *Cousin's first problem on certain locally convex topological spaces,* Acad. Brasil. Cienc. **48**-1(1976), 229-236.

[12] S. Eilenberg and N. Steenrod, Foundations of algebraic topology, Princeton Univ. Press, 1952.

[13] L. Gross, *Potential theory on Hilbert spaces,* J. Funct. Anal. **1** (1967), 123-181.

[14] L. Gruman, *The Levi problem in certain infinite dimensional vector spaces,* Illinois J. Math. **18**(1974), 20-26.

[15] C. J. Henrich, *The $\overline{\partial}$ -equation with polynomial growth on a Hilbert space,* Duke Math. J. **40**-2(1973), 279-306.

[16] S. Hitotumatu, Theorem of analytic functions of several complex variables(in Japanese), Baihuukan, 1960.

[17] J. Kajiwara, *Note on a Cousin-II domain over* $\mathbf{C}^2$, Kōdai Math. Sem. Rep. **17**-1(1965), 44-47.

[18] J. Kajiwara, *Some characterization of Stein manifold through the notion of locally regular boundary points,* Kōdai Math. Sem. Rep. **16**-4(1964), 191-198.

[19] J. Kajiwara, *On the limit of a monotonous sequence of Cousin's domains,* J. Math. Soc. Japan **17**-1(1965), 36-46.

[20] J. Kajiwara, *Some extensions of Cartan-Behnke-Stein's theorem,* Pub. RIMS Kyoto Univ. **2**-1(1966), 133-156.

[21] J. Kajiwara, *Cousin domains over complex projective 2-space,* Mathematica Balkanica **3**(1972), 184-187.

[22] J. Kajiwara, *Domain with many vanishing cohomology sets,* Kōdai Math. Sem. Rep. **26**-2(1975), 258-266.

[23] J. Kajiwara, *La réciproque du théorèm d'annalation et de finitude de cohomologie dans l'espace produit d'une famille dénombrable de sphère de Riemann,* Bull. Soc. Math. France. **103**(1975), 129-139.

[24] J. Kajiwara, *Le principle d'Oka pour certaines espace de dimension infini,* C.R.Paris **16**(1975), 1055-1056.

[25] J. Kajiwara, *Le principle d'Oka pour certaines espaces vectoriets munis de la topologie f-ouvert,* Mem. Fac. Sci. Kyushu. Univ. **26**(1973), 361-37.

[26] J. Kajiwara, *Characterization of Stein subdomains of a complex projective space through Oka's principle,* Mathematica Balkanica **8**(1978), 131-137.

[27] J. Kajiwara, *Equivalence of Steinness and validity of Oka's principle for domains with continous boundaries of a Stein manifold,* Fac. Sci. Kyushu Univ. **33**(1979), 83-93.

[28] J. Kajiwara and H. Kazama, *Oka's principle for relative cohomology sets,* Mem. Fac. Sci. Kyushu Univ. **23**-1(1969), 33-70.

[29] J. Kajiwara and H. Kazama, *Two dimensional complex manifold with vanishing cohomology set,* Math. Ann. **204**(1973),1-12.

[30] J. Kajiwara and M. Nishihara, *Characterisierung der Steinschen Teilgebieten durch Okasches Prinzip in Zwei-dimensionater Mannigfasltigkeit,* Mem. Fac. Sci.Kyushu. Univ. **33**(1979),71-76.

[31] J. Kajiwara and K. H. Shon, *Continuation and vanishing theorem for cohomology of infinite dimensional space*, Pusan Kycheckongnam Math. J. **9**-1(1993) , 65-73.

[32] J. Kajiwara and K. Watanabe, *Domain with a vanishing cohomology set in two dimensional complex projective space*, Mem. Fac. Sci. Kyushu Univ. **32**(1978), 73-80.

[33] P. Mazet, *Un théorèm d'hyperbolicite pour l'opérateur $\overline{\partial}$ sur les espaces de Banach*, C. R. Paris **292**(1981), 31-33.

[34] Ph. Noverraz, Pseudo-Convexité, Convexité Polynomiale et Domaine d'Holomorphie en Dimension Infinie, North-Holland Math. Studies **3**(1973).

[35] S. Ohgai, *Cohomology Vanishing and Validity of Oka's Principle for Infinite Dimensional Domains*, in this Proceedings.

[36] K. Oka, *Sur les fonctions analytiques plusieurs variables: II Domaine d'holomorphie*, J. Sci. Hiroshima Univ. **17**(1937), 115-130.

[37] K. Oka, *Sur les fonctions analytiques plusieurs variables: III Deuxième problème de Cousin*, J. Sci. Hiroshima Univ. **9**(1938), 7-19.

[38] K. Oka, *Sur les fonctions analytiques plusieurs variables: VII Sur quelque notion arithmétiques*, Bull. Soc. Math. France, **78**(1950), 1-27.

[39] K. Oka, *Sur les fonctions analytiques plusieurs variables: IX Domaines finis sans point critique intérieur*, Jap. J. Math. **23**(1953), 1-27.

[40] P. Raboin, *Résolution de l'équation$\overline{\partial} f = F$ sur un espace de Hilbert*, Bull. Soc. Math. France, **107**(1979), 225-240.

[41] G. Scheja, *Riemannsche Hebbarkeitssätze für Cohomologieklassen*, Math. Ann. **144**(1961), 345-360.

[42] J. P. Sèerre, *Dans Séminaire H.Cartan,3*, Ecole Normare Supérieur, Paris, 1951-1952.

[43] J. P. Sèrre, *Quelque problème globaux relatifs aux variété de Stein*, Coll. fonct. plus. var. Bruxelles (1953), 57-68.

[44] R. L. Soraggi, *The $\overline{\partial}$-problem for a (0,2)-form in a D.F.N.space*, J. Func. Anal. **98**(1991), 380-402.

[45] P. Thullen, *Sur deuxième problème de Cousin*, C. R. Paris **200**(1935), 720-721.

7th Int. Coll. on Differential Equations, pp. 181-188
D. Bainov (Ed)

Numerical Analysis on $\bar{\partial}$–Problem and Extension of Holomorphic Functions from Lower Dimensional Subvarieties.

Joji Kajiwara* [1], Champa Lal Parihar**, Vadla Mohammad Raffee*** and Miki Tsuji**** [2]

* Graduate School of Mathematics, Kyushu University 33, Fukuoka, 812-81, Japan
** Department of Mathematics, Holkar Science College, Indore 452 001, India
*** Computer Centre, Centre for Advanced Technology, Indore 452 013, India
**** Graduate School of Mathematics, Kyushu University 33, Fukuoka, 812-81, Japan

Keywords: $\bar{\partial}$–equation, holomorphic extension.

Abstract

Our Indian-Japanese joint work concerns Numerical Analysis on $\overline{\partial}$-problem and aims to solve the inhomogeneous $\overline{\partial}$-equation numerically firstly, using the integral formula by Hörmander[1] and the finite element method. Secondly we apply it to the extension problem.

1 Introduction.

Let Ω be a domain in $\mathbf{C}^n$ and $f = \sum_{j=1}^n f_j d\overline{z_j}$ be a (0,1)-form on Ω such that $\bar{\partial} f = 0$. The equation $\overline{\partial} u = f$ is overdetermined when $n > 1$.

D. C. Spencer[7] formulated the $\overline{\partial}$-Neumann problem and demonstrated why the geometry of $\partial\Omega$ plays an essential role in solving the $\overline{\partial}$-problem. J. J. Kohn[3] discovered how to exploit strong pseudoconvexity to solve the equation and to prove regularity of the solution. L. Hörmander[1] discovered how to suppress the boundary phenomena by working with weighted Sobolev spaces.

In the present paper, using the integral formula by Hörmander[1] and the finite element method, we investigate numerically the above $\overline{\partial}$-problem.

We treat a function with compact support and give an appropriate algorithm to calculate u, using Fortran 77. Next, we apply it to the extension problem. We take a geometric view of the holomorphically extended function.

[1]Supported by Grant-in-Aid for Scientific Research (C) no. 08304012 from the Ministry of Education, Science, Sports and Culture of Japan, 1996.

[2]Supported by Grant-in-Aid for JSPS Research Fellow no.07 2309 from the Ministry of Education, Science, Sports and Culture of Japan, 1996.

2 Numerical Analysis on $\overline{\partial}$-problem.

Theorem 1 *(Hörmander[1]) Let f be a $C^\infty(\mathbf{C})$ (0,1)-form with a compact support. Then the function given by*

$$u(z) := \frac{1}{2\pi i}\int\int \frac{f(\tau)}{\tau - z} d\tau d\overline{\tau}$$

is a solution to the $\overline{\partial}$-equation

$$\overline{\partial} u = f.$$

We use the following example for comparing the algorithms.

Example 1 *For positive numbers c and d with $c < d$, we put*

$$\varphi_1(x) := \exp(-\frac{1}{x^2}) \quad (x > 0), \quad \varphi_2(x; c, d) := \varphi_1(x^2 - c)\varphi_1(d - x^2),$$

and, for $x > 0$, we put

$$\varphi_3(x; c, d) := \frac{\int_c^x \varphi_2(t; c, d)}{\int_c^d \varphi_2(t; c, d)dt} \qquad (x > \sqrt{c}).$$

Then the function $\varphi_3(x; c, d)$ of class C^∞ satisfies $\varphi_3(x; c, d) = 0$ on $x < \sqrt{c}$, $0 < \varphi_3(x; c, d) < 1$ on $\sqrt{c} < x < \sqrt{d}$ and $\varphi_3(x; c, d) = 1$ on $x > \sqrt{d}$.

For $z \in \mathbf{C}$, let

$$\psi(z) := 1 - \varphi_3(|z|; 16, 25)$$

and

$$h(z) =_1 F_1(\frac{1}{2}; \frac{1}{3}, z) = \sum_{k=0}^{\infty} \frac{(\frac{1}{2})_k}{(\frac{1}{3})_k} \frac{z^k}{k!}.$$

We obtain a (0,1)-form f from

$$f := h(z)\overline{\partial}\psi.$$

We have $u(z) = 0$ when $|z| \geq 5$ by the theorem of identity, since there holds $\frac{\partial u(z)}{\partial \bar{z}} = 0$ when $|z| > 5$ and $u(z) \to 0$ as $|z| \to \infty$. Based on the above example, we compare the following four algorithms in double precision computation.

2-1. DE formula for integral.

Our first method is a double integration to τ_1 and τ_2 where $\tau = \tau_1 + i\tau_2$, using the DE formula which is effective when the function is analytic. We take 50 points $z = -5 - 5i + k(0.1 + 0.1i)$ for $k = 1, \cdots, 50$. Then, it takes about 46 hours to calculate the values of $u(z)$ at these 50 points of z. When $4 < |z| < 5$, the value z becomes a singular point of the single integral $\int \frac{f(\tau_1 + \tau_2 i)}{\tau_1 + \tau_2 i - z} d\tau_1$, where $\tau_2 = \text{Im} z$, and many sampling points of τ near the singular point z are necessary. Let M be the maximum of the values $|f|$ on a compact neighborhood of z. For a positive number ϵ sufficiently small, there holds

$$|\frac{1}{2\pi i}\int_{|\tau - z| < \epsilon} \frac{f(\tau)}{\tau - z} d\tau d\overline{\tau}| \leq M\frac{1}{2\pi}\int_0^{\epsilon}\int_0^{2\pi} (1/r) r dr d\theta = M\epsilon \to 0 \quad (\epsilon \to 0).$$

So, we need not take sampling points near z.

2-2. DE formula for modified integral.

For the second double integration, taking the above fact into account, in order to avoid calculating the values near the singular points for $4 < |z| < 5$ and because the DE formula is very efficient and is rather insensitive to singularities that may occur at the end points, the integral intervals for τ_1 are set on $[-5, \mathrm{Re}z - \epsilon]$ and $[\mathrm{Re}z + \epsilon, 5]$ where $\tau_2 \in [\mathrm{Im}z - \epsilon, \mathrm{Im}z + \epsilon]$ and $\epsilon = 10^{-8}$. By the second integration, we get the similar results by the first integration and it takes about 20 minitues for the same points of z.

2-3. Newton-Cotes formula for integral.

As the third integration, after transformation of $\tau = z + t\exp(i\theta)$, we adapt the Newton-Cotes formula of 8 degree to θ and t by the adative automatic integration.

For the adative automatic integration, it is better to use the formula such that sampling points are taken on the same length mesh and hence we choose the Newton-Cotes formula not the DE formula. (See [6])

By the third method, in 18 minutes, we obtain similar results for the same points as those in the first and second integration.

In more detail, for 7 singular points of z $(4 \leq |z| \leq 5)$, it takes 15 minutes by the DE formula for the modified integral vesus 8 minutes by the Newton-Cotes formula and for 28 no singular points of z $(|z| \leq 4)$, it takes 5 minutes by the DE formula vesus 9 minutes by the Newton-Cotes formula.

The results indicates that for singular points of z, the Newton-Cotes formula is best algorithm, on the other hand, for no singular points of z, the DE formula is best one.

2-4. Finite element method.

Finally, by the fourth method, we apply the finite element method to the equation $\Delta u = \frac{1}{4}\partial\bar{\partial}u = \frac{1}{4}\partial f$ and the condition $u = 0$ on the boundary. We point out in [2] that it is impossible to apply the finite element method directly to $\bar{\partial}u = f$.

In Figure 2, we show the graph of $|u|$ by the finite element method with $m = 100$ division as Figure 1, compared with the graph by the Newton-Cotes formula.

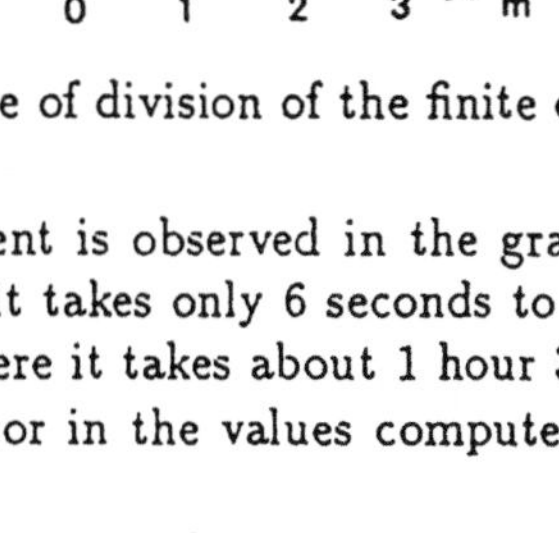

Figure 1: Figure of division of the finite element method.

A reasonably close agreement is observed in the graph of the finite element metod with $m = 100$ division, where it takes only 6 seconds to calculate u and in the graph of the Newton-Cotes formula, where it takes about 1 hour 30 minutes.

However, the maximum error in the values computed by the finite element method are as much as about 0.38.

If the division m should be increased to get more precise values, we would need large memory capacity in hardware. It is a limit in the finite element method.

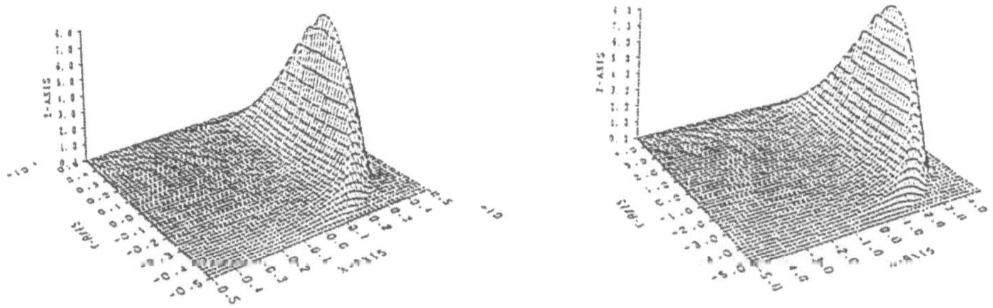

Finite element method with $m = 100$ division Newton-Cotes formula

Figure 2: The graph of $|u|$ by the finite element method with $m = 100$ division as Figure 1 and by the Newton-Cotes formula.

Hence if it is not necessary to get precise values, for example, if you only draw a graph, and your computer has sufficient memory, it is better to use the finite element method and you may obtain a quick result.

In Table 1, we show the comparison the above 4 algorithms by time.

Table 1: Comparison by time

	50 points	Data of graph
DE formular for integral	46 hours	
DE formula for modified integral	20 minuts	
Newton-Cotes formula	18 minuts	1 hour 30 minutes
Finite element method		6 seconds

3 Numerical approach to extension problem.

We put

$$U := \{z = (z_1, z_2) \in \mathbf{C}^2; |z_1| < 1, |z_2| < 1\}$$

and

$$L := \{(z_1, z_2) \in \mathbf{C}^2; z_1 + z_2 + \sqrt{2} = 0\}.$$

Then

$$U \cap L = \{(z_1, z_2) \in \mathbf{C}^2; |z_1| < 1, |z_1 + \sqrt{2}| < 1\}.$$

Let $\pi : \mathbf{C}^2 \to \mathbf{C}^2$ be the Euclidean projection

$$\pi(z_1, z_2) = (z_1, -z_1 - \sqrt{2})$$

and let

$$B = \{z \in U; \pi(z) \notin U \cap L\}.$$

Now, let $h(z_1, z_2)$ be a holomorphic function on $U \cap L$ and ψ be a function of class C^∞ on U so that $\psi = 1$ in a relative neighborhood of $U \cap L$ and $\psi = 0$ on B. $\psi = 0$ on $B = \{z \in U; \pi(z) \notin U \cap L\}$ and, although $h(\pi(z))$ is not defined on B, $\bar{\partial}\psi = 0$ on a neighborhood of $U \cap L = \{(z_1, z_2) \in U; z_1 + z_2 + \sqrt{2} = 0\}$. Hence the funciton $\frac{h(\pi(z))\bar{\partial}\psi}{z_1+z_2+\sqrt{2}}$ is a well defined function of class C^∞ and satisfies

$$\bar{\partial}\left(\frac{h(\pi(z))\bar{\partial}\psi}{z_1 + z_2 + \sqrt{2}}\right) = 0.$$

By the $\overline{\partial}$-theory, we have a solution u in U to the $\overline{\partial}$-equation

$$\bar{\partial}u = \frac{h(\pi(z))\bar{\partial}\psi}{z_1 + z_2 + \sqrt{2}}.$$

Let $\{\sigma_m\}_{m=1}^\infty$ be an increasing sequence of positive numbers such that $\sigma_1 < \sigma_2 < \cdots \sigma_m < \cdots < 1$ and that $\sigma_m \to 1$ as $m \to \infty$. For any positive integer m we put

$$U_m = \{z \in \mathbf{C}^2 : |z_1| < \sigma_m, |z_2| < \sigma_m\}.$$

For any $m \geq 1$, there exits a solution u_m on each U_m to $\overline{\partial}u_m = \frac{h(\pi(z))\bar{\partial}\psi}{z_1+z_2+\sqrt{2}}$.

Let ϵ be a positive number. Assume that we can revise $u_1, u_2, \cdots, u_m$ such that we have

$$|u_{k+1} - u_k| < \epsilon/2^k$$

for $k = 1, 2, \cdots, m$. For the $(m+1)$-th step, the holomorphic function $v_m = u_{m+1} - u_m$ on U_m is approximated by a polynomial P_m on U_{m-1} so that we have

$$|v_m - P_m| < \epsilon/2^m$$

in U_{m-1}. Replacing u_{m+1} by $u_{m+1} - P_m$, we revise the function u_{m+1} so as to satisfy

$$|u_{m+1} - u_m| < \epsilon/2^m.$$

The sequence $\{u_m\}_{m=1}^\infty$, thus revised inductively, converges uniformly to a holomorphic function $u := \lim_{m\to\infty} u_m$ in any compact subset of U. Especially in U_1, there holds

$$|u - u_1| = |\sum_{m=1}^\infty (u_{m+1} - u_m)| \leq \sum_{m=1}^\infty |u_{m+1} - u_m| \leq \sum_{m=1}^\infty \epsilon/2^m = \epsilon.$$

We define a function H of class $\mathbf{C}^\infty$ by putting

$$H(z_1, z_2) := h(\pi(z))\psi - (z_1 + z_2 + \sqrt{2})u(z_1, z_2).$$

Since we have $\overline{\partial}H = 0$, H is holomorphic on the ambiant bidisk U and its restriction satisfies $H|_{U\cap L} = h$. H is the desired holomorphic extension of h. Therefore we can numerically get the approximate value of u by calculating u_1.

The function u_1 is expressed in an exact form by the following theorems.

Theorem 2 *(Hörmander [1]) A function G on $U = \{z \in \mathbf{C} : |z| \leq r\}$ of class C^1 is represented by*

$$G(z) = \frac{1}{2\pi i}\int_U \frac{\partial G(\zeta)}{\partial\bar{\zeta}}\frac{d\zeta \wedge d\bar{\zeta}}{\zeta - z} + \frac{1}{2\pi i}\int_{\partial U} G(\zeta)\frac{d\zeta}{\zeta - z}.$$

Theorem 3 *(Hörmander [1]) Let g be a function of class C^∞ on $U = \{z \in \mathbf{C} : |z| < r\}$ so as to be continuous on $\overline{U}$. Then the function defined by*

$$u(z) = \frac{1}{2\pi i}\int_U g(\zeta)\frac{d\zeta \wedge d\overline{\zeta}}{\zeta - z}$$

satisfies

$$\overline{\partial} u = g d\overline{z}.$$

Theorem 4 *Let $g = g_1 d\overline{z_1} + g_2 d\overline{z_2}$ be a (0,1)-form of class C^∞ on $U = \{z \in \mathbf{C}^2 : |z_1| < R, |z_2| < R\}$ such that $\overline{\partial} g = 0$. Then, for $0 < r < R$, the function*

$$\begin{aligned} u(z) &= \frac{1}{2\pi i}\int_{|\zeta|<r} g_1(\zeta, z_2)\frac{d\zeta \wedge d\overline{\zeta}}{\zeta - z_1} \\ &+ \frac{1}{2\pi i}\int_{|\zeta'|<r}\left\{\frac{1}{2\pi i}\int_{|\zeta|=r} g_2(\zeta, \zeta')\frac{d\zeta}{\zeta - z_1}\right\}\frac{d\zeta' \wedge d\overline{\zeta'}}{\zeta' - z_2} \end{aligned} \tag{1}$$

of class C^∞ on $\{z \in \mathbf{C}^2 : |z_1| < r, |z_2| < r\}$ satisfies the $\overline{\partial}$-equation

$$\overline{\partial} u = g.$$

(Proof) Since the second term of the right hand side in (1) is holomorphic with respect to z_1, we have $\frac{\partial}{d\overline{z_1}} u = g_1$ by Theorem 3.

Next we show $\frac{\partial}{d\overline{z_2}} u = g_2$. Since we have $\frac{\partial g_1}{\partial \overline{z_2}} = \frac{\partial g_2}{\partial \overline{z_1}}$, by Theorem 3 and Theorem 2, we have

$$\begin{aligned} \frac{\partial u}{\partial \overline{z_2}} &= \frac{\partial}{\partial \overline{z_2}}\left(\frac{1}{2\pi i}\int_{|\zeta|<r} g_1(\zeta, z_2)\frac{d\zeta \wedge d\overline{\zeta}}{\zeta - z_1}\right) \\ &+ \frac{\partial}{\partial \overline{z_2}}\left(\frac{1}{2\pi i}\int_{|\zeta'|<r}\left\{\frac{1}{2\pi i}\int_{|\zeta|=r} g_2(\zeta, \zeta')\frac{d\zeta}{\zeta - z_1}\right\}\frac{d\zeta' \wedge d\overline{\zeta'}}{\zeta' - z_2}\right) \\ &= \frac{1}{2\pi i}\int_{|\zeta|<r}\frac{\partial}{\partial \overline{z_2}} g_1(\zeta, z_2)\frac{d\zeta \wedge d\overline{\zeta}}{\zeta - z_1} + \frac{1}{2\pi i}\int_{|\zeta|=r} g_2(\zeta, z_2)\frac{d\zeta}{\zeta - z_1} \\ &= \frac{1}{2\pi i}\int_{|\zeta|<r}\frac{\partial}{\partial \overline{\zeta}} g_2(\zeta, z_2)\frac{d\zeta \wedge d\overline{\zeta}}{\zeta - z_1} + \frac{1}{2\pi i}\int_{|\zeta|=r} g_2(\zeta, z_2)\frac{d\zeta}{\zeta - z_1} \\ &= g_2(z_1, z_2). \end{aligned}$$

q.e.d.

Next we construct the function ψ in Example 1.

The set

$$U \cap L = \{(z_1, z_2) \in \mathbf{C}^2; |z_1| < 1, |z_1 + \sqrt{2}| < 1\}$$

is mapped by

$$v(z_1, z_2) := \sqrt{2} z_1 + 1$$

to the circle-arced di-angular

$$|v - 1| < \sqrt{2}, \qquad |v + 1| < \sqrt{2}$$

which is mapped to the unit disk $|w| < 1$ by the mapping

$$w = F(v) := \frac{2v}{1 - v^2}.$$

Let $G(z) = F(v(\pi(z)))$ for $z \in U$. For positive numbers c and d with $c < d$, we put

$$\phi_1(x) := \exp(-\frac{1}{x}) \qquad (x > 0), \quad \phi_2(x; c, d) := \phi_1(x - c)\phi_1(d - x),$$

and, for $x > 0$, we put

$$\phi_3(x; c, d) := \frac{\int_c^x \phi_2(t; c, d)}{\int_c^d \phi_2(t; c, d)dt} \qquad (x > c).$$

Let $t = |z_1 + z_2 + \sqrt{2}|^2$ and $w = G(z)$. We define one example for ψ by, for $\frac{\pi}{2} > \epsilon > 0$,

$$\rho = \phi_3\Big(\epsilon - \arctan(\frac{t}{1 - |w|^2}), -\frac{\pi}{2} + \epsilon, 0\Big),$$

where $0 \le \arctan x \le 2\pi$.

The function ρ satisfies the condition that $\rho = 1$ in a neighborhood of $U \cap L$ in U, since $\rho = 1$ on

$$\{\arctan(\frac{t}{1 - |w|^2}) \le \epsilon\} \supset \{\arctan(\frac{t}{1 - |w|^2}) = 0\}$$

$$= \{t = 0\} = \{z_1 + z_2 + \sqrt{2} = 0\} = U \cap L$$

and that $\rho = 0$ on B, since $\rho = 0$ on

$$\{\epsilon - \arctan(\frac{t}{1 - |w|^2}) \le -\frac{\pi}{2} + \epsilon\} = \{\arctan(\frac{t}{1 - |w|^2}) \ge \frac{\pi}{2}\}$$

$$= \{|w|^2 \ge 1\} = \{\pi(z) \notin U \cap L\} = B.$$

Example 2 *We define h by*

$$h(z) :=_1 F_1(\frac{1}{2}; \frac{1}{3}, z) = \sum_{k=0}^{\infty} \frac{(\frac{1}{2})_k}{(\frac{1}{3})_k} \frac{z^k}{k!}.$$

By calculating the approximate value of u on $U_1 = \{(z_1, z_2) \in \mathbf{C}^2 : |z_1| < 0.98, |z_2| < 0.98\}$ *we draw the graphs of* $|u(z_1 + z_2 + \sqrt{2})|$, $|h\rho|$ *and* $|h\rho - u(z_1 + z_2 + \sqrt{2})|$ *on* $\{z_1 \in \mathbf{C} : |z_1| \le 0.97\}$ *at* $z_2 = 0$.

Example 3 *We define h by*

$$h(w) = \sum_{k=0}^{\infty} w^{k!},$$

which is unbounded at each boundary point of $U \cap L$.

As the same as Example 2, we draw the graphs of $|u(z_1 + z_2 + \sqrt{2})|$, $|h\rho|$ *and* $|h\rho - u(z_1 + z_2 + \sqrt{2})|$ *on* $\{z_1 \in \mathbf{C} : |z_1| \le 0.97\}$ *at* $z_2 = 0$.

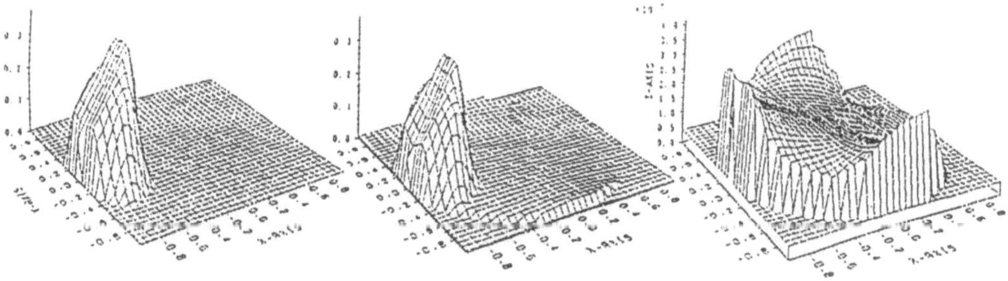

$|u(z_1+z_2+\sqrt{2})|$ $|h\rho|$ $|h\rho-u(z_1+z_2+\sqrt{2})|$,

Figure 3: The graphs in Example 2.

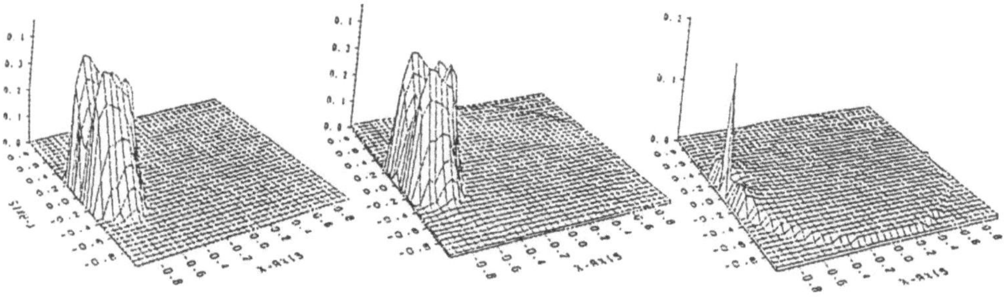

$|u(z_1+z_2+\sqrt{2})|$ $|h\rho|$ $|h\rho-u(z_1+z_2+\sqrt{2})|$

Figure 4: The graphs in Example 3.

References

[1] L. Hörmander, An introduction to Complex Analysis, Northholand (1966), pp. 208.

[2] J. Kajiwara, V. M. Raffee and M. Tsuji, *Numerical Methods for $\bar{\partial}$-problem,* Proceedings of the Beijing Workshop on Finite or Infinite Dimensional Complex Analysis, (to appear) (1996).

[3] J. J. Kohn, *Harmonic integrals on pseudoconvex manifolds,* I and II, Ann. of Math. 78(1963),112-148 and 79(1964), 450-472.

[4] Y. Komatsu and J. Kajiwara, Shoukai-Kansuron enshu, Kyoritsu, (1983) pp. 299.

[5] H.R. Kutt, *The Numerical evaluation of principal value integrals by finite-part integration,* Numer. Math. **24**, (1975) pp.205-210.

[6] M. Mori, *Fortran 77 Numerical calculation programming,* The Iwanami computer science series, (1986) pp.398.

[7] D. C. Spencer, *Overdetermined systems of linear partial differential equations,* Bull. Amer. Math. Soc. 75(1969), 179-239.

7th Int. Coll. on Differential Equations, pp. 189-196
D. Bainov (Ed)

On triple systems and Yang–Baxter equations

Noriaki Kamiya
Department of Mathematics
Shimane University
Matsue,Shimane 690, Japan
and
Susumu Okubo
Department of Physics and Astronomy
University of Rochester
Rochester,NY 14627, USA

Keywords: triple systems, Yang–Baxter equations
AMS Subject Classification: Primary 17A40, 17B60, 82B23

1. Introduction

The concept of a triple system (= vector space equipped with a triple product $< xyx >$) plays an important role in the construction of simple Lie algebras or Lie superalgebras. For this point of view, there exists a construction of the standard embedding Lie algebras or Lie superalgebras associated with the triple systems (cf. [K1, ,K2, K3]).

On the other hand, it is interesting for us to study how the triple system behaves in mathematical physics. For this point of view, there are many approaches to integral models in mathematical physics, and to methods of via a study of classical Yang–Baxter equations (for example, [FT, Ji, O1–O4]).

The purpose of this article is to characterize the (classical) Yang–Baxter equations from the triple system, in particular, called a Freudenthal–Kantor triple system or a Lie triple system.

In this paper, we are concerned with vector space, algebras and triple systems which are finite dimensional over a field Φ of characteristic $\neq 2$.

2. (ε, δ)–Freudenthal–Kantor Triple System

We will sketch some relevant facts about the (ε, δ)–Freudenthal–Kantor triple system [K1]–[K5] in this section so as to facilitate our subsequent discussions.

For $\varepsilon = \pm 1$ and $\delta = \pm 1$, triple system $U(\varepsilon, \delta)$ with the triple product $< -, -, - >$ is called a (ε, δ)–Freudenthal–Kantor triple system if

$$[L(a, b), L(c, d)] = L(< abc >, d) + \varepsilon L(c, < bad >) \tag{2.1}$$

$$K(<abc>,d)+K(c,<abd>)+\delta K(a,K(c,d)b)=0 \tag{2.2}$$

where $L(a,b)c=<abc>$ and $K(a,b)c=<acb>-\delta<bca>$.

In particular, it is said to be a generalized Freudenthal–Kantor triple system if the condition (2.1) alone is satisfied.

In specialy case, we shall consider the case of $K(a,b)=0$ for all $a,b\in U(\varepsilon,\delta)$. We call it a (ε,δ)–Jordan triple system. That is, the definition of (ε,δ)–Jordan triple system is as follows:

$$<abc>-\delta<cba> \tag{2.3}$$

$$<ab<cde>>=<<abc>de>+\varepsilon<c<bad>e>+<cd<abe>>. \tag{2.4}$$

Example 2.1. Let W be a vector space equipped with a symmetric bilinear form $B(x,y)$ for $x,y\in W$. Then

$$<xyz>=B(x,y)z+B(y,z)x-B(z,x)y$$

defines on W a $(-1,1)$–Jordan triple system.

Example 2.2. Let W be a vector space equipped with an anti–symmetric bilinear form $B(x,y)$ for $x,y\in W$. Then

$$<xyz>=B(x,y)z+B(y,z)x-B(z,x)y$$

defines on W a $(1,-1)$–Jordan triple system.

Example 2.3. Let W be a set of matrix $Mat(m,n;\Phi)$. Then $<xyz>=x\,{}^tyz+\delta z\,{}^tyx$ define on W a $(-1,\delta)$–Jordan triple system, where $\delta=\pm1$, ${}^tx$ denote the transpose matrix of $x,x,y,z\in W$.

We recall a Lie triple system (the case of $\delta=1$) or anti–Lie triple system (the case of $\delta=-1$) that satisfy the following identities:

(L1) $[xyz]=-\delta[yxz]$

(L2) $[xyz]+[yzx]+[zxy]=0$

(L3) $[L(x,y),L(z,w)]=L(([xyz],w)+L(z,[xyw])$.

Proposition 2.1.([K2]) *Let $U(\varepsilon,\delta)$–Freudenthal–Kantor triple system. If P is a linear transformation of $U(\varepsilon,\delta)$ such that $P<xyz>=<Px,Py,Pz>$ and $P^2=-\varepsilon\delta Id$, $(U(\varepsilon,\delta),[-,-,-])$ is a Lie triple system (the case of $\delta=1$) or an anti–Lie triple system (the case of $\delta=-1$) with respect to the triple product*

$$[xyz]:=<xPyz>-\delta<yPxz>+\delta<xPzy>-<yPzx>. \tag{2.5}$$

In particular, if $\varepsilon\delta=-1, K(x,y)=0$ and $P=Id$ (that is, $U(\varepsilon,\delta)$ is a (ε,δ) Jordan triple system), then the triple product becomes

$$[xyz]=<xyz>-\delta<yxz>\ (:=S(x,y)z).$$

Corollary. Let $U(\varepsilon,\delta)$ be a (ε,δ)–Freudenthal–Kantor triple system. Then the vector space $U(\varepsilon,\delta)\oplus U(\varepsilon,\delta)$ becomes a Lie triple system (the case of $\delta=1$) or anti–Lie triple system (the case of $\delta=-1$) with respect to the triple product defined by

$$[\begin{pmatrix} a \\ b \end{pmatrix}\begin{pmatrix} c \\ d \end{pmatrix}\begin{pmatrix} e \\ f \end{pmatrix}]=\begin{pmatrix} L(a,d)-\delta L(c,b) & \delta K(a,c) \\ -\varepsilon K(b,d) & \varepsilon(L(d,a)-\delta L(b,c)) \end{pmatrix}\begin{pmatrix} e \\ f \end{pmatrix} \tag{2.6}$$

The Lie triple system obtained from the above theorem is said to be the *standard embedding* associated with the triple system. (cf.[K1, K5]).

We can construct a Lie algebra or Lie superalgebra from any Lie triple system $(\delta=1)$ or anti–Lie triple system $(\delta=-1)$ as follows.

Proposition 2.2. *Let T be a δ–Lie triple system and $L(T,T)$ be the linear space spanned by all $L(x,y),x,y\in T$. Then a vector space $L:=L(T,T)\oplus T$ becomes a Lie algebra $(\delta=1)$ or Lie superalgebra $(\delta=-1)$ with respect to the product*

$$[L(x,y)+z,L(a,b)+c]:=[L(x,y),L(a,b)]+L(z,c)+L(x,y)c-L(a,b)z. \tag{2.7}$$

The Lie algebra (resp. Lie superalgebra) obtained from the above proposition is said to be the *standard embedding* associated with the Lie triple system (resp. anti–Lie triple system).

From Example 2.1 (the case of *dim* $W=2n$), we obtain a Lie triple system

$$[xyz]:=<xyz>-<yxz>=2(B(y,z)x-B(x,z)y).$$

Thus we have the Lie algebra characterized by

$$L=so(2n+1),L(T,T)=so(2n),T=so(2n+1)/so(2n).$$

From Example 2.2 (the case of *dim* $W=2n$), we obtian an anti–Lie triple system

$$[xyz]:=<xyz>+<yxz>=2(B(y,z)x+B(x,z)y).$$

Thus we have the Lie superalgebra characterized by

$$L=osp(1,2n),L(T,T)=sp(2n),T=osp(1,2n)/sp(2n).$$

From Example 2.3 and Corollary of Proposition 2.1 (the case of $\delta=1$), we obtain a Lie triple system T as follows;

$$T=W\oplus W=\{\begin{pmatrix} x_1 \\ x_2 \end{pmatrix}|x_i\in Mat(n,n;\Phi)\ {}^tx_i=x_i\},$$

with respect to the product

$$[\begin{pmatrix} x_1 \\ x_2 \end{pmatrix} \begin{pmatrix} y_1 \\ y_2 \end{pmatrix} \begin{pmatrix} z_1 \\ y_2 \end{pmatrix}] = \begin{pmatrix} L(x_1, y_2) - L(y_1, x_2) & 0 \\ 0 & -L(y_2, x_1) + L(x_2, y_1) \end{pmatrix} \begin{pmatrix} z_1 \\ z_2 \end{pmatrix}$$

where $L(x, y)z = x\ {}^t yz + z\ {}^t yx, x, y, z \in W$.

Thus from the standard embedding, we have the Lie algebra of type C_n characterized by

$$L = sp(2n; S) = \{A \in Mat(2n, 2n) | S^{-1t}AS = -A\},$$
$$L(T, T) := \{\begin{pmatrix} B & 0 \\ 0 & -{}^tB \end{pmatrix} | B \in Mat(n, n; \Phi)\}$$

where $S = \begin{pmatrix} 0 & Id \\ -Id & 0 \end{pmatrix}$.

From Example 2.3 and Corollary of Proposition 2.1 (the case of $\delta = -1$), we obtain an anti-Lie triple system as follows;

$$T = W \oplus W = \{\begin{pmatrix} x_1 \\ x_2 \end{pmatrix} | x_i \in Mat(n, m; \Phi)\}, m \neq n,$$

with respect to the product

$$[\begin{pmatrix} x_1 \\ x_2 \end{pmatrix} \begin{pmatrix} y_1 \\ y_2 \end{pmatrix} \begin{pmatrix} z_1 \\ z_2 \end{pmatrix}] = \begin{pmatrix} L(x_1, y_2) + L(y_1, x_2) & 0 \\ 0 & -L(y_2, x_1) - L(x_2, y_1) \end{pmatrix} \begin{pmatrix} z_1 \\ z_2 \end{pmatrix}$$

where $L(x, y)z = x\ {}^t yz - z\ {}^t yx, x, y, z \in W$.

Thus from the standard embedding, we have the Lie superalgebra of type $A(n-1, m-1)$ characterized by

$$L = spl(n, m), \quad L(T, T) = A_{n-1} + A_{m-1} + Id,$$
$$dim\ L = (n+m)^2 - 1, \quad dim\ L(T, T) = n^2 + m^2 - 1,$$

where A_n is the classical simple Lie algebra of type A_n.

3. Classical Yang–Baxter in Triple Product Form

In order to make this paper as self-contained as possible, we will first sketch the basic formulation of our methods(cf.[O2],[O3]). Let V be a N–dimensional vector space over a field Φ, and we write

$$N = \dim_\Phi V. \tag{3.1}$$

Let $R(\theta)$ for a paramenter θ be a linear mapping in $V \otimes V$, i.e.

$$R(\theta) : V \otimes V \to V \otimes V. \tag{3.2}$$

For given basis vector $e_1, e_2, \ldots, e_N$ of V, we can then expand

$$R(\theta)e_a \otimes e_b = \sum_{c,d=1}^{N} R_{ab}^{dc}(\theta) e_c \otimes e_d \tag{3.3}$$

for scattering matrix elements $R_{ab}^{cd}(\theta)$. Note the order of indices c and d above.

Linear mapping $R_{ij}(\theta)(i,j = 1,2,3)$ in $V \otimes V$ are now defined as usual by

$$R_{12}(\theta)e_a \otimes e_b \otimes e_c = \sum_{a',b'=1}^{N} R_{ab}^{b'a'}(\theta) e_{a'} \otimes e_{b'} \otimes e_c, \tag{3.4a}$$

$$R_{13}(\theta)e_a \otimes e_b \otimes e_c = \sum_{a',c'=1}^{N} R_{ac}^{c'a'}(\theta) e_{a'} \otimes e_b \otimes e_{c'}, \tag{3.4b}$$

$$R_{23}(\theta)e_a \otimes e_b \otimes e_c = \sum_{b',c'=1}^{N} R_{bc}^{c'b'}(\theta) e_a \otimes e_{b'} \otimes e_{'}. \tag{3.4c}$$

However, we will consider hereafter the parameter–free case, i.e. $R(\theta)$ being independent of θ. Similarly, we define $R^* : V \otimes V \to V \otimes V$ by

$$R^* e_a \otimes e_b = \sum_{c,d=1}^{N} R_{ba}^{cd} e_c \otimes e_d, \tag{3.5}$$

and

$$R_{ij}^* : V \otimes V \otimes V \to V \otimes V \otimes V$$

similarly.

Then we have

$$R_{ij}^* = P_{ij} R_{ij} P_{ij}, \tag{3.6}$$

where P_{ij} = permutation operator of vectors in i–th copies of $V(i,j = 1,2,3,$ $i \neq j)$.

Let V be a N–dimensional vector space with a basis vectors $e_1, e_2, \ldots, e_N$ and let V^* = dual of V with dual basis $e^1, e^2 \ldots, e^N$ with

$$< e^a | e_b >= \delta_b^a \ (a, b = 1, 2, \ldots, N). \tag{3.7}$$

We introduced two tripe products:

$$[x, \bar{u}, y] : V \otimes V^* \otimes V \to V \tag{3.8a}$$
$$[\bar{x}, u, \bar{y}]^* : V^* \otimes V \otimes V^* \to V^* \tag{3.8b}$$

for $x, y, u \in V$ and $\bar{x}, \bar{y}, \bar{u} \in V^*$ by

$$[e_a, e^d, e_b] := \sum_{c=1}^{N} e_c R_{ab}^{cd}, \tag{3.9a}$$

$$[d^d, e_a, e^c]^* := \sum_{b=1}^{N} R_{ab}^{cd} e^b. \tag{3.9b}$$

Remark. We have the relation

$$< \bar{u} | [x, \bar{v}, y] > = < [\bar{v}, x, \bar{u}]^* | y > \tag{3.10}$$

since

$$R_{ab}^{cd} = < e^c | [e_a, e^d, e_b] > = < [e^d, e_a, e^c]^* | e_b > .$$

Proposition 3.1. *Let V and $[-,-,-]$ be as above. Then the following relation are equivalent:*

$$1)\ [R_{12}, R_{23}^*] + [R_{12}, R_{13}^*] + [R_{13}.R_{23}^*] = 0 \tag{3.11}$$

$$2)\ [L(x, \bar{u}), L(y, \bar{v})] = L([x, \bar{u}, y], \bar{v}) + L(x, [\bar{v}, y, \bar{u}]^*) - L(y, [\bar{u}, x, \bar{v}]^*) - L([y, \bar{v}, x], \bar{u}) \tag{3.12}$$

where $L(x, \bar{u}) : V \otimes V^* \to gl(V)$ is defined by

$$L(x, \bar{u}) z = [x, \bar{u}, z]. \tag{3.13}$$

Here $gl(V)$ implies the vector space of all general linear transformation in V.

The above identity (3.11) is said to be *classical Yang–Baxter equation* if $R^* = \pm R$ (that is, $R_{ab}^{dc} = \pm R_{ba}^{cd}$). (cf.[Ji],[O4])

Remark. For a Freudenthal–Kantor triple system, we note that if abelian relation holds, i.e.,

$$[L(x, u), L(y, v)] = L([xuy], v) + \varepsilon L(y, [uxv]) = 0,$$

then the above relation (3.12) of Proposition 3.1 is satisfied.

4. Yang–Baxter Equation

If we have

$$R_{12}(\theta) R_{13}(\theta') R_{13}(\theta'') = R_{23}(\theta'') R_{13}(\theta') R_{12}(\theta) \tag{4.1}$$

for parameter θ, θ', and θ'' satisfying

$$\theta' = \theta + \theta'', \tag{4.2}$$

then the relation is called Yang–Baxter equation (e.g. see [Ji],[O2]). Although we can generalize our result to the case of super space, we will consider here only non–super case for simplicity. Suppose that V possesses a non–degenerate bilinear symmetric inner product $< .|. >$ so that we have $< y|x >=< x|y >$. We can then introduce two θ–dependent triple products $[x,y,x]_\theta$ and $[x,y,z]^*_\theta$ satisfying

$$\text{(i)}\ < x|[y,u,v]_\theta >=< y|[x,v,u]^*_\theta > \tag{4.3}$$

$$\text{(ii)}\ R(\theta)(x\otimes y)=\sum_{j=1}^{N}[e^j,y,x]^*_\theta\otimes e_j=\sum_{j=1}^{N}e_j\otimes[e^j,x,y]_\theta . \tag{4.4}$$

Here, e_j and $e^j (j=1,2,\dots,N)$ are basis and its dual basis of V, respectively. Then, the Yang–Baxter equation (hereafter abbreviated as YBE) can be rewritten as a triple product relation

$$\sum_{j=1}^{N}[v,[u,e_j,z]_{\theta'},[e^j,x,y]_\theta]^*_{\theta''}=\sum_{j=1}^{N}[u,[v,e_j,x]^*_{\theta'},[e^j,z,y]^*_{\theta''}]_\theta . \tag{4.5}$$

We are hereafter interested only in the case when we have

$$[x,y,z]^*_\theta=[x,y,z]_\theta \tag{4.6a}$$

or equivalently

$$< y|[x,v,u]_\theta > =< x|[y,u,v]_\theta > . \tag{4.6b}$$

Lemma 4.1. *A neccessary and sufficient condition to have*

$$[R_{ij}(\theta),R_{kl}(\theta')]=0 \tag{4.7}$$

for all $i,j,k,l=1,2,3$ is the validity of

$$[u,v,[x,y,z]_\theta]_{\theta'}=[x,y,[u,v,z]_{\theta'}]_\theta . \tag{4.8}$$

Remark. The validity of Eq.(4.7) implies that the YBE(4.1) as well as classical Yang–Baxter equation

$$[R_{12}(\theta),R_{13}(\theta')]+[R_{12}(\theta),R_{23}(\theta'')]+[R_{13}(\theta'),R_{23}(\theta'')]=0$$

hold valid without assuming the constraint Eq.(4.2).

Proposition 4.2. *Let V be a Lie triple system equipped with a symmetric inner product $< | >$ satisfying*

$$< y|[xvu] > =< x|[yuv] >$$
$$[u,v,[x,y,z]]=[x,y,[u,v,z]].$$

Then, θ-dependent triple product

$$[x, y, z]_\theta = f(\theta)[x, y, z] + g(\theta) < x|y > z$$

for arbitrary functions $f(\theta)$ and $g(\theta)$ of θ gives a solution of Eq.(4.7), and hence of YBE.

References

[FT] L.D.Fadeev and L.A.Takhatajan, Hamiltonian Methods in the Theory of Solitons, Springer-Verlag,Berlin/ Heidelberg/New York /Londn /Paris /Tokyo, 1980

[Ja] N.Jacobson, Structure and Representation of Jordan Algebra, Amer. Math. Soc., Providence,R.I.1968.

[Ji] M.Jimbo,(ed.), Yang-Baxter Equation in Integrable Systems, World Scientific Pub., Singapore 1989.

[K1] N.Kamiya, A Structure Theory of Freudenthal-Kantor Triple Systems, J.Alg.110, (1987)107-123.

[K2] N.Kamiya,A Structure Theory of Freudenthal-Kantor Triple Systems II, Comm. Math. Univ. Sancti Pauli 38, (1989) 41 -60.

[K3] N.Kamiya,The construction of All Simple Lie Algebras Over C from Balanced Freudenthal -Kantor Triple Systems, Contribution to General Algebra 7, (1991) 205 -213, Verlag Hölder-Pichler-Tempsky.

[K4] N.Kamiya,On (ε, δ) Freudenthal -Kantor Triple Systems, International Symposium on a Non-Associative Algebra and Related Topics, Hiroshima, Japan (1991) 65-75, World Scientific Pub.Singapore

[K5] N.Kamiya,On Freudenthal-Kantor Triple Systems and Generalized Structurable Algebras, Mathematics and its Applications 303.(1994) 198-203, Kluwer Academic.Press.

[Me] K.Meyberg,Lectures on Algebras and Triple Systems,Lecture Note, The University of Virginia,1972.

[My] H.C.Myung, Malcev-Admissible Algebras, Birkhäuser, Boston , Basel, Stuttgard 1986.

[O1] S.Okubo,A Generalization of Hurwitz Theorem and Flexible Lie- admissible Algebra,Hadronic. J.3, (1979) 1-52.

[O2] S.Okubo,Triple Products and Yang-Baxter Equations I and II,J. Math. Phys. 34, (1993) 3273-3315.

[O3] S.Okubo, Super-triple Systems,Normal and Classical Yang-Baxter Equations, Mathematics and its Applications 303, (1994) 300-308, Kluwer Academic Press,

[O4] S.Okubo, Introduction to Octonion and Non- Associative Algebras in Physics, Cambridge Univ.Press (1995).

[S1] M.Scheunert, The Theory of Lie Superalgebras, Springer-Verlag, Berline 1979.

7th Int. Coll. on Differential Equations, pp. 197-204
D. Bainov (Ed)

ON CONTINUATION OF GEVREY CLASS SOLUTIONS OF LINEAR PARTIAL DIFFERENTIAL EQUATIONS

AKIRA KANEKO

Graduate School of Mathematical Sciences,
University of Tokyo

ABSTRACT. We survey sufficient or necessary conditions for the removability of thin singularities of Gevrey class solutions of linear partial differential equations.

§0. Introduction

In this report, we gather results on continuation to thin singularity (or removability of thin singularities) of Gevrey class solutions to linear partial differential equations. Most of the results are taken from my recent preprint [Kn4]. For the limited time, however, we restrict ourselves to the case of equations with constant coefficients.

First we shall explain in brief the meaning of our problem. Let K be a locally closed set and let Ω be an open neighborhood. Then the problem is to seek when a solution u of $P(D)u = 0$ defined on $\Omega \setminus K$ can be continued to the whole Ω as a solution of the same equation. In order that this problem is meaningful, we have to treat sufficiently regular solutions. In fact, a (fundamental) solution E of $P(D)E = \delta$ always presents a solution of $P(D)u = 0$ outside $K = \{0\}$ which is easily seen to be non-continuable to K. Moreover, if the solutions are regular but in non-quasianalytic class, K must be thin. For, if K has a non-void interior, for a function φ of this class with support contained in it, the solution u of $P(D)u = \varphi$ on $\boldsymbol{R}^n$ will present a non-continuable solution of $P(D)u = 0$ outside K. Notice that the Gevrey class which we shall treat here belongs to non-quasianalytic class.

The first result to this problem was given by Grushin [G2], who showed the removability of isolated singularities of C^∞-solutions for equations with real simple characteristic directions. Our arguments are based on this pioneering work of him.

For a detailed survey on this subject, we refer to [Kn2] for results until 1992, and [Kn3], where a list of open problems is gathered. The present report includes answers to some of them concerning Gevrey class solutions.

Here is a brief plan of this talk: In §1 we present basic facts on this problem. In §2, we give a sufficient condition for the removability of thin singularities of the Gevrey class solutions. In §3 we discuss the necessity of the condition given in §2. This is to construct non-trivial solutions with irremovable thin singularities under the condition opposite to §2. As an example, the precise Gevrey index for the threshold of existence of solutions with thin singularity is determined for the Schrödinger equation.

§1. Basic facts on continuation of solutions.

Here we present some basic facts on continuation of solutions in our setting. Although we describe with the Gevrey solutions, the results are common to many solution classes. Let $P(D)$ be a linear partial differential operator with constant coefficients, where $P(\zeta)$ is a polynomial in n variables $\zeta = (\zeta_1, ..., \zeta_n)$ and $D = (D_1, ..., D_n)$ with $D_j = -i\partial/\partial x_j$, $j = 1, ..., n$. We define the two spaces of Gevrey class functions of index s as follows:

$$\mathcal{E}^{(s)}(\Omega) := \{f(x) \in C^\infty(\Omega);\ \forall K \subset\subset \Omega,\ \forall h > 0,\ \exists C = C_{K,h} > 0 \quad \sup_{x \in K} |D^\alpha f(x)| \le C h^{|\alpha|} \alpha!^s \ \text{ for } \forall \alpha\}. \tag{1.1}$$

$\mathcal{E}^{\{s\}}(\Omega)$ is defined in a similar way, replacing the quantifiers by $\forall K \subset\subset \Omega$, $\exists h = h(K) > 0$, $\exists C = C(K) > 0$.

We also set

$$\mathcal{E}^{1+}(\Omega) := \bigcap_{s>1} \mathcal{E}^{(s)}(\Omega) = \bigcap_{s>1} \mathcal{E}^{\{s\}}(\Omega). \tag{1.2}$$

This is a very convenient space, still containing enough functions in non-quasianalytic ultra-differentiable class. Following the usage of Komatsu, we shall denote in the sequel by $\mathcal{E}^*(\Omega)$ either of the spaces $\mathcal{E}^{(s)}(\Omega)$, $\mathcal{E}^{\{s\}}(\Omega)$, $\mathcal{E}^{1+}(\Omega)$ when we can state something commonly to these spaces. As usual we let $\mathcal{D}^*(\Omega)$ denote the functions of class $\mathcal{E}^*$ with compact support contained in Ω (together with the obvious topology if the dual space, that is, the space of ultradistributions of this class, is considered).

In general, we denote by $\mathcal{E}^*_P(\Omega)$ the space of solutions in Ω of the equation $P(D)u = 0$ of class $\mathcal{E}^*$. Let $K \subset \Omega$ denote a thin compact subset. Here "thin" means that the interior is void. We assume that it is contained in a hyperplane, say $\nu \cdot x = 0$. (This follows automatically for convex thin set, as we mainly consider in the sequel.) We study the continuation of solutions of $P(D)u = 0$ in $\mathcal{E}^*(\Omega \setminus K)$ to solutions in $\mathcal{E}^*(\Omega)$.

The first easy observation is:

Proposition 1.1. *The canonical map induced by the canonical restriction from Ω to $\Omega \setminus K$:*

$$\mathcal{E}^*_P(\Omega) \to \mathcal{E}^*_P(\Omega \setminus K)$$

is injective. In other words, there are no solutions of $P(D)u = 0$ with compact support.

Thus the quotient space

$$\mathcal{E}^*_P(\Omega \setminus K)/\mathcal{E}^*_P(\Omega) \tag{1.3}$$

will represent the obstruction for continuation of solutions of this class to K. Next we remark that this obstruction space depends only on K and not on Ω. This can be shown in an elementary way, but is related with the local cohomology group with coefficients in the solution sheaf $\mathcal{E}^*_P$ of class $\mathcal{E}^*$ of the equation $Pu = 0$.

Proposition 1.2. *We have the following isomorphism*

$$\mathcal{E}^*_P(\Omega \setminus K)/\mathcal{E}^*_P(\Omega) \cong H^1_K(\Omega, \mathcal{E}^*_P). \tag{1.4}$$

The proof can be made via manipulation of several exact sequences. Note, however, that $H^1(\Omega, \mathcal{E}^*_P)$ does not necessarily vanish. Hence it is a little more complicated than the general nonsense argument.

From this via another manipualtion of exact sequences we obtain the following

Corollary 1.3. *Let $P = Q_1^{m_1} \cdots Q_N^{m_N}$ be the decomposition of $P(\zeta)$ into different irreducible components with their multiplicities counted. Then*

$$H_K^1(\Omega, \mathcal{E}_P^*) \cong \prod_{j=1}^{N} [H_K^1(\Omega, \mathcal{E}_{Q_j}^*)]^{m_j}. \tag{1.5}$$

Hence especially (1.3) *is trivial if and only if the corresponding obstruction is trivial for all the irreducible factors Q_j.*

We only need the last assertion, which can be proved more elementarily.

§2. Continuation of Gevrey class solutions to equations with constant coefficients. Sufficiency part.

Our first main result is the following

Theorem 2.1. *Let K be a compact set contained in a hyperplane $\nu \cdot x = 0$. Assume further that every irreducible component of $P(\zeta)$ has ν as characteristic direction of irregularity μ. Then*

$$\mathcal{E}_P^{(s)}(\Omega \setminus K)/\mathcal{E}_P^{(s)}(\Omega) = 0, \qquad \text{if } s \le \mu/(\mu - 1),$$

$$\mathcal{E}_P^{\{s\}}(\Omega \setminus K)/\mathcal{E}_P^{\{s\}}(\Omega) = 0, \qquad \text{if } s < \mu/(\mu - 1).$$

Here, the irregularity μ of a characteristic direction ν is defined as follows: Consider

$$Q(s,t) := P(t\xi + s\nu) = q_0(\xi)t^m + q_1(s;\xi)t^{m-1} + \cdots + q_m(s;\xi)$$

as a polynomial of the two variables (s,t). For generically fixed ξ, let κ be the minimum value of the dominating power of the Puiseux expansion of the root of Q for t in terms of s representing an irreducible germ of $N(Q)$ passing through the point $(\infty, 0)$ at infinity. Then we set $\mu = (1 - \kappa)^{-1}$. (A more exact definition may be given via the Newton polygon, transforming the point at infinity to the origin.) We have in general $\mu \le$ the multiplicity of the characteristic direction ν.

When we consider the intersection of all Gevrey classes, the result becomes simpler:

Corollary 2.2. *Assume that every irreducible component of $P(\zeta)$ is non-elliptic. Then the isolated singularities of solutions of class $\mathcal{E}^{1+}$ are removable, that is,*

$$\mathcal{E}_P^{1+}(\Omega \setminus \{0\})/\mathcal{E}_P^{1+}(\Omega) = 0.$$

Remark that this last sufficient condition on $P(D)$ is the same as the one for the removablilty of isolated singularities of real analytic solutions given in [Kn1].

Outline of Proof. In view of the Propositions prepared in §1, we can assume without loss of generality that $P(D)$ is irreducible and K, Ω are convex. First recall the Grushin representation: Given $u(x) \in \mathcal{E}_P^*(\Omega \setminus K)$, choose an extension $[u]$ to an element of $\mathcal{E}^*(\Omega)$ with modification in the ε-neighborhood K_ε of K. Then $P(D)[u] \in \mathcal{D}^*(K_\varepsilon)$. Thus its

Fourier transform $F(\zeta) = \widehat{P(D)[u]}$ becomes an entire function and in view of the Paley-Wiener type theorem, it satisfies, for $\forall h > 0$, with $C = C(h) > 0$ in case $* = (s)$, resp. for $\exists h = h(\varepsilon) > 0$, $C > 0$ in case $* = \{s\}$

$$|F(\zeta)| \le C_h e^{-h|\mathrm{Re}\,\zeta|^{1/s} + H_K(\mathrm{Im}\,\zeta) + \varepsilon|\mathrm{Im}\,\zeta|}, \tag{2.1}$$

where H_K denotes the supporting funcion of K. If we restrict $F(\zeta)$ to the complex characteristic variety

$$N(P) := \{\zeta \in C^n;\ P(\zeta) = 0\}$$

of P, it defines a global holomorphic function on $N(P)$ with the same estimate. This does not depend on the choice of the modification $[u]$ of u nor of ε. For, the difference of any two such modifications has the form $P(D)v$ with $v \in \mathcal{D}^*(K_\varepsilon)$, where ε denotes the bigger one. Hence it vanishes by the Fourier transformation and restriction to $N(P)$. Thus we obtain the well defined *Grushin representation* :

$$H^1_K(\Omega, \mathcal{E}^*_P) \to \widehat{\mathcal{E}^*(K)}[N(P)]. \tag{2.2}$$

Here obviously the right-hand side denotes the space of global holomorphic functions on $N(P)$ satisfying the estimate (2.1).

A variant of the so called Fundamental Principle of Ehrenpreis-Palamodov asserts that (2.2) is a topological linear isomorphism. For our class (s) this really takes place, but for $\{s\}$ this holds only partially (see Proposition 2.3 below). To prove the continuation of solutions, however, we only need the injectivity of the mapping (2.2), which is easier to show. Therefore, to prove the theorem it suffices to show that the image of the obstruction $H^1_K(\Omega, \mathcal{E}^*_P)$ in (2.2) is trivial.

Now choose the coordinate system in such a way that $\nu = (0, ..., 0, 1)$ and that $(1, 0, ..., 0)$ is a non-characteristic direction. In the sequel let us employ the abbreviation $\zeta'' = (\zeta_2, ..., \zeta_{n-1})$. By the assumption, for a generic choice of $\zeta'' = \zeta''_0$ and a small $\delta > 0$, the equation $P(\zeta_1, \zeta'', \zeta_n) = 0$ for ζ_1 has a solution $\zeta_1 = \tau(\zeta'', \zeta_n)$ which is multi-valued analytic in ζ_n in $|\zeta_n| \ge 1/\delta$ for each fixed ζ'' in $|\zeta'' - \zeta''_0| \le \delta$, and which satisfies there the following estimate:

$$|\tau(\zeta'', \zeta_n)| \le C|\zeta_n|^{(\mu-1)/\mu}.$$

For the present system of coordinates we have $K \subset \{x_n = 0\}$, hence

$$H_K(\eta) \le A|\eta_1| + A|\eta''|, \qquad H_{K_\varepsilon}(\eta) \le (A+\varepsilon)|\eta_1| + (A+\varepsilon)|\eta''| + \varepsilon|\eta_n|.$$

Thus for each fixed ζ'' we obtain a function

$$G(z) := F(\tau(\zeta'', z), \zeta'', z)$$

of one variable z, which is multi-valued holomorphic in $|z| \ge 1/\delta$ and which satisfies there

$$|G(z)| \le C_\varepsilon e^{-h|z|^{1/s} + (A+\varepsilon)C|z|^{(\mu-1)/\mu} + \varepsilon|\mathrm{Im}\, z|}. \tag{2.3}$$

Here, by the assumption, we have $1/s \ge (\mu-1)/\mu$ but we can choose $h > (A+1)C$ in case $* = (s)$, and we have $1/s > (\mu-1)/\mu$ in case $* = \{s\}$. Thus in either case, in view of the arbitrariness of ε we can apply the Phragmén-Lindelöf principle to $G(z)$ to conclude that $G(z)$ is bounded on $\mathrm{Im}\, z \ge 1/\delta$, and similarly on $\mathrm{Im}\, z \le -1/\delta$, hence on $|z| \ge 1/\delta$. Recall that the multivaluedness of $G(z)$ at $z = \infty$ comes from that of $\tau(\zeta'', z)$. Hence it can be represented by a Puiseux series of z, and we can find an integer $q \ge 1$ such that $H(w) := G(w^q)$ is single valued holomorphic at $w = \infty$. Therefore we can apply Riemann's theorem on removable isolated singularity to $H(w)$ to conclude that it is holomorphic at $w = \infty$. By the estimate (2.3), however, $H(w)$ decreases faster than any inverse power of $|w|$ along the real axis. Thus its Taylor expansion at $w = \infty$ should be trivial. Thus we conclude that $H(w) \equiv 0$, whence $G(z) \equiv 0$.

Now we have shown that $F(\zeta_1, \zeta'', \zeta_n) = 0$ on

$$N(P) \cap \{(\zeta_1, \zeta'', \zeta_n);\ |\zeta'' - \zeta''_0| \le \delta,\ |\zeta_n| \ge 1/\delta,\ \zeta_1 = \tau(\zeta'', \zeta_n)\}.$$

Since this is an open subset of the irreducible algebraic variety $N(P)$, we conclude that $F(\zeta) \equiv 0$ on $N(P)$. This shows the continuation of this solution as is already remarked. □

To discuss the non-triviality of the obstruction in the next section, we need the surjectivity of (2.2):

Proposition 2.3. (2.2) *is an algebraic isomorphism.*

This can be proved in the standard way in the theory of Ehrenpreis-Palamodov on Fundamental Principle.

At the end of this section, we give an analogy of Grushin's result in [G1] concerning the removability of weak singularity. We shall say that an isolated singularity of a solution u of class $\mathcal{E}^*$ is weak if u is prolongeable to a neighborhood of the singularity as an ultradistribution of class $\mathcal{D}^{*\prime}$. Recall that $P(D)$ is called hypoelliptic in the class $\mathcal{E}^*$ if $P(D)$ admits a fundamental solution in $\mathcal{D}^{*\prime}$ which has $\mathcal{E}^*$ regularity outside the origin, or equivalently, if every local solution of $P(D)u = 0$ in $\mathcal{D}^{*\prime}$ becomes regular of class $\mathcal{E}^*$.

Theorem 2.4. *The isolated weak singularity of any solution of $P(D)u = 0$ of class $\mathcal{E}^*$ is always removable if and only if $P(D)$ contains no irreducible factor which is hypoelliptic in the class $\mathcal{E}^*$.*

For the proof see [Kn4].

§3. Equations possessing Gevrey solutions with thin compact singularity. Necessity part.

In this section we present some class of equations with constant coefficients for which the continuation of Gevrey solutions as discussed in the preceding section does not hold. Since the preparatory arguments there apply as well, we can restrict ourselves to consideration of irreducible operators.

First of all, it is well known that hypoelliptic equations have C^∞ solutions with isolated singularity. In our case this corresponds to the following, which makes the necessity part of Theorem 2.1 more concrete:

Proposition 3.1. *Assume that there exist positive constants $q \geq 1$ and δ such that*

$$|\mathrm{Im}\,\zeta| \geq \delta |\mathrm{Re}\,\zeta|^{1/q} \quad \textit{on} \quad \zeta \in N(P) \cap \{|\zeta| \geq 1/\delta\}. \tag{3.1}$$

Then there exist solutions with isolated non-removable singularity of class $\mathcal{E}^{(s)}$ for $s > q$, and of class $\mathcal{E}^{\{s\}}$ for $s \geq q$.

This can be proved either employing the fundamental solution or employing Proposition 2.3 by estimating $F(\zeta) \equiv 1$ on $N(P)$. Note that $q = 1$ corresponds to the elliptic equation.

A remarkable discovery of Grushin was that there exist equations which are non-hypoelliptic but nevertheless allow C^∞ solutions with isolated singularity. Grushin only gave a few examples. Here we shall try to generalize his idea of construction. The following generalizes Grushin's example $D_1^3 - D_2^2$, which is obviously non-hypoelliptic. We first give a result on thin singularity, which is more general than the case of isolated singularity as will be given later:

Theorem 3.2. *Let $P(D)$ be an operator of two independent variables with the following form*

$$P(D) = D_1^m - aD_2^k + \cdots, \qquad k < m, \qquad a \neq 0 \tag{3.2}$$

where $\cdots$ denotes lower order terms in the sense of weighted homogeneity. Then for any prescribed $s > m/k$, $P(D)$ admits a solution in $\mathcal{E}^{\{s\}}$ with a compact thin singularity K contained in $x_2 = 0$.

Note that for this operator $(0, 1)$ is a characteristic direction of multiplicity m and irregularity $\mu = m/(m-k)$, hence $\mu/(\mu - 1) = m/k$.

Outline of Proof. Let $\psi(t) \in \mathcal{D}^{\{q\}}(R^1)$, with $q > 1$ which is supposed to be close to 1. We have

$$|\widehat{\psi}(\tau)| \leq Ce^{-A|\tau|^{1/q}+B|\mathrm{Im}\,\tau|}. \tag{3.3}$$

Set

$$F(\zeta_1, \zeta_2) = \widehat{\psi}(\zeta_1)|_{N(P)}.$$

Employing the asymptotic

$$|\zeta_1| \sim |a|^{1/m}|\zeta_2|^{k/m}, \tag{3.4}$$

we can easily show that $F(\zeta)$ satisfies

$$|F(\zeta)| \leq C_\varepsilon e^{-A'|\zeta|^{k/qm}+B|\mathrm{Im}\,\zeta|}.$$

This gives a solution in $\mathcal{E}^{\{s\}}$ for $s = qm/k$ and in $\mathcal{E}^{(s)}$ for $s > qm/k$ with singularity $|x_1| \leq B$ on the line $x_2 = 0$. Since q can be chosen as close to 1 as we wish, we obtain the assertion. □

Remark that for $s < m/k$ such a solution does not exist because of Theorem 2.1. Notice that the above theorem is not a general nonsence in spite of the simplicity of its proof. Trivial thing is rather that for thin K contained in $a_1x_1 + a_2x_2 = 0$ with $a_1 \neq 0$, we can always show a solution for s in $1 < s < m/k$, employing the weak hyperbolicity:

Proposition 3.3. *Let $P(D)$ be an operator as in the preceding theorem. Then for any prescribed $s > 1$, $P(D)$ admits a solution in $\mathcal{E}^{(s)}$ or in $\mathcal{E}^{\{s\}}$ with a compact thin singularity K contained in $x_1 = bx_2$.*

Next we try to construct solutions with *isolated* singularity. We have rather a partial result on this. We employ the following lemma, which is implicitly contained in Theorems 4.1.1 and 4.1.8 of Boas [Bo1].

Lemma 3.4. *For any $\delta > 0$ we can find $p_\delta > 1$ such that for any prescribed p with $1 < p \leq p_\delta$, we can construct an entire function of one variable $G(\tau)$ which satisfies*

$$|G(\tau)| \leq Ce^{-A|\tau|^{1/p}}, \qquad \text{on } \mathrm{Re}\,\tau \geq \delta|\mathrm{Im}\,\tau|,$$
$$|G(\tau)| \leq Ce^{B|\tau|^{1/p}}, \qquad \text{on } \mathrm{Re}\,\tau \leq \delta|\mathrm{Im}\,\tau|,$$

with some constants $B > A > 0$.

Via scaling of the variable τ, we can let B smaller than any prescribed positive constant or let A larger than any prescribed positive constant, although in each case the remaining constant A resp. B changes proportionally.

Theorem 3.5. *Let $P(D)$ be an operator of two independent variables as in Theorem 3.2. Assume further that $a \neq 0$ is real, m is odd, and k is even. Then for any prescribed $s > m/k$, $P(D)$ admits a solution with isolated singularity in $\mathcal{E}^{\{s\}}$ and in $\mathcal{E}^{(s)}$.*

Outline of Proof. Since m is odd, by the change of sign of the x_1-axis if necessary, we can assume without loss of generality that $a > 0$. Take an entire function of one variable $G(\tau)$ as is given by Lemma 3.4 with $p > 1$ close to 1. Then put

$$F(\zeta_1, \zeta_2) = G(\zeta_1).$$

Recall the asymptotic form (3.4). Thus on the region $|\mathrm{Im}\,\zeta_1| \leq \delta \mathrm{Re}\,\zeta_1$ (actually on $\mathrm{Re}\,\zeta_1 \geq \delta|\mathrm{Im}\,\zeta_1|$) we have

$$|F(\zeta)| \leq Ce^{-A|\zeta_1|^{1/p}} \leq C'e^{-(A/2)|\zeta_1|^{1/p}-(A/3)a^{k/m}|\zeta_2|^{k/pm}}.$$

Next, on the region $|\mathrm{Im}\,\zeta_1| \geq \delta|\mathrm{Re}\,\zeta_1|$, we have $|\zeta_1| \leq \sqrt{1+\delta^{-2}}|\mathrm{Im}\,\zeta_1|$, hence

$$\begin{aligned}|F(\zeta)| &\leq Ce^{B|\zeta_1|^{1/p}} \leq Ce^{-A|\zeta_1|^{1/p}+(A+B)\sqrt{1+\delta^{-2}}|\mathrm{Im}\,\zeta_1|^{1/p}} \\ &\leq C_\varepsilon e^{-(A/2)|\zeta_1|^{1/p}-(A/3)a^{k/m}|\zeta_2|^{k/pm}+\varepsilon|\mathrm{Im}\,\zeta_1|}.\end{aligned}$$

Finally, on the region $|\mathrm{Im}\,\zeta_1| \leq -\delta \mathrm{Re}\,\zeta_1$, we put

$$\zeta_1 = \rho e^{\pi i+\theta i}, \quad \text{where } |\theta| \leq \mathrm{Arctan}\,\delta.$$

Then for $\rho \to \infty$ we have

$$\zeta_2 \sim a^{-m/k}\rho^{m/k}e^{(m/k)\theta i+((2\ell+1)/k)\pi i}, \qquad \ell = 0, ..., k-1.$$

By the assumptions on a, m, k, we see from this that for δ sufficiently small, each branch satisfies

$$\mathrm{Im}\,\zeta_2 \sim a_\ell \rho^{m/k}, \qquad \text{with } a_\ell = a^{-m/k}\sin\left(\frac{m}{k}\theta + \frac{2\ell+1}{k}\pi\right) \neq 0.$$

Thus in this region we obtain, with some $a' > 0$

$$\begin{aligned}|F(\zeta)| &\leq Ce^{B|\zeta_1|^{1/p}} \leq Ce^{-A|\zeta_1|^{1/p}+(A+B)|\zeta_1|^{1/p}} \\ &\leq C_\varepsilon e^{-(A/2)|\zeta_1|^{1/p}-(A/3)a^{k/m}|\zeta_2|^{k/pm}+\varepsilon|\mathrm{Im}\,\zeta_2|}.\end{aligned}$$

The obtained estimate shows that $F(\zeta)$ is the Grushin transform of a solution with isolated singluarity at the origin in the class $\mathcal{E}^{\{s\}}$ for $s = pm/k$ and in the class $\mathcal{E}^{(s)}$ for $s > pm/k$. □

We hope that the additional assumption which we posed in Theorem 3.5 (in comparison to Theorem 3.2) is only technical. Note that for imaginary a or for m, k even and $a < 0$ our operator becomes hypoelliptic and the existence of solutions with isolated singularity becomes trivial. In view of Theorem 2.4 the singularity of such an example of solution is never weak.

We also believe that we will be able to construct solutions with thin singularity in $\mathcal{E}^{(s)}$ with $s = m/k$, but our method does not work for this class.

The equations we have treated in this section are of course very special. Consideration of general equations, especially of more than two variables, will need the study of cohomology groups with bound on the variety $N(P)$, and is left for the future. We conclude this section by showing a solution with thin singularity for the Schrödinger equation for general space dimension.

Proposition 3.6. *Consider the operator $P(D) = D_1^2 + \cdots + D_{n-1}^2 - D_n$ corresponding to the Schrödinger equation. This admits a solution of class $\mathcal{E}^*$ with $* = \{s\}$ or $* = (s)$ for $s > 2$ with compact irremovable singularity contained in $x_n = 0$.*

This can be proved by estimating the function $F(\zeta) = \widehat{\psi}(\sqrt{\zeta_1^2 + \cdots + \zeta_n^2})$ for an even function $\psi(t) \in \mathcal{D}^{\{q\}}(R^1)$, $q > 1$ being close to 1. Note that for this equation with $\nu = (0, ..., 0, 1)$ we have $\mu = \mu/(\mu - 1) = 2$.

It is an interesting problem to know if we can construct similar solution with *isolated singularity* for this equation. In the above proof we can let the "horizontal" size B of the singularity as small as we like, but never to 0.

References

[Bol] Boas R. Ph., *Entire Functions*, Academic Press, 1954.

[G1] Grushin V.V., О Q-гипоэллиптических уравнениях, Мат. Сб. **57** (1962), 233-240.

[G2] ———, О решениях с изолированными особенностями для уравнений в частных производных с постоянными коэффициентами, Труды Москов. Мат. Общ. **15** (1966), 262-278.

[Kn1] ———, *On continuation of regular solutions of partial differential equations to compact convex sets*, J. Fac. Sci. Univ. Tokyo Sec.1A **17** (1970), 567-580.

[Kn2] ———, *On continuation of regular solutios of linear partial differential equations*, Banach Center Publications **27** (1992), 183-195.

[Kn3] ———, *Some problems on continuation of regular solutios of partial differential equations*, Preprint Series UTMS, University of Tokyo **95-43** (1995).

[Kn4] ———, *On continuation of Gevrey class solutions of linear partial differential equations*, Talk at "Workshop on General Theory of Partial Differential Equations and Microlocal Analysis", ICTP, Trieste, September '95. Preprint Series UTMS, University of Tokyo **96-14** (1996).

Graduate School of Mathematical Sciences, University of Tokyo, 3-8-1, Komaba, Meguro-ku, Tokyo 153 Japan

e-mail: kanenko@ms.u-tokyo.ac.jp

7th Int. Coll. on Differential Equations, pp. 205-212
D. Bainov (Ed)

Convergence Properties of Periodic Solutions on a Piecewise Linear Differential Equation

Kaoru KATORI
Faculty of Humanities, Hirosaki University,
Bunkyo-cho 1, Hirosaki-shi, Aomori 036, Japan
Masamitu OTAKE
School of Science and Technology, Meiji University,
Higasi-Mita 1-1-1, Tama-ku, Kawasaki, Kanagawa 214, Japan
Yoichi MANOME
Yokogawa Hewlett Packard,
Takakura-cho 9-1, Hachiouji-shi, Tokyo 192, Japan
Taketoshi MISHIMA
Dept. of Infor. and Comp. science, Saitama University,
255 Simo-okubo, Urawa-shi, Saitama, Japan

Abstract In a previous paper, we had proposed an our new method for solving a piecewise linear ordinary differenntial equation. [1]

That is, supposing that a resistor and a capacitor are linear but an inductor is piecewise linear, the equation is resulted from a series circuit consisting of the resistor, capacitor and inductor connected with an alternating electromotive force. Since the nonlinearity is piecewise linear the solution can be given by the continuation of each component which is an analytic solution in the corresponding linear interval. But such troublesome continuation prevents us from grasping the whole aspect of the solutions. In this paper, we will show some examples ; Approximation of the nonlinear resistance of a superconducting Josephson tunneling device is investigated to obtain one of fundamental data for constructing logic circuits.

AC Josephson oscillations are simulated in two different cases where the V-I characteristic is approximated by a piecewise line and non-approximated method.

Keywords: piecewise linear, exact solution, Josepson circuits

1.Introduction

Josephson devices which cause so-called Josephson effect have attracted wide attention. [2] The devices are characterized by high-speed switching, extremely low power consumption, and high-speed, non-distortion transmission.

Since Rowell found that Josephson junctions would be used as a switching device, various kinds of logic circuits have been proposed. [3]

In this paper, a piecewise linear approximation of the nonlinear resistance of the Josephson device connected with a lossless distributed line is considered.

Computer simulations under a piecewise linear approximation show diffrence in AC Josephson oscillations in comparison with the non-approximation method that we had proposed last year. The nonlinearity is assumed to be a tunneling device. A nonlinear circuit connected with a linear lossless distributed line and the fundamental equations describing the circuit are shown in section 2.

From these equations, a travelling wave and a backward wave are obtained for computer simulations. Figure 1, Figure 2 shows Josephson nonlinear circuit and characteristic of quasi-particle tunnel curent of the junction, respectively.

In section 3, we show some results of computer simulation. In order to demonstrate the validity of our method, the simulation of the circuit is performed. The advantages of our method are as follows:

(1) The solutions of the equations for the circuit with any circuit parameters are easily and acturately obtained.

(2) The program for this method can be built into a general purpose CAD system package.

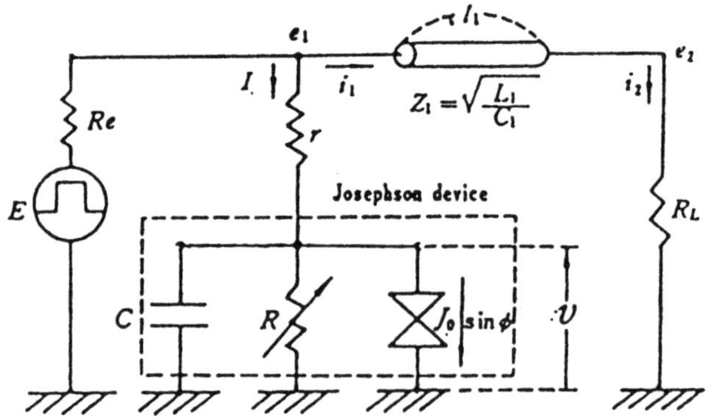

Figure 1. Josephson nonlinear circuit

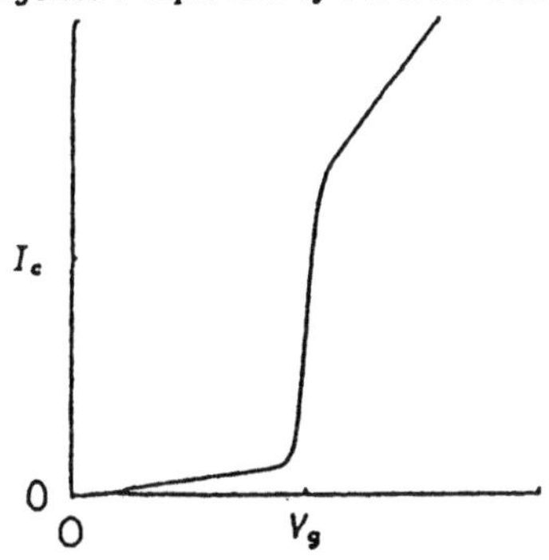

Figure 2. Characteristic of quasi-particle tunnel current

2. Differential Equation and Model

We consider a circuit as shown in Figure 1. The resistively shunted junction model is applied to the Josephson junction as an equivalent circuit, where

E : input voltage given to the Josephson device,
Re : resistance contained in the input port,
r : series resistance to the Josephson device,
Jo : critical current of the Josephson device,
ϕ : phase difference between wave functions of two superconductors,
C : parallel capacitance, including the junction capacitance, of the Josephson device,
R : nonlinear resistance of the Josephsondevice,
l_l : distributed constant line length,
C_l : capacitance per unit length of the line,
L_l : inductance per unit length of the line,
Z_l : characteristic impedance of the line,
R_L : load resistor.

The voltage e1, e2 and the current i1, i2 in Figure 1 are obtainable from the travelling wave F_f and the backward wave F_b on the distributed line as follows :

$$e1 = F_f + F_b (t - l_l / u_l) \tag{1}$$

$$i1 = 1 / Z_1 \{ F_f - F_b (t - l_1 / u_1) \qquad (2)$$

$$e2 = F_f (t - l_1 / u_1) + F_b \qquad (3)$$

$$i2 = 1 / Z_1 \{ F_f (t - l_1 / u_1) - F_b \} \qquad (4)$$

where t and u_1 represent, respectively, time and the propagating velocity of the waves.

The current I and the voltage υ loaded to the Josephson device in Figure 1 are given from the equivalent circuit of the Josephson device.

That is:

$$I = \upsilon / R + C (d\upsilon / dt) + J_0 \sin \phi \qquad (5)$$

$$d\phi / dt = 4\pi e \upsilon / h \qquad (6)$$

where h is Planck's constant.

The current I and the voltage e1 are also obtained as follows:

$$e1 = \upsilon + rI \qquad (7)$$

$$I = (E - e1) / Re - i1 \qquad (8)$$

From the load circuit in Figure 1, we have next relation.

$$e2 = R_L \, i2 \qquad (9)$$

Computer simulations are obtainable by means of equation (1) - (9). Responses of the voltage υ corresponding to the input voltage E are simulated by varing the circuit parameters R_L, C, Z_1 and l_1. The resistance of Re, r, the propagating velocity of the waves and the critical current J_0 are assumed to be 1 kΩ, $C_0 / 2$ (C_0: the light velocity in vacuum), 1 Ω, and 1.0 mA, respectively.

The nonlinear resistance R is approximated by a piecewise line. The approximations of the voltage VS. current characteristic of the Josephson devise as follows:

$I_g = 0.2$ V (when $0 < V < 0.35$)

$I_g = 9.2\ V - 3.6$ (when $0.35 < V < 0.6$)

$I_g = 2\ V$ (when $V > 0.6$)

where I_g and V are represented by [mA] and [mV] , respectively.

3. Results of computer simulations

In Figure 3 , Figure 4 , results of computer simulations are shown. Figure 3 are resuls by piecewise linear approximation , and in Figure 4 , we had not used ' piecewise linear approximation method ' . In these results , there is little difference between two method. (amplitude of the voltage υ and the frequency of the AC Josephson oscillation)

Dr. Motinori Goto solved approximately an extraordinary undamped oscillation in a power supply system. [4]

In conclusion, he showed the next fact ;

--Since even a small term may dominate the dynamics of a nonlinear differential equation, disregarding the small term only because of "small" may cause a serious mistake-- :

that is , we can not obtain any solution caused by the small term even if the approximation of a solution obtained from the equation neglected the small term is increased as much as possible.

In Figure 3-1 and Figure 4-1 , circuit parameter is as follows :

$Z_1 = 0.5\ \Omega$, $R_L = 5.0\ \Omega$, $C = 1$ pF , $l_1 = 10\ \mu$m

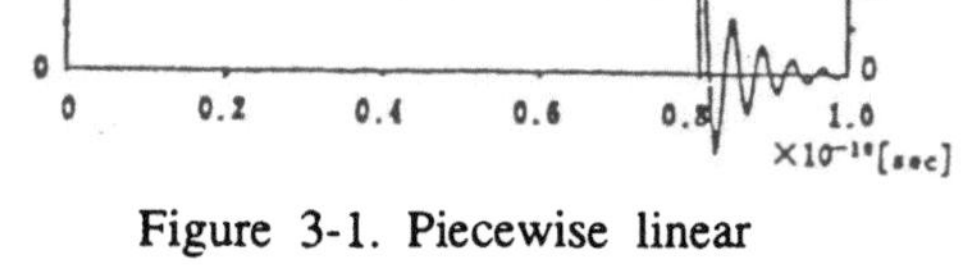

Figure 3-1. Piecewise linear approximation

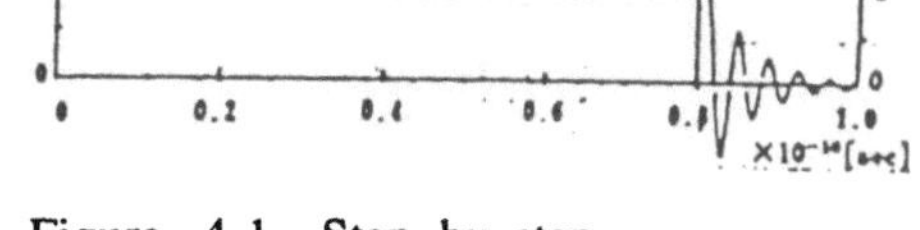

Figure 4-1 . Step by step

Figure 3 and 4 . Voltage responses υ corresponding to input step voltage E

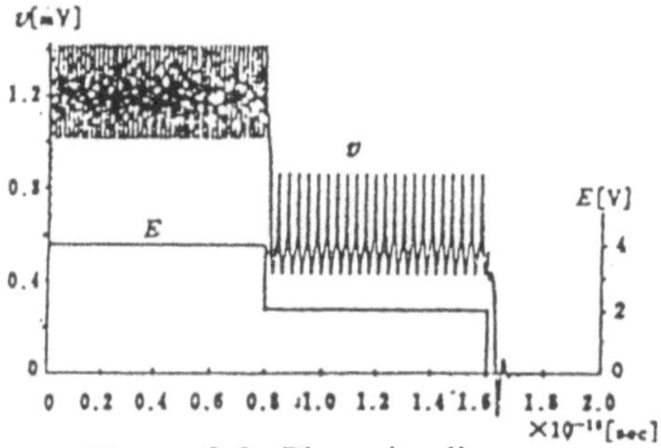

Figure 3-2. Piecewise linear approximation

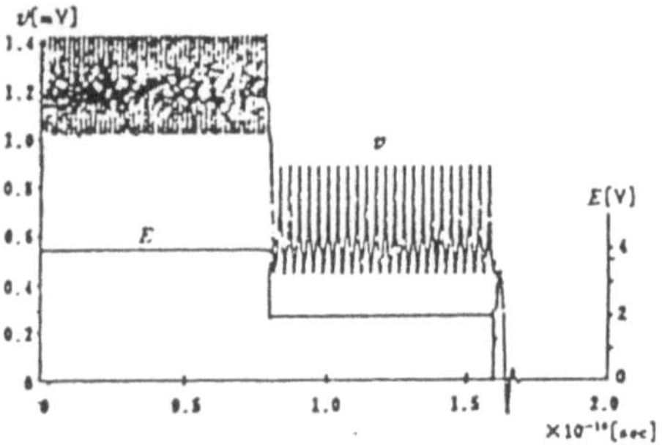

Figure 4-2. Step by step

($Z_1 = 1.0\ \Omega$, $R_L = 1.0\ \Omega$, $C = 1$ pF , $l_1 = 20\ \mu$m)

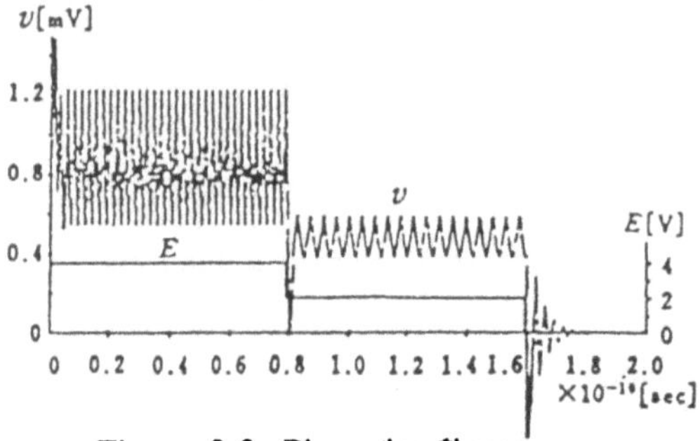

Figure 3-3. Piecewise linear approximation

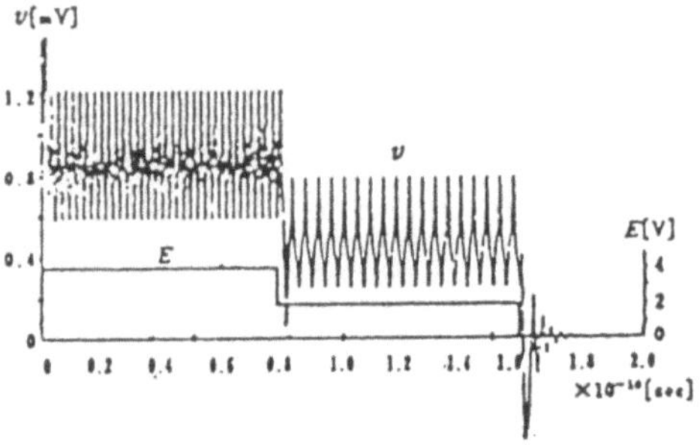

Figure 4-3. Step by step

($Z_1 = 5.0\Omega$, $R_L = 0.5\Omega$, $C = 1$ pF , $l_1 = 20\ \mu$m)

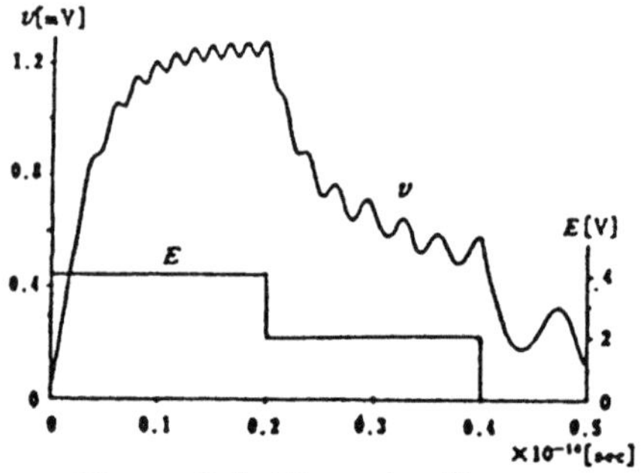

Figure 3-4. Piecewise linear approximation

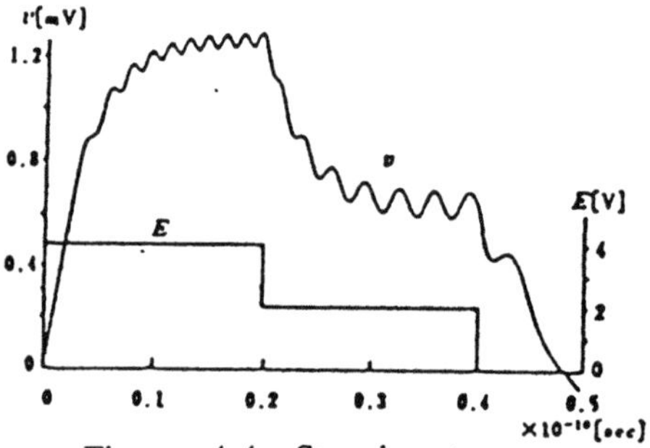

Figure 4-4. Step by step

($Z_1 = 1.0\ \Omega$, $R_L = 1.0\ \Omega$, $C = 10$ pF , $l_1 = 5\ \mu$m)

Figure 3 and 4. Voltage responses υ corresponding to input step voltage E

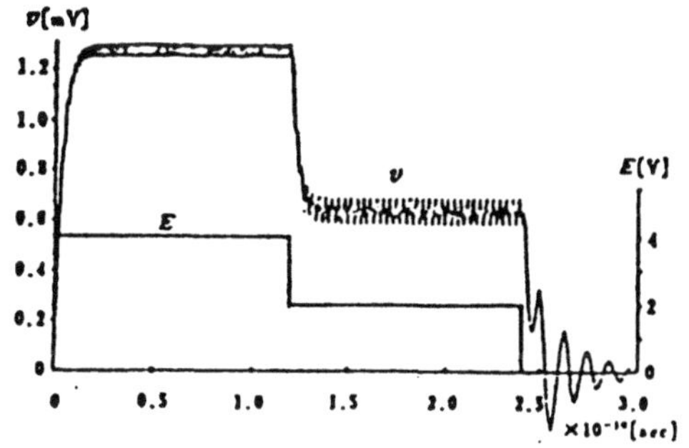

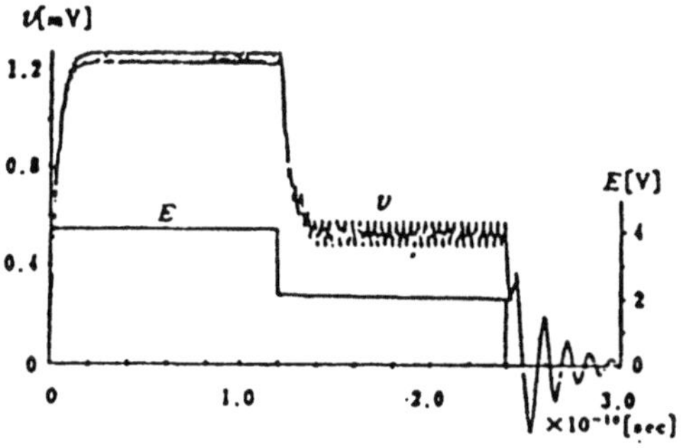

Figure 3-5. Piecewise linear approximation

Figure 4-5. Step by step

($Z_1 = 1.0\Omega$, $R_L = 1.0\ \Omega$, $C = 10\,pF$, $l_1 = 30\ \mu m$)

Figure 3 and 4 . Voltage responses υ corresponding to input step voltage E

Only differences are in the amplitude of the voltage υ in Figure 2-3 , 3-3 and the frequency of the AC Josephson oscillation , but these difference are reducible by using a betterwise linear approximation . For instance , Figure 5 and Figure 6 are obtainable under other linear approximation .

In Figure 5 and Figure 6 , circuit parameter is as follows :

Figure 5 ; $Z_1 = 5.0\Omega$, $R_L = 0.5\ \Omega$, $C = 1\,pF$, $l_1 = 20\ \mu m$

Figure 6 ; $Z_1 = 1.0\Omega$, $R_L = 1.0\ \Omega$, $C = 10\,pF$, $l_1 = 5\ \mu m$

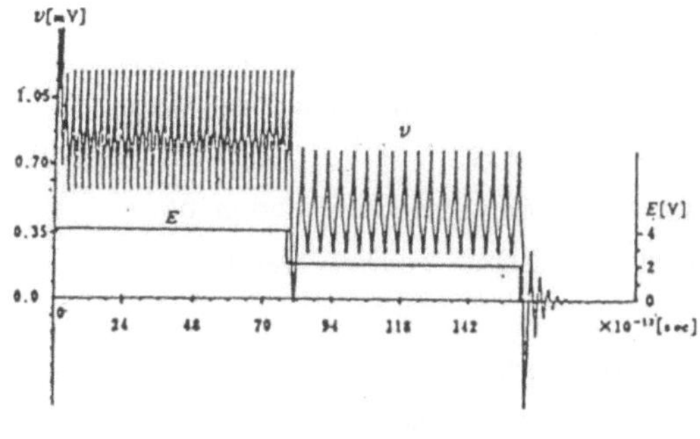

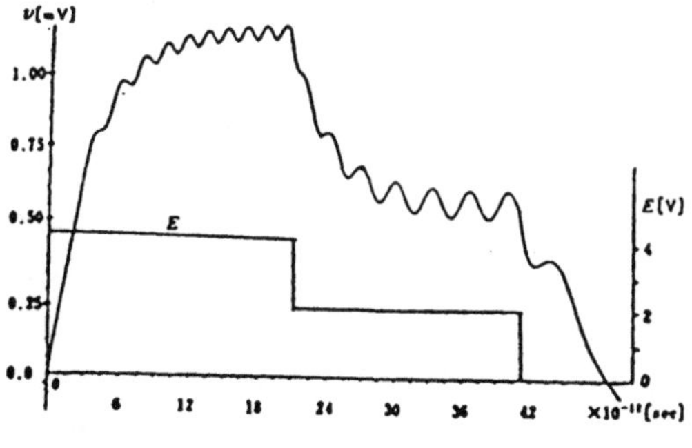

Figure 5. Piecewise linear approximation

Figure 6. Piecewise linear approximation

Figure 5 and 6 . Voltage responses υ corresponding to input step voltage E

4. Conclusion

In this paper, we showed some examples, by using of a piecewise linear approximation method, for solving nonlinear diffrential equations.

These computer simulations tell us that a piecewise linear approximation method can be used.

References

[1] K.Katori, M.Otake, Y.Manome, T.Mishima : One Approach to Exact Solutions of a piecewise Linear Differential Equation, Proc. of the Sixth Intnational Colloquum on Differential Equations, pp. 119-126 (1995).

[2] Josephson, B.d. : Possible new effects in superconductive tunneling, Phys. Lett., 1, 7, pp. 251-253 (1962).

[3] Hasuo, S: Josephson Devices for Computer Applications, J.IECE, Vol 64, No.7, pp. 749-756 (1981).

[4] M.Goto : Undamped Electric Oscillation of Alternator connected to a Transmission Line, Res. of ETL, No281, 1930.

[5] H.D.Block : Periodic Solutions of Forced System Having Hysteresis, IRE Trans. on Circuit Theory, Dec., 1960.

[6] G.Takeuti : Researches on generation of Subharmonics by means of Iron-Core Reactor, Res. of ETL, No 480, 1946.

[7] Y.Kane et al. : Experimental Investigation on Pulse Generating Circuit Composed of Rectangular Hysteresis Core, The Jour. of IEE of Japan, Vol.91, No.11, 1971.

[8] K.Hatta et al. : On Chaotic Responses from a Serial Resonant Circuit including a Saturated Inductor, IEICE Tech. Rept. NLP91-101, 1992.

[9] K.Hatta et al. : Analysis of a Serial Resonant Circuit including a Saturated Inductor, IEICE Tech. Rept. NLP92-110, 1993.

[10] K.Okumura et al. : A Method for Computing Periodic Solution of Non-Linear System with Hysteresis by Discrete Fourier Transform, IEICE Vol.J62-A, No.5, 1979.

[11] M.Otake, K.Katori : Piecewise Linear Resistance of a Superconducting Josephson Tunneling Device, Res.Rept. of the Fac. of ENG., Meiji Univ., No.42, 1982.

[12] K.Katori, M.Otake, M.Morisue : An Analytical Method for Stationary Oscillations in Superconducting Circuit, IEEE Trans. on MAG., MAG-23, No.2, 1987.

7th Int. Coll. on Differential Equations, pp. 213-219
D. Bainov (Ed)

NORM ESTIMATES FOR SCHRÖDINGER SEMIGROUPS WITH MAGNETIC FIELDS

M. LOSS* AND B. THALLER**

*SCHOOL OF MATHEMATICS, GEORGIA INSTITUTE OF TECHNOLOGY, ATLANTA, GA 30332, USA, email: loss@math.gatech.edu

**INSTITUT FÜR MATHEMATIK, UNIVERSITÄT GRAZ, HEINRICHSTR. 36, A-8010 GRAZ, AUSTRIA, email: bernd.thaller@kfunigraz.ac.at

ABSTRACT. We obtain norm estimates for magnetic Schrödinger semigroups in two dimensions under the assumption that the magnetic field is bounded below by some positive constant B_0. The estimates give the exact norm in case of a constant magnetic field.

KEYWORDS: Schrödinger Operator, Magnetic Field, Heat Kernel Estimates

1. INTRODUCTION

Magnetic Schrödinger operators have become a very active field of research. The complexity of the quantum mechanical motion of particles in magnetic fields and the fact that many of the problems cannot be handled by standard techniques contribute to its popularity. To encounter interesting problems it suffices to consider a single charged quantum particle in a magnetic field, which is described by the self-adjoint operator

$$H(B) = \frac{1}{2}(\nabla + \mathrm{i}A)^2 , \tag{1.1}$$

where A is a vector potential for the magnetic field B, i.e., $\operatorname{curl} A = B$.

A good way of studying a self-adjoint operator H is to look at the semigroup e^{tH} (see, e.g., [19] and [6]). In particular, we are interested in smoothing estimates, i.e., estimates on the norm the semigroup acting as an operator from L^p to L^q, $q \geq p$. Only in very rare cases it is possible to obtain these estimates in their sharp form (see, e.g., [15], [2], [3], [12]). As a new example of a sharp smoothing estimate we calculate the norm of the semigroup associated with the Schrödinger operator of a charged particle in a constant magnetic field (see Theorem 2.1).

In order to study more general magnetic fields we used an approach due to Gross [9] (see also Federbush [8]) which is based on an application of logarithmic Sobolev inequalities. The idea is similar to the one in [5] where it was realized that Gross' logarithmic Sobolev inequality when viewed as a family of sharp inequalities on $\mathbb{R}^n$ can be used to obtain sharp smoothing estimates of solutions of diffusion equations with a volume preserving drift. Here we will describe the main results which can be obtained by implementing this method for magnetic Schrödinger semigroups (Theorem 2.2). We refer to the paper [13] for more details.

It is one of the main difficulties that the semigroup generated by the magnetic Schrödinger operator (1.1) is not positivity preserving. This difficulty is usually overcome by the important diamagnetic inequality ([11], [16], [17], [10], [1], [4])

$$\left|\left(e^{tH(B)}u\right)(x)\right| \leq \left(e^{t\Delta/2}|u|\right)(x)\,, \tag{1.2}$$

which relates estimates on magnetic Schrödinger semigroups to estimates on the heat semigroup. The obvious disadvantage is that all effects due to the magnetic field are not seen. Thus, the natural problem to consider is how the magnetic field affects the behaviour of the semigroup.

The first paper that addressed this question is presumably [14], where it is proved that for a magnetic Schrödinger operator in $\mathbb{R}^3$,

$$\lim_{t\to\infty} \frac{1}{t}\, ln\, \|e^{tH(B)}\|_{L^1\to L^\infty} \leq -\Phi(C)B_0\,, \tag{1.3}$$

for all curl-free magnetic fields B that satisfy $0 < B_0 \leq |B| \leq CB_0$. Here $|B| = \sqrt{B_1^2+B_2^2+B_3^2}$ with B_1, B_2, B_3 being the components and the function Φ is explicitly given in [14]. It is noteworthy that the estimates were obtained by probabilistic techniques.

This work was subsequently improved in [7], again using stochastic methods. It was shown that the estimate (1.3) holds in any dimension and without the condition that B be curl-free. Similar results can be found in [20], where techniques from differential geometry were used.

In [13] we considered the simpler problem for a magnetic field of constant direction, or what amounts to the same, a magnetic field in two dimensions. However, we tried to retain as much information as possible about the magnetic field in the estimate on the semigroup. In the following we describe the main results obtained in [13].

2. MAIN RESULTS

A magnetic field in two dimensions is a scalar function B which can be expressed in terms of a (non-unique) vector potential A,

$$B(x) = \frac{\partial A_2(x)}{\partial x_1} - \frac{\partial A_1(x)}{\partial x_2}\,, \qquad A(x) = \left(A_1(x), A_2(x)\right)\,. \tag{2.1}$$

For example, if $B(x) = B_0$ is a constant magnetic field then

$$A(x) = \frac{B_0}{2}\left(-x_2, x_1\right)\,. \tag{2.2}$$

We want to estimate (for suitable p and q with $1 \le p \le q \le \infty$)

$$C(t;p,q) := \|\exp(tH)\|_{L^p\to L^q} = \sup_{u\in L^p} \frac{\|u(t)\|_q}{\|u\|_p} \quad \text{where } u(t) = e^{tH}\,u. \tag{2.3}$$

This norm is finite if $p \le q$, because Eq. (1.2) implies that

$$\|e^{tH}\|_{L^p\to L^q} \le \|e^{t\Delta/2}\|_{L^p\to L^q} < \infty \,. \tag{2.4}$$

If $p > q$ the norm is infinite, as can be seen from an appropriately chosen sequence u_n for which $\|u_n(t)\|_q/\|u_n\|_p \to \infty$ (see [13]).

The following theorem is essentially a corollary of a a theorem in [12].

Theorem 2.1. *Let $B(x) = B_0 > 0$, all $x \in \mathbb{R}^2$. For $1 < p \le q < \infty$, there is a centered Gaussian function*

$$u_0(x) = N_0 \exp\left\{-\frac{B_0}{4a_0}\,x^2\right\} \tag{2.5}$$

(where N_0 is an arbitrary constant and $a_0 > 0$ is given below) such that

$$C_0(t;p,q) := \|e^{tH(B_0)}\|_{L^p\to L^q} = \frac{\|u_0(t)\|_q}{\|u_0\|_p}. \tag{2.6}$$

The constant a_0 is given explicitly by

$$a_0 = d_{t,p,q} + \sqrt{d_{t,p,q}^2 + p - 1}\,, \tag{2.7}$$

$$d_{t,p,q} = \frac{1}{2c_t s_t}\left(\frac{p}{q} - 1 + s_t^2(p-2)\right)\,, \tag{2.8}$$

$$c_t = \cosh\left(\frac{B_0 t}{2}\right)\,, \qquad s_t = \sinh\left(\frac{B_0 t}{2}\right)\,. \tag{2.9}$$

This theorem implies that the norm of the semigroup with a constant magnetic field can be calculated explicitly, because in this case the time evolution operator has an explicitly known Gaussian integral kernel. If the vector potential A is chosen as in Eq. (2.2) above, the time evolution is given by

$$u(t,x) = \int G(t,x,y)\,u(y)\,d^2y\,, \tag{2.10}$$

where $G(t,x,y)$ is the Mehler kernel ([18], p. 168),

$$G(t,x,y) = \frac{B_0}{4\pi c_t}\exp\left\{-\frac{B_0}{4}\frac{c_t}{s_t}(x-y)^2 + \mathrm{i}\,\frac{B_0}{2}(x_1y_2 - x_2y_1)\right\}\,, \tag{2.11}$$

which is a complex, degenerate, centered Gaussian kernel (in the terminology of [12]). The time evolution of the initial function $u_0(x)$ is given by

$$u_0(t,x) = N(t)\exp\left\{-\frac{B_0}{4a(t)}\,x^2\right\}\,, \tag{2.12}$$

where

$$N(t) = \frac{N_0\,a_0}{a_0\,c_t + s_t}\,, \qquad a(t) = \frac{a_0\,c_t + s_t}{a_0\,s_t + c_t}\,. \tag{2.13}$$

The function $a(t)$ is the unique solution of the initial value problem

$$\dot{a}(t) = \frac{B_0}{2}\left(1 - a(t)^2\right), \qquad a(0) = a_0 . \tag{2.14}$$

For $a_0 = 1$, the function $u_0(x) = N_0 \exp(-B_0 x^2/4)$ is the eigenfunction of H corresponding to the eigenvalue $-B_0/2$.

Using the notation above, we obtain for the norm

$$C_0(t;p,q) = \frac{N(t)}{N_0}\left(\frac{B_0}{4a_0}\frac{p}{\pi}\right)^{1/p}\left(\frac{B_0}{4a(t)}\frac{q}{\pi}\right)^{-1/q} \tag{2.15}$$

In case of a non-constant field in two dimensions, we obtain the following result.

Theorem 2.2. *Assume that $B(x) \geq B_0 > 0$ is continuous. Then the estimate*

$$C(t;p,q) := \|e^{tH(B)}\|_{L^p \to L^q} \leq C_0(t;p,q), \tag{2.16}$$

holds for

a) all $t > 0$, if $p \leq 2$, $q \geq 2$,

b) $t > 0$ and $2 < p < q \leq \infty$, if $\frac{q-p}{q(p-2)} \geq s_t^2$,

c) $t > 0$ and $1 \leq p < q < 2$, if $\frac{q-p}{p(2-q)} \geq s_t^2$.

Remark 2.1. In the case where neither a), b) nor c) holds but $p \leq q$ we get a nontrivial bound but it is not sharp in the constant field case, see Eq. (3.10).

A simple consequence of Theorem 2.2 is the following bound

$$\|e^{tH(B)}\|_{L^1 \to L^\infty} \leq \frac{B_0}{4\pi \sinh\left(\frac{B_0 t}{2}\right)}, \tag{2.17}$$

which is best possible. In fact there is equality if $B(x) \equiv B_0$ as can be seen from the Mehler kernel. Using a technique due to Davies [6] this result can be extended to a pointwise bound on the integral kernel of $\exp(tH(B))$.

Theorem 2.3. *Assume that $B(x) \geq B_0 > 0$ is continuous. Then the integral kernel of* $\exp(tH(B))$ *satisfies the bound*

$$|e^{tH(B)}(x,y)| \leq \frac{B_0}{4\pi \sinh\left(\frac{B_0 t}{2}\right)} e^{-\frac{(x-y)^2}{2t}} . \tag{2.18}$$

Remark 2.2. The Gaussian decay on the right side of (2.18) is the one of the heat kernel, which is not optimal. Is it true that $|e^{tH}(x,y)|$ is bounded by the Mehler kernel $|G(t,x,y)|$? This is an open problem.

3. Discussion

The main idea in the proof of Theorem 1.2 is to consider

$$\frac{d}{ds}\ln\|u(s)\|_{r(s)} = \frac{\dot r(s)}{r(s)^2}\int |u(s,x)|^{r(s)}\ln|u(s,x)|^{r(s)}\,d^2x \tag{3.1}$$

$$+\frac{1}{2}\int |u(s,x)|^{(r(s)-2)}\frac{d}{ds}|u(s,x)|^2\,d^2x\,. \tag{3.2}$$

where $r:(0,t]\to\mathbb{R}$ is a twice differentiable function with $r(s)\geq 2$ and $\dot r(s)\geq 0$ for all $s\in(0,t]$. Here we assumed that $\|u(s)\|_{r(s)}=1$. We can use the "heat equation"

$$\frac{d}{ds}u = H\,u = \frac{1}{2}(\nabla+\mathrm{i}A)^2 u\,. \tag{3.3}$$

to rewrite and estimate the expression (3.2) in a simple and gauge invariant way. In order to estimate (3.1), we use the following family of logarithmic Sobolev inequalities for $g=|u|^{r/2}$

$$\int g^2\ln g^2 - \frac{\lambda}{\pi}\int(\nabla g)^2 \leq -2-\ln\lambda\,,\qquad \text{all } \lambda>0, \tag{3.4}$$

where we assumed that $\int g^2 = 1$. Equality holds if g is a normalized Gaussian,

$$g(x)=\frac{1}{\sqrt{\lambda}}\exp\left\{-\frac{\pi}{2\lambda}x^2\right\}\,. \tag{3.5}$$

This family is essentially Gross' logarithmic Sobolev inequality rewritten as an inequality on $\mathbb{R}^2$ (see, e.g., [5]).

We obtain the following result

Theorem 3.1. *Let $r:(0,t]\to\mathbb{R}$ be a twice differentiable function with $r(s)\geq 2$ and $\dot r(s)\geq 0$ for all $s\in(0,t]$. Then*

$$\frac{d}{ds}\ln\|u(s)\|_{r(s)} \leq -L(r(s),\dot r(s))\,, \tag{3.6}$$

for all $s\in(0,t]$, where

$$L(r,\dot r)=\frac{\dot r}{r^2}\left\{2+\ln\frac{4\pi a(r,\dot r)}{B_0 r}\right\}+\frac{a(r,\dot r)B_0}{r}\,, \tag{3.7}$$

and

$$a(r,\dot r)=\frac{1}{B_0 r}\left(\sqrt{\dot r^2+B_0^2r^2(r-1)}-\dot r\right). \tag{3.8}$$

Equality holds in Eq (3.6) for $B(x)=B_0$ and

$$u(s,x)=\exp\left(-\frac{B_0}{4a\big(r(s),\dot r(s)\big)}x^2\right)\,. \tag{3.9}$$

We note that if for some $s \in [0, t]$, the derivative $\dot{r}$ is negative, then $d/ds\ \ln \|u\|_r$ is unbounded. This can be seen already in case of a constant magnetic field.

Theorem 3.1 can be used to prove Theorem 2.2 in the following way. As an example, we consider only the case b), i.e., $2 < p < q < \infty$. c) follows from b) by duality, the cases $p = 1$ and $q = \infty$ follow by a limiting argument.

To prove b) for $q < \infty$ we integrate the inequality (3.2) and obtain

$$\frac{\|u(t)\|_q}{\|u(0)\|_p} \leq \exp\left\{-\int_0^t L(r(s), \dot{r}(s))\, ds\right\}. \tag{3.10}$$

Recall that the only assumption on $r(s)$ is that $r(0) = p$, $r(t) = q$ and that $\dot{r}(s)$ be nonnegative. Thus, we can optimize the right side of (4.1) over $r(s)$ which leads to an interesting problem in the calculus of variations. The Euler-Lagrange equation is

$$\ddot{r} = 2\frac{\dot{r}^2}{r} + \frac{1}{2}B_0^2 r(r-2). \tag{3.11}$$

It can be verified that $r(s)$ defined by

$$r(s) = p\Big(a_0 c_s + (1-p)s_s\Big)^{-1}\Big(\frac{c_s}{a_0} + s_s\Big)^{-1}, \tag{3.12}$$

(with c_t and s_t as in Eq. (2.9)) is a solution of Eq. (3.11) and satisfies $r(0) = p$. Moreover, if we choose a_0 as in (2.7), we see that $r(t) = q$. Using the function $r(s)$ as a trial function in (3.10) we obtain b) after some lengthy computations.

Remark 3.1. We note that the differential equation (3.11) for r and the equation (2.14) for a are equivalent, if we choose $a(s) = a(r(s), \dot{r}(s))$ according to Eq. (3.8). Hence the Gaussian function u_0 from Theorem 2.1 is the optimizer of the differential inequality (3.6) for all s, if $r(s)$ is chosen according to (3.12).

Remark 3.2. We finally explain, why our method fails to give an optimal bound if neither a), b) nor c) holds in Theorem 2.2. From

$$C_0(t; p, q) = \frac{\|u_0(t)\|_q}{\|u_0\|_p} = \frac{\|u_0(s)\|_r}{\|u_0\|_p}\frac{\|u_0(t)\|_q}{\|u_0(s)\|_r}, \tag{3.13}$$

($p \leq r \leq q$, $0 \leq s \leq t$, u_0 as in Theorem 2.1) it is obvious that

$$C_0(t; p, q) \leq C_0(s; p, r)\, C_0(t - s, r, q). \tag{3.14}$$

Our method can only work, if for all $s \in (0, t)$ there exists a number r with $p < r < q$ such that the above formula is an equality, i.e.,

$$C_0(t; p, q) = \min_{p \leq r \leq q}\{C_0(s; p, r)\, C_0(t - s, r, q)\}. \tag{3.15}$$

However, it is easy to see that Eq. (3.15) cannot hold in general. For example, it is wrong in the specific case $q = p > 2$, $0 < s < t$. From the explicit form of $C_0(t; p, q)$ we find that

$$C_0(t; p, q) = C_0(s; p, r(s))\, C_0(t - s; r(s), q) \tag{3.16}$$

holds for some $r(s) \in [p, q]$ if and only if the optimizer u_0 giving $C_0(t; p, q)$ is also the optimizer for $C_0(s; p, r(s))$ (i.e., $C_0(s; p, r(s)) = \|u(s)\|_{r(s)}/\|u_0\|_p$).

Eq. (3.16) can only hold for all $s \in (0, t)$ if the function $r(s)$ is monotonic nondecreasing, with $r(0) = p$ and $r(t) = q$. Given u_0 (and hence a_0) as the optimizer for $C_0(t;p,q)$, we can easily determine $r(s)$ using Theorem 1.1 by solving

$$a_0 = d_{s,p,r(s)} + \sqrt{d^2_{s,p,r(s)} + p - 1}\,. \tag{3.17}$$

This gives again the function (3.12) which is monotonic nondecreasing precisely in the cases a), b) and c) of Theorem 2.2. The identity

$$a(s) = d_{t-s,r(s),q} + \sqrt{d^2_{t-s,r(s),q} + r(s) - 1} \tag{3.18}$$

(with $a(s)$ as in (1.6)) shows that if u_0 is the optimizer for $C_0(s;p,r(s))$, then $u_0(s)$ is the optimizer for $C_0(t-s;r(s),q)$, with $r(s)$ as in Eq. (3.12).

References

1. J.E.Avron, I.Herbst, and B.Simon, Duke Math. J., 45, 847–883 (1978),
2. W.Beckner, Ann. of Math., 102, 159–182 (1975),
3. H.J.Brascamp and E.H.Lieb, Adv. in Math., 20, 151–172 (1976),
4. H.L.Cycon, R.G.Froese, W.Kirsch, and B.Simon, Schrödinger operators, Springer Verlag (1987).
5. Carlen, E. and Loss, M., Duke Math. J., 81, 135–157 (1996),
6. Davies, E.B., Heat kernels and spectral theory, Cambridge tracts in Mathematics Vol. 92, Cambridge (1989),
7. L.Erdös, Duke Math. J., 76, 541–566 (1994),
8. P.Federbush, J. Math. Phys., 10, 50–52 (1969),
9. L.Gross, Amer.J. Math., 97, 1061–1083 (1976),
10. H.Hess, R.Schrader, and D.A.Uhlenbrock, Duke Math. J., 44, 893-904 (1977),
11. T.Kato, Isr. J. Math., 13, 135–148 (1972),
12. E.Lieb, Inv. math., 102, 179–208 (1990),
13. M.Loss and B.Thaller, Optimal heat kernel estimates for Schrödinger operators with magnetic fields in two dimensions, Preprint (1996),
14. P.Malliavin, C.R.Acad.Sci.Paris Ser.I.Math., 302, 481–486 (1986),
15. E.Nelson, in: Mathematical Theory of Elementary Particles, R.Goodman and I.Segal, (Eds.), M.I.T. Press (1966),
16. B.Simon, Ind. Math. J., 26, 1067–1073 (1977),
17. B.Simon, J. Funct. Anal., 32, 97–101 (1979),
18. B.Simon, Functional integration and Quantum Physics, Academic Press (1979).
19. B.Simon, Bull. Amer. Math. Soc. (N.S.), 7, 447–526 (1982),
20. N.Ueki, J. Func. Anal., 120, 344–379 (1994), and 127, 257–258 (1995),

7th Int. Coll. on Differential Equations, pp. 221-229
D. Bainov (Ed)

On Linear Differential Equations of Fractional Order

Francesco MAINARDI

Department of Physics, University, Via Irnerio 46, I-40126 Bologna, Italy
Fax : *+39-51-247244* E-mail : *mainardi@bo.infn.it*

The processes involving basic phenomena of relaxation, diffusion, oscillations and wave propagation are of great relevance in physics; from a mathematical point of view they are known to be governed by simple differential equations of order 1 and 2 in time. The introduction of fractional derivatives of order α in time, with $0 < \alpha < 1$ or $1 < \alpha < 2$, leads to processes that, in mathematical physics, we may refer to as fractional phenomena. Our analysis, carried out by the Laplace transform, leads to certain special functions in one variable, the Mittag-Leffler and the Wright functions, which generalize in a straightforward way the characteristic functions of the basic phenomena, namely the exponential and the gaussian.

1991 Mathematics Subject Classification: 26A33, 33E20, 45E10, 70J99.
Key Words: Fractional calculus, Mittag-Leffler function, Wright function.

INTRODUCTION

In recent years considerable interest has been shown in the so called *fractional calculus*, which allows one to consider integration and differentiation of any order, not necessarily integer. Such interest has been stimulated by the applications that this calculus finds in different areas of physics and engineering, possibly including fractal phenomena.

It is the purpose of this note to provide an approach to fractional calculus that is as simple as possible. Here we adopt the following definition for the *fractional derivative of order* $\alpha > 0$ of a causal function $f(t)$ (*i.e.* vanishing for $t < 0$)

$$\frac{d^\alpha}{dt^\alpha} f(t) := \begin{cases} f^{(n)}(t) & \text{if} \quad \alpha = n \in \mathbb{N}, \\ \dfrac{1}{\Gamma(n-\alpha)} \displaystyle\int_0^t \frac{f^{(n)}(\tau)}{(t-\tau)^{\alpha+1-n}}\, d\tau & \text{if} \quad n-1 < \alpha < n, \end{cases} \tag{1}$$

where $f^{(n)}(t)$ denotes the derivative of order n and Γ is the Gamma function. We apply the above notion in order to generalize some basic topics of classical mathematical physics, which are treated by simple, linear, ordinary or partial, differential equations.

We start to consider the *fractional* ordinary differential equation

$$\frac{d^\alpha u}{dt^\alpha} + \omega^\alpha\, u(t;\alpha) = 0\,, \quad 0 < \alpha \le 2\,, \tag{2}$$

where ω is a positive constant with the dimensions T^{-1} as a frequency, and the field variable $u(t;\alpha)$ is assumed to be a casual function of time. According to the cases $0 < \alpha \le 1$ and $1 < \alpha \le 2$ the equation (2) can be referred to as the *fractional relaxation* or the *fractional oscillation* equation, respectively.

Then, we consider the following *fractional* partial differential equation

$$\frac{\partial^{2\beta} u}{\partial t^{2\beta}} = D \frac{\partial^2 u}{\partial x^2}, \qquad -\infty < x < +\infty, \quad 0 < \beta \le 1, \tag{3}$$

where D denotes a positive constant with the dimensions $L^2 T^{-2\beta}$, and the field variable $u = u(x,t;\beta)$ is assumed to be a causal function of time with $u(\mp\infty, t; \beta) = 0$. It should be noted that here, for the sake of convenience, we agree to denote the order of the time differentiation as 2β rather than α. As before, we have to distinguish two cases, namely $0 < \beta \le 1/2$ and $1/2 < \beta \le 1$, for which the equation (3) can be referred to as the *fractional diffusion* or the *fractional wave-propagation* equation, respectively.

The solutions of the fractional differential equations (2) and (3) will be obtained by applying the classical technique of the Laplace transform. In this respect, using the following notation for the Laplace transform pairs

$$f(t) \div \tilde{f}(s) = \mathcal{L}\,[f(t)] := \int_0^{\to\infty} e^{-st} f(t)\,dt, \quad t \in \mathbb{R}^+, \quad s \in \mathbb{C}, \tag{4}$$

we need to recall the Caputo rule [1]

$$\mathcal{L}\left[\frac{d^\alpha}{dt^\alpha} f(t)\right] = s^\alpha \tilde{f}(s) - \sum_{k=0}^{n-1} s^{\alpha-1-k} f^{(k)}(0^+), \quad n-1 < \alpha \le n. \tag{5}$$

The reader, who is interested in extensive treatments of the fractional calculus and some of its applications, can consult *i.e.* the references [1-49]. For more details about the results here resumed, one can refer to our original papers [39-42], [45-47] and our recent lecture notes [49].

THE FRACTIONAL RELAXATION-OSCILLATION EQUATION

According to our plan outlined in the introduction, we start to consider the *fractional* ordinary differential equation (2) distinguishing the cases $0 < \alpha \le 1$ (*fractional relaxation*) and $1 < \alpha \le 2$ (*fractional oscillation*). In the former case, Eq. (2) must be equipped with a single intial condition, say $u(0^+;\alpha) = u_0$, and in the latter with two intial conditions, say $u(0^+;\alpha) = u_0$, and $\dot{u}(0^+;\alpha) = v_0$. We assume $v_0 \equiv 0$ in order to ensure the continuous dependence of the solution of (2) on the parameter α also in the transition from $\alpha = 1^-$ to $\alpha = 1^+$.

The solution of (2) can be obtained by applying the Laplace transform technique and using the rule (5) with $n = 1, 2$; in both cases, since $v_0 \equiv 0$, the image solution turns out to be

$$\tilde{u}(s;\alpha) = u_0 \frac{s^{\alpha-1}}{s^\alpha + \omega^\alpha}. \tag{6}$$

To invert (6) we can use the series expansion theorem or the complex Bromwich formula; we obtain

$$u(t;\alpha) = u_0 \, E_\alpha[-(\omega t)^\alpha], \tag{7}$$

where E_α denotes the Mittag-Leffler function [50]. In the complex plane this function is defined by the following series and integral representations

$$E_\alpha(z) := \sum_{n=0}^{\infty} \frac{z^n}{\Gamma(\alpha\, n+1)}\,, \quad \alpha > 0\,, \quad z \in \mathbf{C}\,, \tag{8}$$

$$E_\alpha(z) := \frac{1}{2\pi i}\int_{Ha} \frac{\zeta^{\alpha-1} e^{\zeta}}{\zeta^\alpha - z}\, d\zeta\,, \quad \alpha > 0\,, \quad z \in \mathbf{C}\,. \tag{9}$$

In (9) Ha denotes the Hankel path, a loop which starts from $-\infty$ along the lower side of the negative real axis, encircles the circular disc $|\zeta| \le |z|^{1/\alpha}$ in the positive sense and ends at $-\infty$ along the upper side of the negative real axis.

From the series representation (8) we can easily recognize that $E_\alpha(z)$ is an *entire function* (of order $1/\alpha$), which provides a simple generalization of the exponential function ($\alpha = 1$). Other particular cases in which the Mittag-Leffler function reduces to known functions are

$$E_{1/2}\left(\pm z^{1/2}\right) = \exp(z)\,\mathrm{erfc}\,(\mp z^{1/2})\,, \tag{10}$$

$$E_2(+z^2) = \cosh(z)\,, \qquad E_2(-z^2) = \cos(z)\,. \tag{11}$$

The most interesting properties of the Mittag-Leffler function are associated with its asymptotic expansions as $z \to \infty$ in various sectors of the complex plane and with its connection with the Laplace integral [50]. For the present purposes, our interest in this function is limited to negative argument and $0 < \alpha \le 2$. Taking for the sake of simplicity $\omega \equiv 1$ and putting

$$e_\alpha(t) := E_\alpha(-t^\alpha) = \sum_{n=0}^{\infty} (-1)^n \frac{t^{\alpha n}}{\Gamma(\alpha\, n+1)}\,, \tag{12}$$

the main properties are hereafter resumed.

- For $0 < \alpha \le 1$, $e_\alpha(t)$ turns out to be *positive* and *completely monotonic* i.e.

$$(-1)^n \frac{d^n}{dt^n} e_\alpha(t) > 0\,, \qquad t > 0\,, \quad 0 < \alpha \le 1\,. \tag{13}$$

- For $1 < \alpha < 2$, $e_\alpha(t)$ exhibits an odd number of zeros, while for $\alpha = 2$, $e_2(t) = \cos(t)$ exhibits an infinite number of zeros.
- For $0 < \alpha < 1$, and $1 < \alpha < 2$, the following asymptotic expansion is valid

$$e_\alpha(t) \sim -\sum_{n=1}^{\infty} (-1)^n \frac{t^{-\alpha n}}{\Gamma(1-\alpha\, n)}\,, \qquad t \to +\infty\,, \quad 0 < \alpha < 1\,, \quad 1 < \alpha < 2\,. \tag{14}$$

Other relevant properties of $e_\alpha(t)$ can be derived in detail for $0 < \alpha \le 2$, directly from its representation as a Laplace inversion integral (Bromwich formula), without detour on the general theory of Mittag-Leffler functions in the complex plane. The essential step consists in decomposing $e_\alpha(t)$ into two parts according to $e_\alpha(t) = f_\alpha(t) + g_\alpha(t)$, as indicated below. The part f_α, present for α not integer, turns out to be a *completely monotonic* function, which tends to zero as t tends to infinity; the other part turns out to be of *oscillatory* character if $1 < \alpha < 2$, but identically vanishing if $0 < \alpha < 1$.

We separately discuss the cases: (a) $0 < \alpha < 1$, (b) $1 < \alpha < 2$, recalling that for $\alpha = 1, 2$, we know $e_\alpha(t)$ as elementary function, namely $e_1(t) = \exp(-t)$, $e_2(t) = \cos(t)$. In order to obtain the desired decomposition of e_α we bend the Bromwich path of integration Br into the equivalent Hankel path $Ha(1^+)$, a loop which starts from $-\infty$ along the lower side of the negative real axis, encircles the circular disc $|s| = 1$ in the positive sense and ends at $-\infty$ along the upper side of the negative real axis. One obtains in all cases (a), (b),

$$e_\alpha(t) = f_\alpha(t) + g_\alpha(t), \quad t \geq 0, \tag{15}$$

with

$$f_\alpha(t) := \frac{1}{2\pi i} \int_{Ha(\epsilon)} e^{st} \frac{s^{\alpha-1}}{s^\alpha + 1} ds, \tag{16}$$

where now the Hankel path $Ha(\epsilon)$ denotes a loop constituted by a small circle $|s| = \epsilon$ with $\epsilon \to 0$ and by the two borders of the cut negative real axis, and

$$g_\alpha(t) := \sum_h e^{s'_h t} Res \left[\frac{s^{\alpha-1}}{s^\alpha + 1}\right]_{s'_h}, \tag{17}$$

where s'_h are the relevant poles of $s^{\alpha-1}/(s^\alpha + 1)$. In fact the poles turn out to be $s_h = \exp[i(2h+1)\pi/\alpha]$ with unitary modulus; they are all simple but relevant are only those situated in the main Riemann sheet, *i.e.* the poles s'_h for which $-\pi < \arg s'_h < \pi$.

If $0 < \alpha < 1$, there are no such poles, since for all integer h we have $|\arg s_h| = |2h+1|\,\pi/\alpha > \pi$; as a consequence,

$$g_\alpha(t) \equiv 0, \quad \text{hence} \quad e_\alpha(t) = f_\alpha(t), \quad \text{if} \quad 0 < \alpha < 1. \tag{18}$$

If $1 < \alpha < 2$, then there exist precisely two relevant poles, namely $s'_0 = \exp(i\pi/\alpha)$ and $s'_{-1} = \exp(-i\pi/\alpha) = \overline{s'_0}$, and one obtains

$$g_\alpha(t) = \frac{2}{\alpha} e^{t \cos(\pi/\alpha)} \cos\left[t \sin\left(\frac{\pi}{\alpha}\right)\right], \quad \text{if} \quad 1 < \alpha < 2. \tag{19}$$

We note that this function exhibits oscillations with circular frequency $\sin(\pi/\alpha)$ and with an amplitude which exponentially decays with the rate $|\cos(\pi/\alpha)|$. Of course, in the classical case $\alpha = 2$, we recover the sinusoidal behaviour with unitary frequency.

It is now an exercise in complex analysis to show that the contribution from the Hankel path $Ha(\epsilon)$ as $\epsilon \to 0$ is provided by

$$f_\alpha(t) := \int_0^\infty e^{-rt} K_\alpha(r)\, dr, \quad K_\alpha(r) = \frac{1}{\pi} \frac{r^{\alpha-1} \sin(\alpha\pi)}{r^{2\alpha} + 2r^\alpha \cos(\alpha\pi) + 1}. \tag{20}$$

This function $K_\alpha(r)$ vanishes identically if α is an integer, it is positive for all r if $0 < \alpha < 1$, negative for all r if $1 < \alpha < 2$. In fact, in (20) the denominator is, for α not integer, always positive being $> (r^\alpha - 1)^2 \geq 0$. Hence $f_\alpha(t)$ in case (a), $-f_\alpha(t)$ in case (b) has the aforementioned property of complete monotonicity, decreasing towards zero for $t \to \infty$. We also note that, in order to satisfy the initial condition $e_\alpha(0^+) = 1$, we find $\int_0^\infty K_\alpha(r)\, dr = 1$ if $0 < \alpha < 1$, $\int_0^\infty K_\alpha(r)\, dr = 1 - 2/\alpha$ if $1 < \alpha < 2$.

THE FRACTIONAL DIFFUSION-WAVE EQUATION

Let us now consider the *fractional* partial differential equation (3) distinguishing the cases $0 < \beta \le 1/2$ (*fractional diffusion*) and $1/2 < \beta \le 1$ (*fractional wave*). In the former case, Eq. (3) must be equipped with a single intial condition, say $u(x, 0^+; \beta) = f(x)$, while in the latter with two intial conditions, say $u(x, 0^+; \beta) = f(x)$, and $\dot{u}(x, 0^+; \beta) = g(x)$. We assume $g(x) \equiv 0$ in order to ensure the continuous dependence of the solution of (3) on the parameter β also in the transition from $\beta = (1/2)^-$ to $\beta = (1/2)^+$.

Let us now solve the above intial value problem (or *Cauchy problem*) recurring to the *Green function* $\mathcal{G}_c(x, t; \beta)$ which represents the solution for $f(x) = \delta(x)$. This method allows one to obtain the required solution as

$$u(x, t; \beta) = \int_{-\infty}^{+\infty} \mathcal{G}_c(\xi, t; \beta)\, f(x - \xi)\, d\xi\,. \tag{21}$$

It should be noted that $\mathcal{G}_c(x, t; \beta) = \mathcal{G}_c(|x|, t; \beta)$ since the Green function turns out to be an even function of x.

The Green function will be determined by using the technique of the Laplace transform. This technique allows us to obtain the transformed functions $\widetilde{\mathcal{G}_c}(x, s; \beta)$ by solving an ordinary differential equations of the 2-nd order in x and then, by inversion, $\mathcal{G}_c(x, t; \beta)$. In fact, the application of the Laplace transform to (3) with $f(x) = \delta(x)$, leads to the following non homogeneous differential equation satisfied by the image of the Green function, $\widetilde{\mathcal{G}_c}(x, s; \beta)$,

$$s^{2\beta}\, \widetilde{\mathcal{G}_c} - \delta(x)\, s^{2\beta - 1} = D\, \frac{d^2 \widetilde{\mathcal{G}_c}}{dx^2}\,. \tag{22}$$

Because of the singular term $\delta(x)$, we have to consider the above equation separately in the two intervals $x < 0$ and $x > 0$, imposing the vanishing conditions at $x = \mp\infty$ and the necessary matching conditions at $x = 0^\mp$. We obtain

$$\widetilde{\mathcal{G}_c}(x, s; \beta) = \frac{1}{2\sqrt{D}\, s^{1-\beta}}\, \mathrm{e}^{-(|x|/\sqrt{D})s^\beta}\,. \tag{23}$$

We are going to show how the inversion of the Laplace transform in (23) leads us to the *auxiliary function* of the *similarity variable*

$$z = \frac{|x|}{\sqrt{D}\, t^\beta}\,, \qquad 0 < \beta < 1\,. \tag{24}$$

In fact, applying the inversion Bromwich formula to (23), introducing the *similarity variable* z as in (24), and putting $\sigma = st$, we just obtain

$$|x|\, \mathcal{G}_c(x, t; \beta) = \frac{z}{2}\, M(z; \beta)\,, \quad M(z; \beta) := \frac{1}{2\pi i} \int_{Br} \mathrm{e}^{\sigma - z\sigma^\beta}\, \frac{d\sigma}{\sigma^{1-\beta}}\,, \quad 0 < \beta < 1\,, \tag{25}$$

where $M(z; \beta)$ is our *auxiliary function* and Br denotes the Bromwich path. The above definition of $M(z; \beta)$ (*Bromwich representation*) can be analytically continued from $z > 0$ to $\forall z \in \mathbb{C}$, adopting suitable integral and series representations valid in all of $\mathbb{C}$. For this purpose, let us deform the Bromwich path Br into the Hankel path Ha [a contour that begins at $\sigma = -\infty - ia$ $(a > 0)$, encircles the branch cut that lies along the negative real axis, and ends up at $\sigma = -\infty + ib$ $(b > 0)$], which is equivalent to the original path (at least) for $z > 0$. We therefore obtain the *Hankel representation*

$$M(z;\beta) := \frac{1}{2\pi i}\int_{Ha} e^{\sigma - z\sigma^{\beta}} \frac{d\sigma}{\sigma^{1-\beta}}, \quad 0 < \beta < 1. \tag{26}$$

Writing

$$\begin{aligned} 2\pi i\, M(z;\beta) &= \int_{Ha} e^{\sigma} \left[\sum_{n=0}^{\infty} \frac{(-1)^n z^n}{n!} \upsilon^{\beta n}\right] \frac{d\sigma}{\sigma^{1-\beta}} \\ &= \sum_{n=0}^{\infty} \frac{(-1)^n z^n}{n!} \left[\int_{Ha} e^{\sigma}\, \sigma^{\beta n+\beta-1}\, d\sigma\right], \end{aligned} \tag{27}$$

and using the Hankel representation of the reciprocal of the Gamma function, we obtain the following *series representation*

$$M(z;\beta) := \sum_{n=0}^{\infty} \frac{(-1)^n z^n}{n!\,\Gamma[-\beta n + (1-\beta)]}, \quad 0 < \beta < 1. \tag{28}$$

Since the radius of convergence of the power series (28) can be proven to be infinite for $0 < \beta < 1$, our function turns out to be entire in z and therefore the exchange between the series and the integral in (27) is legitimate. Consequently, (26) and (28) provide the integral and series representations of $M(z;\beta)$, valid on all of $\mathbf{C}$, respectively. In the theory of special functions (see, *e.g.* [50]) we find an entire function of z, referred to as the *Wright function*, which reads (in our notation)

$$W(z;\lambda,\mu) := \sum_{n=0}^{\infty} \frac{z^n}{n!\,\Gamma(\lambda n + \mu)} := \frac{1}{2\pi i}\int_{Ha} e^{\sigma + z\sigma^{-\lambda}} \frac{d\sigma}{\sigma^{\mu}}, \tag{29}$$

where $\lambda > -1$, $\mu > 0$. We thus recognize that

$$M(z;\beta) = W(-z;-\beta, 1-\beta). \tag{30}$$

We can prove that for $\beta = 1/q$, where $q \geq 2$ is a positive integer, the auxiliary function can be expressed as a sum of $(q-1)$ simpler entire functions. In the particular cases $q = 2$ and $q = 3$ we find

$$M(z;1/2) = \frac{1}{\sqrt{\pi}} \sum_{m=0}^{\infty} (-1)^m \left(\frac{1}{2}\right)_m \frac{z^{2m}}{(2m)!} = \frac{1}{\sqrt{\pi}} \exp\left(-z^2/4\right), \tag{31}$$

and

$$\begin{aligned} M(z;1/3) &= \frac{1}{\Gamma(2/3)} \sum_{m=0}^{\infty} \left(\frac{1}{3}\right)_m \frac{z^{3m}}{(3m)!} - \frac{1}{\Gamma(1/3)} \sum_{m=0}^{\infty} \left(\frac{2}{3}\right)_m \frac{z^{3m+1}}{(3m+1)!} \\ &= 3^{2/3}\,\mathrm{Ai}\left(z/3^{1/3}\right), \end{aligned} \tag{32}$$

where $(a)_m := a(a+1)\ldots(a+m-1)$ denotes the Pochhammer symbol and Ai the *Airy function*. Thus, we recognize that for $\beta = 1/2$ the *Gaussian auxiliary function* of the classical diffusion equation is recovered. Furthermore, always for $\beta = 1/q$, it can be proved that our function satisfies the following differential equation of order $q-1$,

$$\frac{d^{q-1}}{dz^{q-1}} M(z;1/q) + \frac{(-1)^q}{q} z\, M(z;1/q) = 0\,, \tag{33}$$

subjected to the $q-1$ initial conditions at $z = 0$, derived from (28),

$$M^{(h)}(0;\beta) = \frac{(-1)^h}{\pi} \Gamma[\beta(h+1)] \sin[\pi\beta(h+1)]\,, \quad h = 0,\, 1,\, \ldots\, q-2\,. \tag{34}$$

We note that, for $q \geq 4$, Eq. (33) is akin to the *hyper-Airy differential equation* of order $q-1$. Consequently, in view of the above considerations, the auxiliary function can be considered as a *generalized hyper-Airy function.*

For the fractional diffusion-wave equation (3), only real positive values of z are of interest. As a matter of fact, it turns out that, for $z \in \mathbb{R}^+$, the function $M(z;\beta)$ is positive; in particular, it is *monotonically decreasing* for $0 < \beta < 1/2$, while for $1/2 < \beta < 1$ it is *unimodal, i.e.* it exhibits only one maximum, $M_0(\beta)$, at a certain point $z_0(\beta)$. It should be noted that, as $\beta \to 1^-$, $M_0(\beta) \to +\infty$ and $z_0(\beta) \to 1$. The case $\beta = 1$, for which (3) reduces to the classical wave equation, is of course singular since $M(z;1) \propto \delta(z-1)$.

REFERENCES

1. M. Caputo, *Elasticità e Dissipazione*, Zanichelli, Bologna (1969). [in Italian]
2. H.T. Davis, *The Theory of Linear Operators*, The Principia Press, Bloomington, Indiana (1936).
3. A. Erdélyi (Editor), *Tables of Integral Transforms*, Vol. 2, Ch. 13, McGraw-Hill, New York (1954).
4. I.M. Gel'fand and G.E. Shilov, *Generalized Functions*, Vol. 1, Academic Press, New York (1964).
5. K.B. Oldham and J. Spanier, *The Fractional Calculus*, Academic Press, New York (1974).
6. B. Ross (Editor), *Fractional Calculus and its Applications*, Lecture Notes in Mathematics # 457, Springer Verlag, Berlin (1975).
7. A.C. McBride, *Fractional Calculus and Integral Transforms of Generalized Functions*, Pitman Research Notes in Mathematics # 31, Pitman, London (1979).
8. A.C. McBride and G.F. Roach (Editors), *Fractional Calculus*, Pitman Research Notes in Mathematics # 138, Pitman, London (1985).
9. Yu. I. Babenko, *Heat and Mass Transfer*, Chimia, Leningrad (1986). [in Russian]
10. S.G. Samko, A.A. Kilbas and O.I. Marichev, *Fractional Integrals and Derivatives, Theory and Applications*, Gordon and Breach, Amsterdam (1993). [Engl. Transl. from the Russian edition, Nauka i Tekhnika, Minsk (1987)]
11. H.M. Srivastava and S. Owa (Editors), *Univalent Functions, Fractional Calculus, and their Applications*, Ellis Horwood, Chichester (1989).
12. K. Nishimoto (Editor), *Fractional Calculus and its Applications*, Nihon University, Tokyo (1990).
13. K. Nishimoto, *An Essence of Nishimoto's Fractional Calculus*, Descartes Press, Koriyama (1991).

14. R. Gorenflo and S. Vessella, *Abel Integral Equations: Analysis and Applications*, Lecture Notes in Mathematics # 1461, Springer-Verlag, Berlin (1991).

15. R.N. Kalia (Editor), *Recent Advances in Fractional Calculus*, Global Publ., Sauk Rapids, Minnesota (1993).

16. K.S. Miller and B. Ross, *An Introduction to the Fractional Calculus and Fractional Differential Equations*, Wiley, New York (1993).

17. V. Kiryakova, *Generalized Fractional Calculus and Applications*, Pitman Research Notes in Mathematics # 301, Longman, Harlow (1994).

18. P. Rusev, I. Dimovski and V. Kiryakova (Editors), *Transform Methods and Special Functions, Sofia 1994*, Science Culture Technology, Singapore (1995).

19. M. Caputo and F. Mainardi, A new dissipation model based on memory mechanism, *Pure Appl. Geophys.*, **91**, 134-147 (1971).

20. M. Caputo and F. Mainardi, Linear models of dissipation in anelastic solids, *Riv. Nuovo Cimento* (Ser. II), **1**, 161-198 (1971).

21. M. Caputo, Vibrations of an infinite viscoelastic layer with a dissipative memory, *J. Acoust. Soc. Am.*, **56** (3), 897-904 (1974).

22. R.L. Bagley and P.J. Torvik, A theoretical basis for the application of fractional calculus, *J. Rheology*, **27**, 201-210 (1983).

23. P.J. Torvik and R.L. Bagley, On the appearance of the fractional derivatives in the behavior of real materials, *J. Appl. Mech.* (Trans. ASME), **51**, 294-298 (1984).

24. R. C. Koeller, Applications of fractional calculus to the theory of viscoelasticity, *J. Appl. Mech.* (Trans. ASME), **51**, 299-307 (1984).

25. R.R. Nigmatullin, On the theory of relaxation with "remnant" memory, *Phys. Stat. Sol. B*, **124**, 389-393 (1984). [English transl. from Russian]

26. R.R. Nigmatullin, The realization of the generalized transfer equation in a medium with fractal geometry, *Phys. Stat. Sol. B*, **133**, 425-430 (1986). [English transl. from Russian]

27. F. Mainardi and E. Bonetti, The application of real-order derivatives in linear viscoelasticity, *Rheol. Acta*, **26** Suppl., 64-67 (1988).

28. W.R. Schneider and W. Wyss, Fractional diffusion and wave equations, *J. Math. Phys.*, **30**, 134–144 (1989).

29. A.N. Kochubei, Fractional order diffusion, *J. Diff. Eqs*, **26** (4), 485–492 (1990). [English transl. from Russian]

30. W.R. Young, A. Pumir and Y. Pomeau, Anomalous diffusion of tracer in convection rolls, *Phys. Fluids A*, **1** (3), 462-469 (1989).

31. U.J. Choi and R.C. MacCamy, Fractional order Volterra equations with applications to elasticity, *J. Math. Anal. Applics.*, **139**, 448-464 (1989).

32. A. LeMéhauté, *Les Géometries Fractales*, Hermes, Paris (1990).

33. T.F. Nonnenmacher, Fractional integral and differential equations for a class of Levy-type probability densities, *J. Phys A: Math. Gen.*, **23**, L697-L700 (1990).

34. N. Sugimoto, Burgers equation with a fractional derivative; hereditary effects of nonlinear acoustic waves, *J. Fluid Mech.*, **225**, 631-653 (1991).

35. M. Giona and H.E. Roman, Fractional diffusion equation for transport phenomena in random media, *Physica A*, **185**, 82-97 (1992).

36. M. Caputo, The splitting of the seismic rays due to dispersion in the Earth's interior, *Rend. Fis. Acc. Lincei* (Ser. IX), **4**, 279-286 (1993).

37. G.M. Zaslavsky, Fractional kinetic equation for Hamiltonian chaos, *Physica D*, **76** (1-3), 110-122 (1994).

38. F. Mainardi, Fractional relaxation in anelastic solids, *J. Alloys and Compounds*, **211/212**, 534-538 (1994).

39. F. Mainardi, On the initial value problem for the fractional diffusion-wave equation, in *Waves and Stability in Continuous Media*, edited by S. Rionero and T. Ruggeri, pp. 246-251, World Scientific, Singapore (1994).

40. F. Mainardi, Fractional diffusive waves in viscoelastic solids, in *Nonlinear Waves in Solids*, edited by J.L. Wegner and F.R. Norwood, pp. 93-97, ASME/AMR, Fairfield (1995). [Abstract in *Appl. Mech. Rev.*, **46**, 549 (1993)]

41. F. Mainardi, The time fractional diffusion-wave equation, *Radiofisika* [*Radiophysics & Quantum Electronics*], **38** (1-2), 20-36 (1995).

42. F. Mainardi and M. Tomirotti, On a special function arising in the time fractional diffusion-wave equation, in ref. 18 (1995), pp. 171-183.

43. R. Gorenflo and R. Rutman, On ultraslow and on intermediate processes, in ref. 18 (1995), pp. 61-81.

44. I. Podlubny, Solutions of linear fractional differential equations, in ref. 18 (1995), pp. 227-230.

45. F. Mainardi, Fractional relaxation-oscillation and fractional diffusion-wave phenomena, *Chaos, Solitons & Fractals* **7** (9), 1461-1477 (1996).

46. F. Mainardi, The fundamental solutions for the fractional diffusion-wave equation, *Appl. Maths Lett.*, **9** No 6, 23-28 (1996).

47. R. Gorenflo and F. Mainardi, Fractional oscillations and Mittag-Leffler functions, Preprint No. A-14/96, Freie Universität Berlin, Serie A Mathematik (1966).

48. F. Mainardi and P. Pironi, The fractional Langevin equation: Brownian motion revisited, *Extracta Mathematicae*, **11** (1), 140-154 (1996).

49. F. Mainardi, Fractional Calculus: An Introduction with Applications, in *Scaling, Fractals and Fractional Calculus in Continuum Mechanics*, edited by A. Carpinteri and F. Mainardi, Springer Verlag, Wien and New York (1997). [Lecture Notes of the Advanced School held at CISM, Udine, Italy, 23-27 September 1996]

50. A. Erdélyi (Editor), *Higher Transcendental Functions*, Vol. 3. Ch. 18, McGraw-Hill, New York (1955).

7th Int. Coll. on Differential Equations, pp. 231-236
D. Bainov (Ed)

DIFFERENTIABILITY FOR NONLINEAR PARABOLIC SYSTEMS

MARIO MARINO AND ANTONINO MAUGERI
Dipartimento di Matematica, Università di Catania, Viale A. Doria, 6, 95125 Catania, ITALY

Abstract. Gagliardo-Nirenberg estimates in generalized Sobolev spaces are proved in order to achieve the differentiability of the weak solutions to the system (1.1).

Keywords: Nonlinear parabolic systems, Gagliardo–Nirenberg estimates, Interpolation theory, Besov Spaces.

1. Introduction

Let us consider a weak solution $u \in L^2\left(-T, 0, H^1(\Omega)\right) \cap C^{0,\lambda}(\overline{Q})$, $0 < \lambda < 1$, to the nonlinear parabolic system of second order with natural growth

$$(1.1) \quad -\sum_{j=1}^{n} D_j\, a^j(X,u,Du) + \frac{\partial u}{\partial t} = B^0(X,u,Du), \quad X=(x,t), \;\; Q = \Omega \times (-T,0),$$

where $a^j(X,u,p)$, $j = 1,2,\dots,n$, and $B^0(X,u,p)$ are Carathéodory vectors of $\mathbb{R}^N$, N integer ≥ 1) defined in $\Lambda = \overline{Q} \times \mathbb{R}^N \times \mathbb{R}^{nN}$ such that

i) $a^j(X,u,p) \in C^1(\Lambda)$ *and there exists* $\nu > 0$ *such that*

$$\sum_{h,k=1}^{N} \sum_{i,j=1}^{n} \frac{\partial a_h^j(X,u,p)}{\partial p_k^i} \xi_h^j \xi_k^i \geq \nu \sum_{i=1}^{n} \|\xi^i\|^2,$$
$$\forall \xi = (\xi^1,\dots,\xi^n), \xi^i \in \mathbb{R}^N, \; \forall (X,u,p) \in \Lambda;$$

ii) $\forall (X,u,p) \in \Lambda$ *there exists a constant* $M > 0$ *such that*

$$\sum_{k=1}^{N} \sum_{i=1}^{n} \left\| \frac{\partial a^j}{\partial p_k^i} \right\| \leq M, \quad j = 1,\dots,n,$$

$$\|a^j\| + \sum_{s=1}^{n} \left\| \frac{\partial a^j}{\partial x_s} \right\| + \sum_{k=1}^{N} \left\| \frac{\partial a^j}{\partial u_k} \right\| \leq M \left(1 + \sum_{i=1}^{n} \|p^i\| \right), \; j = 1,\dots,n;$$

iii) $\forall (X,u,p) \in \Lambda$ *it results*

$$\|B^0(X,u,p)\| \leq M \left(1 + \sum_{i=1}^{n} \|p^i\|^2 \right).$$

In [1] it was shown the following differentiability result:

Theorem 1.1. *Under the assumptions above, if* $u \in L^2\left(-T, 0, H^1(\Omega)\right) \cap C^{0,\lambda}(\overline{Q}), 0 < \lambda < 1$, *is a solution to the system* (1.1), *then, for each cube* $B(3\sigma) = B(x^0, 3\sigma) \subset\subset \Omega$, $\forall a, b \in (0, T)$, $a < b$, *we have*

$$u \in L^2\left(-a, 0, H^{1+\vartheta}(B(\sigma))\right), \quad \forall \vartheta \in (0,1), \tag{1.2}$$

and

$$\int_{-a}^{0} | Du |^2_{\vartheta, B(\sigma)}\, dt \leq c \left\{ \int_{-b}^{0} | u |^2_{1, B(3\sigma)}\, dt + 1 \right\}. \tag{1.3}$$

The result above does not allow us to reach $\vartheta = 1$, i.e., to obtain $u \in L^2\left(-a, 0, H^2(B(\sigma))\right)$ since to achieve this further result we need

$$Du \in L^4\left(B(\sigma) \times (-a, 0)\right).$$

Then a question naturally arises: if $u \in L^2\left(-a, 0, H^{1+\vartheta}(B(\sigma))\right) \cap C^{0,\lambda}(\overline{Q}), \forall \vartheta \in (0,1)$, it is possible to show that

$$Du \in L^4\left(B(\sigma) \times (-a, 0)\right)?$$

In other words, are the Gagliardo-Nirenberg estimates true also in the generalized Sobolev spaces?

A positive answer to this question is presented in the next section. By the way, we recall that the Hölder-continuity results for the solutions $u \in L^2\left(-T, 0, H^1(\Omega)\right)$ to the systems (1.1) are shown in [2], [3], [4], [5].

2. Gagliardo-Nirenberg estimates in the generalized Sobolev spaces

It is well known the following important result (see [9])

Theorem 2.1. *Let* Ω *be a bounded open subset of* R^n *with cone property. If* $u \in H^{m,r}(\Omega) \cap C^{s,\lambda}(\Omega)$, *with* m *integer* ≥ 2, $1 < r < \infty$, s *integer* ≥ 0, $0 < \lambda < 1, s < m-1$, *then, for each integer* j *with* $s + \lambda < j < m$, *there exist two constants* c_1 *and* c_2 *(depending on* $\Omega, m, r, s, \lambda, j$*) such that:*

$$| D^j u |_p \leq c_1 \, | D^m u |_r^a \, | u |^{1-a}_{-\frac{n}{s+\lambda}} + c_2 \, | u |_{-\frac{n}{s+\lambda}}, \tag{2.1}$$

where $\frac{1}{p} = \frac{j}{n} + a\left(\frac{1}{r} - \frac{m}{n}\right) - (1-a)\frac{s+\lambda}{n}, \forall a \in \left[\frac{j-s-\lambda}{m-s-\lambda}, 1\right[$.

Then the following results have been just proved in the forth coming work [7]:

Theorem 2.2. *Let* Ω *be a bounded open subset of* $\mathbb{R}^n$ *with a* C^∞ *boundary* $\partial\Omega$. *If* $u \in H^{m+\vartheta,r}(\Omega) \cap C^{s,\lambda}(\Omega)$, *with* m *integer* ≥ 1, $0 < \vartheta < 1$, $1 < r < \infty$, s *integer* ≥ 0, $0 < \lambda < 1$, $s < m$, *then, for each* j *with* $\max\{s + \lambda, m + \vartheta - \frac{n}{r}\} < j < m + \vartheta$, *there exists a constant* c *(depending on* $\Omega, m, \vartheta, r, s, \lambda, j, n, a$*) such that:*

$$\|u\|_{H^{j,p}(\Omega)} \leq c \|u\|^a_{H^{m+\vartheta,r}(\Omega)} \|u\|^{1-a}_{C^{s,\lambda}(\Omega)} \tag{2.2}$$

where $\frac{1}{p} = \frac{j}{n} + a\left(\frac{1}{r} - \frac{m+\vartheta}{n}\right) - (1-a)\frac{s+\lambda}{n}, \forall a \in \left[\frac{j-s-\lambda}{m+\vartheta-s-\lambda}, 1\right[$ *with* $(1-a)(s+\lambda) + a(m+\vartheta) \neq$ *integer.*

Proof. It is well known (see [10], Remark 4.4.2/2) that

$$H^{m+\vartheta,r}(\Omega) = B^{m+\vartheta}_{r,r}(\Omega) \tag{2.3}$$

and that (see [10], Remark 4.5.2/3 and (3.4.2/2) in [11])

$$C^{s,\lambda}(\Omega) = B^{s+\lambda}_{\infty,\infty}(\Omega). \tag{2.4}$$

Then it results:

$$u \in B^{m+\vartheta}_{r,r}(\Omega) \cap B^{s+\lambda}_{\infty,\infty}(\Omega)$$

and also, for each $a \in]\,0,1\,[$:

$$u \in \left(B^{s+\lambda}_{\infty,\infty}(\Omega), B^{m+\vartheta}_{r,r}(\Omega)\right)_a = B^h_{q,q}(\Omega) \tag{2.5}$$

and

$$\|u\|_{B^h_{q,q}(\Omega)} \leq c\|u\|^a_{B^{m+\vartheta}_{r,r}(\Omega)}\|u\|^{1-a}_{B^{s+\lambda}_{\infty,\infty}(\Omega)} \tag{2.6}$$

with $h = (1-a)(s+\lambda) + a(m+\vartheta)$, $\frac{1}{q} = \frac{a}{r}$ (see [10], formula (1.9.3/3) and [11], Theorem 3.3.6).

Now fixed $j \in]\max(s+\lambda, m+\vartheta-\frac{n}{r}),\, m+\vartheta[$, let us consider (2.5) and (2.6) for $a \in \left[\frac{j-s-\lambda}{m+\vartheta-s-\lambda)}, 1\right[$ with $(1-a)(s+\lambda) + a(m+\vartheta)$ non integer.

For such values of a, setting

$$\frac{1}{p} = \frac{j}{n} + a\left(\frac{1}{r} - \frac{m+\vartheta}{n}\right) - (1-a)\frac{s+\lambda}{n}$$

it results:

$$1 < q \leq p < \infty, \quad 0 < j \leq h < \infty, \quad h - \frac{n}{q} = j - \frac{n}{p}.$$

Thanks to the Remark 4.4.2/2 in [10] and to (4.6.1/8) in [10], it follows

$$B^h_{q,q}(\Omega) = H^{h,q}(\Omega) \subset H^{j,p}(\Omega), \tag{2.7}$$

with algebraic and topological inclusion.

From (2.6), (2.7), (2.3) and (2.4) the assert follows.

From Theorem 2.2 we reach the following

Corollary 2.1. *Under the some assumptions on Ω, if $u \in H^{1+\vartheta}(\Omega) \cap C^{0,\lambda}(\Omega)$, with $0 < \vartheta < 1$ and $0 < \lambda < 1$, then $u \in H^{1,p}(\Omega)$ and there exists a constant c (depending on $\Omega, \vartheta, \lambda, n, a$) such that:*

$$\|u\|_{H^{1,p}(\Omega)} \leq c\|u\|^a_{H^{1+\vartheta}(\Omega)}\|u\|^{1-a}_{C^{0,\lambda}(\Omega)}, \tag{2.8}$$

where $\frac{1}{p} = \frac{1}{n} + a\left(\frac{1}{2} - \frac{1+\vartheta}{n}\right) - (1-a)\frac{\lambda}{n}$, $\forall a \in]\frac{1-\lambda}{1+\vartheta-\lambda}, 1[$.

In particular, if $1-\lambda < \vartheta < 1$, for $a = \frac{1}{2}$, we get

$$u \in H^{1,p}(\Omega)$$

and there exists a constant c (depending on $\Omega, \vartheta, \lambda, n$) such that:

$$\|u\|_{H^{1,p}(\Omega)} \leq c\|u\|^{\frac{1}{2}}_{H^{1+\vartheta}(\Omega)}\|u\|^{\frac{1}{2}}_{C^{0,\lambda}(\Omega)}, \tag{2.9}$$

where

$$p = 4 + \frac{8(\vartheta+\lambda-1)}{n-2(\vartheta+\lambda-1)} \; (> 4).$$

3. Differentiability of solutions to system (1.1)

Let u be a solution to the system (1.1) and let us observe that by Theorem 1.1 u belongs to $L^2(-a,0,\ H^{1+\vartheta}(B(\sigma)))\cap C^{0,\lambda}(\overline{Q})$, for each $\vartheta\in(0,1)$. Then for t a.e. in $(-a,0)\,u$ belongs to $H^{1+\vartheta}(B(\sigma))\cap C^{0,\lambda}(\Omega)$ with $\vartheta=1-\frac{\lambda}{2}\in(0,1)$ for instance. Therefore the estimate (2.9), with $\Omega=B(\sigma),\vartheta=1-\frac{\lambda}{2},\ p=4+\frac{4\lambda}{n-\lambda}$, is true.

Hence we deduce:

$$\left(\int_{-a}^{0}\|u\|^4_{H^{1,4}(B(\sigma))}dt\right)^{\frac14}\leq c\|u\|^{\frac12}_{C^{0,\lambda}(\overline{Q})}\left(\int_{-a}^{0}\|u\|^2_{H^{1+\vartheta}(B(\sigma))}dt\right)^{\frac14}$$

that is $Du\in L^4\left(B(\sigma)\times(-a,0)\right)$. Now we are able to show the following

Theorem 3.1. *If $u\in L^2(-T,0,H^1(\Omega))\cap C^{0,\lambda}(\overline{Q})$, $0<\lambda<1$, is a solution to the system* (1.1), *if the assumptions* i)-iii) *hold, then, $\forall B(3\sigma)=B(x^0,3\sigma)\subset\subset\Omega$, $\forall a,b\in(0,T)$, $a<b$, it results:*

$$u\in L^2(-a,0,H^2(B(\sigma)))\cap H^1(-a,0,L^2(B(\sigma)))$$

and the following estimate holds:

$$\int_{-a}^{0}\left\{|u|^2_{2,B(\sigma)}+\left|\frac{\partial u}{\partial t}\right|^2_{0,B(\sigma)}\right\}dt\leq c(\nu,\sup_Q\|u\|,[u]_{\lambda,\overline{Q}},\lambda,\sigma,a,b,n)\left\{1+\int_{-b}^{0}|u|^2_{1,B(3\sigma)}dt\right\}.$$

Proof. We give an outline of the proof; for more details see [7]. Fixed $B(3\sigma)\subset\subset\Omega$, $a,b\in(0,T)$, $a<b$, $b^*=\frac{a+b}{2}$ and $|h|<\frac{\sigma}{2}$, setting $\tau_{i,h}v(x,t)=v(x+he^i,t)-v(x,t)$, $i=1,2,\ldots,n$, with $\{e^i\}_{i=1,2,\ldots,n}$ the canonic base of $\mathbb{R}^n$ and denoted by $\psi(x)$, $\rho_m(t)(m\in N)$ cut off functions, one can obtain the following estimate

$$\frac{\nu}{2}\int_{-b^*}^{-\frac1m}dt\int_{B(2\sigma)}\psi^2\rho_m^2\|\tau_{i,h}Du\|^2dx\leq c\mid h\mid^2\int_{-b^*}^{-\frac1m}dt\int_{B(3\sigma)}(1+\|Du\|^2)dx+$$

$$+c\int_{-b^*}^{-\frac1m}dt\int_{B(2\sigma)}\psi^2\rho_m^2\|\tau_{i,h}u\|^2\|Du\|^2dx+$$

$$+c\int_{-b^*}^{-\frac1m}\rho_m^2dt\int_{B(\frac52\sigma)}(1+\|Du\|^2)\|\tau_{i,-h}(\psi^2\tau_{i,h}u)\|dx.$$

In this estimate the crucial term is:

$$\int_{-b^*}^{-\frac1m}\rho_m^2dt\int_{B(\frac52\sigma)}(1+\|Du\|^2)\|\tau_{i,-h}(\psi^2\tau_{i,h}u)\|dx$$

which one can estimate in the following way:

$$\int_{B(\frac52\sigma)}(1+\|Du\|^2)\|\tau_{i,-h}(\psi^2\tau_{i,h}u)\|dx\leq\left(\int_{B(\frac52\sigma)}|h|^2(1+\|Du\|^2)^2dx\right)^{\frac12}\cdot$$

$$\left(\int_{B(\frac{5}{2}\sigma)} |h|^{-2}\|\tau_{i,-h}(\psi^2\tau_{i,h}u)\|^2 dx\right)^{\frac{1}{2}}$$

only if one knows that $Du \in L^4(B(\frac{5}{2}\sigma))$ for a.e. $t \in (-b^*, 0)$.
The condition $Du \in L^4(B(\frac{5}{2}\sigma))$ is a consequence of Theorem 1.1 and Corollary 2.1, as showed above.
The previous estimates allow us to get, after non trivial calculations,

$$\int_{-a}^{0} dt \int_{B(\sigma)} \|\tau_{i,h}Du\|^2 dx \le c|h|^2\Big\{1 + \int_{-b^*}^{0}(|u|^2_{1,B(3\sigma)} + |Du|^2_{1-\frac{\lambda}{2},B(\frac{5}{2}\sigma)})dt\Big\} \le$$

$$\le c|h|^2\Big\{1 + \int_{-b}^{0} |u|^2_{1,B(3\sigma)}dt\Big\}$$

having taken into account that

$$\int_{-b^*}^{0} |Du|^2_{1-\frac{\lambda}{2},B(\frac{5}{2}\sigma)}dt \le c\left\{1 + \int_{-b}^{0} |u|^2_{1,B(3\sigma)}dt\right\}.$$

From the estimate

$$\frac{1}{|h|^2}\int_{-a}^{0} dt \int_{B(\sigma)} \|\tau_{i,h}Du\|^2 dx \le c\Big\{1 + \int_{-b}^{0} |u|^2_{1,B(3\sigma)}dt\Big\}$$

by means of a well known lemma by L. Nirenberg, we get

$$\int_{-a}^{0} |u|^2_{2,B(\sigma)}dt \le c\Big\{1 + \int_{-b}^{0} |u|^2_{1,B(3\sigma)}dt\Big\}.$$

Making again use of the result $Du \in L^4\,(B(\sigma) \times (-a, 0))$ and taking into account the result above and the assumptions ii) and iii) on the growth with respect the gradient, it is possible now to show that $u \in H^1(-a, 0, L^2\,(B(\sigma)))$.

From the result of Theorem 3.1 it follows that

Theorem 3.2. *Under the same assumptions of Theorem* 3.1, *the gradient Du of a solution to the system* (1.1) *is partial Hölder continuous* (see [5], [6]).

Remark 3.1. Theorems 1.1, 3.1 and 3.2 hold if, instead of the assumptions i), ii) we assume that a.e. in Q, $\forall u \in R^N$, $\forall p, \bar{p} \in \mathbb{R}^{nN}$

i')
$$(a(X, u, p) - a(X, u, \bar{p})|p - \bar{p}) \ge \nu\|p - \bar{p}\|^2$$

and that a.e. in Q, $\forall u, v \in \mathbb{R}^N$, $\forall p, \bar{p} \in \mathbb{R}^{nN}$

ii')
$$\|a(X, u, p) - a(X, u, \bar{p})\| \le M\|p - \bar{p}\|,$$
$$\|a(X, u, p)\| \le M(1 + \|p\|),$$
$$\|a(x, t, u, p) - a(y, t, v, p)\| \le M(\|x - y\| + \|u - v\|)(1 + \|p\|).$$

Remark 3.2. A result similar to our one on differentiability has been stated in the paper [8] by means of a different technique.

REFERENCES

[1] L. Fattorusso, *Sulla differenziabilità delle soluzioni di sistemi parabolici non lineari del secondo ordine ad andamento quadratico,* Boll. U.M.I., (7) 1-B (1987), pp. 741-764.

[2] M. Marino and A. Maugeri, *Partial Hölder continuity of solutions of nonlinear parabolic systems of second order with quadratic growth,* Bollettino U.M.I. 3-B 1989, pp. 397-435.

[3] M. Marino and A. Maugeri, *Régularité et principe du maximum pour la solution du probléme de Cauchy-Dirichlet inhomogéne pour des systémes paraboliques,* C.R. Acad. Sc. 312 (1991), pp. 277-279.

[4] M. Marino and A. Maugeri, $L^{2,\lambda}$ *regularity of the spatial derivatives of the solutions to parabolic systems in divergence form,* Ann. di Matematica Pura e Applicata, 164 (1993), pp. 275-298.

[5] M. Marino and A. Maugeri, *Partial Hölder continuity of the spatial derivatives of the solutions to nonlinear parabolic systems with quadratic growth,* Rend. Sem. Mat. Univ. Padova, 76 (1986), pp. 219-245.

[6] M. Marino and A. Maugeri, *A Remark on the Note: "Partial Hölder continuity of the spatial derivatives of the solutions to nonlinear parabolic systems with quadratic growth",* Rend. Sem. Mat. Univ. Padova, 95 (1996).

[7] M. Marino and A. Maugeri, *Differentiability of weak solutions of nonlinear parabolic systems with quadratic growth,* Le Matematiche, 50 (1995), pp. 361-377.

[8] J. Naumann and J. Wolf, *Interior Differentiability of Weak Solutions to Parabolic Systems with Quadratic Growth Nonlinearities,* Rend. Sem. Mat. Univ. Padova, to appear.

[9] L. Nirenberg, *An extended interpolation inequality,* Ann. Scuola Norm. Sup. Pisa, (3) 20 (1966), pp. 733-737.

[10] H. Triebel, *Interpolation Theory, Function Spaces, Differential Operators,* North-Holland Publ. Comp., Amsterdam, New York, Oxford, 1978.

[11] H. Triebel, *Theory of Function Spaces,* Birkhäuser Verlag, Basel, Boston, Stuttgart, 1983.

7th Int. Coll. on Differential Equations, pp. 237-242
D. Bainov (Ed)

Extension of Holomorphic Mappings into Manifolds with Weak Disc Property

Yasuo MATSUDA* [1] ,

Graduate School of Mathematics, Kyushu University 33, Fukuoka, 812-81, Japan, FAX : +81-93-881-2863 or +81-92-632-2737

Abstract

Let Ω be a Riemann domain over a Banach space with Shauder basis.@$\tilde{\Omega}$ be its envelope of holomorphy and X be a Banach manifold with weak disc property. Ban[2] and [3] proved that any holomorphic mapping $f : \Omega \to X$ can be extended to a holomorphic mapping $\tilde{f} : \tilde{\Omega} \to X$.

The present paper generalizes the above theorem Ban[2] to a Riemann domain over a D. F. N. space.

1. Introduction.

J. Kajiwara, L. Li, M. Nishihara and M. Yoshida [14] discussed extension of holomorphic mappings from a separable Fréchet space, with the bounded approximation property, into a Banach complex Lie group. M. Yoshida[34] investigated the extension of holomorphic maps from domains over locally convex spaces where the Levi problem was solved.

On the other hand, P. K. Ban[2] proved that a Banach manifold with the weak disc property has the holomorphic extension property. It seems to the author that this is the most general result concerning image spaces.

Let Ω be a Riemann domain over a Banach space with Shauder basis. $\tilde{\Omega}$ be its envelope of holomorphy and X be a Banach manifold with weak disc property. Ban[2] and [3] proved that any holomorphic mapping $f : \Omega \to X$ can be extended to a holomorphic mapping $\tilde{f} : \tilde{\Omega} \to X$.

The main results of the present paper is the following generalization of the Ban's theorem[2]:

Let E be a D. F. N. space, (Ω, φ) be a Riemann's domain over E, $(\tilde{\lambda}, \tilde{\Omega}, \tilde{\varphi})$ be the envelope of holomorphy of the Riemann's domain (Ω, φ) over E, X be a Banach manifold with weak disc property and $f : \Omega \to X$ be a holomorphic mapping. Then there exists a holomorphic continuation $(\tilde{\lambda}, \tilde{f})$ of f to $(\tilde{\lambda}, \tilde{\Omega}, \tilde{\varphi})$.

[1]Supported by Grant-in-Aid for Scientific Research (C) no. 08304012 from the Ministry of Education, Science, Sports and Culture of Japan, 1996.
Key Words: Holomorphic extension, weak disc property
AMS Classifications: 32D10, 32D15

2. Ban' theory

In this section, we discuss the extension of holomorphic maps with values in a Banach manifold X with the weak disc property.

Throughout the present paper, we use the notations

$$D = \{z \in C; |z| < 1\}, \tag{1}$$

and

$$D^* = \{z \in D; z \neq 0\}. \tag{2}$$

Let X be a Banach manifold and let $\mathrm{H}(D, X)$ be the space of holomorphic maps from D into X equipped with compact open topology. The manifold X is said to satisfy the *weak disc condition,* if every sequence $\{f_n\}$ of $\mathrm{H}(D, X)$ which converges in $\mathrm{H}(D^*, X)$, converges in $\mathrm{H}(D, X)$ too.

Let (Ω, φ) be a Riemann domain over a Banach space with a Schauder basis and let $(\tilde{\Omega}, \tilde{\varphi})$ be the envelope of holomorphy of the domain (Ω, φ) in the sense of Malgrange[18].

The following theorem was proved by P. K. Ban[2] and, we dare hesitate to expose his theory, because we use repeatedly his method in the proof of our Main Theorem.

THEOREM Let X be a Banach manifold satisfying the weak disc condition. Then every holomorphic map from Ω to X can be extended holomorphically to $\tilde{\Omega}$.

Proof. Let

$$f : \Omega \longrightarrow X \tag{3}$$

be a holomorphic map, where Ω is a Riemann domain over a Banach space B with a Schauder basis. By S_X we denote the sheaf of germs of holomorphic maps on $\hat{\Omega}$ with values in X. By S_X^f we denote the domain of existence of f. Then S_X^f is the component of S_X containing the set $\{(x, f_x) : x \in \Omega\}$, where f_x denotes the germ of f at x.

At first we show that S_X^f satisfies the weak disc condition.

Given a sequence $\{u_n\} \subset \mathrm{H}(D, S_X^f)$ convergent to u in $\mathrm{H}(D^*, S_X^f)$. By hypothesis $\{\tilde{f}u_n\}$ converges to h in $\mathrm{H}(D, X)$. Take a neighbourhood U of $h(0)$in X which is isomorphic to an open subset of a Banach space E. Then we can assume that $\tilde{f}u_n(\delta D) \subset U$ for all $n \geq 1$.

Let K be the closure of convex set

$$\bigcup_{n\geq 1} \tilde{f}u_n((\frac{\delta}{2})D). \tag{4}$$

Let F be the canonical Banach space spanned by K. Then

$$\tilde{f}u_n : \partial((\frac{\delta}{2})D) \longrightarrow F \tag{5}$$

is continuous for $n \geq 1$. Hence, by the Cauchy's integral formula

$$\tilde{f}u_n(z) = \frac{1}{2\pi i}\int_{|z|=\frac{\delta}{2}} \tilde{f}u_n(\lambda)(\lambda - z)^{-1}d\lambda \tag{6}$$

it follows that

$$\tilde{f}u_n : (\frac{\delta}{2})D \longrightarrow F \tag{7}$$

is holomorphic for all $n \geq 1$.

Now for each $n \geq 1$, consider the map

$$\tilde{u}_n : (\frac{\delta}{2})D \longrightarrow \mathrm{limind}_{0\in W}\mathrm{H}^\infty(W,F) \tag{8}$$

where $\mathrm{H}^\infty(W,F)$ is the Banach space of bounded holomorphic functions on W with value in F, defined by

$$\tilde{u}_n(\lambda) = [u_n(\lambda) \circ \theta_{\gamma u_n(\lambda)}]_0 \tag{9}$$

where $\gamma = \hat{p}\beta, \hat{p} : \hat{\Omega} \longrightarrow B$, $\beta : S_X^f \longrightarrow \hat{\Omega}$ are canonical map, and $\theta_v(r) = v + r$

Then $\{\tilde{u}_n\}$ converges to $\tilde{u}$ in $\mathrm{H}((\frac{\delta}{2})D^*, \mathrm{limind}_{0\in W}\mathrm{H}^\infty(W,F))$, and hence in $\tilde{u}$ in $H((\frac{\delta}{2})D, \mathrm{limind}_{0\in W}\mathrm{H}^\infty(W,F))$. Since the above inductive limit is regular, there exists a neighbourhood W of zero in B, such that $\{\tilde{u}_n\}$ is contained and bounded in $H^\infty(W,F)$. Extend u holomorphically to $0 \in D$ by setting

$$u(0) = \tilde{u}(0) \circ \theta^{-1}_{\gamma u(0)}. \tag{10}$$

It remains to check that $\{u_n\}$ converges to u in $\mathrm{H}(D, S_X^f)$.

Now we consider the neighbourhood $\tilde{W}$ of $u(\partial((\frac{\delta}{2})D))$ given by

$$\tilde{W} = \bigcup\{(\gamma u(\lambda) + x, [u(\lambda)]_{\gamma u(\lambda)+x}) : |\lambda| = \frac{\delta}{2}, x \in W\}. \tag{11}$$

Take n_0 such that $u_n(\partial((\frac{\delta}{2})D)) \subset \tilde{W}$ for all $n \geq n_0$. Now for each $|\lambda| = \frac{\delta}{2}$ and $n \geq n_0$, there exists a neighbourhood $W(n,\lambda)$ of $\gamma u(\lambda)$ in $\gamma u(\lambda) + W$, such that $u_n(\lambda)x = u(\lambda)x$ for all $x \in \gamma u(\lambda) + W$ and $|\lambda| = \frac{\delta}{2}$. This shows that the above equality holds for all $|\lambda| \leq \frac{\delta}{2}$. Hence $u_n \longrightarrow u$ in $\mathrm{H}(D, S_X^f)$.

Since B has a Schauder basis, S_X^f is the domain of existence of a holomorphic function. This implies that the canonical map $\alpha : \Omega \longrightarrow S_X^f$ can be extended holomorphically to $\hat{\Omega}$. Hence the canonical map $\beta : S_X^f \longrightarrow \hat{\Omega}$ is a biholomorphism and $\tilde{f}\beta^{-1}$ is a holomorphic extension of f to $\hat{\Omega}$.

The theorem is proved.

3. Main theorem

Let (Ω, ψ) be a Riemann domain over a locally convex space E. Schottenloher[26] proved that there exists a regular cover of Ω satisfying the assumption of Theorem 3.1 in the sense of [14], if E is hereditary *Lindelöf* locally convex space with a infinite dimensional Schauder decomposition including a *Fréchet* space with a Schauder basis. Pelcznski[23] proved that a separable *Fréchet* space with the bounded approximation property is a complement subspace of a *Fréchet* space with a Schauder basis. Colombeau-Mujica[4], Mujica[19], Nachbin[21], and Paques-Zaine[22] proved that there exists a continuous Hilbert norm α of E such that (Ω, ψ) is a Riemann domain over the separable pre-Hilbert space (E, α).

Now we prove the next main theorem by the method of Ban's[2] and Soraggi's[31].

MAIN THEOREM Let E be a D. F. N. space, (Ω, φ) be a Riemann's domain over E, $(\tilde{\lambda}, \tilde{\Omega}, \tilde{\varphi})$ be the envelope of holomorphy of the Riemann's domain (Ω, φ) over E, X be a Banach manifold with weak disc property and $f : \Omega \rightarrow X$ be a holomorphic mapping. Then there exists a holomorphic continuation $(\tilde{\lambda}, \tilde{f})$ of f to $(\tilde{\lambda}, \tilde{\Omega}, \tilde{\varphi})$.

Proof. The D. F. N. space can be regarded as the inductive limit of a sequence of nuclear maps $E_\nu \to E_\mu$ between Banach spaces satisfying Ban[2]'s conditions. For any ν, there exists a domain $(\Omega_\nu, \varphi_\nu)$ such that Ω is the inductive limit of the sequence $\{\Omega_\nu\}_{\nu\in\Lambda}$ and $\tilde{\Omega}$ is the inductive limit of the sequence $\{\tilde{\Omega}_\nu\}_{\nu\in\Lambda}$, $(\tilde{\lambda}_\nu, \tilde{\Omega}_\nu, \tilde{\varphi}_\nu)$ being the envelope of holomorphy of the Riemann's domain $(\Omega_\nu, \varphi_\nu)$ over E_ν.

Let $(\tilde{\lambda}_f, \tilde{\Omega}_f, \tilde{\varphi}_f)$ be the envelope of holomorphy of the mapping f and F be the holomorphic extension of f to it.

Let n be a positive integer and ϵ be a positive number with $\epsilon < 1$. In the complex n-space C^n, we put

$$K = \{(w_1, w_2, \cdots, w_n) \in C^n; |w_1| \leq 1, |w_j| \leq \epsilon, (j = 2, 3, \cdots, n)\} \tag{12}$$

$$\cup\{(w_1, w_2, \cdots, w_n) \in C^n; 1 - \epsilon \leq |w_1| \leq 1, |w_j| \leq 1, (j = 2, 3, \cdots, n)\}$$

and

$$\tilde{K} = \{(w_1, w_2, \cdots, w_n) \in C^n; |w_k| \leq 1(k = 1, 2, 3, \cdots, n)\}. \tag{13}$$

We assume that the compact set $K \subset \Omega$. Then we may assume that there exists ν such that φ is a local homeomorphism of the compact set K in the Banach space E_ν.

Connecting the results of Colombeau-Mujica[4] with Soraggi[31], we find an element $\Lambda \ni \mu > \nu$ so as to regard $f : K \to X$ as a mapping of an open neighborhood of K in the Banach space E_μ. By the theorem of Ban[2], the holomorphic map $f : K \to X$ can be extended to a holomorphic map $\tilde{f}_\mu : \tilde{\Omega}_\mu \to X$. Since $\tilde{\Omega}_\mu$ is pseudoconvex, we may assume that $\tilde{K} \subset \tilde{\Omega}_\mu$. Combining these arguments with the arguments in Kajiwara-Kazama[11], we can prove the pseudoconvexity of $\tilde{\Omega}_f$ of the domain of existence of f. Since $\tilde{\Omega}$ is the smallest pseudoconvex domain over E, by Kajiwara-Li-Nisihara-Yoshida[14], ther exists a local homeomorphis $\psi : \tilde{\Omega} \to \tilde{\Omega}_f$ with $\tilde{\varphi} = \tilde{\varphi}_f \circ \psi$ such that $(\tilde{\Omega}_f, \psi)$ is a domain over $\tilde{\Omega}$.

Then the compostion $\tilde{f} := F \circ \psi$ is the desired holomorphic extension of f to the envelope of holomorphy $(\tilde{\Omega}, \tilde{\varphi})$, what was to be proved.

REFERENCES

[1] K. Adachi, M. Suzuki and M. Yoshida, *Continuation of holomorphic mappings. With value in a complex Lie group.* Pacific J. Math. **47**(1973), 1-4.

[2] P. K. Ban, *Some remarks on holomorphic extension in infinite dimensions,Colloquium Mathematicum,* **167**(1994), 155-159.

[3] P. K. Ban, *Banach hyperbolicity and extension of holomorphic maps,* Acta Math. Vietnam, **16**(1991), 187-200.

[4] J. F. Colombeau and J. Mujica, *The Levi problem in nuclear Silva spaces,* Arkiv ör Math. ,**18**(1980), 117-123.

[5] S. Dineen, *Sheaves of holomorphic functions on infinite dimensional spaces,* Math. Ann., **202**-2(1973), 337-345.

[6] F. Docquier und H.Grauert, *Levisches und Rungescher Satz für Teilgebiete Steinscher Mannigfaltigkeiten,* Math. Ann., **140**(1960), 94-123.

[7] L. Gruman, *The Levi problem in certain infinite dimensional vector spaces,* Illinois J. Math. , **18**-2(1974), 20-26.

[8] M. Harita, *Continuation of meromorphic functions in a domain of the cartesian product of denumberable family of complex planes*, Mem. Fac. Sci. Kyushu Univ. Ser. A, **29**-2(1975).

[9] J. Kajiwara, *On the limit of a monotonous sequence of Cousan's domains*, J. Math. Soc. Japan **17**-1(1965), 36-46.

[10] J. Kajiwara, *Some results on the equivalence of complex-analytic fiber bundles*, Mem. Fac. Sci. Kyushu Univ. Ser.A, **19**-1(1959), 37-48.

[11] J. Kajiwara and H. Kazama, *Oka's principle for relative cohomology sets, Mem. Fac. Sci. Kyushu Univ.* **23**-1(1969),33-70.

[12] J. Kajiwara and H. Kazama, *Two dimensional complex manifold with vanishing cohomology set,* Math. Ann. **204**(1973),1-12.

[13] J. Kajiwara, L. Li, M. Nishihara, and M. Yoshida, *Continuation of holomorphic sections of holomorphic fiber bundles over domains of infinite dimensional spaces equipped with the finite open topology*, Fukuoka Univ. Sci. Rep. , **23**(2)(1993), 69-79.

[14] J. Kajiwara, L. Li, M. Nishihara, and M. Yoshida, *On the Levi problem for holomorphic mappings of Riemann domains over infinite dimensional spaces into a complex Lie group*, Res. Bull. Fukuoka Inst. Tech. , **26**-2(1994), 151-161.

[15] J. Kajiwara and M. Nishihara, *Characterisierung der Steinschen Teitgebieten durch Okasches Prinzip in Zwei-dimensionater Mannigfasltigkeit,* Mem. Fac. Sci.Kyushu. Univ. **33**(1979),71-76.

[16] J. Kajiwara and K. H. Shon, *Continuation and vanishing theorem for cohomology of infinite dimensional spaces,* Kyonbuk J. Mat.(1993).

[17] H. Kazama, *On psedoconvexity of complex Lie groups,* Mem. Fac. Sci. Kyushu Univ. Series A, Math. **27**-2(1973), 241-247.

[18] B. Malgrange, *On the theory of functions of several complex valiables,* Tata Inst. Fund. Res. Bombay, **13**(1958).

[19] J. Mujica, *Domains of holomorphy in DFN spaces,* Functional Analysis, Holomorphy and Approximation Theory, Lecture Notes in Math. Springer, **843**(1981),500-533.

[20] J. Mujica, *Complex analysis in Banach spaces,* North-Holland Math. Studies, 120(1986).

[21] L. Nachbin, *On pure uniform holomorphy in spaces of holomorphic germs,* Results in Math. , **8**(1985), 117-122.

[22] O. W. Paques and M. C. F. Zaine, *Uniformly holomorphic continuation in locally convex spaces,* J. Math. Anal. and Appl., **123**(1987), 448-454.

[23] A. Pelcznski, *Any separable Banach space with the bounded approximation property is a complemented subspace of a Banach space with a basis,* Studia Math. **40**(1971), 149-243.

[24] P. Raboin, *Résolution de l'équation* $\overline{\partial} f = F$ *sur un espace de Hilbert,* Bull. Soc. Math. France, **107**(1979), 225-240.

[25] G. Scheja, *Riemannsche Hebbarkeitssätze für Cohomologieklassen,* Math. Ann. **144**(1961), 345-360.

[26] M. Schottenloher, *The Levi problem for domains spread over locally convex spaces with a finite dimensional Schauder decomposition,* Ann. Inst. Fourier, **26**(1976), 207-237.

[27] J. P. Serre, *Dans Séminaire H.Cartan,3,* École Normare Supérieur, Paris, 1951-1952.

[28] M. Schottenlocher, *The Levi problem for domains spread over locally convex space with a finite dimensional Schauder docomposition* Ann. Inst. Fourier, **26**(1976), 207-237.

[29] B. Shiffman, *Extension of Holomorphic Maps into Hermitian Maniford,* Math. Ann. **194**(1971), 249-258.

[30] B. Shiffman, *Extension of holomorphic maps into Hermitian manifolds,* Math.Ann. Vietnam. **16**(1971), 241-253.

[31] R. L. Soraggi, *The $\overline{\partial}$-problem for a (0,2)-form in a D. F. N. space,* J. Func. Anal. **98**(1991), 380-402.

[32] B. D. Tac, *Extending holomorphic maps in infinite dimensions,* Ann. Polon. Math. **54**(1991), 241-253.

[33] D. D. Thai, *On the D^*-extension and the Hartogs extension,* Ann. Scuola Norm. Sup. Pisa **18**(1991), 13-38.

[34] M. Yoshida, *On the Levi Priblem for Holomorphic Mappings of Riemann Domains over Infinite Dimensional Spaces into a Complex Lie Group,* Reasearch Bulletin of Fukuoka Institute of Technology, Vol. **26**, Mar.(1994).

[35] M. Yoshida, *On the Continuation of Holomorphic Mapping,* to appear in Irish Mathematical Society Bulletin, **33**, December,(1995).

7th Int. Coll. on Differential Equations, pp. 243-252
D. Bainov (Ed)

Global Instability of Natural Convection in an Elliptic Cavity, Using a Fourier Spectral Difference Method

Yoshihiro MOCHIMARU
Dept. of International Development Engineering
Tokyo Institute of Technology, Tokyo 152, Japan
Key words: Instability, Natural Convection, Spectral Difference Method
AMS Classification: 76E Hydrodynamic stability

ABSTRACT

A Fourier spectral finite difference scheme shows that onset of a strong steady-state Bénard-cell-like cellular flow pattern at natural convection in an elliptic cavity takes place in cooperation with a finite magnitude of disturbance existing at an initial stage for a given Grashoh number and a Prandtl number.

1. INTRODUCTION

Natural convection in an elliptic (two-dimensional) cavity heated from below was analyzed numerically by Mochimaru [1], using a Fourier spectral finite difference method, the effectiveness of which was shown as in Ref. [2]. Here treated is a problem of a possibility of transition to different types of steady-state flow fields corresponding to given boundary conditions (from different initial conditions) in a viewpoint of global stability, whereas linear stability theory on a stratified fluid was reviewed by Chandrasekhar [3]. For example, concrete numerical analysis of onset of bifurcation on a stratified fluid in a rectangular enclosure was treated by Lee and Schultz [4], as an eigen value problem based on a linear theory.

2. ANALYSIS

2.1. Basic Assumption

It is assumed that a fluid is enclosed in an elliptic cavity placed horizontally, and that the fluid and thermal field is laminar. Numerical analysis is based on a two-dimensional vorticity transport equation under a Boussinesq approximation, supplemented with an energy equation neglecting dissipation terms, which is expressed in an elliptic coordinate system (α, β) relating to a Cartesian coordinate (x, y) [the x-coordinate is horizontal, the y-coordinate being vertically upward] as

$$x + iy = \begin{cases} -i\sinh(\alpha + i\beta)/\cosh\alpha_0 & (a > b) \\ -i\cosh(\alpha + i\beta)/\sinh\alpha_0 & (a < b) \end{cases} , \tag{1}$$

where a : 1/2 of the horizontal principal axis, b : 1/2 of the vertical principal axis [the circular case $(a = b)$ is not included], and all the length are made dimensionless with respect to the length a, and

$$-\pi < \beta \leq \pi \quad \text{or} \quad (0 \leq \beta < 2\pi) \ , \tag{2}$$

$$0 \leq \alpha \leq \alpha_0 \tag{3}$$

Here the elliptic surface is specified by $\alpha = \alpha_0$, $\alpha_0 \equiv \tanh^{-1}\{\min(a/b, b/a)\}$

2.2. Basic Equations

Dimensionless forms of governing equations (the equation of vorticity transport, the relation between vorticity and a stream function, and the energy equation) are

$$K\frac{\partial \zeta}{\partial t} + \frac{\partial(\zeta,\psi)}{\partial(\alpha,\beta)} = \frac{1}{\sqrt{Gr}}\left(\frac{\partial^2}{\partial \alpha^2} + \frac{\partial^2}{\partial \beta^2}\right)\zeta + \frac{\partial(T,y)}{\partial(\alpha,\beta)} \quad , \tag{4}$$

$$K\zeta = -\left(\frac{\partial^2}{\partial \alpha^2} + \frac{\partial^2}{\partial \beta^2}\right)\psi \quad , \tag{5}$$

$$K\frac{\partial T}{\partial t} + \frac{\partial(T,\psi)}{\partial(\alpha,\beta)} = \frac{1}{Pr\sqrt{Gr}}\left(\frac{\partial^2}{\partial \alpha^2} + \frac{\partial^2}{\partial \beta^2}\right)T \quad , \tag{6}$$

$$K \equiv \frac{\partial(x,y)}{\partial(\alpha,\beta)} = \left(\frac{\partial y}{\partial \alpha}\right)^2 + \left(\frac{\partial y}{\partial \beta}\right)^2 \quad , \tag{7}$$

where ζ : vorticity, ψ : stream function [in the Cartesian system, velocity $\boldsymbol{v}$ is given by $\boldsymbol{v} = \nabla \times (0,0,\psi)$, and $(0,0,\zeta) = \nabla \times \boldsymbol{v}$], T : dimensionless temperature $\equiv$ [temperature $-(T_l+T_u)/2]/(T_l-T_u)$, Gr : Grashof number based on the temperature difference $(T_l - T_u)$ and length a, Pr : Prandtl number, T_u : constant temperature on the upper wall ($y > 0$), T_l : constant temperature on the lower wall ($y < 0; T_l > T_u$). Reference velocity is specified as $(\nu/a)\sqrt{Gr}$, (ν : kinematic viscosity of the fluid); reference time is defined as $(a^2/\nu)/\sqrt{Gr}$.

2.3. Boundary Conditions

Assumed are no slip flow on the wall

$$\psi(\alpha_0,\beta,t) = 0 \quad , \tag{8}$$

$$\frac{\partial}{\partial \alpha}\psi(\alpha_0,\beta,t) = 0 \tag{9}$$

and the following (for dimensionless temperature)

$$T(\alpha_0,\beta,t) = \mp(1/2) \quad (\, y \gtrless 0\,) \quad . \tag{10}$$

2.4. Initial Conditions

The following two types of initial conditions are treated. (I) All the fluid is at rest with an initial temperature of $T_i = \pm 1/2$ or $\delta \cos 2\beta$ ($|\delta| \leq 1/2$). At least $\delta \neq 0$ is a source for asymmetry leading to a Bénard-cell-like pattern. (II) The field is assumed to be identical to the steady one corresponding to the type (I) of $\delta = 0$ except a superposed non-uniform temperature of the form $\delta \cos 2\beta$.

3. NUMERICAL SOLUTION

Numerical solution procedure is based on a spectral finite difference scheme [2]. Hereafter only symmetrical solution with respect to the y-axis is sought, so that ζ, ψ , and T are of the form

$$\begin{bmatrix} \zeta \\ \psi \end{bmatrix} = \sum_{n=1}^{\infty} \begin{bmatrix} \zeta_n(\alpha, t) \\ \psi_n(\alpha, t) \end{bmatrix} \sin n\beta \quad , \tag{11}$$

$$T = \sum_{n=0}^{\infty} T_n(\alpha, t) \cos n\beta \tag{12}$$

respectively. If the expressions (11),(12) for the unknowns are substituted into Eqs.(4) - (7), then Eqs.(4) - (6) can be decomposed into each sine or cosine components of β, using addition formulae of trigonometric functions to separate the variable β. Resulting decomposed differential equations are discretized in time and space in the interior region by finite difference approximations. For discretization of spatial partial derivatives, the following non-uniform grid spacing may be used:

$$\alpha_n = \alpha_0 - h \left(\frac{\sinh(n-1)\gamma}{\sinh \gamma} + 1 \right) \quad , \tag{13}$$

where α_n is a coordinate at the n-th grid point counted from the wall along the α coordinate ; γ : a suitable constant, with $\gamma \to 0$ corresponding to a uniform grid spacing in α; h : a constant to be determined.
Auxiary conditions on $\alpha = 0$ (between foci) are estimated as follows:
At the points on $\alpha = 0$ all the scalar quantities and vector quantities are continuous, that is, in case of $a > b$ for any scalar function ϕ(either ζ, ψ, or T)

$$\phi(0, \pi - \beta, t) = \phi(0, \beta, t) \quad , \tag{14}$$

$$\frac{\partial}{\partial \alpha}\phi(0, \pi - \beta, t) = -\frac{\partial}{\partial \alpha}\phi(0, \beta, t) \tag{15}$$

which reduce to

$$\zeta_n(0, t) = \psi_n(0, t) = \frac{\partial}{\partial \alpha}T_n(0, t) = 0 \quad (n : \text{even}) \quad , \tag{16}$$

$$\frac{\partial}{\partial \alpha}\zeta_n(0, t) = \frac{\partial}{\partial \alpha}\psi_n(0, t) = T_n(0, t) = 0 \quad (n : \text{odd}) \quad . \tag{17}$$

On the other hand, in case of $a < b$ for any scalar function ϕ(either ζ, ψ, or T)

$$\phi(0, 2\pi - \beta, t) = \phi(0, \beta, t) \quad , \tag{18}$$

$$\frac{\partial}{\partial \alpha}\phi(0, 2\pi - \beta, t) = -\frac{\partial}{\partial \alpha}\phi(0, \beta, t) \tag{19}$$

which reduce to

$$\zeta_n(0, t) = \psi_n(0, t) = 0 \quad (n \geq 1) \quad , \tag{20}$$

$$\frac{\partial}{\partial \alpha}T_n(0, t) = 0 \quad (n \geq 0) \quad . \tag{21}$$

Treatment of the boundary conditions on the wall is the same as in Ref. [2]. Special care should be taken to the second boundary condition (9) even for a spectral finite difference method, Ref.[2].
Time discretization with respect to terms involving $(\partial/\partial t)$ is based on a forward finite difference approximation of the order δt (time increment), which is compatible to evaluate, at the current time, nonlinear terms and other linear terms than possessing the same component as in the corresponding decomposed differential equations. Thus, the difference expressions are coupled each other only if they have been derived from the same Fourier components, and constitute a set of simultaneous linear algabraic equations for the future time stage. In case of no initial disturbance ($\delta = 0$ for the first and the second kind of the type of the initial condition), at the initial stage $\partial\zeta/\partial t = 0, \partial T/\partial t = 0$ inside the flow domain (excluding the boundary), whereas in case of $\delta \neq 0$ the values will be slightly different from zero according to Eqs.(4) and (6). To uncouple discretized equations of partial differential equations of different components each other, $\partial\zeta/\partial t, \partial T/\partial t$ have to be known at each time step; at each time the values of partial time derivatives can be computed using backward difference approximation. (As $a/b \to 1$, equations tend to be uncoupled automatically with respect to terms $\partial\zeta/\partial t, \partial T/\partial t$). Thus, to avoid ambiguity only at the initial stage the approximation $\partial\zeta/\partial t = 0, \partial T/\partial t = 0$ is used for uncoupling irrespectively of the value of δ (the absolute value of which under discussion is usually small as shown later.) So numerical time integration is based on a semi-implicit scheme (independently of each component) as an initial value problem to get a steady-state solution for moderate combinations of Gr and Pr.

4. NUMERICAL RESULTS

4.1. General Feature

If $\zeta(x,y,t), \psi(x,y,t)$, and $T(x,y,t)$ in a Cartesian coordinate system is a set of solution of the proposed problem, then $\zeta^*(x,y,t), \psi^*(x,y,t)$, and $T^*(x,y,t)$ can be a solution for a modified initial condition if

$$\zeta^*(x,y,t) \equiv -\zeta(x,-y,t) \quad , \tag{22}$$

$$\psi^*(x,y,t) \equiv -\psi(x,-y,t) \quad , \tag{23}$$

$$T^*(x,y,t) \equiv -T(x,-y,t) \quad . \tag{24}$$

Under the current situation, boundary conditions are consistent with Eqs.(22) - (24). Thus if $\delta = 0$ for the case of (I), a four cell solution (as shown in Fig.1, symmetric with respect to the x-axis and y-axs) is always possible. A Fourier spectral solution of the type Eqs.(11) and (12) is compatible with this solution. The conjugate solution denoted with an asterisk can be obtained merely by changing the sign of δ.
The stream function for the four cell steady-state solution is given (in case of $a > b$) by (for small values of Gr and a fixed Pr)

$$\begin{aligned}\psi &= -\sqrt{Gr}\left[\sum_{n=1}^{\infty} A_{2n+1}\left\{\sinh(2n+4)\alpha\ \sin 2n\beta + \sinh 2n\alpha\ \sin(2n+4)\beta\right\}\right. \\ &\quad + \sum_{n=1}^{\infty} B_{2n}\left\{\sinh(2n+2)\alpha\ \sin 2n\beta - \sinh 2n\alpha\ \sin(2n+2)\beta\right\}\end{aligned}$$

$$- \sum_{n=1}^{\infty} C_{2n} \sinh 2n\alpha \, \sin 2n\beta \Bigg] + O(Gr) \ , \tag{25}$$

where

$$\{\sinh(4n+2)\alpha_0 - (2n+1)\sinh 2\alpha_0\} B_{2n}$$

$$+ \{\sinh(4n-2)\alpha_0 - (2n-1)\sinh 2\alpha_0\} B_{2n-2}$$

$$+ \{2\sinh(4n+4)\alpha_0 - (2n+2)\sinh 4\alpha_0\} A_{2n+1}$$

$$- \{2\sinh(4n-4)\alpha_0 - (2n-2)\sinh 4\alpha_0\} A_{2n-3} = 0 \quad (n \geq 1) \ , \tag{26}$$

$$\{\sinh(4n+6)\alpha_0 - (2n+3)\sinh 2\alpha_0\} A_{2n+1}$$

$$- \{3\sinh(4n-2)\alpha_0 - (2n-1)\sinh 6\alpha_0\} A_{2n-3}$$

$$+ \{2\sinh 4n\alpha_0 - 2n\sinh 4\alpha_0\} B_{2n-2}$$

$$+ \{\sinh(4n+2)\alpha_0 - (2n+1)\sinh 2\alpha_0\} C_{2n} = 0 \quad (n \geq 1) \ , \tag{27}$$

$$A_{2n-1} \equiv \frac{(-1)^{(n-1)}}{256\, n\pi \cosh^3 \alpha_0} \left\{ \frac{1}{(2n-1)\sinh(2n-1)\alpha_0} - \frac{1}{(2n+1)\sinh(2n+1)\alpha_0} \right\}$$

$$(n \geq 1) \ , \tag{28}$$

$$A_{-1} \equiv 0, \quad B_0 \equiv 0 \ . \tag{29}$$

4.2. Possibility of Transition to Steady-State Cellular Flow Patterns from Initial Disturbance Superposed on a Uniform (Zero) Temperature

In case of the boundary condition (I) with $\delta = 0$ (no disturbance), mathematically the unique four cell solution exists at any time (even at a transient state), which can be rigorously (algorithmically) supported by the current Fourier spectral difference scheme. For a given b/a and Pr, if Grashof number is less than a certain critical value corresponding to b/a, the steady-state flow pattern (streamlines) is the same as obtained with $\delta = 0$ independently of $\delta(|\delta| \leq 0.5)$. However, if Gr is greater than a certain critical value corresponding to b/a, the steady-state flow pattern is the same pattern as obtained with $\delta = 0$ if $|\delta|$ is less than a certain critical value depending on Gr, whereas if $|\delta|$ is greater than the critical value, then steady-state flow pattern corresponds to a cellular pattern (either of the two cases antisymmetric each other with respect to the x-axis is possible: the final state depends on the initial value of δ).

An example of a steady four cell flow pattern is shown in Fig.1 ($b/a = 1/2, Gr = 4400, Pr = 0.7$).

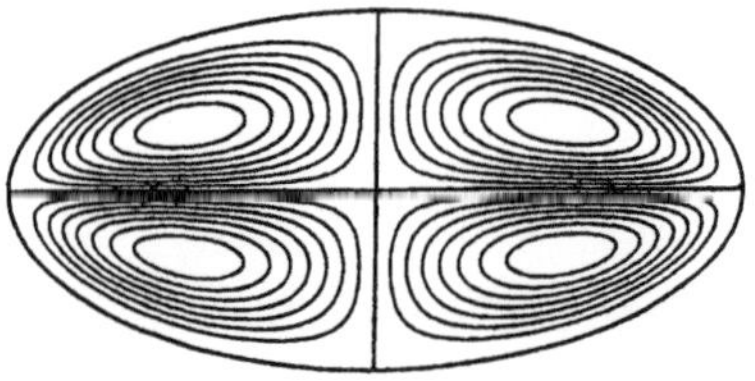

Fig.1 Steady-state streamlines for
$b/a = 0.5, Gr = 4400, \delta = 0, Pr = 0.7.[\Delta\psi = 0.00025] < + >$.

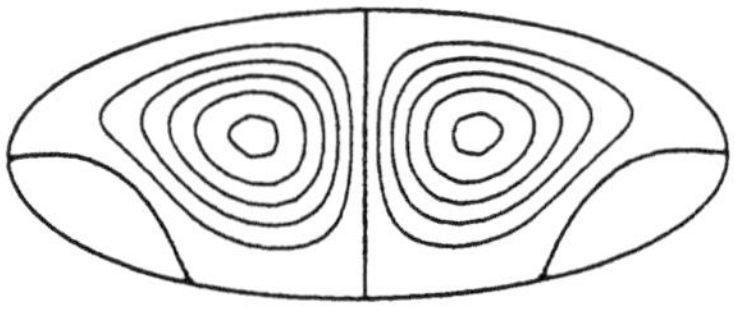

Fig.2 Steady-state streamlines for $b/a = 0.4, Gr = 7450, Pr = 0.7, \delta = 0.5.[\Delta\psi = 0.001] < + >$.

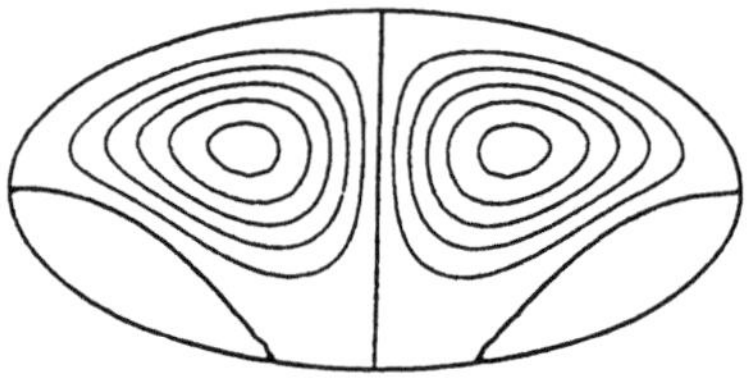

Fig.3 Steady-state streamlines for $b/a = 0.5, Gr = 4400, Pr = 0.7, \delta = 0.1.[\Delta\psi = 0.001] < + >$.

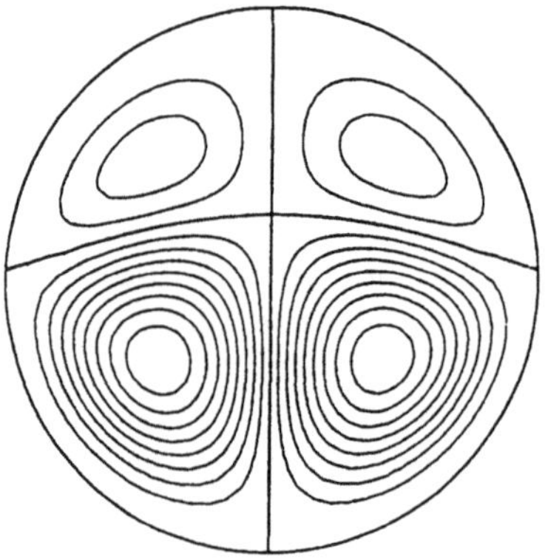

Fig.4 Steady-state streamlines for $b/a = 1.03, Gr = 1700, Pr = 0.7, \delta = 0.01.$ $[\Delta\psi = 0.0025] < + >$.

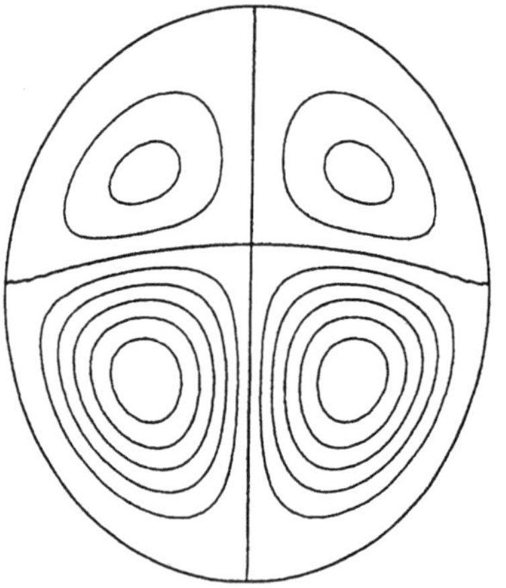

Fig.5 Steady-state streamlines for $b/a = 1.2, Gr = 1630, Pr = 0.7, \delta = 0.5.[\Delta\psi = 0.005] < + >$.

Steady-state cellular flow patterns (the shape of which depends on Gr, Pr, and b/a) are shown in Figs.2,3,4, and 5. The sign $(+,-)$ enclosed by angle brackets shown in captions in Figs.1 – 5 denotes the sign of streamlines ψ in the lower circulating region in the right half plane $(x > 0)$.
Physical shape of distribution of streamlines may vary according to $b/a \lessgtr 1$, since the loci are on the x-axis or y-axis. The steady-state isotherms corresponding to the cellular flow pattern are shown in Figs.6 and 7.

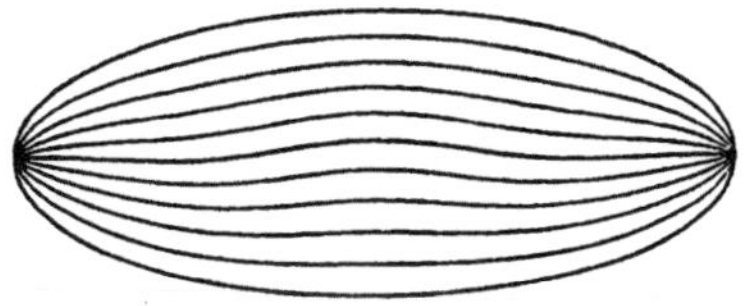

Fig.6 Steady-state isotherms for $b/a = 0.4, Gr = 7450, Pr = 0.7, \delta = 0.5.[\Delta T = 0.1.]$

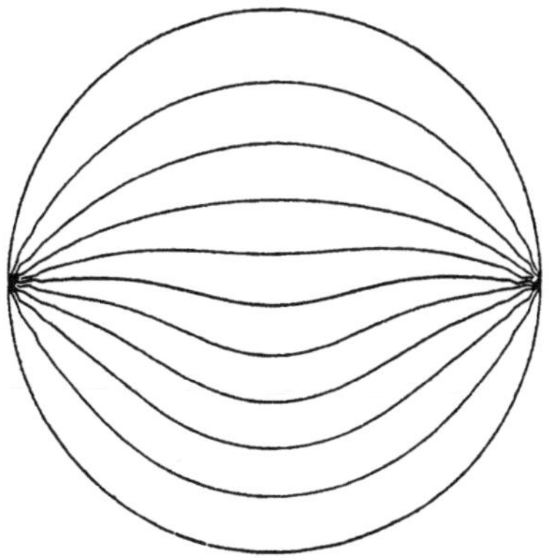

Fig.7 Steady-state isotherms for $b/a = 1.03, Gr = 1700, Pr = 0.7, \delta = 0.01.$ $[\Delta T = 0.1]$.

Classification of steady-state flow pattern by initial disturbance δ in case of the type (I) are shown in Figs. 8 – 13 (o : Bénard-cell-like cellular pattern, • : symmetric four cell pattern). Here z is the measure of the disturbance δ :

$$z \equiv \ln\left\{1 + \ln(1 + 10^4|\delta|)\right\} \quad . \tag{30}$$

$z = 0$ corresponds to $\delta = 0$, and $z = 2.253$ corresponds to $|\delta| = 0.5$.

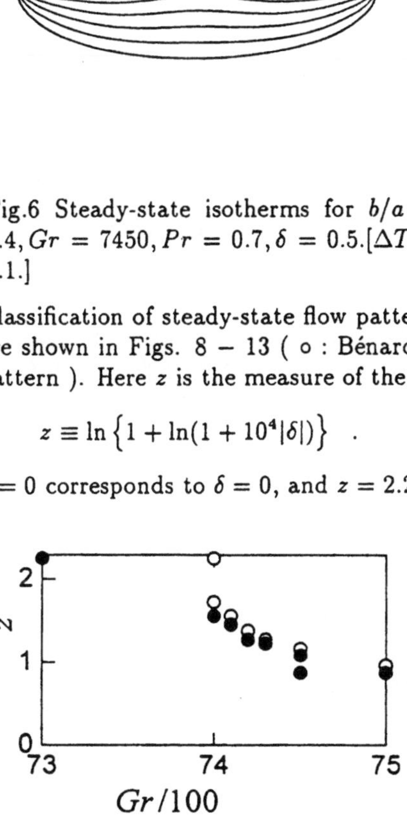

Fig.8 Classification of flow pattern for $b/a = 0.4, Pr = 0.7$.

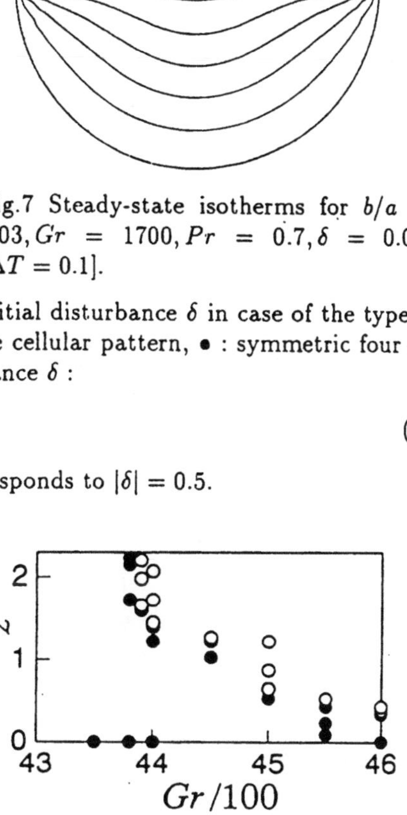

Fig.9 Classification of flow pattern for $b/a = 0.5, Pr = 0.7$.

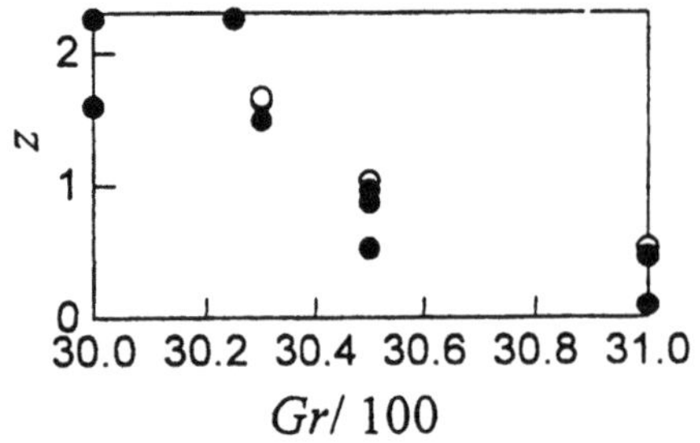

Fig.10 Classification of flow pattern for $b/a = 0.6, Pr = 0.7$.

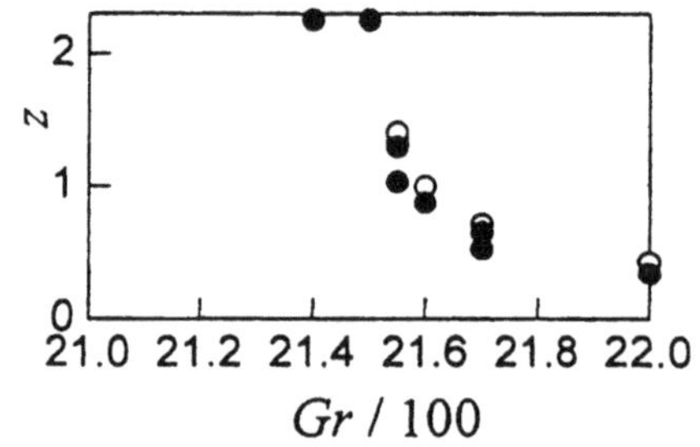

Fig.11 Classification of flow pattern for $b/a = 0.75, Pr = 0.7$.

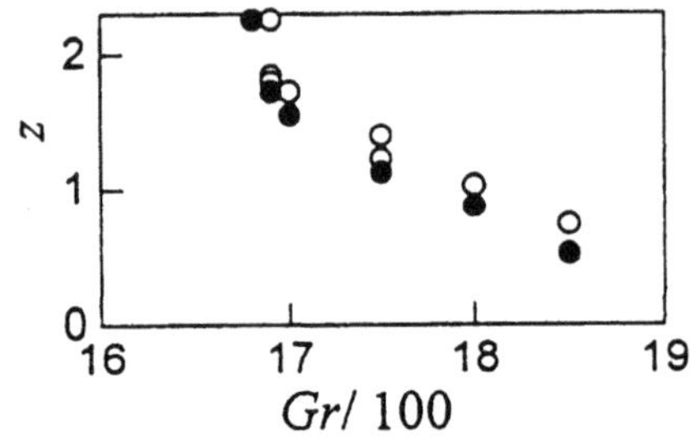

Fig.12 Classification of flow pattern for $b/a = 1.03, Pr = 0.7$.

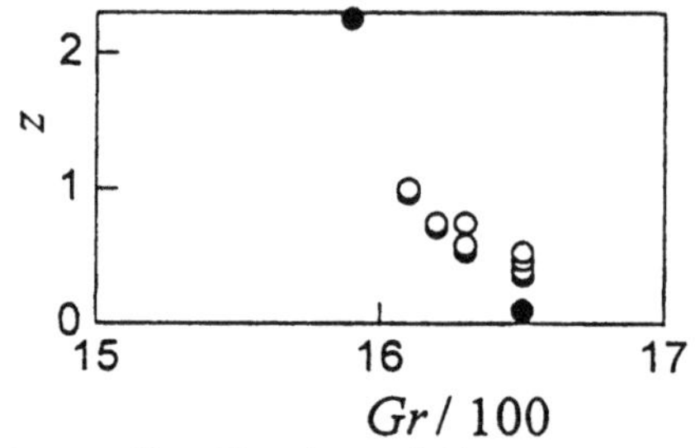

Fig.13 Classification of flow pattern for $b/a = 1.2, Pr = 0.7$.

In case of the type (I) of $T_i = \pm 1/2$ at $b/a = 1/2$, the steady-state flow pattern corresponds to those of $\delta = 1/2 (T_i = 1/2), \delta = -1/2 (T_i = -1/2)$ as long as $Gr = 4400, Pr = 0.7$. Characteristics of mean Nusselt number along the lower half wall for $b/a = 1/2, Pr = 0.7$ is shown in Fig.14.

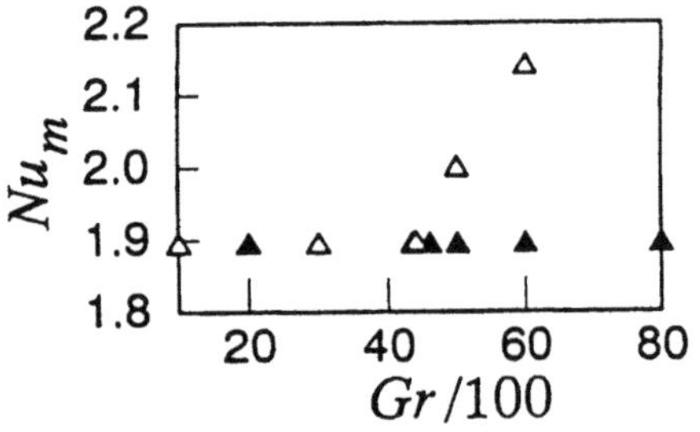

Fig.14 Mean Nusselt number Nu_m along the lower wall (up to the 50th component) for $b/a = 0.5, Pr = 0.7$. ▲ : $\delta = 0$, △ : $T_i = \pm 0.5$.

4.3. Effect of a Prandtl Number

Previous numerical results are all concerned with $Pr = 0.7$ (air). As an example for a different Prandtl number, classification of steady-state flow pattern by initial disturbance for $Pr = 7$ (water) is shown in Fig.15; for larger values of $|\delta|$ at $Pr = 7$, kinds of the pattern (not shown) of streamlines may be dependent on the magnitude of $|\delta|$.

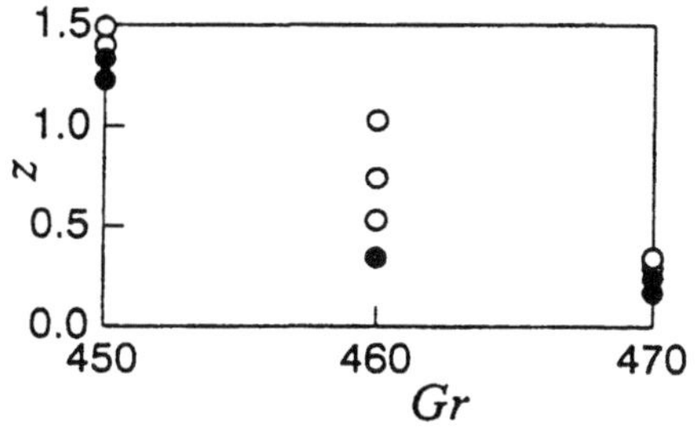

Fig.15 Classification of flow pattern for $b/a = 0.5, Pr = 7$.
Symbols are the same as in Figs.8 - 13.

4.4. Posibility of Transition to Steady-State Cellular Patterns from Initial Disturbance Superposed on the Steady-State Four Cell Pattern Flow

If the disturbance $|\delta|$ is greater than a certain value corresponding to $b/a, Gr$, and Pr, then the initial four cell flow will shift to a steady-state cellular flow as shown in Fig.16, whereas if $|\delta|$ is less than the critical value corresponding to $b/a, Gr$, and Pr, then the initial disturbance will decay to zero as time tends to infinity.

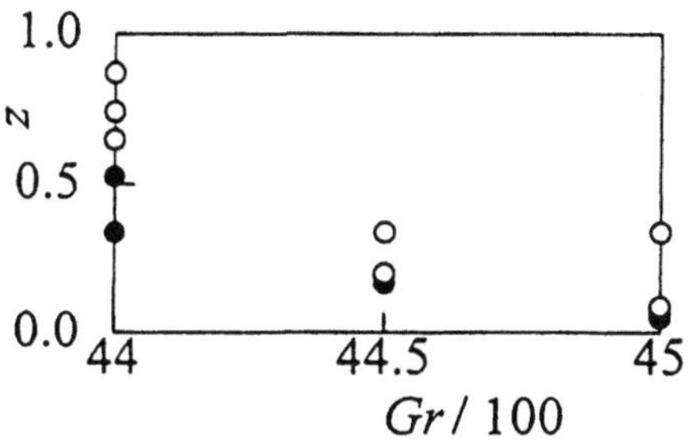

Fig.16 Classification of flow pattern for $b/a = 0.5, Pr = 0.7$.

5. CONCLUSIONS

Using a Fourier spectral finite difference scheme expressed in the boundary fitted coordinate system, which is compatible to an antisymmetrical solution automatically under a certain initial condition, it is found that transition to an asymmetric steady-state solution will take place if the magnitude of the initial disturbance is greater than a certain finite value corresponding to $b/a, Gr$, and Pr.

REFERENCES

[1] Mochimaru, Y., 2nd ISAIF, Vol.2, pp.387-392(1993).
[2] Mochimaru, Y., 26th Ann. Iranian Mathematics Conference, Vol.2, pp.245-254(1995)
[3] Chandrasekhar,S., *Hydrodynamics and Hydrodynamic Stability*, (1961).
[4] Lee, N.Y. and Schultz, W.W., Int.J.Heat Mass Transfer, Vol.32, pp.513-520(1989)

7th Int. Coll. on Differential Equations, p.p. 253-260
D. Bainov (Ed)

Capacity associated with an α-parabolic operator

MASAHARU NISHIO

Department of Mathematics, Osaka City University, Osaka 558, Japan

Abstract. We consider a parabolic operator of fractional order. We generalize an α-parabolic capacity and give some applications.

Keywords: capacity, heat equation, parabolic operator of fractional order

INTRODUCTION

Let $\mathbf{R}^{n+1} = \mathbf{R}^n \times \mathbf{R}$ be the $(n+1)$-dimensional Euclidean space with $n \geq 1$. We denote by $X = (x, t)$ a point in $\mathbf{R}^{n+1}$ with $x \in \mathbf{R}^n$ and $t \in \mathbf{R}$. In this note, we consider capacities associated with the α-parabolic operator and its adjoint operator:

$$L^{(\alpha)} = \frac{\partial}{\partial t} + (-\Delta)^\alpha \text{ and } \widetilde{L}^{(\alpha)} = -\frac{\partial}{\partial t} + (-\Delta)^\alpha,$$

where $0 < \alpha \leq 1$ and $\Delta = \sum_{i=1}^n \frac{\partial^2}{\partial x_i^2}$ denotes the Laplace operator on $\mathbf{R}^n$.

The purpose of this note is to generalize the α-parabolic capacity stated in [1] and [2] and to apply them to α-parabolic morphisms.

1. PRELIMINALIES

We shall recall the notion and notations with respect to the potential theory of $L^{(\alpha)}$. For those of the adjoint operator, we use "$\sim$" like $\widetilde{L}^{(\alpha)}$.

Denote by $W^{(\alpha)}$ the fundamental solution of $L^{(\alpha)}$, that is,

$$W^{(\alpha)}(x,t) = \begin{cases} (2\pi)^{-n} \int_{\mathbf{R}^n} \exp(-t|\xi|^{2\alpha} + \sqrt{-1}x \cdot \xi)\, d\xi & t > 0 \\ 0 & t \leq 0. \end{cases}$$

When $\alpha = 1$, $L^{(1)}$ is the heat operator and $W(x,t) = (4\pi t)^{-\frac{n}{2}} \exp(-\frac{|x|^2}{4t})$ for $t > 0$.

For any $s > 0$, we put $P_s^{(\alpha)} := (W^{(\alpha)}(x,s)dx) \otimes \varepsilon_s$ and $P_0^{(\alpha)} := \varepsilon$, where ε_s and ε denote the Dirac measure at s and the origin, respectively. Then $(P_s^{(\alpha)})_{s \geq 0}$ is the convolution semi-group whose infinitesimal generator is $-L^{(\alpha)}$.

In ths paper, we consider a strip domain

$$D = \mathbf{R}^n \times (a, b),$$

where $-\infty \leq a < b \leq +\infty$.

DEFINITION. For a lower semi-continuous function $u > 0$, we write $u \in S_\alpha(D)$ if $u < \infty$ a.e. and for every $s > 0$, $u \geq P_s^{(\alpha)} * u$ on D. Here "$*$" denotes the convolution in $\mathbf{R}^{n+1}$ and u is considered to be extended to $\mathbf{R}^n \times (-\infty, b)$ by putting 0.

REMARK. For a positive Radon measure μ on D, we consider a convolution $u = W^{(\alpha)} * \mu$. If $u < \infty$ a.e., then $u \in S_\alpha(D)$, which is said to be a potential.

Proposition 1. (see cf. [3, Proposition 2.10]) *Let $u \in S_\alpha(D)$. Then there exist a measure μ on D and $h \in S_\alpha(D)$ such that*

$$u = W^{(\alpha)} * \mu + h$$

and $P_s^{(\alpha)} * h(x, t) = h(x, t)$ for every $0 < s < t - a$. Here μ and h are determined uniquely. The measure μ is called the associated measure of u.

For a finite continuous function $u \in S_\alpha(D)$ and a compact set K in D, we put

$$Q_K^{(\alpha)} u(Y) = \inf\{v(Y); v \in S_\alpha(D), v \geq u \text{ on } K\}$$

and

$$R_K^{(\alpha)} u(X) = \liminf_{Y \to X} Q_K^{(\alpha)} u(Y).$$

Furthermore for any $u \in S_\alpha(D)$ and any set A in D, we write

$$R_A^{(\alpha)} u(X) = \sup\{R_K^{(\alpha)} v(X); v \in S_\alpha(D) : \text{finite continuous}, v \leq u, K \subset A : \text{compact}\},$$

$$\overline{Q}_A^{(\alpha)} u(X) = \inf\{R_O^{(\alpha)} u(X); O \supset A : \text{open}\}$$

and

$$\overline{R}_A^{(\alpha)} u(X) = \liminf_{Y \to X} \overline{Q}_A^{(\alpha)} u(Y).$$

DEFINITION. Let $u = W^{(\alpha)} * \mu$, $w = \widetilde{W}^{(\alpha)} * \nu$ be potentials and A be a set in D. Then we denote by μ'_A and ν''_A the associated measure of $R_A^{(\alpha)} u$ and of $\widetilde{R}_A^{(\alpha)} w$, respectively.

We refer to [1] and [3] for the potential theoretic properties with respect to the α-parabolic operator.

2. Capacity

In this section, let $u \in S_\alpha(D)$ and $w \in \widetilde{S}_\alpha(D)$ be fixed. For a set A, we denote by μ_A and ν_A the associated measure of $R_A^{(\alpha)}u$ and of $\widetilde{R}_A^{(\alpha)}w$, respectively. For an open set Ω in D, we denote by $\mathcal{K}(\Omega)$ the totality of compact sets in Ω.

Lemma 1. *For $K \in \mathcal{K}(D)$, we have*

$$\int w\, d\mu_K = \int u\, d\nu_K.$$

Proof. Let ω be a relatively compact open set containing K. Then by [3, Propositions 3.7, 3.8], we have

$$\int w\, d\mu_K = \int \widetilde{W} * \nu_\omega\, d\mu'_{\omega,K} = \int W * \mu\, d\nu''_{\omega,K} = \int u\, d\nu_K.$$

Definition. For a set A in D, we put

$$\underline{C}_{u,w}(A) = \underline{C}(A) := \sup\{\int w\, d\mu_K; K \subset A : \text{compact}\},$$

$$\overline{C}_{u,w}(A) = \overline{C}(A) := \inf\{\underline{C}(O); O \supset A : \text{open}\}.$$

If $\underline{C}(A) = \overline{C}(A)$, then we denote the common value by $C_{u,w}(A)$ or $C(A)$ and say that A is capacitable with respect to (u, w).

Lemma 2. *For $K_1, K_2 \in \mathcal{K}(D)$, we have*

$$R^{(\alpha)}_{K_1 \cup K_2}u + R^{(\alpha)}_{K_1 \cap K_2}u \leq R^{(\alpha)}_{K_1}u + R^{(\alpha)}_{K_2}u$$

and

$$\widetilde{R}^{(\alpha)}_{K_1 \cup K_2}w + \widetilde{R}^{(\alpha)}_{K_1 \cap K_2}w \leq \widetilde{R}^{(\alpha)}_{K_1}w + \widetilde{R}^{(\alpha)}_{K_2}w.$$

Proof. We may assume that u is continuous. For every open sets ω_1 and ω_2 with $\omega_1 \supset K_1$ and $\omega_2 \supset K_2$, we have

$$R^{(\alpha)}_{K_1 \cup K_2}u + R^{(\alpha)}_{K_1 \cap K_2}u \leq R^{(\alpha)}_{\omega_1}u + R^{(\alpha)}_{\omega_2}u \quad \text{on} \quad \omega_1 \cup \omega_2,$$

which implies

$$R^{(\alpha)}_{K_1 \cup K_2}u + R^{(\alpha)}_{K_1 \cap K_2}u \leq R^{(\alpha)}_{\omega_1}u + R^{(\alpha)}_{\omega_2}u \quad \text{on} \quad D$$

by the domination principle [3, Proposition 3.11]. Then the first inequality follows from [3, Proposition 3.5]. The second one is shown similarly.

Proposition 2. *The capacity $\underline{C}_{u,w}$ on $\mathcal{K}(D)$ is strongly subadditive.*

Proof. Taking a relatively compact open set ω with $K_1 \cup K_2 \subset \omega$, we have

$$\begin{aligned}\underline{C}(K_1 \cup K_2) + \underline{C}(K_1 \cap K_2) &= \int w\, d\mu_{K_1\cup K_2} + \int w\, d\mu_{K_1\cap K_2} \\ &= \int (R^{(\alpha)}_{K_1\cup K_2}u + R^{(\alpha)}_{K_1\cap K_2}u)\, d\nu_\omega \\ &\leq \int (R^{(\alpha)}_{K_1}u + R^{(\alpha)}_{K_2}u)\, d\nu_\omega = \int w\, d\mu_{K_1} + \int w\, d\mu_{K_2} \\ &= \underline{C}(K_1) + \underline{C}(K_2).\end{aligned}$$

Theorem 1. *Let Ω be an open set in D. If both u and w are continuous on D, then the capacity $\underline{C}_{u,w}$ defined on $\mathcal{K}(\Omega)$ is a Choquet capacity. Consequently every analytic set, especially every Borel set in Ω is capacitable.*

Proof. By the above proposition, we have only to show the right continuity. Let $(K_k)_{k=1}^{\infty}$ be a decreasing sequence in $\mathcal{K}(\Omega)$ and put $K = \cap_{k=1}^{\infty} K_k$. Then by [3, Lemmas 2.7, 3.5], we have w*-$\lim_{k\to\infty} \mu_{K_k} = \mu_K$. Since w is continuous, $\lim_{k\to\infty} \int w\, \mu_{K_k} = \int w\, \mu_K$. This completes the proof.

Theorem 2. *Let A be a set in D.*

(1)If $R^{(\alpha)}_A u$ is a potential, then $\underline{C}(A) = \int w\, d\mu_A$.

(2) If $w = \widetilde{W}^{(\alpha)} * \nu$, then $\underline{C}(A) = \int R^{(\alpha)}_A u\, d\nu$.

Assume that u and w are continuous on a neighborhood of A.

(3) If $\overline{R}^{(\alpha)}_A u$ is a potential $W^{(\alpha)} * \overline{\mu}_A$, then $\overline{C}(A) = \int w\, d\overline{\mu}_A$.

(4) If $w = \widetilde{W}^{(\alpha)} * \nu$, then $\overline{C}(A) = \int \overline{R}^{(\alpha)}_A u\, d\nu$.

Proof. Let $(\omega_k)_{k=1}^{\infty}$ be an exhaustion of D. Then for any compact set $K \subset A$,

$$\underline{C}(K) = \lim_{k\to\infty} \int \widetilde{W}^{(\alpha)} * \nu_{\omega_k}\, d\mu_K = \lim_{k\to\infty} \int R^{(\alpha)}_K u\, d\nu_{\omega_k}.$$

By the Dini lemma (cf. [4, Lemma 2.23]), we have

$$\underline{C}(A) = \lim_{k\to\infty} \int R^{(\alpha)}_A u\, d\nu_{\omega_k} = \lim_{k\to\infty} \int \widetilde{W}^{(\alpha)} * \nu_{\omega_k}\, d\mu_A = \int w\, d\mu_A,$$

which shows (1). Next let $(O_k)_{k=1}^{\infty}$ be a decreasing sequence of open sets including A such that u and w are continuous on O_1, $\lim_{k\to\infty} C(O_k) = \overline{C}(A)$ and the lower regularization of $\lim_{k\to\infty} R^{(\alpha)}_{O_k} u$ is equal to $\overline{R}^{(\alpha)}_A u$. Putting $B := \cap_{k=1}^{\infty} O_k$, we have $\overline{C}(B) = \overline{C}(A)$ and $\overline{R}^{(\alpha)}_B u = \overline{R}^{(\alpha)}_A u$, that is, $\overline{\mu}_B = \overline{\mu}_A$. Then remarking that B is Borel, we have

$$\overline{C}(A) = C(B) = \lim_{k\to\infty} \int R^{(\alpha)}_B u\, d\nu_{\omega_k} \leq \lim_{k\to\infty} \int \overline{R}^{(\alpha)}_B u\, d\nu_{\omega_k} = \int w\, d\overline{\mu}_B = \int w\, d\overline{\mu}_A.$$

On the other hand,

$$\begin{aligned}\overline{C}(A) &= \inf_j C(O_j) = \inf_j \sup_k \int R_{O_j}^{(\alpha)} u \, d\nu_{\omega_k} \geq \sup_k \inf_j \int R_{O_j}^{(\alpha)} u \, d\nu_{\omega_k} \\ &\geq \sup_k \int \overline{R}_A^{(\alpha)} u \, d\nu_{\omega_k} = \sup_k \int \widetilde{R}_{\omega_k}^{(\alpha)} u \, d\overline{\mu}_A = \int w \, d\overline{\mu}_A,\end{aligned}$$

which shows (3). (2) and (4) are shown in a similar way without taking an exhaustion.

REMARK. In the above theorems, it is necessary to assume that u and w are continuous. In fact, we choose $u := 1$, $w := \widetilde{W}^{(\alpha)}$ and $A := \{0\}$. Then $\underline{C}(A) = 0$ and $\overline{R}_A^{(\alpha)} u = 0$ nevertheless $\overline{C}(A) = 1$.

Corollary 1. *For a compact set M in $\mathbf{R}^n$ and $t \in \mathbf{R}$,*

$$\underline{C}_{u,w}(M \times \{t\}) = \int_M u(x,t) w(x,t) \, dx.$$

When we take $u = w = 1$, the capacity $\underline{C}_{1,1}$ is said to be the α-parabolic capacity and for a compact set K in $\mathbf{R}^{n+1}$, μ_K is called the equilibrium measure of K. Then we have

Proposition 3. *If for a set A in $\mathbf{R}^{n+1}$, $R_A^{(\alpha)} 1$ (resp. $\overline{R}_A^{(\alpha)} 1$) is not a potential, then $\underline{C}_{1,1}(A) = \infty$ (resp. $\overline{C}_{1,1}(A) = \infty$).*

Proof. Assume that $R_A^{(\alpha)} 1 = 1$. Then taking an exhaustion $(\omega_k)_{k=1}^\infty$ of $\mathbf{R}^{n+1}$, in the same manner in the proof of Theorem 2(1), we have

$$\underline{C}(A) = \lim_{k \to \infty} \int R_A^{(\alpha)} 1 \, d\nu_{\omega_k} = \lim_{k \to \infty} C_{1,1}(\omega_k) = C_{1,1}(\mathbf{R}^{n+1}) = \infty$$

by Corollary 1. Next assume that $\overline{R}_A^{(\alpha)} 1 = 1$. Then for every open set $O \supset A$, $C_{1,1}(O) = \infty$, because $R_O^{(\alpha)} 1 = 1$. This shows $\overline{C}_{1,1}(A) = \infty$.

As an application of Theorems 1 and 2 to the reducibility (cf. [3, Lemma 3.5]), we have

Proposition 4. *Let $u \in S_\alpha(D)$ and A be an analytic set in D. If u is continuous on some open set $\omega \supset A$, then $R_A^{(\alpha)} u = \overline{R}_A^{(\alpha)} u$.*

Proof. Let $\varphi \geq 0$ be a continuous function on D with compact support and put $\nu := \varphi(X) dX$. Then since $\widetilde{W}^{(\alpha)} * \nu$ is continuous, by Theorems 1 and 2(2),(4), we have

$$\int R_A^{(\alpha)} u \, d\nu = \underline{C}(A) = \overline{C}(A) = \int \overline{R}_A^{(\alpha)} u \, d\nu.$$

This shows $R_A^{(\alpha)}u = \overline{R}_A^{(\alpha)}u$ a.e. on D, which implies

$$R_A^{(\alpha)}u = \overline{R}_A^{(\alpha)}u$$

by [3, Lemma 2.4(3)]. This is the required result.

3. Parabolic morphism

Let $D = \mathbf{R}^n \times (a,b)$ and $D' = \mathbf{R}^n \times (a',b')$ be strip domains. We consider a homeomorphism f from D to D' which preserve superharmonic functions.

Definition. Let $\varphi \in S_\alpha(D)$ and $\widetilde{\varphi} \in \widetilde{S}_\alpha(D)$. f is said to be an α–parabolic morphism with $(\varphi, \widetilde{\varphi})$ when maps $f^* : u \mapsto \varphi \cdot (u \circ f)$ and $\widetilde{f}^* : u \mapsto \widetilde{\varphi} \cdot (u \circ f)$ are bijections from $S_\alpha(D')$ to $S_\alpha(D)$ and from $\widetilde{S}_\alpha(D')$ to $\widetilde{S}_\alpha(D)$, respectively.

Example. We give two typical examples.

(1) Parabolic dilations: For $\lambda > 0$, we put $\tau_\lambda^{(\alpha)}(x,t) := (\lambda x, \lambda^{2\alpha}t)$. $\tau_\lambda^{(\alpha)}$ is an α–parabolic morphism with $(1,1)$.

(2) The Appell transformation: Let $\alpha = 1$, $D = \mathbf{R}^n \times (0,\infty)$ and $D' = \mathbf{R}^n \times (-\infty, 0)$. We put $\mathcal{A}(x,t) := (x/t, -1/t)$. Then $\mathcal{A}$ is an α–parabolic morphism with $(W^{(1)}, 1/(\widetilde{W}^{(1)} \circ \mathcal{A}^{-1}))$.

Theorem 3. *Let $f : D \to D'$ be an α-parabolic morphism. Then for $u \in S_\alpha(D')$, $w \in \widetilde{S}_\alpha(D')$ and a set A in D, we have*

$$R_A^{(\alpha)} f^* u = f^* R_{f(A)}^{(\alpha)} u$$

and

$$\widetilde{R}_A^{(\alpha)} \widetilde{f}^* w = \widetilde{f}^* \widetilde{R}_{f(A)}^{(\alpha)} w.$$

We apply Theorem 3 to parabolic dilations and the Appell transformation. Since $\tau_\lambda^{(\alpha)}$ is linear and

$$W^{(\alpha)} \circ \tau_\lambda^{(\alpha)} = \lambda^{-n} \cdot W^{(\alpha)},$$

we have

Corollary 2. *Let $\lambda > 0$ and K be a compact set in $\mathbf{R}^{n+1}$ and denote by μ and μ_λ the equilibrium measure of K and $\tau_\lambda^{(\alpha)}(K)$, respectively. Then we have*

$$\mu_\lambda = \lambda^n \cdot \tau_{\lambda *}^{(\alpha)} \mu$$

and

$$C_{1,1}(\tau_\lambda^{(\alpha)}(K)) = \lambda^n C_{1,1}(K),$$

where $\tau_{\lambda *}^{(\alpha)}\mu$ *is the image measure of* μ *by* $\tau_\lambda^{(\alpha)}$.

Theorem 4. *Let* K *be a compact set in* $\mathbf{R}^n \times (0,\infty)$ *and denote by* ν_K *and* $\mu_{\mathcal{A}(K)}$ *the associated measure of* $\widetilde{R}_K^{(1)}1$ *and* $R_{\mathcal{A}(K)}^{(1)}1$, *respectively. Then we have*

$$\mathcal{A}_* \varepsilon'_K = \widetilde{W}^{(1)} \cdot \mu_{\mathcal{A}(K)}, \quad \mathcal{A}_*^{-1}\varepsilon''_{\mathcal{A}(K)} = W^{(1)} \cdot \nu_K$$

and

$$C_{W^{(1)},1}(K) = C_{1,\widetilde{W}^{(1)}}(\mathcal{A}(K)).$$

Proof. Since $\mathcal{A}^*1 = W^{(1)}$, by Theorem 3, we have

$$\begin{aligned} W^{(1)} * \varepsilon'_K(x,t) &= \mathcal{A}^*(W^{(1)} * \mu_{\mathcal{A}(K)})(x,t) \\ &= W^{(1)}(x,t)\int W^{(1)}(\mathcal{A}(x,t) - \mathcal{A}(y,s))\, d(\mathcal{A}_*^{-1}\mu_{\mathcal{A}(K)})(y,s). \end{aligned}$$

By a direct calculation, we have

$$\begin{aligned} W^{(1)}(x,t)W^{(1)}(\mathcal{A}(x,t) - \mathcal{A}(y,s)) &= W^{(1)}(x-y,t-s)\left(\frac{4\pi}{s}\right)^{-\frac{n}{2}} \exp\left(-\frac{|y|^2}{4s}\right) \\ &= W^{(1)}(x-y,t-s)\cdot(\widetilde{W}^{(1)} \circ \mathcal{A})(y,s), \end{aligned}$$

which shows the first relation. We can obtain the second relation in the same manner. Integrating this relation, we have the last equality.

4. Minimal thinness for the heat equation

Finally, we give an application of the Appell transformation to the minimal thinness. In this section, we assume $\alpha = 1$. For a closed set F in $\mathbf{R}^{n+1}$, F is said to be 1-thin if $R_F^{(\alpha)}1 \neq 1$. By Theorems 3, 4 and the Wiener criterion [5], we have the following

Theorem 5. *Let* F *be a closed set in* $\mathbf{R}^{n+1}$. F *is* 1*-thin if and only if*

$$\sum_{k=1}^{\infty} C_{1,\widetilde{W}^{(1)}}(F_k) < \infty,$$

where

$$F_k = \{(x,t) \in F; t < 0, kc \leq \log(-t) - \frac{|x|^2}{2nt} \leq (k+1)c\}$$

for some fixed $c > 0$.

K.Uchiyama also mentioned the Wiener criterion of the 1–thinness in [6], where he uses another cutting process. It is an open question to obtain this type of criterion of the 1–thinness for $\alpha < 1$.

ACKNOWLEDGEMENT

This paper was written while the author was visiting Katholische Universität Eichstätt with the support of DAAD foundation. The author wishes to express his gratitude for the support received.

REFERENCES

[1] M.Nishio, The Wiener criterion of regular points for the parabolic operator of order α, Nagoya Math. J., 116(1989), 163–179.

[2] N.A.Watson, Thermal capacity, Proc. London Math. Soc., 37(1978), 342–362.

[3] M.Itô and M.Nishio, Poincaré type conditions of the regularity for the parabolic operator of order α, Nagoya Math. J., 115(1989), 1–22.

[4] L.L.Helms, Introduction to potential theory, Wiley-Interscience, New York, 1969.

[5] L.Evans and R.Gariepy, Wiener criterion for the heat equation, Arch. Rational Mech. Anal., 78(1981), 293–314.

[6] K.Uchiyama, A probabilistic proof and applications of Wiener's test for the heat operator, Math. Ann., 283(1989), 65–86.

7th Int. Coll. on Differential Equations, pp. 261-267
D. Bainov (Ed)

Martin Boundary at Infinity for the Heat Equation

MASAHARU NISHIO

Department of Mathematics, Osaka City University, Osaka 558, Japan

Abstract. We formulate the notion of the Martin boundary at infinity and apply it to some concrete domeins.

Keywords: heat equation, uniqueness of positive solutions, Martin boundary

INTRODUCTION

We consider positive solutions of the heat equation

$$Hu := (\frac{\partial}{\partial t} - \Delta)u = 0$$

on an unbounded domain Ω of the $(n+1)$–dimensional Euclidean space $\mathbf{R}^{n+1}$. Here we denote by X or (x,t) a point of $\mathbf{R}^{n+1}$ and by Δ the Laplace operator with respect to x. Our main interest is the convex cone

$$\mathcal{H}_\infty(\Omega) := \{u \geq 0; Hu = 0 \text{ on } \Omega, u = 0 \text{ on } \partial\Omega\},$$

which is closely related to the uniqueness of positive solutions. In fact, for example, if $\mathcal{H}_\infty(\Omega) = 0$, then positive solutions are determined by their boundary values. The Martin boundary theory is useful for treating a convex cone of positive solutions.

In this note, we construct the Martin boundary for the heat equation and examin $\dim \mathcal{H}_\infty(\Omega)$ for some concrete domains Ω by their geometric properties.

1. PRELIMINARIES

In this section, we recall the basic properties for the heat equation.

For a domain Ω in $\mathbf{R}^{n+1}$, we denote by $G(X,Y) = G_\Omega(X,Y)$ the Green function on Ω and by ω_Ω^X the heat measure at $X \in \Omega$, that is, for a bounded continuous function f on $\partial\Omega$,

$$h(X) := \int f \, d\omega_\Omega^X$$

is the solution of the Dirichlet problem for the heat equation. When Ω is a strip domain of form $\Omega = \mathbf{R}^n \times (a,b)$, we can write $G_\Omega(X,Y) = W(X-Y)$, where

$$W(x,t) = \begin{cases} (4\pi t)^{-\frac{n}{2}} \exp(-\frac{|x|^2}{4t}) & t > 0 \\ 0 & t \leq 0 \end{cases}$$

and

$$\omega_\Omega^{(x,y)} = (W(x-y, t-a)dy) \otimes \varepsilon_a,$$

where ε_a denotes the Dirac measure at a.

We call a solution h of $Hh = 0$ a temperature. A lower semi-continuous function u on Ω is said to be a supertemperature if $Hu \geq 0$ in the sense of distribution. We denote by $\mathcal{S}^+(\Omega)$ the totality of nonnegative supertemperatures on Ω. Since $HG(\cdot, Y) = \varepsilon_Y$, for every measure $\mu \geq 0$ on Ω,

$$G\mu := \int G(\cdot, Y)\, d\mu(Y) \in \mathcal{S}^+(\Omega),$$

whenever $G\mu < \infty$ a.e. on Ω. Then we call $G\mu$ a potential. Conversely, we have the following Riesz decomposition.

Proposition 1. (see cf. [1]) *For $u \in \mathcal{S}^+(\Omega)$, there exist a measure $\mu \geq 0$ and a temperature h on Ω such that*

$$u = G\mu + h.$$

2. Martin boundary

We shall introduce a topology in $\mathcal{S}^+(\Omega)$. We denote

$$\Psi(\Omega) := \{1 - \varphi; \varphi \in C_0^\infty(\Omega), 0 \leq \varphi \leq 1\},$$

where $C_0^\infty(\Omega)$ is the set of all infinitely differentiable functions on Ω with compact support. For $u = G\mu + h \in \mathcal{S}^+(\Omega)$ and $\psi \in \Psi$, we put

$$u_{(\psi)} := G(\psi \cdot \mu) + h,$$

which is a temperature on $\Omega \setminus \mathrm{supp}(\psi)$. We consider the following semi-norms on the vector space $\mathcal{V}(\Omega) = \mathcal{S}^+(\Omega) - \mathcal{S}^+(\Omega)$ which consists of all formal differences of elements in $\mathcal{S}^+(\Omega)$.

DEFINITION. Let $\varphi \in C_0^\infty(\Omega)$, $\psi \in \Psi$ and K be a compact set in $\Omega \setminus \mathrm{supp}(\psi)$. For $u - u' = (G\mu + h) - (G\mu' + h') \in \mathcal{V}(\Omega)$, we put

$$\|u - u'\|_\varphi := \left| \int \varphi \, d\mu - \int \varphi \, d\mu' \right|$$

and

$$\|u - u'\|_{\psi,K} := \sup_K |u_{(\psi)} - u'_{(\psi)}|,$$

which are semi-norms on $\mathcal{V}(\Omega)$. We consider $\mathcal{S}^+(\Omega)$ to be endowed with the relative topology in $\mathcal{V}(\Omega)$. We remark that $\mathcal{S}^+(\Omega)$ is metrizable.

We fix a measure $\lambda \geq 0$ on Ω such that

$$G^*\lambda := \int G(X, \cdot) \, d\lambda(X)$$

is strictly positive and finite continuous on Ω. We consider

$$\mathcal{S}_\lambda^{+1}(\Omega) := \{u \in \mathcal{S}^+(\Omega); \int u \, d\lambda \leq 1\}$$

and

$$\Omega_\lambda := \left\{ k_Y := \frac{G(\cdot, Y)}{G^*\lambda(Y)}; Y \in \Omega \right\} \subset \mathcal{S}_\lambda^{+1}(\Omega).$$

Then we have

Lemma 1. *Let K be a compact set in Ω. Then $\{\mu(K); G\mu \in \mathcal{S}_\lambda^{+1}(\Omega)\}$ is bounded.*

Proof. Put $\alpha := \inf_K G^*\lambda > 0$. Then we have

$$1 \geq \int G\mu \, d\lambda = \int G^*\lambda \, d\mu \geq \alpha\mu(K),$$

which shows Lemma 1.

Lemma 2. *Let $\psi \in \Psi$ and K be a compact set in $\Omega \setminus supp(\psi)$. Then a family $\{u_{(\psi)}; u \in \mathcal{S}_\lambda^{+1}(\Omega)\}$ is uniformly bounded and equicontinuous on K.*

Proof. Since $Hu_{(\psi)} = 0$ on $\Omega \setminus \mathrm{supp}(\psi)$, we have only to show the uniform boundedness. By the Harnack inequality, $\{h; G\mu + h \in \mathcal{S}_\lambda^{+1}(\Omega)\}$ is uniformly bounded on K. On the other hand, $\sup_{K \times \mathrm{supp}(\psi)} G < \infty$. Hence Lemma 1 shows Lemma 2.

Proposition 2. *$\mathcal{S}_\lambda^{+1}(\Omega)$ is compact.*

Proof. Let $(u_k = G\mu_k + h_k)_{k=1}^{\infty}$ be a sequence in $\mathcal{S}_{\lambda}^{+1}(\Omega)\}$. By Lemma 1, we may assume that $(\mu_k)_{k=1}^{\infty}$ converges to some measure μ. Since $\mathcal{S}_{\lambda}^{+1}(\Omega)$ is metrizable, by using the diagonal method and Lemma 2, we obtain a convergent subsequence. This shows Proposition 2.

Proposition 3. *Ω is homeomorphic to Ω_λ by the mapping $\iota_\lambda : Y \mapsto k_Y$.*

Proof. First we shall show the continuity. Let $Y_0 \in \Omega$. For $\varphi \in C_0^{\infty}(\Omega)$,

$$\int \frac{\varphi}{G^*\lambda(Y)}\,d\varepsilon_Y = \frac{\varphi(Y)}{G^*\lambda(Y)}$$

is continuous. Let $\psi \in \Psi$ and K be a compact set in $\Omega \setminus \mathrm{supp}(\psi)$.

$$k_{Y(\psi)}(X) = \frac{G(X,Y)\psi(Y)}{G^*\lambda(Y)}$$

is equicontinuous functions of Y at Y_0 for $X \in K$. Therefore the mapping ι_λ is continuous. Next, taking $\varphi \in C_0^{\infty}(\Omega)$, we have

$$\left\{Y \in \Omega;\, \frac{\varphi(Y)}{G^*\lambda(Y)} > 0\right\} \subset \mathrm{supp}(\varphi).$$

This shows Proposition 3.

DEFINITION. We call $\overline{\Omega_\lambda}$ the Martin compactification of Ω and put $\partial_\lambda\Omega := \overline{\Omega_\lambda} \setminus \Omega_\lambda$, which is said to be the Martin boundary of Ω.

Theorem 1. *Let $h \geq 0$ be a temperature on Ω. If $\int h\,d\lambda < \infty$, there exists a measure $\tilde{\mu}$ on $\partial_\lambda\Omega$ such that*

$$h(X) = \int_{\partial_\lambda\Omega} v(X)\,d\tilde{\mu}(v).$$

Proof. Let $(\Omega_k)_{k=1}^{\infty}$ be an exhaustion of Ω. Then the function

$$h_k(X) := \begin{cases} h(X) & X \in \Omega_k \\ \liminf_{Y \to X} \int h\,d\omega^Y_{\Omega\setminus\overline{\Omega_k}} & X \in \partial\Omega_k \\ \int h\,d\omega^X_{\Omega\setminus\overline{\Omega_k}} & X \in \Omega \setminus \overline{\Omega_k} \end{cases}$$

is a posential, that is, $h_k = G\mu_k$ for some measure μ_k on Ω supported by $\Omega \setminus \Omega_k$. We put

$$\tilde{\mu}_k := (G^*\lambda \circ \iota_\lambda^{-1}) \cdot \iota_{\lambda *}\mu_k,$$

where $\iota_{\lambda *}\mu_k$ is the image measure of μ_k by ι_λ. Hence we have

$$\int v(X)\,d\tilde{\mu}_k(v) = \int k_Y(X) G^*\lambda(Y)\,d\mu_k(Y) = G\mu_k(X) = h_k(X)$$

and

$$\int d\tilde{\mu}_k = \int G^*\lambda \, d\mu_k = \int h_k \, d\lambda \leq \int h \, d\lambda < \infty,$$

so that the measure $\tilde{\mu}_k$ can be regarded as a measure on $\mathcal{S}^{+1}_{\lambda}(\Omega)$ and, if necessary, by taking a subsequence, we obtain a measure $\tilde{\mu}$ which is the limit of $(\tilde{\mu}_k)_{k=1}^{\infty}$. Then $\tilde{\mu}$ is supported by $\partial_\lambda \Omega$ and

$$\int v(X) \, d\tilde{\mu}(v) = h(X).$$

This completes the proof.

3. Martin boundary at infinity

For a domain Ω in $\mathbf{R}^{n+1}$, we consider

$$\mathcal{H}_\infty(\Omega) = \{u \geq 0; Hu = 0 \text{ on } \Omega, u = 0 \text{ on } \partial\Omega\},$$

where the assertion "$u = 0$ on $\partial\Omega$" means that for every relatively compact open set O in $\mathbf{R}^{n+1}$,

$$u(X) = \int_\Omega u \, d\omega^X_{\Omega \cap O}.$$

Definition. For a domain Ω in $\mathbf{R}^{n+1}$, we put

$$\partial^\infty_\lambda \Omega := \partial_\lambda \Omega \cap \mathcal{H}_\infty(\Omega),$$

which is called the Martin boundary at infinity.

For the heat equation, the Martin boundary depends on the choice of the measure λ. Hereafter we restrict domains to treat. Kemper [2] deals with a spatially bounded (1,1/2)–Lipschitz domain andshow the uniqueness of kernel functions for finite boundary points. Our interest is kernel functions at infinity.

A domain Ω in $\mathbf{R}^{n+1}$ is said to be spacially bounded if for every $t \in \mathbf{R}$, $\{x \in \mathbf{R}^n; (x,t) \in \Omega\}$ is bounded. A domain Ω in $\mathbf{R}^{n+1}$ is called a $(1, 1/2)$–Lipschitz domain with the Lipschitz constant m if for every boundary point $(y,s) \in \partial\Omega$, there exist a coodinate system $(x_1, \cdots, x_n)$ of $\mathbf{R}^n$, a function f on $\mathbf{R}^{n-1} \times \mathbf{R}$ and a neighborhood U of (y,s) such that for every $x^*, \xi^* \in \mathbf{R}^{n-1}$ and every $t, \tau \in \mathbf{R}$,

$$|f(x^*,t) - f(\xi^*,\tau)| \leq m(|x^* - \xi^*| + |t - \tau|^{\frac{1}{2}})$$

and

$$\Omega \cap U = \{(x^*, x_n, t) \in U; x_n > f(x^*,t)\}.$$

REMARK. Every boundary point Y of a (1,1/2)–Lipschitz domain Ω is regular with respect to the Dirichlet problem:

$$\text{w}^*\text{-}\lim_{X\to Y}\omega_\Omega^x = \varepsilon_Y$$

(see cf. [3]).

For a spatially bounded (1,1/2)–Lipschitz domain, the following Harnach inequality of elliptic type holds.

Proposition 4. (see cf. [3, Proposition 1])) *Let Ω be a spatially bounded (1,1/2)–Lipschitz domain in $\mathbf{R}^{n+1}$, $\tau \in \mathbf{R}$ and K a compact set in $\Omega_\tau := \Omega \cap (\mathbf{R}^n \times (,\tau,\infty))$. Then there exists a constant $C > 0$ auch that for every temperature $h \geq 0$ on Ω_τ which vanishes continuously on $\partial\Omega \cap (\mathbf{R}^n \times (,\tau,\infty))$,*

$$\max_K h \leq \min_K h.$$

From this proposition, we can choose a good measure as λ.

Theorem 2. *Let Ω be a spatially bounded (1,1/2)–Lipschitz domain in $\mathbf{R}^{n+1}$. Then there exists a measure $\lambda \geq 0$ such that $G^*\lambda$ is strictly positive and continuous ans that*

$$\mathcal{H}_\infty(\Omega) \subset L^1(\Omega, \lambda).$$

Corollary 1. *Let Ω be as above and $X_0 \in \Omega$. Then*

$$\mathcal{H}_\infty^1(\Omega, X_0) := \{h \in \mathcal{H}_\infty(\Omega); h(X_0) = 1\}$$

is compact.

Moreover we know

Theorem 3. ([4, Proposition 2]) *Let Ω and X_0 be as above. Then*

$$\mathcal{H}_\infty^1(\Omega, X_0) \neq \emptyset.$$

Let D be a bounded Lipschitz domain which is star–like at the origin in the sense of Hunt–wheeden [5]. For $\alpha \in \mathbf{R}$, put

$$\Omega_\alpha(D) := \{(x,t); t < 0, (-t)^{-\alpha}x \in D\}.$$

Then we have

Theorem 4. **(cf. [4],[6])** *Let $X_0 \in \Omega_\alpha(D)$. If $\alpha < 1$, then*

$$^{\#}\mathcal{H}^1_\infty(\Omega_\alpha(D), X_0) = 1,$$

where for a set A, $^{\#}A$ denotes the number of elements in A. Hence the Martin boundary is the following:

$$\begin{aligned} \partial^\infty_\lambda \Omega_\alpha(D) &\cong \{0\} \cup \{-\infty\}, \\ \partial_\lambda \Omega_\alpha(D) &\cong (\partial\Omega_\alpha(D) \cap (\mathbf{R}^n \times (-\infty, 0))) \cup \{0\} \cup \{-\infty\}. \end{aligned}$$

Here "$-\infty$" means the unique element $h \in \mathcal{H}_\infty(\Omega\alpha(D))$ such that $\int h\, d\lambda = 1$.

On the other hands, for $\alpha \geq 1$, we have

Theorem 5. *if $\alpha \geq 1$, then*

$$\dim \mathcal{H}_\infty(\Omega\alpha(D)) := {}^{\#}ex\, \mathcal{H}^1_\infty(\Omega_\alpha(D), X_0) = {}^{\#}\partial^\infty_\lambda \Omega_\alpha(D) = \infty,$$

where $ex\, \mathcal{H}^1_\infty(\Omega_\alpha(D), X_0)$ denotes the set of all extremal elements of $\mathcal{H}^1_\infty(\Omega_\alpha(D), X_0)$.

REMARK. For $\alpha = 1$, $\partial^\infty_\lambda \Omega_1(D)$ is homeomorphic to $\overline{D}$ disjointly unioned by one point, which corresponds to the constant function 0. Hence we have

$$\partial_\lambda \Omega_1(D) \cong \partial\Omega_1(D) \cup \overline{D}.$$

ACKNOWLEDGEMENT

This paper was written while the author was visiting Katholische Universität Eichstätt with the support of DAAD foundation. The author wishes to express his gratitude for the support received.

REFERENCES

[1] N.A.Watson, Green functions, potentials and the Dirichlet problem for the heat equation, Proc. London Math. Soc., (3)33(1976), 251–298.

[2] J.T.Kemper, Temperatures in several variables: kernel functions, representations, and parabolic boundary values, Trans. Amer. Math. Soc., 167(1972), 243-262.

[3] E.Efrros and J.Kazdan, On the Dirichlet problem for the heat equation, Indiana U. Math. J., 20(1971), 683–693.

[4] M.Nishio, The uniqueness of positive solutions of parabolic equations of divergence form on an unbounded domain, Nagoya Math. J., 130(1993), 111–121.

[5] R.R.Hunt and R.L.Wheeden, Positive harmonic functions on Lipschitz domains, Trans. Amer. Math. Soc., 147(1970), 507–527.

[6] M.Nishio and N.Suzuki, Minimal thickness and uniqueness of kernel functions for the heat equation in several variables, Osaka J. Math., 31(1994), 331–339.

7th Int. Coll. on Differential Equations, pp. 269-276
D. Bainov (Ed)

A general form of a mean value property for poly-temperatures on a strip domain

MASAHARU NISHIO, KATSUNORI SHIMOMURA and NORIAKI SUZUKI

Department of Mathematics, Osaka City University, Osaka 558, Japan,
Department of Mathematical Sciences, Ibaraki University Mito, Ibaraki 310, Japan and
Graduate School of Polymathematics, Nagoya University, Chikusa Nagoya 464-01, Japan

Abstract. We show that the mean value property of poly-temperature, the solution of the iterate of the heat operator, also holds for more general mean value than that of [1].
Keywords: heat equation, mean value property

Introduction

In our previous paper [1], we investigated a mean value property of poly-temperatures and of poly-supertemperatures on a strip domain

$$D = \{(X,t); X \in \mathbf{R}^n, 0 < t < T\}.$$

In [1] we used the mean value defined by the integrals over parallel m hyperplanes placed with same intervals. In this note, we shall show that the mean value property also holds without the same interval assumption.

Let

$$H = \sum_{k=1}^{n} \frac{\partial^2}{\partial x_k^2} - \frac{\partial}{\partial t}$$

be the heat operator on $\mathbf{R}^{n+1} = \{(X,t); X = (x_1, \cdots, x_n) \in \mathbf{R}^n, t \in \mathbf{R}\}$. For an integer $m \geq 1$, a lower semi-continuous and locally integrable function u on D is called a *poly-supertemperature* of degree m, if

$$(-H)^m u \geq 0 \text{ on } D \text{ (in the sense of distribution).}$$

If u and $-u$ are both poly-supertemperatures of degree m, then u is called a *poly-temperature* of degree m. Since H is hypoelliptic, every poly-temperature belongs to $C^\infty(D)$ and hence u satisfies the equation

$$(-H)^m u(X,t) = 0 \quad (\forall (X,t) \in D) \text{ (in the usual sense).}$$

In the case $m = 1$, we simply call the function supertemperature and temperature.

1. Mean value property

Let W denote the fundamental solution for the heat equation on $\mathbf{R}^{n+1}$, that is,

$$W(X,t) = \begin{cases} \frac{1}{(4\pi t)^{n/2}} e^{\frac{-|X|^2}{4t}} & \text{if } t > 0, \\ 0 & \text{if } t \leq 0. \end{cases}$$

If a measurable function v on D satisfies the growth condition

$$|v(X,t)| \leq M e^{a|X|^2} \quad \forall (X,t) \in D \tag{1}$$

with some constants $M > 0$ and $a \geq 0$ (Here $|X|^2 = x_1^2 + \cdots + x_n^2$), we can define the integral mean $W[v,c](X_0,t_0)$ for $(X_0,t_0) \in D$ and $0 < c < \min\{t_0, 1/4a\}$ by

$$W[v,c](X_0,t_0) := \int_{\mathbf{R}^n} v(X_0 - X, t_0 - c) W(X,c) dX.$$

Now we define a mean value $A[v, c_1, \cdots, c_m](X_0,t_0)$ for $(X_0,t_0) \in D$ and $0 < c_1 < \cdots < c_m < \min\{1/4a, t_0\}$ by

$$A[v, c_1, \cdots, c_m](X_0,t_0) := \sum_{k=1}^{m} A_k \, W[v, c_k](X_0,t_0),$$

where the coefficients A_k are given by

$$A_k := \prod_{j=1, j\neq k}^{m} \frac{c_j}{c_j - c_k}. \qquad (A_1 := 1 \text{ when } m = 1) \tag{2}$$

REMARK 1. (a). The above mean value $A[v, c_1, \cdots, c_m]$ gives one of the generalization of our previous mean value $A[u,c]$ in [1]. In fact, putting $c_k = ck$, we have $A_k = (-1)^{k-1} {}_mC_k$, and hence $A[v, c, 2c, \cdots, mc] = A[u,c]$, where the right hand side denotes the mean value in [1].

(b). We have already encountered a similar quantity as A_k in our previous paper. Introducing new parameters $t_1, \cdots, t_n$ by $t_k = t_0 - c_k$, we have

$$A_k = \prod_{j=1, j\neq k}^{m} \frac{t_0 - t_j}{t_k - t_j}.$$

The right hand side equals to $A^{m,k}_{(t_0,\cdots,t_m)}$ which is used in Lemma 8 of [1].

(c). If v is a temperature on D satisfying (1), then for $0 < c < \min\{1/4a, t_0\}$,

$$v(X_0,t_0) = W[v,c](X_0,t_0)$$

(cf. [2, p.85]) and hence $v(X_0,t_0) = A[v,c](X_0,t_0)$ (cf. Theorem 1 below).

Then we can extend the mean value property to $A[u,c_1,\cdots,c_m]$ as follows.

Theorem 1. *Let u be a poly-temperature of degree m on D. We assume that there exist constants $M>0, a\geq 0$ such that for all $(X,t)\in D$*

$$|H^k u(X,t)| \leq Me^{a|X|^2} \quad for \quad k=0,1,\cdots,m-1. \tag{3}$$

Then we have

$$u = A[u,c_1,\cdots,c_m] \quad on \quad \mathbf{R}^n \times (c_m, T) \tag{4}$$

for every $c_1,\cdots,c_m$ with $0<c_1<\cdots<c_m<\min\{1/4a, T\}$.

For the proof, we use the following Almansi type formula:

Proposition 1. (see [1] Proposition 1) *A function u on D is a poly-temperature of degree m if and only if there are temperatures h_k on D $(k=0,1,\cdots,m-1)$ such that*

$$u(X,t) = h_0(X,t) + th_1(X,t) + \cdots + t^{m-1}h_{m-1}(X,t).$$

Moreover, if u satisfies the growth condition (3), then every h_k also satisfies

$$|h_k(X,t)| \leq M'e^{a|X|^2}$$

on D with some constant $M'>0$.

Proof of Theorem 1. Let $(X_0,t_0)\in D$ be fixed. By Proposition 1, there exist temperatures $h_0,h_1,\cdots,h_{m-1}$ on D satisfying the growth condition (1) such that

$$u(X,t) = h_0(X,t) + (t_0-t)h_1(X,t) + \cdots + (t_0-t)^{m-1}h_{m-1}(X,t).$$

Then, by Remark 1 (c), we have for $k=1,\ldots,m$,

$$W[u,c_k](X_0,t_0) = h_0(X_0,t_0) + c_k h_1(X_0,t_0) + \cdots + c_k^{m-1}h_{m-1}(X_0,t_0),$$

which implies, by the Cramer formula, (by $\check{c}_k$ we indicate that the factor c_k is missing)

$$\begin{aligned} u(X_0,t_0) &= h_0(X_0,t_0) = \begin{vmatrix} 1 & c_1 & \cdots & c_1^{m-1} \\ \vdots & \vdots & & \vdots \\ 1 & c_m & \cdots & c_m^{m-1} \end{vmatrix}^{-1} \begin{vmatrix} W[u,c_1](X_0,t_0) & c_1 & \cdots & c_1^{m-1} \\ \vdots & \vdots & & \vdots \\ W[u,c_m](X_0,t_0) & c_m & \cdots & c_m^{m-1} \end{vmatrix} \\ &= \frac{1}{\prod_{i<j}(c_j-c_i)} \sum_{k=1}^{m} (-1)^{k-1} \prod_{i<j, i\neq k, j\neq k} (c_j-c_i) c_1\cdots \check{c}_k \cdots c_m W[u,c_k](X_0,t_0) \\ &= \sum_{k=1}^{m} \prod_{j=1, j\neq k}^{m} \frac{c_j}{c_j-c_k} W[u,c_k](X_0,t_0) \\ &= \sum_{k=1}^{m} A_k W[u,c_k](X_0,t_0). \end{aligned}$$

This completes the proof.

It is an open question whether the conditions (1) for $k = 1, \cdots, m-1$ are really necessary or not.

2. The fundamental solution and poly-supertemperatures

Next we deal with poly-supertemperatures. If $m > 1$, the mean value $A[\cdot, \cdots]$ is no more positive mean, which means that a nonnegative function may have a negative mean value. Nevertheless poly-supertemperatures satisfy the following super-mean value property.

Theorem 2. *Let u be a C^{2m-2}-function on D satisfying the growth condition (3). If u is a poly-supertemperature of degree m on D, then we have*

$$u \geq A[u, c_1, \cdots, c_m] \quad \text{on } \mathbf{R}^n \times (c_m, T)$$

for all $c_1, \cdots, c_m$ with $0 < c_1 < \cdots < c_m < \min\{1/4a, T\}$.

For the proof, we first recall the Riesz decomposition of poly-supertemperatures. For two functions f and g on $\mathbf{R}^{n+1}$, we denote by $f * g$ the convolution of f and g, that is,

$$f * g(X,t) := \iint_{\mathbf{R}^n \times \mathbf{R}} f(X-Y, t-s) g(Y,s) dY ds$$

provided that the integral exists. We set inductively

$$W^1 := W, \qquad W^k(X,t) := W^{k-1} * W(X,t), \quad k \geq 2.$$

Then we have $W^m(X,t) = \frac{1}{(m-1)!} t^{m-1} W(X,t)$, which is the fundamental solution of the equation $(-H)^m u = 0$.

Proposition 2. *For $\varphi \in C_0^\infty(D)$, we have*

$$(-H)^m (W^m * \varphi) = W^m * ((-H)^m \varphi) = \varphi.$$

Proof. The equality $(-H)^m(W^m * \varphi) = \varphi$ follows from the fact that W is the fundamental solution of $-Hu = 0$. We shall show the second equality. Note that $-H(W * (-H\varphi)) = -H\varphi$, $h := W * (-H\varphi)) - \varphi$ is a temperature, and h satisfies the growth condition (1) with $a = 0$. Then by the mean value property in Remark 1 (c), we have for every $(X_0, t_0) \in D$,

$$h(X_0, t_0) = \lim_{c \uparrow t_0} W[h, c](X_0, t_0) = 0,$$

that is, $W*(-H\varphi) = \varphi$. Inductively, we have the second equality for every integer $m \geq 1$, which shows Proposition 2.

Using this notation, we have the following Riesz decomposition for poly-supertemperatures

Proposition 3. (see [1] Proposition 2) *Let $u \in C^{2m-2}(D)$ be a poly-supertemperature of degree m satisfying the growth condition (3). Then for any $d > 0$, there exist temperatures h_k on $D_d := \{(X,t) \in D; d < t < d + \frac{1}{4a}\}$ $(k = 0, 1, \cdots, m-1)$ such that*

$$u = h_0 + W * h_1 + \cdots + W^{m-1} * h_{m-1} + W^m * ((-H)^m u)$$

on D_d and that $|h_k| \leq M' m_{a,d}$ on D_d with some constant M'. Here

$$m_{a,d}(X,t) = \int_{\mathbf{R}^n} W(X-Y, t-d) e^{a|Y|^2} dY \quad ((X,t) \in D_d)$$

and to take the convolution, we consider the functions $(-H)^m u$ and h_k to be extended to $\mathbf{R}^{n+1}$ by putting 0 outside D_d.

In the above proposition, we do not know whether $h_0 + W * h_1 + \cdots + W^{m-1} * h_{m-1}$ is the greatest poly-temperature minorant of u on D_d or not.

Next we consider the auxiliary functions ϕ_p $(p = 0, \cdots, m)$ defined by

$$\phi_p(t) = \phi_p(t_0, \cdots, t_m; t) := (t_0 - t)^{m-1} - \sum_{k=1}^{p} A^{m,k}_{(t_0,\cdots,t_m)} (t_k - t)^{m-1}, \tag{5}$$

where the coefficients $A^{m,k}_{(t_0,\cdots,t_m)}$ are defined in Remark 1 (b).
The following nonnegativity of ϕ_p is the key point for the proof of Theorem 2.

Lemma 1. (see [1] Lemma 8) *Let $t_0, t_1, \cdots, t_m$ be real numbers such that*

$$t_m < t_{m-1} < \cdots < t_0.$$

Then we have for $0 \leq p \leq m-1$

$$\phi_p(t) \geq 0 \quad on \quad t_{p+1} \leq t \leq t_p, \tag{6}$$

and

$$\phi_m = 0. \tag{7}$$

Now we give a proof of Theorem 2.

Proof of Theorem 2. Let $u \in C^{2m-2}(D)$ be a poly-supertemperature of degree m. Take a point $(X_0, t_0) \in D$ and $c_1, \cdots, c_m$ with $0 < c_1 < \cdots < c_m$. Choose $d > 0$ with $c_m < t_0 - d < 1/4a$. Then by the Riesz decomposition

$$u = h_0 + W * h_1 + \cdots + W^{m-1} * h_{m-1} + W^m * ((-H)^m u) \text{ on } D_d.$$

Since $h := h_0 + W * h_1 + \cdots + W^{m-1} * h_{m-1}$ is poly-temperature of degree m, the mean value property implies $A[h, c_1, \cdots, c_m](X_0.t_0) = h(X_0, t_0)$. Then we put $\mu := (-H)^m u$. We may assume that $W^m * \mu(X_0, t_0) < \infty$. For $0 < c < t_0$, we have

$$\begin{aligned}
&W[W^m * \mu, c](X_0, t_0) \\
&= \int_{\mathbf{R}^n} W^m * \mu(X_0 - Y, t_0 - c) W(Y, c)\, dY \\
&= \int_{\mathbf{R}^n} \int_{\mathbf{R}^{n+1}} \frac{(t_0 - c - t)^{m-1}}{(m-1)!} W(X_0 - Y - X, t_0 - c - t) W(Y, c)\, d\mu(X, t) dY \\
&= \int_{\mathbf{R}^n \times (0, t_0 - c)} \frac{(t_0 - c - t)^{m-1}}{(m-1)!} W(X_0 - X, t_0 - t)\, d\mu(X, t).
\end{aligned}$$

Then putting $t_p = t_0 - c_p$ for $p = 0, 1, \cdots, m-1$, $t_{m+1} = 0$ and

$$\mu_p := \mu|_{\mathbf{R}^n \times [t_{p+1}, t_p)}$$

for $0 \le p \le m$, we obtain

$$\begin{aligned}
&W^m * \mu(X_0, t_0) - A[W^m * \mu, c_1, \cdots, c_m](X_0, t_0) \\
&= \int_{\mathbf{R}^{n+1}} \frac{(t_0 - t)^{m-1}}{(m-1)!} W(X_0 - X, t_0 - t)\, d\mu(X, t) \\
&\quad - \sum_{k=1}^m A_k \int_{\mathbf{R}^n \times (0, t_k)} \frac{(t_k - t)^{m-1}}{(m-1)!} W(X_0 - X, t_0 - t)\, d\mu(X, t) \\
&= \sum_{p=0}^m \int_{\mathbf{R}^{n+1}} \frac{(t_0 - t)^{m-1}}{(m-1)!} W(X_0 - X, t_0 - t)\, d\mu_p(X, t) \\
&\quad - \sum_{k=1}^m A_k \sum_{p=0}^m \int_{\mathbf{R}^{n+1}} \frac{(t_k - t)^{m-1}}{(m-1)!} W(X_0 - X, t_0 - t)\, d\mu_p(X, t) \\
&= \frac{1}{(m-1)!} \sum_{p=0}^m \int_{\mathbf{R}^{n+1}} \phi_p^m(t) W(X_0 - X, t_0 - t)\, d\mu_p(X, t), \\
&\ge 0
\end{aligned}$$

because of Lemma 1. This completes the proof of Theorem 2.

We give a remark on Lemma 1.

REMARK 2. For $0 < c_1 < \cdots < c_m$, we consider a C^{m-2}-function of τ

$$\psi_m(\tau) = \psi_m(c_1, \cdots, c_m; \tau) := \sum_{p=0}^{m-1} \sum_{k=p+1}^m A_k (\tau - c_k)^{m-1} \chi_{(c_p, c_{p+1}]}(\tau),$$

where $c_0 = 0$, and χ_K denotes the characteristic function of a set K. Then by (7), Lemma 1 is equivalent to $\psi_m(\tau) \geq 0$. If we take $c_k = kc$ with some constant $c > 0$, then we have

$$\psi_m(c, 2c, \cdots, mc; \tau) = \sum_{p=0}^{m-1} \sum_{k=0}^{p} (-1)^k {}_mC_k \, (\tau - kc)^{m-1} \chi_{(pc,(p+1)c]}(\tau),$$

which can be directly proved to be nonnegative by showing that $\psi_m(c, 2c, \cdots, mc; \tau)$ is an increasing function of c. In fact, for $m = 1$,

$$\psi_1(c; \tau) = \chi_{(0,c]}(\tau)$$

is increasing with respect to c and obviously nonnegative. Next take $m \geq 2$ and assume that $\psi \geq 0$. If $mc \leq \tau$, then $\psi(c, \cdots, mc; \tau) = 0$. If $pc < \tau < (p+1)c$ for some $p = 0, 1, \cdots, m-1$, then we have

$$\begin{aligned} \frac{\partial \psi_m}{\partial c}(c, \cdots, mc; \tau) &= \sum_{k=1}^{p} (-1)^{k-1} (m-1) k \, {}_mC_k (\tau - kc)^{m-2} \\ &= m(m-1) \sum_{k=1}^{p} (-1)^{k-1} \, {}_{m-1}C_{k-1}((\tau - c) - (k-1)c)^{m-2} \\ &= \psi_{m-1}(c, \cdots, (m-1)c; \tau) \\ &\geq 0. \end{aligned}$$

From this, we conjecture that $\phi^m(c_1, \cdots, c_m; \tau)$ is increasing with respect to each variable c_k $(k = 1, \cdots, m)$.

3. The characterization of poly-temperatures and poly-supertemperatures

As in [1], the mean value property characterize poly-temperature and super-mean value property characterize poly-supertemperature. For the proof, see [3].

Theorem 3. *Let u be a measurable function on D satisfying the growth condition (1). Then*

$$\lim_{c_1, \cdots, c_m \to 0,\ c_1 < \cdots < c_m} \frac{m!}{c_1 \cdots c_m} (u - A[u, c_1, \cdots, c_m]) = (-H)^m u, \tag{8}$$

in the distribution sense.

Theorem 4. *Let u be a measurable function on D satisfying the growth condition (1). If u satisfies*

$$u = A[u, c_1, \cdots, c_m] \tag{9}$$

almost everywhere on $\mathbf{R} \times (c_m, T)$ for all $0 < c_1 < \cdots < c_m < \min\{1/4a, T\}$, then u is equal to a poly-temperature of degree m almost everywhere on D.

Theorem 5. *Let u be a measurable function on D satisfying the growth condition (1). If u satisfies*

$$u \geq A[u, c_1, \cdots, c_m] \quad on \quad \mathbf{R}^n \times (c_m, T)$$

almost everywhere on $\mathbf{R} \times (c_m, T)$ *for all* $c_1, \cdots, c_m$ *with* $0 < c_1 < \cdots < c_m < \min\{1/4a, T\}$, *then u is equal to a poly-supertemperature of degree m almost everywhere on D.*

Acknowledgment

This paper was written while the first author M.Nishio was visiting Katholische Universität Eichstät with the support of DAAD foundation. The author wishes to express his gratitude for the support received.

References

[1] M.Nishio, K.Shimomura and N.Suzuki, A mean value property of poly-temperatures on a strip domain, J. London Math. Soc., in press.

[2] N.A.Watson, Parabolic equations on an infinite strip, Marcel Dekker, New York, 1988.

[3] M.Nishio, K.Shimomura and N.Suzuki, A characterization of poly-supertemperatures, preprint.

7th Int. Coll. on Differential Equations, pp. 277-282
D. Bainov (Ed)

Cohomology vanishing and q-convex functions on infinite dimensional domains

Seiko Ohgai * [1] ,

Graduate School of Mathematics, Kyushu University 33, Fukuoka, 812-81, Japan

Abstract

We give a theorem of cohomology vanishing with values in a Fréchet space for a q-concave domain in the infinite dimensional locally convex space E equipped with the finite open topology.

1 Introduction

In the previous paper[22], the author extended Kajiwara-Shon[20]'s infinite dimensional cohomology vanishing theorem, which asserts that

$$\mathrm{H}^p(D-A,\mathcal{O})=0 \tag{1}$$

for the cohomology of degree $p \leq \operatorname{codim} A - 2$ of the complement $D-A$ of an analytic subset A of a pseudoconvex open set D of a vector space E equipped with the finite open topology.

Let F be a Fréchet space and $\mathcal{O}_{\mathcal{F}}$ be the sheaf of germs of holomorphic mappings into F. By Fujimoto[9] , the study of F-valued holomorphic functions is not only to generalize the results on ordinary holomorphic functions , but also to contribute to the study of ordinary holomorphic functions on a product space.

The first aim of the present paper is to generalize the result of the previous paper[22] to the category of the F-holomorphic functions over infinite dimensional domains as follows:

Let E be a **C**-vector Hausdorff space and $\mathcal{O}$ be its structural sheaf. Let D be a domain of E. A real valued C^∞ function φ on D is said to be $q-convex$ if there exists a positive integer n_0 such that, for any integer n with $n \geq n_0$ and for any n-dimensional **C**-linear subspace H of E, the Levi form of the restriction $\varphi|D\cap H$ of the function φ to the n-dimensional domain $D\cap H$ has $n-q+1$ positive eigen values at every point of $D\cap H$.

The first aim of the present paper is to establish the cohomology vanishing theorem

$$\mathrm{H}^p(\{x\in D;\varphi(x)>c\},\mathcal{O}_{\mathcal{F}})=0 \tag{2}$$

[1]Supported by Grant-in-Aid for Scientific Research (C) no. 08304012 from the Ministry of Education, Science, Sports and Culture of Japan, 1996.
Keywords: q-convexity, cohomology vanishing, Oka's principle, infinite dimension
A.M.S. Classification: 32F10-32L10

for any pseudoconvex domain D, for any positive integers p and q, and for any q-convex function φ on D.

Kajiwara-Kazama[17] established the validity of the Oka's principle for relative cohomology sets for pairs of Stein spaces X and analytic subset A for complex Lie group L modelled with Banach spaces.

The second aim of the present paper is to infinite dimensionalize the above results[17] as follows:

Let Ω be a pseudoconvex domain in E, A be an analytic subset of Ω, L be a complex Lie group modelled with a Banach space. (B, L, Ω) be an analytic fibre bunble with base space Ω and structure group L and, $I_B^a(A)$ and $I_B^c(A)$ be, respectively, the sheaves of all germs of holomorphic and continuous sections over Ω. Then the canonical mapping

$$\iota : H^1(\Omega, I_B^a(A)) \to H^1(\Omega, I_B^c(A)) \tag{3}$$

is injective.

2 q-convexity and cohomology vanishing theorem

Let E be a **C**-linear Hausdorff space and Λ be the set of finite dimensional **C**-linear subspaces of E. A complex valued function f on an open subset D of E is said to be *Gâtaux holomorphic* if, for any $L \in \Lambda$, the restriction of f to $D \cap L$ is holomorphic on the open subset $D \cap L$ of the finite dimensional complex space L. A complex valued Gâtaux holomorphic function f on an open subset D of E is said to be *holomorphic* if f is continuous on D. The sheaf $\mathcal{O}$ of germs of holomorphic function over E is called the *structural sheaf* of E. An open subset D of E is said to be *pseudoconvex* if, for any $L \in \Lambda$, the intersection $D \bigcap L$ is pseudoconvex domain of the **C**-linear space L of finite dimension for any $L \in \Lambda$. A topology on E is said to be *finite open* if the family of open sets consists of subsets O of E such that, for any $L \in \Lambda$, the intersections $O \cap L$ are open in the finite dimensional Hausdorff **C**-linear spaces L.

As a **C**-linear vector space E equipped with the finite open topology, we consider exclusively in this section the space $\Sigma\mathbf{C}$ which is a unique locally convex space euipped with the finite open topology, which pulls therefore the definite article over it and which is defined as follows:

For any integers m and n with $m < n$, we regard the complex m-space $\mathbf{C}^m$ as a subspace

$$\mathbf{C}^m = \{z = (z_1, z_2, \cdots, z_n) \in \mathbf{C}^n; z_{m+1} = z_{m+2} = \cdots = z_n = 0\} \tag{4}$$

of the superspace $\mathbf{C}^n$. Let $\pi_{m,n} : \mathbf{C}^m \to \mathbf{C}^n$ be the canonical injection. We put

$$\Sigma\mathbf{C} := \bigcup_{n \geq 1} \mathbf{C}^n \tag{5}$$

and, for any $n \geq 1$, denote by $\pi_n : \mathbf{C}^n \to \Sigma\mathbf{C}$ the canonical injection. We induce the strongest topology in $\Sigma\mathbf{C}$ so as each injection π_n is continuous.

The **C**-linear space $\Sigma\mathbf{C}$ is the locally convex space with the finite open topology. Its structural sheaf is denoted by $\mathcal{O}$ which is the sheaf of germs of holomorphic functions over

$\Sigma\mathbf{C}$.

Since the space $\Sigma\mathbf{C}$ is paracompact, cohomology groups over $\Sigma\mathbf{C}$ with coefficients in any sheaf of abelian groups coincides with those of Čech cohomology groups. Since we have a partition of unity by C^∞-functions subordinate to an open covering of an open set in $\Sigma\mathbf{C}$, the Čech cohomology coincides with that of the $\bar{\partial}$-cohomology.

By Dineen[7], Colombeau-Perrot[5]-[6] and Kajiwara-Shon[20], cohomology groups, with coefficients in the structural sheaf $\mathcal{O}$, of a pseudoconvex covering of a domain D in the locally convex space $\Sigma\mathbf{C}$ equipped with the finite open topology coincides with the cohomology groups of the domain D. An F-valued function f on D is called to be F-holomorphic on D if f is continuous and uf is holomorphic on D for each u in F^*, where F^* is the dual of F. For a domain D with the structure sheaf $\mathcal{O}$ and for a Fréchet space F, we consider the sheaf of germs of F-holomorphic functions over D and denote it by $\mathcal{O}_\mathcal{F}$. $\mathcal{O}_\mathcal{F}$ is an analytic sheaf by Fujimoto[9]. Under these notations, we have the following theorem:

Theorem 1 *Let D be a pseudoconvex domain in the locally convex $\mathbf{C}$-linear space $\Sigma\mathbf{C}$ equipped with the finite open topology, p and q be positive integers and φ be a q-convex fumction on D. Assume moreover the existence of a positive integer n_0 such that, for positive numbers c_1 and c_2 with $c_1 < c_2$, and for any integer $n \geq n_0$ and the set $\{x \in D\cap\mathbf{C}^n; c_1 < \varphi(x) < c_2\}$ is relatively compact in $D\cap\mathbf{C}^n$. Let c be a negative number. We put $Y = \{x \in D; \varphi(x) > c\}$. Then we have*

$$H^p(Y, \mathcal{O}_\mathcal{F}) = 0. \tag{6}$$

(Proof) By the assumption of the q-convexity of the function φ on D, there exists a positive integer n_0 such that, for any $n \geq n_0$ and for any n-dimensional $\mathbf{C}$-linear subspace L of E, the Levi form of φ has $n - q + 1$ positive eigen values at every point of $D \cap L$. We may identify L with $\mathbf{C}^n$.

Let $\mathcal{U} = \{U_i; i \in I\}$ be an open covering of D such that each open set U_i is convex and

$$\beta = \{\beta_{i_0 i_1\cdots i_p}; i_0, i_1, \cdots, i_p \in I\} \tag{7}$$

is an element of $Z^p(\mathcal{U}, \mathcal{O}_\mathcal{F})$, i.e., a p-cocycle of the covering $\mathcal{U}$ with coefficients in the structural sheaf $\mathcal{O}_\mathcal{F}$. We prove by induction with respect to a positive integer $n \geq n_0$ the proposition P_n which asserts that, for any positive integer m with $n_0 \leq m \leq n$, the restriction of the p-cocycle β to the covering $\mathcal{U}(m) := \{U_i \cap \mathbf{C}^m; i \in I\}$ is the coboundary $\delta\gamma(m)$ of a $(p-1)$-cochain $\gamma(m) = \{\gamma(m)_{i_0 i_1\cdots i_{p-1}}; i_0, i_1, \cdots, i_{p-1} \in I\}$ of the covering $\mathcal{U}(m)$ and that, for any $m < n$, each rear $\gamma(m+1)$ is an extension of the preceding $\gamma(m)$.

We assume the validity of the proposition P_n and prove P_{n+1}. Since $p < n + 1 - q$, by uniting the results of Fujimoto[9] with Theorem 15 of Andreotti-Grauert [1], we have the isomorphism

$$\mathrm{H}^p(\mathbf{C}^{n+1} \cap Y, \mathcal{O}_\mathcal{F}) \cong \mathrm{H}^p(D \cap \mathbf{C}^{n+1}, \mathcal{O}_\mathcal{F}) = 0. \tag{8}$$

Since the Stein covering $\mathcal{U}(n+1)$ is a Leray covering, we have the isomorphism

$$\mathrm{H}^p(\mathcal{U}(n+1), \mathcal{O}_\mathcal{F}) \cong \mathrm{H}^p(D \cap \mathbf{C}^{n+1} \cap Y, \mathcal{O}_\mathcal{F}) = 0. \tag{9}$$

Hence the cocycle

$$\beta(n+1) \in \mathrm{Z}^p(\mathcal{U}(n+1), \mathcal{O}_\mathcal{F}) \cong \mathrm{B}^p(\mathcal{U}(n+1), \mathcal{O}_\mathcal{F}) \tag{10}$$

is the coboundary of a cochain $\zeta(n+1) = \{\zeta(n+1)_{i_0 i_1 \cdots i_{p-1}}; i_0, i_1, \cdots, i_{p-1} \in I\}$ of the covering $\mathcal{U}(n+1)$, i. e. we have

$$\beta(n+1) = \delta\zeta(n+1). \tag{11}$$

For each $i_0, i_1, \cdots, i_{p-1} \in I$, the open set

$$U_{i_0 i_1 \cdots i_{p-1}} := U_{i_0} \cap U_{i_1} \cap \cdots \cap U_{i_{p-1}} \tag{12}$$

is convex. For any positive integer m, we put

$$U_{i_0 i_1 \cdots i_{p-1}}(m) := U_{i_0 i_1 \cdots i_{p-1}} \cap \mathbf{C}^m. \tag{13}$$

For each $i_0, i_1, \cdots, i_{p-1} \in I$, the preceding function $\gamma(n)_{i_0 i_1 \cdots i_{p-1}}$ is holomorphic in the analytic subset

$$U_{i_0 i_1 \cdots i_{p-1}}(n) = U_{i_0 i_1 \cdots i_{p-1}}(n+1) \cap \mathbf{C}^n \tag{14}$$

of the Stein domain open set $U_{i_0 i_1 \cdots i_{p-1}}(n+1)$, $\mathbf{C}^n$ being regarded as a hyperplane of $\mathbf{C}^{n+1}$. Hence the preceding function $\gamma(n)_{i_0 i_1 \cdots i_{p-1}}$ is extended to a holomorphic function $\eta(n+1)_{i_0 i_1 \cdots i_{p-1}}$ in the convex domain $U_{i_0 i_1 \cdots i_{p-1}}(n+1)$.

We construct a $(p-1)$-cochain by putting

$$\eta(n+1) := \{\eta(n+1)_{i_0 i_1 \cdots i_{p-1}}; i_0, i_1, \cdots, i_{p-1} \in I\}. \tag{15}$$

We have two cochains $\zeta(n+1)$ and $\eta(n+1) \in \mathrm{C}^{p-1}(\mathcal{U}, \mathcal{O}_{\mathcal{F}})$. We consider the coboundary $\vartheta(n+1)$ of their difference:

$$\vartheta(n+1) = \{\vartheta(n+1)_{i_0 i_1 \cdots i_p}; i_0, i_1, \cdots, i_p \in I\} := \delta\eta(n+1) - \delta\zeta(n+1). \tag{16}$$

Since $\vartheta(n+1)$ vanishes on the hyperplane $z_{n+1} = 0$, we have

$$\vartheta \in \mathrm{Z}^p(\mathcal{U}(n+1), \mathcal{O}_{\mathcal{F}}/z_{n+1}\mathcal{O}_{\mathcal{F}}) \tag{17}$$

which is isomorphic to $\mathrm{B}^p(\mathcal{U}(n+1), \mathcal{O}_{\mathcal{F}}/z_{n+1}\mathcal{O}_{\mathcal{F}})$ because, by Andreotti-Grauert[1], we have

$$\mathrm{H}^p(\mathcal{U}(n+1), \mathcal{O}/z_{n+1}\mathcal{O}) \cong \mathrm{H}^p(D \cap Y \cap \mathbf{C}^{n+1}, \mathcal{O}/z_{n+1}\mathcal{O}) = 0 \tag{18}$$

for the analytic coherent sheaf $\mathcal{O}/z_{n+1}\mathcal{O}$ of homological dimension $n+1-1=n$. Uniting this with Fujimoto[9], we get a cochain

$$\kappa(n+1) = \{\kappa(n+1)_{i_0 i_1 \cdots i_p}; i_0, i_1, \cdots, i_{p-1} \in I\} \in \mathrm{C}^{p-1}(\mathcal{U}(n+1), \mathcal{O}_{\mathcal{F}}/z_{n+1}\mathcal{O}_{\mathcal{F}}) \tag{19}$$

such that

$$\delta\kappa(n+1) = \vartheta(n+1). \tag{20}$$

Now we define a rear cochain $\gamma(n+1)$ by putting

$$\gamma(n+1) := \zeta(n+1) + \kappa(n+1) \tag{21}$$

modifying the cochain $\zeta(n+1)$ by adding the cochain $\kappa(n+1)$ which vanishes on the hyperplane $z_{n+1} = 0$. Hence the rear cochain $\gamma(n+1)$ coincides with the preceding cochain $\gamma(n)$ on the hyperplane $z_{n+1} = 0$ and the coboundary of the rear cochain $\gamma(n+1)$ coincides with the target cocycle $\beta(n+1)$.

Thus we have proved the proposition P_n for any integer $n \geq n_0$ and we have constructed cochains $\gamma(n)$ for any integer $n \geq n_0$. Since our space is $\Sigma\mathbf{C}$ and its topology is finite open, our construction of a cochain $\gamma \in \mathrm{C}^{p-1}(\mathcal{U}, \mathcal{O}_{\mathcal{F}})$, the coboundary of which is the given cocycle β is completed. q.e.d.

3 Validity of Oka's principle

Theorem 2 *Let Ω be a pseudoconvex domain in E, A be an analytic subset of Ω, L be a complex Lie group modelled with a Banach space, $B(L,\Omega)$ be an analytic fibre bundle with base space Ω and structure group L and, $I^a_B(A)$ and $I^c_B(A)$ be, respectively, the sheaves of all germs of holomorphic and continuous section over Ω. Then the canonical mapping*

$$\iota : H^1(\Omega, I^a_B(A)) \to H^1(\Omega, I^c_B(A)) \tag{22}$$

are injective.

We proof the theorem by arguments in the previous paper[22] using theorems and lemmma in [17].

As a collorary, we have the following theorem:

Theorem 3 *A holomorphic mapping $f : A \to L$ is extended to a holomorphic mapping of Ω in L if and only if it is extended to a continuous mapping of Ω in L.*

References

[1] A. Andreotti et H. Grauert, *Théorèmes de finitude pour la cohomologie des espaces complexs,* Bull. Soc. Math. France **90**(1962), 193-259.

[2] H. Cartan, *Idéaux de fonctions analytiques de n complexes variables,* Bull. Soc. Math. France. **78**(1950), 28-64.

[3] H. Cartan, Séminaire E.N.S., École Normare Supérieur, Paris, 1951/1952.

[4] G. Coeuré, *L'équation $(\overline{\partial})u = F$ en dimension infinie,* université Lille Publ. Int. **131**(1968), 6-9.

[5] J. Colombeau and B. Perri, *The $(\overline{\partial})$-equation in D. F. N. spaces,* J. Math. Anal. Appl. **78**-2(1980), 466-87.

[6] J.Colombeau and B.Perri, *L'équation $(\overline{\partial})$ dans les ouverts pseudo-convexes des espaces D. F. N.,* Bull. Soc. Math. France. **110**(1982), 15-26.

[7] S.Dineen, *Sheaves of holomorphic functions on infinite dimensional vector spaces,* Math. Ann. **202**(1973), 337-345.

[8] S. Dineen, *Cousin's first problem on certain locally convex topological spaces,* Acad. Brasil. Cienc. **48**-1(1976), 229-236.

[9] H. Fujimoto, *Vector-valued holomorphic functions on complex space,* J. Math. Soc. Japan. Vol. **17**. No. 1(1965), 52-66.

[10] L. Gross, *Potential theory on Hilbert spaces,* J. Funct. Anal. **1**(1967), 123-181.

[11] L. Gruman, *The Levi problem in certain infinite dimensional vector spaces,* Illinois J. Math. **18**(1974), 20-26.

[12] C. J. Henrich, *The $\overline{\partial}$ -equation with polynomial growth on a Hilbert space,* Duke Math. J. **40**-2(1973), 279-306.

[13] S. Hitotumatu, *Theorem of analytic functions of several complex variables(in Japanese)*, Baihuukan, 1960.

[14] J. Kajiwara, *La réciproque du théorème d'annalation et de finitude de cohomologie dans l'espace produit d'une famille dénombrable de sphère de Riemann*, Bull. Soc. Math. France. **103**(1975), 129-139.

[15] J. Kajiwara, *Le principle d'Oka pour certaines espace de dimension infini*, C. R. Paris **16**(1975), 1055-1056.

[16] J. Kajiwara, *Le principle d'Oka pour certaines espaces vectoriets munis de la topologie f-ouvert*, Mem. Fac. Sci. Kyushu. Univ. **26**(1973), 361-37.

[17] J. Kajiwara and H. Kazama, *Oka's principle for relative cohomology sets, Mem. Fac. Sci. Kyushu Univ.* **23**-1(1969),33-70.

[18] J. Kajiwara and H. Kazama, *Two dimensional complex manifold with vanishing cohomology set*, Math. Ann. **204**(1973),1-12.

[19] J. Kajiwara and M. Nishihara, *Characterisierung der Steinschen Teitgebieten durch Okasches Prinzip in Zwei-dimensionater Mannigfasltigkeit*, Mem. Fac. Sci.Kyushu. Univ. **33**(1979),71-76.

[20] J. Kajiwara and K. H. Shon, *Continuation and vanishing theorem for cohomology of infinite dimensional space*, Pusan Kyŏngnam Math. J. **1**(1993), 65-73.

[21] P. Mazet, *Un théorème d'hyperbolicité pour l'opérateur $\overline{\partial}$ sur les espaces de Banach*, C. R. Paris **292**(1981), 31-33.

[22] S. Ohgai, *Cohomology vanishing and validity of Oka's principle for infinite dimensional domains, Proceedings of the Third International Colloquium on Finite or Infinite Dimensional Complex Analysis, Seoul Korea, July 31-August 2,1995,283-288*

[23] K. Oka, *Sur les fonctions analytiques plusieurs variables: VII Sur quelques notions arithmétiques*, Bull. Soc. Math. France, **78**(1950), 1-27.

[24] P. Raboin, *Résolution de l'équation $\overline{\partial} f = F$ sur un espace de Hilbert*, Bull. Soc. Math. France, **107**(1979), 225-240.

[25] G. Scheja, *Riemannsche Hebbarkeitssätze für Cohomologieklassen*, Math. Ann. **144**(1961), 345-360.

[26] J. P. Serre, *Dans Séminaire H.Cartan,3*, École Normare Supérieur, Paris, 1951-1952.

[27] R. L. Soraggi, *The $\overline{\partial}$-problem for a (0,2)-form in a D. F. N. space*, J. Func. Anal. **98**(1991), 380-402.

7th Int. Coll. on Differential Equations, pp. 283-287
D. Bainov (Ed)

The Existence and the Continuation of Holomorphic Solutions for Convolution Equations in a Half-Space in C^n

JUN-ICHI OKADA
Department of Mathematics and Informatics, Institute of Science and Technology, Chiba University, 1-33 Yayoi-cho, Inage-ku, Chiba, 263 JAPAN
E-mail : mokada@math.s.chiba-u.ac.jp

Abstract. We study holomorphic solutions for convolution equations (E) $T * f = g$ in a half-space in C^n. Under a natural condition (the condition (S)), we will prove the existence of solutions of (E) and the analytic continuation of homogeneous equation (E′) $T * f = 0$.

Keywords: convolution equations, entire functions, the condition (S)

1. INTRODUCTION

Let Ω be a convex domain and let K be a compact convex set in C^n. We denote by $\mathcal{O}(\Omega)$ the space of holomorphic functions on Ω provided with the topology of uniform convergence on compact subsets of Ω, and by $\mathcal{O}(K)$ the space of germs of functions holomorphic on K, endowed with the usual topology of the inductive limit. Then each nonzero analytic functional $T \in \mathcal{O}'(C^n)$ carried by K (or equivalently, $T \in \mathcal{O}'(K)$) defines a continuous linear convolution operator

$$T* \ : \ \mathcal{O}(\Omega + K) \longrightarrow \mathcal{O}(\Omega)$$

which is given by

$$(T * f)(z) = T_\zeta(f(z+\zeta)), \ z \in \Omega.$$

If $K = \{0\}$, the convolution operator $T*$ is a linear partial differential operator of infinite order with constant coefficients on $\mathcal{O}(\Omega)$. The convolution equation has been historically studied by many authors, especially the equation in the category of holomorphic functions defined on a complex domain. For example, using the notion of an entire function of completely regular growth on a fixed ray, Morzhakov [1] established sufficient condition for $T*$ to be surjective in the general case, and gave a criterion for the solvability for three classes of domains: smooth domains, products of smooth planar domains, and domains whose boundaries consist of smooth points and vertices. On the other hand, under the condition (S) due to Kawai [2], Ishimura - Y. Okada [3] studied the existence and the continuation problem of holomorphic solutions for convolution equations of hyperfunction kernels in the tube domain. In [4], Ishimura and the author proved that the property of completely regular growth is equivalent to the condition (S) for entire functions, in more general case, *i.e.* for sub-harmonic functions.

In this paper, we consider the convolution equation in the case where Ω is a half-space, and under the condition $(S)_{\zeta_0}$, we will prove the existence of solutions of (E) and the analytic continuation of homogeneous equation (E′).

Most of results is based on the paper [3], and refer to it for the details of proofs.

2. PRELIMINARIES

Let

$$|z|^2 = z_1\overline{z_1} + \cdots + z_n\overline{z_n}, \qquad \langle z, w\rangle = z_1\overline{w_1} + \cdots + z_n\overline{w_n}$$

for

$$z = (z_1, \cdots, z_n),\ w = (w_1, \cdots, w_n) \in \boldsymbol{C}^n.$$

For $\zeta_0 \in \boldsymbol{C}^n$ and $|\zeta_0| = 1$, we put

$$\Omega = \{\, \zeta \in \boldsymbol{C}^n \mid \mathrm{Re}\langle \zeta_0, \zeta\rangle < 0 \,\}$$

and take a compact convex set K as $K \subset \Omega$, *i.e.* $\Omega + K = \Omega$. As it is well-known, the properties of convolution operator are reflected in the properties of the Fourier-Borel transform of T, namely

$$\widehat{T}(\zeta) = T_z(\exp\langle z, \zeta\rangle),$$

which is an entire function of exponential type satisfying the following estimate:

Theorem 2.1.(Polyà-Ehrenpreis-Martineau) If $T \in \mathcal{O}'(\boldsymbol{C}^n)$ and T is carried by a compact set $K \subset \boldsymbol{C}^n$, then $\widehat{T}(\zeta)$ is an entire function and for every $\varepsilon > 0$, there is a constant $C_\varepsilon > 0$ such that

$$|\widehat{T}(\zeta)| \le C_\varepsilon \exp(H_K(\zeta) + \varepsilon|\zeta|), \quad \zeta \in \boldsymbol{C}^n \tag{2.1}$$

where $H_K(\zeta) = \sup_{z\in K} \mathrm{Re}\langle z, \zeta\rangle$ is the supporting function of K.
Conversely, if K is a compact convex set and $f(\zeta)$ an entire function satisfying (2.1) for every $\varepsilon > 0$, there exists an analytic functional $T \in \mathcal{O}'(\boldsymbol{C}^n)$ carried by K such that $\widehat{T}(\zeta) = f(\zeta)$.

In this paper, we suppose the following condition for the entire function $\widehat{T}(\zeta)$.

Definition 2.2. We say that $\widehat{T}(\zeta)$ satisfies the condition (S) to the direction ζ_0 or simply it satisfies the condition $(S)_{\zeta_0}$ if we have

$$\begin{cases} \text{For every } \varepsilon > 0, \text{ there exists } N > 0 \text{ such that} \\ \text{for any } r \in \boldsymbol{R} \text{ with } r > N, \\ \text{we can find } \zeta \in \boldsymbol{C}^n, \text{ which satisfies} \\ |\zeta - \zeta_0| < \varepsilon r, \\ |\widehat{T}(\zeta)| \ge \exp(-\varepsilon r). \end{cases}$$

3. THE EXISTENCE OF HOLOMORPHIC SOLUTIONS

We will prove the surjectivity theorem under the condition $(\mathrm{S})_{\zeta_0}$.

Theorem 3.1. Let $T \in \mathcal{O}'(\boldsymbol{C}^n)$ carried by K. Assume that $\widehat{T}(\zeta)$ satisfies the condition $(\mathrm{S})_{\zeta_0}$. Then the convolution operator

$$T* \; : \; \mathcal{O}(\Omega + K) \longrightarrow \mathcal{O}(\Omega)$$

is surjective.

Proof. The transpose of

$$P = T* \; : \; \mathcal{O}(\Omega + K) \longrightarrow \mathcal{O}(\Omega)$$

is

$${}^tP = \check{T}* \; : \; \mathcal{O}'(\Omega) \longrightarrow \mathcal{O}'(\Omega + K)$$

with $\widehat{\check{T}}(\zeta) = \widehat{T}(-\zeta)$. By the standard argument, it is enough to prove that ${}^tP$ is injective and that the image of ${}^tP$ is weakly closed. In fact, the injectivity of ${}^tP$ shows that the image of P is dense, and the closedness of the image of ${}^tP$ shows the closedness of the image of P. Because T is not 0, the injectivity of ${}^tP$ is clear. Then we will show that the image of ${}^tP$ is weakly closed. To do this, we use the following division lemma, which we can prove in an analoguous way to the proof of Lemma 2.1 in [3].

Lemma 3.2. Let f, g and h be entire functions satisfying $fg = h$, and K and L be two compact convex sets in $\boldsymbol{C}^n$ with $K,\ L \subset \Omega$. We suppose that for every $\varepsilon > 0$, f and h satisfy the following estimates with constants $A_\varepsilon > 0$ and $B_\varepsilon > 0$,

$$\begin{cases} \log|f(\zeta)| \le A_\varepsilon + H_K(\zeta) + \varepsilon|\zeta|, \\ \log|h(\zeta)| \le B_\varepsilon + H_L(\zeta) + \varepsilon|\zeta|, \end{cases}$$

for any $\zeta \in \boldsymbol{C}^n$. We also assume that f satisfies the condition $(\mathrm{S})_{\zeta_0}$. Then for any $\varepsilon > 0$, there exists a compact convex set $M = M_\varepsilon \subset \boldsymbol{C}^n$ and $C_\varepsilon > 0$ such that

$$\begin{cases} M \subset \Omega \\ \log|g(\zeta)| \le C_\varepsilon + H_M(\zeta). \end{cases}$$

End of the proof of the theorem. — Let $\{T_\nu\}$ be a sequence in $\mathcal{O}'(\Omega)$ and assume that $\{{}^tPT_\nu\}$ converges to $S \in \mathcal{O}'(\Omega+K)$ in $\mathcal{O}'(\Omega+K)$. By taking the Fourier-Borel transform, $\widehat{T}(-\zeta)\widehat{T}_\nu(\zeta)$ converges to $\widehat{S}(\zeta)$. Then it is well-known that $G(\zeta) = \dfrac{\widehat{S}(\zeta)}{\widehat{T}(-\zeta)}$ becomes an entire function. By Lemma 3.2 and Theorem 2.1, there exists a compact convex set M and $\mu \in \mathcal{O}'(\boldsymbol{C}^n)$ carried by M such that $\widehat{\mu}(\zeta) = G(\zeta)$ and ${}^tP\mu = \check{T}*\mu = S$, *i.e.* $S \in \mathrm{Im}\,{}^tP$. □

4. THE CHARACTERISTIC SET AND THE CONTINUATION OF HOMOGENEOUS SOLUTIONS

Under the condition $(\mathrm{S})_{\zeta_0}$, we shall now solve the problem of continuation for the solutions of the homogeneous equation (E′). For any open set $U \subset \boldsymbol{C}^n$, we set:

$$\mathcal{N}(U) = \{ f \in \mathcal{O}(U) \mid T * f = 0 \}.$$

For an open set $V \subset \boldsymbol{C}^n$ with $U \subset V$, the problem is formulated as to get the condition so that the restriction map

$$r : \mathcal{N}(V) \longrightarrow \mathcal{N}(U)$$

is surjective.

In order to describe the theorem of continuation, we will prepare the notion of characteristics which is a natural generalization of the case of usual differential operators of finite order with constant coefficients. We define the sphere at infinity S_∞^{2n-1} by $(\boldsymbol{C}^n \setminus \{0\})/\boldsymbol{R}_+$ and consider the compactification with directions $\boldsymbol{D}^{2n} = \boldsymbol{C}^n \sqcup S_\infty^{2n-1}$ of $\boldsymbol{C}^n$. For $\zeta \in \boldsymbol{C}^n \setminus \{0\}$, we denote by $\zeta\infty \in S_\infty^{2n-1}$ the equivalence class of ζ, *i.e.*

$$\{\zeta\infty\} = (\text{the closure of } \{ t\zeta \mid t > 0 \} \text{ in } \boldsymbol{D}^{2n}) \cap S_\infty^{2n-1}.$$

For $\varepsilon > 0$, we put:

$$\begin{cases} V_{\widehat{T}}(\varepsilon) = \{ \zeta \in \boldsymbol{C}^n \mid \exp(\varepsilon|\zeta|)|\widehat{T}(\zeta)| < 1 \}, \\ W_{\widehat{T}}(\varepsilon) = (\text{the closure of } V_{\widehat{T}}(\varepsilon) \text{ in } \boldsymbol{D}^{2n}) \cap S_\infty^{2n-1}. \end{cases}$$

Now we define the characteristic set of $T*$.

Definition 4.1. With the above notation, we define the characteristics of $T*$ (at infinity)

$$\mathrm{Char}_\infty(T*) = \text{the closure of } \bigcup_{\varepsilon>0} W_{\widehat{T}}(\varepsilon).$$

Under the above situation, we can state the theorem of the continuation without proof.

Theorem 4.2. Let $T \in \mathcal{O}'(\boldsymbol{C}^n)$ carried by K and $f \in \mathcal{O}(\Omega + K)$ be a solution of $T * f = 0$. Assume that $\widehat{T}(\zeta)$ satisfies the condition $(\mathrm{S})_{\zeta_0}$. If $\zeta_0 \notin \mathrm{Char}_\infty(T*)$, then the restriction map

$$r : \mathcal{N}(\boldsymbol{C}^n) \longrightarrow \mathcal{N}(\Omega + K)$$

is surjective, that is, f can be analytically continued to the whole $\boldsymbol{C}^n$.

REFERENCES

[1] V. V. Morzhakov, *Convolution equations in convex domains of $\boldsymbol{C}^n$*, Compl. Anal. and Appl. '87, Sofia, pp.360–364 (1989).

[2] T. Kawai, *On the theory of Fourier hyperfunctions and its applications to partial differential equations with constant coefficients*, J. Fac. Sci. Univ. Tokyo, Sect. IA Math., **17**, pp.467–517 (1970).

[3] R. Ishimura and Y. Okada, *The existence and the continuation of holomorphic solutions for convolution equations in tube domains*, Bull. Soc. Math. France, **122**, pp.413–433 (1994).

[4] R. Ishimura et J. Okada, *Sur la condition (S) de kawai et la propriété de croissance régulière d'une fonction sous-harmonique et d'une fonction entière*, Kyushu J. Math. **48**, pp.257–263 (1994).

7th Int. Coll. on Differential Equations, pp. 289-294
D. Bainov (Ed)

SOLVABILITY AND PROPAGATION OF SINGULARITIES FOR SOME CLASS OF MICRODIFFERENTIAL EQUATIONS IN THE SPACES OF MICRO-DISTRIBUTIONS

Yasunori Okada*
Department of Mathematics and Informatics,
Faculty of Science, Chiba University,
1-33 Yayoi-cho, Inage-ku, Chiba, 263 Japan

Abstract

We study a class of microdifferential equations with double involutive characteristics in the spaces of micro-distributions. We treat the operators of the form $P = P_1P_2 +$ *lower order* with simple characteristic operators P_j's at a double characteristic point $\dot{q} \in \mathrm{Char}(P_1) \cap \mathrm{Char}(P_2)$. Under additional assumptions of Levi condition, the regular involutiveness, and the micro-hyperbolicity, we get the solvability of the equation $Pu = v$ in the spaces of tempered microfunctions and differentiable microfunctions, and the relation between the micro-differentiable singularities of solutions of $Pu = 0$, and the Hamiltonian flows of $\sigma(P_1)$ and $\sigma(P_2)$.

1 Introduction

Let M be a real analytic manifold with complexification X. We consider the operator $P \in \mathcal{E}_X$ of the form $P = P_1P_2 +$ *lower order* with simple characteristic operators P_j's. Here P_j's have principal symbols p_j's which are real valued on T^*_MX and satisfy $dp_1 \wedge dp_2 \wedge \omega \neq 0$ for the canonical 1-form ω.

Note that the structure of the microdifferential equation generated by P is completely studied outside of $\mathrm{Char}(P_1) \cap \mathrm{Char}(P_2)$ by M. Sato et al. [6]. On the double characteristic points $\mathrm{Char}(P_1) \cap \mathrm{Char}(P_2)$, N. Tose [7, 8] treats the analytic singularities by using the method of second microlocalization, and G. Uhlmann [9] treats the differentiable singularities by constructing a parametrix.

In this paper, under additional assumption of Levi condition, we study micro-differential equations

$$Pu = v, \tag{1}$$

and

$$Pu = 0, \tag{2}$$

in the spaces of micro-distributions at double characteristic points. In the analysis of our problem, we use Bony-Schapira's calculus (see [2]), the notion of simple sheet operators (see [1]), and the relation between microfunctions with holomorphic parameters and their defining holomorphic functions (see [5] and [4]).

2 Main theorems

For a real analytic manifold M and its complexification X, we denote by π_M, the projection $T_M^*X \to M = T_M^*M$. Using the spectrum map $\mathrm{Sp}_M : \pi_M^{-1}\mathcal{B}_M \to \mathcal{C}_M$, we introduce the subsheaves $\mathcal{C}_M^f$ and $\mathcal{C}_M^d$ of $\mathcal{C}_M$ by

$$\mathcal{C}_M^f = \mathrm{Sp}_M(\pi_M^{-1}\mathcal{D}'_M),$$
$$\mathcal{C}_M^d = \mathrm{Sp}_M(\pi_M^{-1}C_M^\infty),$$

where $\mathcal{B}_M$, $\mathcal{C}_M$, $\mathcal{D}'_M$ and C_M^∞ are the sheaf of Sato's hyperfunctions, that of microfunctions, that of distributions and that of differentiable functions respectively.

Sections of $\mathcal{C}_M^f$ (resp. $\mathcal{C}_M^d$) are called tempered microfunctions (resp. differentiable microfunctions). $\mathcal{C}_M^f$ and $\mathcal{C}_M^d$ are conically soft.

Let V be a regular involutive submanifold in $\dot{T}_M^*X = T_M^*X \setminus T_M^*M$. Then for any $u \in \mathcal{C}_M^f|_V$, we can define the tempered second wave front set $\mathrm{WF}_V^2(u)$ of u along V, which is a closed biconic subset of $\dot{T}_V\dot{T}_M^*X = T_V\dot{T}_M^*X \setminus T_VV$. Refer to G. Lebeau [3] for the definition. We also define subsheaves $\mathcal{A}_V^{2,f}$ and $\mathcal{A}_V^{2,d}$ of $\mathcal{C}_M^f|_V$ by

$$\mathcal{A}_V^{2,f}(U) = \{u \in \mathcal{C}_M^f|_V(U); \mathrm{WF}_V^2(u) = \emptyset\},$$
$$\mathcal{A}_V^{2,d}(U) = \{u \in \mathcal{C}_M^d|_V(U); \mathrm{WF}_V^2(u) = \emptyset\},$$

for any open subset $U \subset V$.

Let $\dot{q}$ be a point in $\dot{T}_M^*X$. We consider a microdifferential equation

$$Pu = v \tag{3}$$

and

$$Pu = 0 \tag{4}$$

at $\dot{q}$, where P is a microdifferential operator satisfying the following conditions.

(A.1) There exist two regular involutive submanifolds $V_1^{\mathbf{C}}$ and $V_2^{\mathbf{C}}$ of T^*X, which satisfy

- (A.1.1) the characteristic variety $\mathrm{Char}(P) \subset T^*X$ coincides with $V_1^{\mathbf{C}} \cup V_2^{\mathbf{C}}$,
- (A.1.2) $\mathrm{codim}_{\mathbf{C}} V_1^{\mathbf{C}} = \mathrm{codim}_{\mathbf{C}} V_2^{\mathbf{C}} = 1$,
- (A.1.3) $V_1^{\mathbf{C}}$ and $V_2^{\mathbf{C}}$ intersect transversally along $V^{\mathbf{C}} = V_1^{\mathbf{C}} \cap V_2^{\mathbf{C}}$, which is a regular involutive submanifold of complex codimension 2 containing $\dot{q}$,
- (A.1.4) P is simple characteristic along $\mathrm{Char}(P) \setminus V^{\mathbf{C}}$, i.e., the principal symbol $p = \sigma(P)$ vanishes exactly up to order 1 along $V_1^{\mathbf{C}} \setminus V^{\mathbf{C}}$ and $V_2^{\mathbf{C}} \setminus V^{\mathbf{C}}$,
- (A.1.5) $V_1^{\mathbf{C}}$ and $V_2^{\mathbf{C}}$ are real, i.e., we can take defining functions p_1 of $V_1^{\mathbf{C}}$ and p_2 of $V_2^{\mathbf{C}}$ which are real valued on T_M^*X, and satisfy $dp_1 \neq 0$ and $dp_2 \neq 0$.

(A.2) P satisfies the Levi condition along $V^{\mathbf{C}}$. That is, in this case, the symbol of $(ord\,P - 1)$'th order part of P vanishes up to order 1 along $V^{\mathbf{C}}$.

Let us set

$$V_1 = V_1^{\mathbf{C}} \cap T_M^*X, \quad V_2 = V_2^{\mathbf{C}} \cap T_M^*X, \quad V = V^{\mathbf{C}} \cap T_M^*X.$$

Then, from (A.1.2), (A.1.3) and (A.1.5), we find V_1, V_2 and V are regular involutive submanifolds in T_M^*X of real codimension 1, 1 and 2 respectively.

Let q be a point of V. We denote by b_q the bicharacteristic leaf of V through q, and by b_q^j $(j = 1, 2)$ the bicharacteristic curve of V_j through q. We also set, for a subset $K \subset b_{\dot{q}}$:

$$\hat{K} = \{q \in b_{\dot{q}};\ \text{both } b_q^1 \text{ and } b_q^2 \text{ intersect with } K\},$$

and call it the (V_1, V_2)-bicharacteristic convex hull of K.

Our main results are:

Theorem 2.1 *Let P be as above. Then the equation* (3) *is solvable for $\mathcal{C}_M^f$ and $\mathcal{C}_M^d$. That is, if v belongs to $\mathcal{C}_{M,\dot{q}}^f$ (resp. $\mathcal{C}_{M,\dot{q}}^d$), there exists a solution u of* (3) *belonging to $\mathcal{C}_{M,\dot{q}}^f$ (resp. $\mathcal{C}_{M,\dot{q}}^d$).*

Theorem 2.2 *Let P be as above. There exists a neighborhood $U \subset b_{\dot{q}}$ of $\dot{q}$ with the following property. Assume that φ is a continuous map from the unit square $I \times I$ ($I = [0, 1]$) to U satisfying*

- *$\varphi(I, t)$ is bicharacteristic curve of V_1 for any t,*
- *$\varphi(s, I)$ is bicharacteristic curve of V_2 for any s,*

and that $u \in \mathcal{C}_M^f(\varphi(I \times I))$ is a solution of (4). *Then there exist solutions $u_1 \in \mathcal{A}_{V_1}^{2,f}(\varphi(I \times I))$ and $u_2 \in \mathcal{A}_{V_2}^{2,f}(\varphi(I \times I))$ of* (4) *which satisfy*

$$u = u_1 + u_2.$$

Moreover if $\dot{q} \notin \mathrm{WF}(u)$, the preceding u_1 and u_2 can be chosen with the property $\dot{q} \notin \mathrm{WF}(u_1)$ and $\dot{q} \notin \mathrm{WF}(u_2)$.

Theorem 2.3 *Let P be as above. There exists a neighborhood $U \subset b_{\dot{q}}$ of $\dot{q}$ with the following property. Assume that φ is a continuous map from the unit square $I \times I$ ($I = [0, 1]$) to U satisfying*

- *$\varphi(I, t)$ is bicharacteristic curve of V_1 for any t,*
- *$\varphi(s, I)$ is bicharacteristic curve of V_2 for any s,*

and that $u \in \mathcal{C}_M^f(\varphi(I \times I))$ is a solution of (4). *Then we have*

$$\varphi(0,0), \varphi(1,0), \varphi(0,1) \notin \mathrm{WF}(u) \Longrightarrow \varphi(1,1) \notin \mathrm{WF}(u).$$

Corollary 2.4 *Let P be as above. There exists a neighborhood $U \subset b_{\dot{q}}$ of $\dot{q}$ satisfying the following property. Let K be an arbitrary connected compact subset of U with $\hat{K} \subset U$. Then, for any solution $u \in \mathcal{C}_M^f(\hat{K})$ of* (4), *we have*

$$\mathrm{WF}(u) \cap K = \emptyset \Longrightarrow \mathrm{WF}(u) \cap \hat{K} = \emptyset.$$

3 Goursat problem in complex flat domains

We prove the theorems in the previous section by the well-posedness of the Goursat problem of micro-differential equation in suitable flat domains, by using the Bony-Schapira's actions of micro-differential operators to holomorphic functions. See Bony and Schapira [2].

In this section we assume that $M = \mathbf{R}^n$ and $X = \mathbf{C}^n$ with coordinates

$$\begin{aligned} &x = (x_1, x_2, \dots, x_n) \in M, \\ &z = (z_1, z_2, \dots, z_n) = x + \sqrt{-1}y \in X, \\ &(x, \sqrt{-1}\xi \cdot dx) \in T^*_M X \text{ with } \xi = (\xi_1, \xi_2, \dots, \xi_n). \end{aligned}$$

We take, using these coordinates, V_j, V and $\dot{q}$ as

$$\begin{aligned} &V_j = \{(x, \sqrt{-1}\xi \cdot dx) \in T^*_M X; \xi_j = 0\} \quad (j = 1, 2), \\ &V = V_1 \cap V_2 = \{\xi_1 = \xi_2 = 0\}, \\ &\dot{q} = (0; \sqrt{-1}dx_n) \in V. \end{aligned}$$

Let $P = D_1D_2 - D_1A - D_2B - C$ be a microdifferential operator defined in a neighborhood of $\dot{q} = (0; \sqrt{-1}dx_n) \in T^*_M X$, where A, B and C are microdifferential operators of order ≤ 0.

We consider the Goursat problem

$$\begin{cases} (D_1D_2 - D_1A_\Sigma - D_2B_\Sigma - C_\Sigma)f = g \\ f|_{z_1=0} = 0, \quad f|_{z_2=0} = 0 \end{cases} \tag{5}$$

for a given function $g \in \mathcal{O}(\Omega)$ and an unknown function $f \in \mathcal{O}(\Omega)$, where Σ is a hyperplane $\{z_n = \dot{\sigma}\}$, Ω is a suitable flat domain, and A_Σ, B_Σ and C_Σ are the Bony-Schapira's action of A, B and C.

Moreover, for a holomorphic function $f \in \mathcal{O}(\Omega)$, non-negative integer ℓ, and an arbitrary positive valued function λ defined in Ω depending only on z_n, we set

$$\begin{aligned} d'(z) &= \inf_{(\tilde{z}', z'', z_n) \notin \Omega} \sup_{i=1,2} |\tilde{z}_i - z_i|, \\ d''(z) &= \inf_{(z', \tilde{z}'', z_n) \notin \Omega} \sup_{3 \leq i \leq n-1} |\tilde{z}_i - z_i|, \\ \hat{\lambda}(z_n) &= \sup_{0 \leq t \leq 1} \lambda(tz_n + (1-t)\dot{\sigma}), \\ \|f\|_\lambda &= \sup_{z \in \Omega} \frac{|f(z)|}{\hat{\lambda}(z_n) d'(z)^{-1} d''(z)^{-1}}, \\ \|f\|_{\lambda,\ell} &= \sup_{|\alpha| + \alpha_n \leq \ell} \|D^\alpha f\|_\lambda. \end{aligned}$$

Theorem 3.1 *Let P be as above, Ω be sufficiently small, and λ be an arbitrary positive valued function defined in Ω depending only on z_n. Then the Goursat problem (5) is well-posed for class G_λ and G^∞_λ. Here $G_\lambda(\Omega)$ and $G^\infty_\lambda(\Omega)$ are vector spaces defined by*

$$\begin{aligned} &G_\lambda(\Omega) = \{f \in \mathcal{O}(\Omega); \|f\|_\lambda < \infty\}, \\ &G^\infty_\lambda(\Omega) = \{f \in \mathcal{O}(\Omega); \|f\|_{\lambda,\ell} < \infty \text{ for any } \ell\}. \end{aligned}$$

In the proof of this theorem, we use the notion of simple sheet operators due to Bony [1]. We omit the details.

4 Proof of the results

In this section, we give a sketch of proof of our results.

By a suitable real (quantized) contact transformation, we can reduce our problem to the case

$$M = \mathbf{R}^n \hookrightarrow X = \mathbf{C}^n,$$
$$V_1 = \{\xi_1 = 0\}, V_2 = \{\xi_2 = 0\}, V = \{\xi_1 = \xi_2 = 0\},$$
$$\dot{q} = (0; \sqrt{-1}dx_n).$$

Moreover from the assumption of Levi condition, we deduce

$$P = Q(D_1D_2 - D_1A - D_2B - C)$$

with an elliptic operator Q and operators A, B, C of order ≤ 0. Taking account that an elliptic operator operates bijectively on $\mathcal{C}_M^f$, $\mathcal{C}_M^d$ and $\mathcal{A}_{\#}^{2,*}$, we may assume, from the beginning, that

$$P = D_1D_2 - D_1A - D_2B - C.$$

The solvability (Theorem 2.1) and the decomposition of solutions of homogeneous equations (the first half of Theorem 2.2) is a easy consequence of solvability of the Goursat problem in the previous section.

The last half of Theorem 2.2 can be proved by using the uniqueness of the Goursat problem.

Now we give the proof of Theorem 2.3. From Theorem 2.2, we can take u_j's $(j = 1, 2)$ with

$$u = u_1 + u_2,$$
$$u_j \in \mathcal{A}_{V_j}^{2,f}(L) \quad (j = 1, 2),$$
$$\varphi(0,0) = (0; \sqrt{-1}dx_n) \notin \mathrm{WF}(u_j) \quad (j = 1, 2).$$

We give a remark that each u_j is a restriction of a tempered microfunction with a holomorphic parameter z_j to $\{Im\, z_j = 0\}$. Thus, from the propagation theorem of differentiable singularities along holomorphic parameters (refer to [1]. See also [4]), we have

$$\varphi(1,0) \notin \mathrm{WF}(u_1),$$
$$\varphi(0,1) \notin \mathrm{WF}(u_2).$$

Since these two points are also outside of $\mathrm{WF}(u)$, the estimates

$$\varphi(1,0) \notin \mathrm{WF}(u - u_1) = \mathrm{WF}(u_2),$$

and

$$\varphi(0,1) \notin \mathrm{WF}(u - u_2) = \mathrm{WF}(u_1),$$

follow. Applying the propagation theorem again, we get

$$\varphi(1,1) \notin \mathrm{WF}(u_2),$$
$$\varphi(1,1) \notin \mathrm{WF}(u_1),$$

which concludes the desired result.

REFERENCES

[1] Bony, J.M., *Propagation des singularités différentiables pour une classe d'opérateurs différentiels à coefficients analytiques*, Astérisque, Soc. Math. de France, **34-35** (1976) 43-92.

[2] Bony, J.M. and P. Schapira, *Propagation des singularités analytiques pour les solutions des équations aux dérivées partielles*, Ann. Inst. Fourier, Grenoble, **26(1)** (1976) 81-140.

[3] Lebeau G., *Deuxième microlocalisation à croissance*, Séminaire Goulaouic-Meyer-Schwartz 1982-83, Exposé 15.

[4] Okada, Y. *Second microlocal singularities of tempered and Gevrey classes*, J. Fac. Sci., Univ. Tokyo, Sect. IA, Math., **39** (1992), 475-505.

[5] Okada, Y., and N. Tose, *FBI transformation and second microlocalization – Equivalence between second analytic wave front sets and second singular spectrums*, J. Math. Pures Appl. **70** (1991), 427-453.

[6] Sato, M., T. Kawai, and M. Kashiwara, *Microfunctions and pseudo-differential equations* Lect. Notes in Math., **287** Springer, (1973) 265-529.

[7] Tose, N., *On a class of microdifferential equations with involutory double characteristics*, J. Fac. Sci., Univ. Tokyo, Sect. IA, Math., **33** (1986), 619-634.

[8] ———, *The 2-microlocal canonical form for a class of microdifferential equations and propagation of singularities*, Publ. of RIMS, Kyoto Univ., **23** (1987), 101-116.

[9] Uhlmann, G., *Pseudodifferential operators with double involutive characteristics*, Comm. in P.D.E. **2** (1977), 713-779.

7th Int. Coll. on Differential Equations, pp. 295-303
D. Bainov (Ed)

Rapidly Varying Behaviour of the Solutions of a Second Order Linear Differential Equation

EDWARD OMEY
EHSAL, Stormstraat 2, 1000 Brussels, BELGIUM

Abstract. In this paper we consider the nonnegative solutions of the linear differential equation $y''(x) = f(x)y(x)$ where $f(x) \geq 0$ is continuous. In many cases such solutions are regularly varying or rapidly varying. The regularly varying case has been studied by several authors before. In the present paper we mainly study the rapidly varying case. We provide several estimates of the precise asymptotic behaviour of $y(x)$ as x tends to infinity. We define and use properties of the subclass $\Gamma(a(x))$ of the rapidly varying functions and use other classes of functions related to this class.

Keywords: Regularly varying functions, O-regular variation, rapidly varying functions, Beurling varying functions, selfcontrolled functions

1. INTRODUCTION

In this paper we consider the nonnegative solutions of the following second-order linear differential equation

$$y''(x) = f(x)y(x) \tag{1.1}$$

where $f(x) \geq 0$ and continuous. The asymptotic behaviour of nonnegative solutions of (1.1) has been studied by many authors, see e.g. [1], [4], [6], [7], [8], [12], [14], [15].

If $y(x) \geq 0$ is such a solution it follows from (1.1) that $y(x)$ is convex and that two possibilities occur:

(a) $\lim y(x) = \infty$ and $\lim y'(x) = A$, $0 < A \leq \infty$

or (b) $\lim y(x) = A$, $0 \leq A < \infty$ and $\lim y'(x) = 0$.

The cases (a) and (b) complement each other in an obvious way: if $y_1(x)$ denotes a solution of type (a), then $y_2(x)$ defined by

$$y_2(x) = ky_1(x) \int_x^\infty (y_1(s))^{-2} ds \qquad (k > 0) \tag{1.2}$$

is a solution of type (b).

Solutions of the type (a) (resp. (b)) are usually called principal solutions (resp. non-principal solutions). We shall mainly restrict attention to solutions of type (a). In many cases it turns out that the nonnegative solutions of (1.1) are connected with the class of *regularly varying functions* and related classes of function. Recall that a nonnegative measurable function $f : \mathbf{R} \to \mathbf{R}^+$ is *regularly varying* at infinity and with (finite) index α (=: RV(α)) if as $t \to \infty$,

$$\lim \frac{f(tx)}{f(t)} = x^{\alpha}, \quad \text{for all } x > 0. \tag{1.3}$$

If $\alpha = 0$, the function $f(x)$ is called slowly varying (SV).

The function $f(x)$ is *rapidly varying* (=: RV(∞) or RV($-\infty$)) if (1.3) holds with $\alpha = +\infty$ or $\alpha = -\infty$.

A function $f(x)$ is called *O-regularly varying* (=: ORV) if as $t \to \infty$,

$$\limsup \frac{f(tx)}{f(t)} < \infty, \quad \text{for all } x > 0. \tag{1.4}$$

RV- and ORV-functions behave like power functions. One can prove that if $f(x) \in$ RV(α) (resp. ORV), then (1.3) (resp. (1.4)) holds locally uniformly (notation: l.u.) in x.

Further elementary properties of RV and ORV can be found in [2], [5], [9], [18]. Rapidly varying functions have a behaviour close to that of exponential functions. In connection with equation (1.1), the following result motivates the use of these classes of functions.

Proposition 1.1. [15] *Suppose $x^2 f(x) \to c$ $(0 \le c \le \infty)$ and let $y(x)$ denote a nonnegative solution of* (1.1) *tending to* ∞. *Then $y(x) \in$ RV(α) and $xy'(x)/y(x) \to \alpha$ where $\alpha \ge 1$ satisfies $\alpha(\alpha - 1) = c$.*

If $c < \infty$, it follows that $y(x)$ has the form $y(x) = x^{\alpha} L(x)$ where $\alpha \ge 1$ and where $L(x)$ is SV. *If $c = \infty$, then $y(x)$ is rapidly regularly varying.*

If $c < \infty$ and if $y_1(x)$ denotes a solution such as in Proposition 1.1, then $y_2(x)$ defined by (1.2) is a decreasing solution of (1.1). Basic properties of RV show that in this case $y_2(x) \in$ RV($1 - \alpha$) and that

$$\frac{y_1(x) y_2(x)}{x} \to \frac{k}{2\alpha - 1}. \tag{1.5}$$

Remarks. (1) In [14] the authors gave the necessary and sufficient condition $\lim_{x\to\infty} \int_x^{\infty} f(s)ds = c$ in order to ensure the regular variation of the nonnegative solutions of (1.1). Additional and related results were proved in [7], [8].

(2) In [6], Geluk studied into detail the second order behaviour of $y(x)/x^{\alpha}$ in the case where $c < \infty$.

(3) The discrete analogue of Prop. 1.1 also holds, cf. [16].

In the present paper we want to provide more information in the case where $c = \infty$, i.e. the case where the solutions of (1.1) are rapidly varying. In sections 2 and 3 we define and use the class of functions $\Gamma(a(x))$ to distinguish different types of rapidly varying behaviour.

2. THE CLASS $\Gamma(a(x))$ AND RELATED CLASSES

If $f(x) \in \mathrm{RV}(\infty)$, then (1.3) shows that for each fixed $x < 1$, $f(tx) = o(1)f(t)$. If we replace x by a function $a(t)$ tending to 1, we may be able to specify the $o(1)$-term.

Definition. ([2], [3]) A positive and measurable function $a(x)$ is called *Beurling-varying* (=: BV) if

$$a(x) = o(1)x, \quad \text{and} \tag{2.1}$$

$$\text{for all real } y, \ \frac{a(x + ya(x))}{a(x)} \to 1. \tag{2.2}$$

If (2.1) holds and if (2.2) holds l.u. in y, then $a(x)$ is called *self-neglecting* (=: SN).

The class of functions $\Gamma(a(x))$ is defined as follows.

Definition. Let $a(x) \in \mathrm{SN}$. A measurable and positive function $f : \mathbf{R} \to \mathbf{R}^+$ is in the class $\Gamma(a(x))$ with auxiliary function $a(x)$ if for all real y, as $x \to \infty$,

$$\frac{f(x + ya(x))}{f(x)} \to e^y. \tag{2.3}$$

A function $f(x) \in \Gamma_-(a(x))$ if $1/f(x) \in \Gamma(a(x))$.

These classes have been useful in a variety of problems in analysis and probability theory, cf. [2], [9], [10], [11], [17]. If $f(x) \in \Gamma(a(x))$, then automatically (2.3) holds l.u. in y (cf. [11]). Also, if $f(x)$ is monotone and if (2.3) holds, then relation (2.3) alone implies that $a(x) \in \mathrm{SN}$ ([5]). It turns out that $\Gamma(a(x))$ is a proper subclass of $\mathrm{RV}(\infty)$.

If (2.3) holds, Proposition 2.1 below shows that one can replace $a(x)$ by $b(x)$ where $b(x) \sim a(x)$ and $b'(x) = o(1)$.

Proposition 2.1. ([2], [3]) *Suppose that $a(x) \in \mathrm{BV}$ is continuous. Then $a(x) \in \mathrm{SN}$ and $a(x) = c(x)b(x)$ where $c(x) \to 1$ and where $b'(x) \to 0$ is continuous.*

Note that the behaviour of a function $a(x) \in \mathrm{BV}$ is controlled by the function $a(x)$ itself. The class of functions controlled by another function $b(x)$ is defined as follows.

Definition. Suppose that $b(x) \in \mathrm{SN}$. A positive and measurable function $a(x)$ is controlled by $b(x)$ if for all real y,

$$\frac{a(x+yb(x))}{a(x)} \to 1.$$

Notation: $a(x) \in \mathrm{SC}(b(x))$.

In the next result we summarize some of the properties of the classes defined before, see [5], [9], [10], [11].

Proposition 2.2. (i) *If $f'(x) \in \Gamma(a(x))$, then $f(x) \sim a(x)f'(x)$ and $f(x) \in \Gamma(a(x))$.*

(ii) *If $f'(x) \in \Gamma_-(a(x))$, then $\lim f(x) = A$ exists, $A - f(x) \sim a(x)f(x)$ and $A - f(x) \in \Gamma_-(a(x))$.*

(iii) *If $a'(x) = o(1)$, then $a(x) \in \mathrm{SN}$.*

(iv) *If $a'(x) \in \mathrm{SC}(b(x))$, then for all real y,*

$$a(x+yb(x)) - a(x) = \big(1+o(1)\big)yb(x)a'(x), \quad \textit{l.u. in } y.$$

3. DIFFERENTIAL EQUATIONS AND $\Gamma(a(x))$

3.1. First order results

Our first result is a refinement of Proposition 1.3 in the case where $c = \infty$.

Proposition 3.1. *Let $y(x) \geq 0$ denote an increasing solution of (1.1) and suppose that $f(x) \geq 0$ is continuous. Let $a(x) \geq 0$ be defined by $f(x)a^2(x) = 1$ and assume that $a(x) \in \mathrm{BV}$. Then*

(i) $\lim \dfrac{a(x)y'(x)}{y(x)} = \lim \dfrac{a(x)y''(x)}{y'(x)} = 1$;

(ii) $y(x) \in \Gamma(a(x))$, $y'(x) \in \Gamma(a(x))$, $y''(x) \in \Gamma(a(x))$.

The next Corollary 3.2 generalizes a result of Grimms and Hall [8, p. 4] and is the analogue of (1.5).

Corollary 3.2. *If $y_1(x)$ is a solution as in Proposition 3.1 and if $y_2(x)$ is defined as in (1.2), then $y_2(x) \in \Gamma_-(a(x))$.*
Moreover, $\dfrac{a(x)y_2'(x)}{y_2(x)} \to -1$ *and* $\dfrac{y_1(x)y(x)}{a(x)} \to \dfrac{k}{2}$.

Before proving Proposition 3.1, we first prove the next Lemma 3.3, interesting in its own right. We use the notation $u(x) \approx v(x)$ when $u(x) = O(v(x))$ and $v(x) = O(u(x))$.

Lemma 3.3. *Let $b(x) \geq 0$ and let $y(x) \geq 0$ denote a nonnegative increasing solution of* (1.1).

(i) *If $b'(x) = O(1)$ and $b^2(x)f(x) \approx 1$, then $b(x)y'(x) \approx y(x)$.*

(ii) *If $b'(x) \approx 1$ and $b^2(x)f(x) \approx 1$, then $xy'(x) \approx y(x)$ and hence $y(x) \in$ ORV.*

(iii) *If $b'(x) \to \beta \geq 0$ and $b^2(x)f(x) \to 1$, then $\dfrac{b(x)y'(x)}{y(x)} \to A$ where $A \geq 0$ satisfies $A^2 = 1 + \beta A$.*

Remark. Lemma 3.3 and Proposition 3.1 show that the results complete the results of Proposition 1.1 as follows.

If $b'(x) = o(1)$, then $y(x) \in \Gamma(b(x))$; if $b'(x) \to \beta > 0$, then $y(x) \in \mathrm{RV}(A/\beta)$; if $b'(x) \approx 1$, then $y(x) \in$ ORV.

Proof of Lemma 3.3. (i) Let $K(x) := \dfrac{b(x)y'(x)}{y(x)}$. We shall prove the $K(x) \approx 1$. Obviously $K(x) \geq 0$ and $K'(x)$ satisfies

$$b(x)K'(x) = b^2(x)f(x) + b'(x)K(x) - K^2(x). \tag{3.1}$$

First assume $K'(x) > 0$ for $x \geq c$. In this case $\lim K(x) = A$ exists and $0 < A \leq \infty$. If $A = \infty$, then we would have $b(x)K'(x) \sim -K^2(x)$ which is not possible since $K'(x) > 0$. Hence $A < \infty$ and the result follows.

Next assume $K'(x) < 0$ for $x \geq c$. In this case $\lim K(x) = A$ exists and $0 \leq A < \infty$. If $A = 0$, then it would follow that $b(x)K'(x)$ ultimately has the same sign as $b^2(x)f(x)$. Since the latter is positive, this is not possible. Hence $A > 0$ and again the result follows.

Finally suppose that $K'(x) = 0$ infinitely often. Using (3.1) and $K(x) \geq 0$ it follows that

$$2K(x) = b'(x) + \sqrt{b'^2(x) + 4b^2(x)f(x)} \tag{3.2}$$

in each zero of $K'(x)$.

Since $K(x)$ is monotone between consecutive zeros of $K'(x)$ and since $b'(x) + \sqrt{b'^2(x) + 4b^2(x)f(x)} \approx 1$, it follows again that $K(x) \approx 1$. This proves (i).

(ii) Since $b'(x) \approx 1$ implies $b(x) \approx x$, the first part follows from (i). Integrating the relation $\dfrac{y'(s)}{y(s)} \approx \dfrac{1}{s}$ between x and xt shows that $\log\left(\dfrac{y(xt)}{y(x)}\right) \approx \log(t)$. Hence $y(x) \in$ ORV.

(iii) As in part (i) we use (3.1) and consider three cases.

First assume $K'(x) > 0$ for all $x \geq c$. In this case $\lim K(x) = A$ exists and $0 < A < \infty$. Using (3.1) it follows that

$$\lim b(x)K'(x) = B =: 1 + \beta A - A^2. \tag{3.3}$$

Since $K'(x) > 0$ we have $B \geq 0$ and we have to show that $B = 0$. Suppose that $B > 0$. Since $\frac{b(x)}{x} \to \beta$ it follows from (3.2) that $\lim xK'(x) = \frac{B}{\beta}$ ($= \infty$ if $\beta = 0$). This would imply that $\lim K(x) = \infty$, a contradiction. Hence $B = 0$ and $\lim K(x) = A$ with A as in the statement of the Lemma. If $K'(x) < 0$ for all $x \geq c$, the result follows in a similar way. Finally suppose that $K'(x_n) = 0$ infinitely often. As in part (i) it follows that (3.2) holds in each zero x_n of $K'(x)$. Since $K(x)$ is monotone between consecutive zeros of $K'(x)$ and since (cf. (3.2)) $\lim K(x_n)$ exists, also in this case the result follows. □

Proof of Proposition 3.1. (i) Since $a(x) \in$ BV and $a(x)$ is continuous, Proposition 2.1 shows that we can find a function $b(x)$ such that $b(x) \sim a(x)$ and such that $b'(x) \to 0$. An application of Lemma 3.3 (iii) shows that $\frac{a(x)y'(x)}{y(x)} \to 1$.

Using (1.1) it also follows that $\frac{a(x)y''(x)}{y'(x)} \to 1$.

(ii) To prove (ii), choose $e > 0$ and $c > 0$ so that for $x \geq c$, $\frac{1-e}{a(x)} \leq \frac{y'(x)}{y(x)} \leq \frac{1+e}{a(x)}$.

Let $z > 0$ and take the integral between x and $x + za(x)$. Since by local uniform convergence,

$$\int_x^{x+za(x)} \frac{1}{a(s)} ds = \int_0^z \frac{a(x)}{a(x+sa(x))} ds \to z$$

it follows that $(1-e)z \leq \liminf \log\left(\frac{y(x+za(x))}{y(x)}\right)$ and $\limsup \log\left(\frac{y(x+za(x))}{y(x)}\right) \leq (1+e)z$.

Now let $e \to 0$ to see that $y(x) \in \Gamma(a(x))$. Using $a(x) \in$ SN, (i) and (1.1), it follows that also $y'(x)$ and $y''(x) \in \Gamma(a(x))$. □

Proof of Corollary 3.2. Use (1.2) and Proposition 2.2 (ii). □

3.2. Second-order results

In the next result we compare the solutions of two equations of the type (1.1). We consider the solutions of the equations $y''(x) = f(x)y(x)$ and $u''(x) = h(x)u(x)$. One is tempted to believe that $y(x) \approx u(x)$ if $r(x) = f(x) - h(x)$ is close to zero. If $r(x)$ satisfies a certain integrability condition, this is indeed the case.

Proposition 3.4. (cf. [13, Theorem 9.1]) *Let $y_1(x)$ and $y_2(x)$ be as in Corollary 3.2 and assume that for some $c > 0$,*

$$\int_c^\infty a(x)\left|h(x) - f(x)\right| dx < \infty.$$

Consider the equation $v''(x) = h(x)v(x)$. We have:

(i) $\lim \frac{v'(x)}{y_1'(x)}$ *exists;*

(ii) *The general solution of* $v''(x) = h(x)v(x)$ *is asymptotic to* $d_1y_1(x) + d_2y_2(x)$.

Example. Assume $y''(x) = f(x)y(x)$ where $f(x) = g^2(x) + (g(x))'$ and $\frac{g'(x)}{g^2(x)} \to 0$. In this case $y_1(x) = \exp\left(\int_a^x g(s)ds\right)$ and $\lim y_2(x)y_1(x)f^{1/2}(x) = 1$.

If $v''(x) = h(x)v(x)$ with $\int_a^\infty \frac{|h(x) - f(x)|}{\sqrt{f(x)}}dx < \infty$, then $v(x) \sim d_1y_1(x) + d_2y_2(x)$.
If for example $g(x) = 1$, then $v(x) \sim d_1e^x + d_2e^{-x}$.

In our next result we obtain a second-order result by taking a closer look at $L(x) := \frac{y(x)}{a(x)} - y'(x)$.

Proposition 3.5. *Let* $a(x) \geq 0$ *defined by* $a^2(x) = f(x)$. *Suppose that* $a'(x) = o(1)$ *and that* $a'(x) \in \mathrm{SC}(a(x))$. *Then*

(i) $\frac{a(x)y'(x)}{y(x)} = 1 - \frac{(1 + o(1))a'(x)}{2}$.

(ii) $y(x) \in \Gamma(a(x))$ *with remainder term as follows:*

$$\frac{y(x + za(x))}{y(x)} - e^z = -a'(x)(1 + o(1))e^z\left(z + \frac{z^2}{2}\right).$$

Proof. (i) Let $u(x) := \exp\left(\int_{x_0}^x a^{-1}(s)ds\right)$, $L(x) := \frac{y(x)}{a(x)} - y'(x)$ and $C(x) := \frac{L(x)}{u(x)}$.
It is easy to verify that $C(x)$ satisfies $C'(x) = -u(x)a'(x)y(x)$.

Using $u(x) \in \Gamma(a(x))$, $y(x) \in \Gamma(a(x))$ and $a'(x) \in \mathrm{SC}(a(x))$ it follows that $C'(x) \in \Gamma\left(\frac{a(x)}{2}\right)$. Proposition 2.2 (i) shows that $C(x) \sim \frac{C'(x)a(x)}{2}$ so that $L(x) \sim -\frac{a'(x)y(x)}{2a(x)}$.
This proves (i).

(ii) Using (i) we integrate between x and $x + za(x)$ to obtain

$$\log\left(\frac{y(x + za(x))e^{-z}}{y(x)}\right) = I - II$$

where $I = \int_0^z \left(\frac{a(x)}{a(x + ta(x))} - 1\right) dt$

and $II = \int_0^z (1 + o(1))\frac{a'(x + ta(x))a(x)}{a(a + ta(x))}dt$.

Using $a'(x) \in \mathrm{SC}(a(x))$ and l.u. convergence we have $II = z(1 + o(1))a'(x)$. Also, since $a'(x) \in \mathrm{SC}(a(x))$, Proposition 2.5 (iv) shows that $I = -\frac{(1 + o(1))a'(x)z^2}{2}$. It follows that

$$\log\left(\frac{y(x + za(x))e^{-z}}{y(x)}\right) = -a'(x)(1 + o(1))\left(z + \frac{z^2}{2}\right).$$

Since $a'(x) \to 0$, the result follows. □

Remark. (1) Using stronger assumptions on $f(x)$ (or $a(x)$) one can obtain different types of asymptotic results and expansions. If for example $f(x) = 1 + r(x)$ where $r(x) \in \mathrm{RV}(\alpha)$ and $\alpha < -1$ then $\log(y(x)) = \text{Const.} + x + \frac{(1+o(1))xr(x)}{-1+\alpha}$. If $\alpha > -1$, then $\log(y(x)) = x + \frac{(1+o(1))xr(x)}{1+\alpha}$. This means that $\log(y(x))$ is regularly varying with a remainder term. This and other cases are discussed in [16].

(2) For a discussion of the class $\Gamma(a(x))$ with remainder term, we refer to [11].

REFERENCES

[1] R. Bellman, *Stability Theory of Differential Equations,* Mc. Graw-Hill, New York, 1953.

[2] N. H. Bingham, C. M. Goldie and J. L. Teugels, *Regular Variation,* Cambridge University Press, 1987.

[3] S. Bloom, A characterization of B-slowly varying functions, *Proc. Am. Math. Soc.,* **54** (1976), 243–250.

[4] M. S. P. Eastham, *Theory of Ordinary Differential Equations,* Van Nostrand Reinhold Comp., London, 1970.

[5] J. L. Geluk and L. de Haan, *Regular Variation, Extensions and Tauberian Theorems,* CWI tract 40, M.C. Amsterdam, 1987.

[6] J. L. Geluk, On slowly varying solutions of the linear second order differential equation, *Publ. Inst. Math. Nouvelle Série (Béograd)* **48** (62) (1990), 52–60.

[7] L. J. Grimm and L. M. Hall, Self-adjoint differential equations and Karamata functions, In: *Proceedings of the Third International Colloquium on Differential Equations* (Ed. D. Bainov and V. Covachev), 77–87, VSP, Utrecht, 1993.

[8] L. J. Grimm and L. M. Hall, Regular variation and solutions of second order differential equations, In: *Trends and Developments in Ordinary Differential Equations,* World Scientific Publishing Co., Singapore, 1994.

[9] L. de Haan, *On Regular Variation and its Applications to the Weak Convergence of Sample Extremes,* M.C. Tracts, Math. Centre Amsterdam, 1970.

[10] L. de Haan, Equivalence classes of regularly varying functions, *Stoch. Proc. Appl.,* **2** (3) (1974), 243–259.

[11] L. de Haan and E. Omey, On a subclass of Beurling varying functions, preprint, EHSAL / EUR Rotterdam, 1989.

[12] M. Hacik and E. Omey, On the zeros of oscillatory solutions of linear second order differential equations, *Publ. Inst. Math. Nouvelle Série (Béograd)* **49** (63) (1991), 189–200.

[13] Ph. Hartman, *Ordinary Differential Equations,* J. Wiley & Sons, New York, London, Sidney, 1964.

[14] V. Maric and M. Tomic, A classification of solutions of second order linear differential equations by means of regularly varying functions, *Publ. Inst. Math. Nouvelle Série (Béograd)* **48** (62) (1990), 199–207.

[15] E. Omey, Regular variation and its applications to second order linear differential equations, *Bull. Soc. Math. Belg.,* **33**, Ser. B, (1981).

[16] E. Omey, Asymptotic behaviour of the solutions of a second order linear differential equation, preprint, EHSAL, August 1996.

[17] G. E. Peterson, Tauberian theorems for integrals II, *J. London Math. Soc.,* **5** (2) (1972), 182–190.

[18] E. Seneta, *Regularly Varying Functions,* Lecture Notes in Math., No. 508, Springer, 1976.

7th Int. Coll. on Differential Equations, pp. 305-311
D. Bainov (Ed)

Stability of Lie Groups of the Perturbed Non-linear wave equation $u_{tt} + \epsilon u_t = [f(x,u)u_x]_x$

N. Omolo-Ongati and E. E. Rosinger
Department of Mathematics and Applied Mathematics
University of Pretoria, 0002 Pretoria, South Africa

Abstract

The groups of the perturbed non-linear wave equation $u_{tt} + \epsilon u_t = [f(x,u)u_x]_x$ are discussed for arbitrary f ($f \in C^2(\mathbf{R} \times \mathbf{R})$, $f > 0$, $f_u \neq 0$). Further the theory of approximate groups of transformations developed in [1] is applied to construct Approximate Lie Groups of the same wave equation.

1. Introduction

In this paper we study the properties of Lie Groups and algebras associated with the perturbed non-linear wave equation

$$u_{tt} + \epsilon u_t = [f(x,u)u_x]x, \; f \in C^2(\mathbf{R} \times \mathbf{R}), \; f > 0, \; f_x \neq 0, \; f_u \neq 0, \; \epsilon \in \mathbf{R}. \tag{1}$$

We aim to construct approximate symmetries of first order, i.e. $p = 1$, for (1), see [1] for definitions and Appendix [A] for notations and assumptions. So far, the most general perturbed equation studied from the point of view of approximate symmetries was, see [1]

$$u_{tt} + \epsilon u_t = [f(u)u_x]_x.$$

In order to obtain the approximate symmetries of a perturbed equation, it is known that, one has to start with the usual Lie symmetries of the unperturbed equation. For convenience we mention that the unperturbed cases

$$u_{tt} = [f(u)u_x]_x \tag{2}$$
$$u_{tt} = [f(x,u)u_x]_x \tag{3}$$

have been studied, respectively, by Ames, Lohner and Adams [2] and Torrisi and Valenti [3]. We note that the computation of the Lie groups of (3) can be obtained in a more simple and straightforward manner than in [3]. This helps further in the computation of the approximate symmetries of (1).

2. Symmetry Groups of (3)

We shortly recall the group properties for (3) as presented in [3].
Using the notations in Appendix [A] and (3), we have

$$F = u_{11} - f_x u_2 - f_u (u_2)^2 - f u_{22}. \tag{4}$$

The invariant condition $V^{(2)}F = 0$ when $F = 0$ gives

$$\begin{aligned}
0 \;=\; & [-\eta_x f_x + \eta_{tt} - \eta_{xx} f] + u_1[\xi_x f_x + 2\eta_{tu} - \xi_{tt} + \xi_{xx} f] \\
& + u_2[-\theta f_{xx} - \eta f_{xu} + \theta_x f_x - 2\eta_x f_u - \theta_{tt} - 2\xi_t f_x - 2\eta_{xu} f + \theta_{xx} f] \\
& + u_1 u_2[\xi_u f_x + 2\xi_x f_u - 2\theta_{tu} - 3\xi_u f_x + 2\xi_{xu} f] \\
& + (u_1)^2[\eta_{uu} - 2\xi_{tu}] \\
& + (u_2)^2[-\theta f_{xu} - \eta f_{uu} + \theta_u f_x - \eta_u f_u + 2\theta_x f_u - 2\xi_t f_u - \theta_u f_x - \eta_{uu} f + 2\theta_{xu} f] \\
& - (u_1)^3 \xi_{uu} + (u_2)^2[\theta_u f_u - \theta_{uu} f] - (u_1)^2 u_2 \theta_{uu} \\
& + u_1 (u_2)^2 [\xi_{uu} - \xi_u f_u] + u_{12}[2\xi_x f - 2\theta_t] \\
& + u_{22}[-\theta f_x - \eta f_u - 2\xi_t f + 2\theta_x f] - 2u_1 u_{12}\theta_u - 2u_1 u_{22}\xi_u f \\
& + 2u_2 u_{12}\xi_u f + 2u_2 u_{22}\theta_u f.
\end{aligned} \tag{5}$$

The coefficients of u_1u_{22}, u_2u_{22}, u_1u_{12}, $(u_1)^2u_2$, $(u_1)^3$, u_1, u_1u_2 and u_{12} in (5) give the relations

$$\xi_u = \xi_{uu} = \theta_u = \theta_{uu} = 0 \quad \text{and} \quad \xi_x = \theta_t = 0. \tag{6}$$

It is easy to see that the coefficients of $(u_1)^2$ in (5) together with (6) give

$$\xi(t,x,u) = a(t), \quad \theta(t,x,u) = b(x) \quad \text{and} \quad \eta(t,x,u) = c(t,x)u + d(t,x) \tag{7}$$

where a, b, c and d are arbitrary functions.
Considering (6), (7) and the remaining independent conditions supplied by (5) we have the following:

$$-\eta_x f_x + \eta_{tt} - \eta_{xx} f = 0 \tag{8}$$

$$2\eta_{tu} - \xi_{tt} = 0 \tag{9}$$

$$\eta f_u + \theta f_x - 2\theta_x f + 2\xi_t f = 0 \tag{10}$$

$$\eta_x f_u + \theta_{xx} f + 2\eta_{xu} f = 0. \tag{11}$$

In case we want to solve (8) – (11) for arbitrary f, we obtain from (7) that

$$\xi(t,x,u) = c = \text{const}, \quad \theta(t,x,u) = \eta(t,x,u) = 0 \tag{12}$$

After solving the O.D.E. corresponding to the above infinitesimals (12) we finally get the Lie Groups of transformation admitted by (3) for *arbitrary f*, as

$$\begin{aligned} t^* &= t + ca \\ x^* &= x \\ u^* &= u \end{aligned} \tag{13}$$

where c is an arbitrary constant and a is the group parameter.

3. Stability symmetries of (1)

An algorithm for constructing approximate groups of the form

$$z^* \approx f_0(z,a) + \epsilon f_1(z,a) \tag{14}$$

determined by the infinitesimal generator

$$V = [\xi_0(z) + \epsilon\xi_1(z)]\frac{\partial}{\partial z} \tag{15}$$

is developed in detail in [1]. It is known that the construction of (14) leaving the equation

$$F(z,\epsilon) \approx 0 \tag{16}$$

invariant reduces to the solution of

$$VF(z,\epsilon)\Big|_{F\approx 0} \approx 0. \tag{17}$$

For equation (1), with $p = 1$, (17) yields, see [1]

$$\sum_{k=1}^{8} \xi_0^k(z)\frac{\partial F_0(z)}{\partial z^k} = 0 \tag{18}$$

$$\sum_{k=1}^{8} \xi_1^k(z)\frac{\partial F_0(z)}{\partial z^k} + \sum_{k=1}^{8} \xi_0^k(z)\frac{\partial F_1(z)}{\partial z^k} + \sum_{l=1}^{8}\sum_{k=1}^{8} y_1^l \frac{\partial}{\partial z^l}\left(\xi_0^k(z)\frac{\partial F_0(z)}{\partial z^k}\right) = 0 \tag{19}$$

under the conditions

$$F_0(z) = 0, \quad F_1(z) + \sum_{l=1}^{8} \frac{\partial F_0(z)}{\partial z^l} = 0 \tag{20a, b}$$

where from (1)

$$\begin{aligned} F_0 &= u_{11} - f_x u_2 - f_u (u_2)^2 - f u_{22}; \ F_1 = u_1; \\ z &= (t, x, u, u_1, u_2, u_{11}, u_{12}, u_{22}). \end{aligned} \tag{21}$$

and from the infinitesimals of (3) which correspond to (18) and (20a), we have

$$\begin{aligned}
\xi_0^1(z) &= \xi = c; \quad \xi_0^2(z) = \theta = 0; \quad \xi_0^3(z) = \eta = 0 \\
\xi_0^4(z) &= \eta_1^{(1)} = 0; \quad \xi_0^5(z) = \eta_2^{(1)} = 0 \\
\xi_0^6(z) &= \eta_{11}^{(2)} = 0; \quad \xi_0^7(z) = \eta_{12}^{(2)} = 0 \\
\xi_0^8(z) &= \eta_{22}^{(2)} = 0
\end{aligned} \tag{22}$$

We turn now to solving (19) and (20b). In view of (22), equation (19) becomes

$$\begin{aligned}
&(f_{x^2}u_2 + f_{xu}(u_2)^2 + f_x u_{22})\xi_1^2(y_0) + \\
&+(f_{xu}u_2 + f_{u^2}(u_2)^2 f_u u_{22})\xi_1^3(y_0) + \\
&+(f_x + 2f_u u_2)\xi_1^5(y_0) - \xi_1^6(y_0) + f\xi_1^8(y_0) = 0
\end{aligned}$$

By identifying with zero the coefficients of 1, u_2, $(u_2)^2$ and u_{22} in the above equation, we obtain

$$\begin{aligned}
&f_x\xi_1^5(y_0) - \xi_1^6(y_0) + f\xi_1^8(y_0) = 0 \\
&f_{x^2}\xi_1^2(y_0) + f_{xu}\xi_1^3(y_0) + 2f_u\xi_1^5(y_0) = 0 \\
&f_{xu}\xi_1^2(y_0) + f_{u^2}\xi_1^3(y_0) = 0 \\
&f_x\xi_1^2(y_0) + f_u\xi_1^3(y_0) = 0
\end{aligned} \tag{23}$$

The last two equations in (23) give

$$\xi_1^2(y_0) = \xi_1^3(y_0) = 0 \tag{24}$$

unless

$$f_u f_{xu} = f_x f_{u^2} \tag{25}$$

Putting (24) in the second equation (23), we obtain

$$\xi_1^5(y_0) = 0 \tag{26}$$

unless

$$f_u = 0 \tag{27}$$

We note here that condition (27) can be assumed not to hold, otherwise equation (1) becomes linear. On the other hand, (25) is equivalent with the relation

$$f_u(x,u) = g(x) f_x(x,u) \tag{28}$$

for a suitable function g. Certainly, condition (28) is not satisfied for an arbitrary f. For instance, it is not satisfied for

$$f(x,u) = f_1(x) + f_2(u) \tag{29}$$

unless f_2 is of the particular form

$$f_2(u) = \alpha u + \beta \tag{30}$$

with α, β constants.

Therefore, in the sequel, we shall assume that none of the conditions (25) or (27) holds.

On the other hand, in case

$$f(x,u) = f(u)$$

as is assumed in [1], [2] and (2), it is obvious that (25) holds. Therefore, the results in this paper do not contain as a particular case those in [1], although the equation in (1) is more general than that studied in [1].
Now, the first equation (23) gives

$$\xi_1^6(y_0) = f\xi_1^8(y_0) \tag{31}$$

In this way we computed $\xi_1^2(y_0)$, $\xi_1^3(y_0)$ and $\xi_1^5(y_0)$, with $\xi_1^8(y_0)$ being *arbitrary*. We also note that in view of (21) and (19), $\xi_1^1(y_0)$, $\xi_1^4(y_0)$ and $\xi_1^7(y_0)$ will also remain *arbitrary* since their coefficients in (19) vanish.
Finally we now find $f_1 = (f_1^1, f_1^2, \ldots, f_1^8)$. For this we solve the system, see [1] for details,

$$\frac{df_1(z,a)}{da} = \sum_{k=1}^{8} \frac{\partial \xi_0(f_0)}{\partial z^k} f_1^k(z,a) + \xi_1(f_0) \tag{32}$$

under the initial conditions $f_1^k(z,a)|_{a=0} = 0$ where $f_0 = (f_0^1, f_0^2, \ldots, f_0^8)$ are known from (13).
First we note that in view of (21), the functions $\xi_0(z)$ and $\xi_1(z)$ in (15) are only supposed to depend on the first three variables in z, namely t, x and u. Consequently in (14), only the first three components of f_0 and f_1, that is, $f_0^1, f_0^2, f_0^3, f_1^1, f_1^2, f_1^3$ have to be determined, since the other components follow through prolongation. Furthermore, these components can also be assumed to depend on t, x and u only. In this way, in the system (32), we are only interested in the first three equations and the respective initial conditions. Now (13) will give

$$\begin{aligned} f_0^1(t,x,u,a) &= t^* = t + ca \\ f_0^2(t,x,u,a) &= x^* = x \\ f_0^3(t,x,u,a) &= u^* = u \end{aligned} \tag{33}$$

In view of (22), (24), (26), (31) and (33), the system (32) becomes

$$\frac{df_1(z,a)}{da} = \xi_1(f_0(z,a)) \tag{34}$$

and its first three equations are

$$\begin{aligned} \frac{df_1^1(t,x,u,a)}{da} &= h(t+ca,x,u,a) \\ \frac{df_1^2(t,x,u,a)}{da} &= 0 \\ \frac{df_1^3(t,x,u,a)}{da} &= 0 \end{aligned} \tag{35}$$

where h is an arbitrary function, while $a \in \mathbf{R}$ is the group parameter. In view of (32), the system (35) has to be solved for the initial conditions

$$f_1^1(t,x,u,0) = f_1^2(t,x,u,0) = f_1^3(t,x,u,0) = 0 \tag{36}$$

With the notation in (21), the equations (35) and (36) give

$$\begin{aligned} f_1^1(t,x,u,a) &= \int_0^a h(t+c\alpha,x,u,\alpha)d\alpha \\ f_1^2(t,x,u,a) &= f_1^3(t,x,u,a) = 0 \end{aligned} \tag{37}$$

In this way, the first three components in the approximate groups transformation (14) become

$$\begin{aligned} t^* &\approx t + ca + \epsilon \int_0^a h(t+c\alpha,x,u,\alpha)d\alpha \\ x^* &\approx x \\ y^* &\approx u \end{aligned} \tag{38}$$

and they are those which give the approximate Lie groups of the equation (1).

Appendix A

Notation

The following notation is used throughout this paper.
$z = (z^1, z^2, \ldots, z^N)$ — independent variable; $\theta_p(z,\epsilon)$ denotes an infinitesimally small function of order ϵ^{p+1}, $p \geq 0$ i.e. $\theta_p(z,\epsilon) = 0(\epsilon^p)$.
A one-parameter Lie Group of transf. acting on (t,x,u)-space will be denoted by:

$$t^* = T(t,x,u;a); \; x^* = X(t,x,u;a); \; u^* = U(t,x,u;a).$$

The infinitesimal generator

$$V = \xi(t,x,u)\frac{\partial}{\partial t} + \theta(t,x,u)\frac{\partial}{\partial x} + \eta(t,x,u)\frac{\partial}{\partial u}$$

will have its once and twice extended generators as

$$
\begin{aligned}
V^{(1)} &= V + \eta_1^{(1)}(t,x,u,u_1,u_2)\frac{\partial}{\partial u_1} + \eta_2^{(1)}(t,x,u,u_1,u_2)\frac{\partial}{\partial u_2} \\
V^{(2)} &= V^{(1)} + \eta_{11}^{(2)}(t,x,u,u_1,u_2,u_{11},u_{12},u_{22})\frac{\partial}{\partial u_{11}} + \eta_{12}^{(2)}(\ldots)\frac{\partial}{\partial u_{12}} + \ldots \\
&\quad + \eta_{22}^{(2)}(\ldots)\frac{\partial}{\partial u_{22}}
\end{aligned}
$$

where $u_t = u_1,\ u_x = u_2,\ u_{tt} = u_{11},\ u_{tx} = u_{xt} = u_{12},\ u_{xx} = u_{22}$.

Assumptions

We make the following assumptions throughout:

(a) All functions are jointly analytic in their arguments.

(b) $\exists$ a constant $c > 0$ such that $|\theta_p(z,\epsilon)| \leq c|\epsilon|^{p+1}$ where c does not depend on z and c does not depend on ϵ when ϵ is bounded.

(c) The approximate equality $f \approx g$ means the equality $f(z,\epsilon) = g(z,\epsilon) + 0(\epsilon^p)$ for some fixed value of $p \geq 0$.

REFERENCES

[1] N.Kh. Ibragimov, V.A. Baikov, R.K. Gazizov, *Approximate Symmetries*, Math USSR Sbornik, Vol. 64, (1989), No. 2, pp 427—440.

[2] W.F. Ames, R.J. Lohner, E.Adams, *Group Properties of* $u_{tt} = [f(x)u_x]_x$, Int. J. Non-linear Mech., Vol. 16, No. 5/6, (1981), pp 439–447.

[3] M. Torrisi and A.Valenti, *Group properties and invariant solutions for infinitesimal transformation of a non-linear wave equation*, Int. J. Non-linear Mechanics, Vol. 20, No. 3, (1985), pp 135–144.

[4] G.W. Bluman and S. Kumei, *Symmetries and Differential Equations*, Springer-Verlag, New York, Inc, (1989).

[5] P.J. Olver, *Applications of Lie Groups to Differential Equations*, Springer-Verlag, New York, Inc, (1986).

7th Int. Coll. on Differential Equations, p.p. 313-320
D. Bainov (Ed)

Application of Weighted Ranking Methods for Tie-Break Simulations

Kunio Oshima*, Ichiro Hofuku**, *and* Katsuhisa Horimoto*
**Science University of Tokyo in Yamaguchi Onoda 756, Japan*
***Tokyo Metropolitan College of Technology, Shinagawa-ku 1-10-40, Japan*

Abstract

In general, to each participant in a contest we want to assign a score that is based on the interactions with other participants. The assigned score should depend on both the outcome of the interaction and the strength of its opponents. In this paper, the participants are the students and the problems in the examination, and the outcome and the strength are those scores that the student took from the problem and that the problem does not give to the students. We will give a new weighted rule how to assign each element to a matrix which will be a tool of ranking those students.

Keywords: Perron-Frobenius theorem, weighted ranking, playnig off

1 Introduction

There are many games or competitions to end up in a tie, for instance, golf, football, and examination as well. Even though they end up in a tie, we still want to make a decision on which is the best. If a mathematically based ranking scheme would make use of such a case and there is room for subjectivity in the sense of a professional skill, we may be able to make an appropriate ranking without playing off. The mathematical foundation of this ranking method comes from Perron-Frobenius theorem for a non-negative irreducible matrix.

The purpose of an examination is that a teacher wants to know how deep his students understand the subject which he has taught. When he determines the examination result, he first finds out the highest score, the mean, and the standard diviation then he ranks all students based on their scores. Even if he curves all the scores, the ranking of the students is unchanged without changing their original scores. This result simply tells us only one aspect of the examinees'understanding, since one could make 30 points working on relatively easy problems and another could make the same points working on relatively good problems. It may not be fair to decide that their comprehension is the same. In order to be fairer, we want to impose some kind of a relative difficulty on each problem or a relative influence on each student. The ranking scheme which we will discuss may have some weight which is not objective but rather subjective, and this is the reason why an examiner who gave an examination may be able to give some weight depending on the result. The mathematical foundation of this ranking method comes from Perron-Frobenius theorem for a non-negative irreducible matrix.

In this paper we will exclusively treat the examination result which contains three students having same score and will make a new fundamental but very important weighted ranking scheme based upon their achievement of each problem in the examination. We will give two different examples, one can be ranked and the other can not be ranked.

In section 2, the well-known Perron-Frobenius Theorem and the related propositions are stated.

In section 3, we will construct an irreducible matrix by assigning each weighted value to each element from the examination result that three students made the same score, and then apply Perron-Frobenius theorem to the matrix in order to determine the ranking in the group. The construction of such a matrix and its application will be explained in detail.

In section 4, a generalized equally weighted irreducible matrix will be introduced and the influence on the ranking will also be discussed, when the weighted values on the matrix have been changed.

In section 5, we will discuss the case that this method does not work, especially, an irreducible matrix can not make a decision of ranking the students who made the same score on an examination.

In section 6, some comments will be given and be mentioned another possible applications of this method for other fields.

We will start the following section with the definition of an irreducible matrix:

2 Perron-Frobenius Theorem and its properties

Definition 2.1 *A square matrix A is said to be reducible if there is a permutation matrix P such that*

$$PAP^T = \begin{pmatrix} B_{11} & B_{12} \\ 0 & B_{22} \end{pmatrix}$$

where B_{11} and B_{22} are square matrices.

A matrix A is said to be irreducible if it is not reducible.

The following two theorems are fundamental properties of an irreducible matrix.

Theorem 2.2 *A square matrix A is irreducible if and only if for any two distinct indices $1 \leq i, j \leq n$, there is a sequence of nonzero elements of A of the form*

$$\{a_{i,i_1}, a_{i_1,i_2}, \cdots, a_{i_m,j}\}$$

Theorem 2.3 *If the inequality $Ar > 0$ is satisfied for any nonnegative matrix A and a vector $r \geq 0$ then A is irreducible.*

Theorem 2.4 [**Perron-Frobenius theorem**]

If the matrix A has nonnegative entries, then there exists an eigenvector r with nonnegative entries, corresponding to a positive eigenvalue λ. Furthermore if the matrix A is irreducible, the eigenvector r has strictly positive entries, is unique and simple, and the corresponding eigenvalue is the largest eigenvalue of A in absolute value which is the spectral radius of A.

a(i)	a(4)	a(5)	a(6)	total
a(1)	8	7	5	20
a(2)	7	6	7	20
a(3)	5	6	9	20
mean	6.7	6.3	7	—

Table 1: each student's score for each problem and the mean.

Notice that we refer to a vector with nonnegative entries as a nonnegative vector, and a vector with positive entries as a positive vector and that a partial order on the set of nonnegative vectors by saying that $p > q$ whenever $p - q$ is a positive vector and $p \geq q$ whenever $p - q$ is nonnegative.

The proof of these theorems will be found in a number of books [1] or [2], so we recommend to see these references.

3 Generating procedure of an irreducible matrix on the examination result

In this section we will construct an irreducible matrix for the examination result that three students made the same score, and then apply Perron-Frobenius theorem to the matrix in order to determine the ranking in the group. In the following application we will assume that problems are independent one another, another words, students could answer each problem independently.

Suppose that three students, say $a(1)$, $a(2)$ *and* $a(3)$, answered three problems which are 10 points each, say $a(4)$, $a(5)$ *and* $a(6)$. The first comes the question of how to make an irreducible matrix A. Referring to the result in Table 1, each total score of the student is 20 and the mean which each student made in each problem have little significant difference. So we may be able to consider that the relative difficulty among problems is little. It is very difficult to rank those students without giving another examination. In this paper we will introduce a ranking method for such a case.

Step-1

Let S be the set of the all students, say,

$$S = \{\ a(1),\ a(2),\ a(3)\ \}$$

and T be the set of the all problems, say,

$$T = \{\ a(4),\ a(5),\ a(6)\ \}.$$

Referring Table 1. we will create a matrix A $(= a[i,j])$ according to the following rule;

(2.1) $\cdots$ If the students, $a(i)$, $(1 \leq i \leq 3)$ makes k−points from the problems, $a(j)$, $(4 \leq j \leq 6)$, then $a[i,j] = k,\ \ a[j,i] = 10 - k$

(2.2) $\cdots$ $a[i,j] = 0.1 \quad (1 \leq i,j \leq 3 \ or \ 4 \leq i,j \leq 6,\ where\ i \neq j)$

(2.3) $\cdots$ $a[i,j] = 0 \quad (1 \le i, j \le 6,\ where\ i = j)$

The following matrix A is created based on the above three steps. We will call this matrix as an evaluated matrix. The important property of this matrix is that each $a[i,j]$ expresses the relative difficulty (the outcome and the strength) between $a(i)$ and $a[i,j]$ as a value.

$$A = \begin{pmatrix} 0 & 0.1 & 0.1 & 8 & 7 & 5 \\ 0.1 & 0 & 0.1 & 7 & 6 & 7 \\ 0.1 & 0.1 & 0 & 5 & 6 & 9 \\ 2 & 3 & 5 & 0 & 0.1 & 0.1 \\ 3 & 4 & 4 & 0.1 & 0 & 0.1 \\ 5 & 3 & 1 & 0.1 & 0.1 & 0 \end{pmatrix}$$

We will explain the above three criteria (2.1),(2.2) and (2.3): The elements in the matrix A, $a[i,j]$ $(1 \le \mathrm{i} \le 3, 4 \le \mathrm{j} \le 6)$, mean that the students took the points from the problems, on the other hand, the elements in A, $a[i,j]$ $(4 \le \mathrm{i} \le 6, 1 \le \mathrm{j} \le 3)$, mean that the problem does not give the points to the students. Therefore we will make use of the scores as the evaluating value between each student and each problem (the interactions with other participants) .

The expression in (2.2) means evaluated value among students or among problems. For example, the element $a[1,2]$ the evaluated value of the student $a(1)$ with the student $a(2)$ and the element $a[2,1]$, on the other hand, the evaluated value of the student $a(2)$ with the student $a(1)$. There exist some interactions to evaluate two objects respectively and so it is not realistic to express $a[i,j] = 0$ but it is not easy to give a definite value to each element $a[i,j]$. Hence we will cosider that the values (the outcome and the strength)in the all elements

$$a[i,j] \quad (1 \le i, j \le 3 \ or\ 4 \le i, j \le 6\ except\ i = j\ )$$

will be 0.1. The more about this argument will be given in the section 3.

The expression in (2.3) means the evaluated value of themselves. Since there is no self comparison,the value will be

$$a[i,j] = 0\ for\ all\ i = j.$$

Step-2 Let

$$y = [\ 1,\ 1,\ 1,\ 1,\ 1,\ 1\ ]^T$$

be an initial vector which means that the ranking of the initial vector is even, and we compute A^1y, A^2y, A^3y, and A^4y, respectively, then

$$\begin{aligned} & A^1y = [\ 20.2,\ 20.2,\ 20.2,\ 10.2,\ 11.2,\ 9.2\ ]^T \\ (2.4)\cdots\quad & A^2y = [\ 210.04,\ 207.04,\ 205.04,\ 204.04,\ 224.14,\ 183.94\ ]^T \\ & A^3y = [\ 4162.21,\ 4102.21,\ 4062.21,\ 2107.21,\ 2307.24,\ 1919.18\ ]^T \\ & A^4y = [\ 43490.7,\ 42910.6,\ 42538.5,\ 41365.7,\ 45546.9,\ 37622.3\ ]^T \end{aligned}$$

Referring Table 1, the values of $a(1)$ through $a(6)$ in the vector A^1y indicate the new values which added 0.2 to the original scores, respectively.

Definition 3.1 *The normalized vector*

$$z^n = \frac{A^n y}{||A^n y||}$$

of $A^n y$ $(1 \leq n < \infty)$ is said to be an n-th ranking vector at each step n. The subvector z_1^n consisted of the elements

$$\{z(1), z(2), z(3)\} \text{ in the ranking vector } z^n$$

is called a ranking of the students who took the examination.
The subvector z_2^n consisted of the elements

$$\{z(4), z(5), z(6)\} \text{ in the ranking vector } z^n$$

is called an n-th ranking of the problems in the examination.

We will explain the meaning of the above definition with the vectors $A^n y$ and z_i^n. In fact, if we normalize the vector $A^1 y$ then,

$$\frac{A^1 y}{||A^1 y||} = \begin{bmatrix} 0.515038 \\ 0.515112 \\ 0.515112 \\ 0.260069 \\ 0.285566 \\ 0.234572 \end{bmatrix}.$$

Hence the ranking of the subvector z_1^1 of the students who took the examination will be

$$z(2) = z(3) \simeq z(1)$$

and the ranking of the subvector z_2^1 of the problems in the examination will be

$$z(5) > z(4) > z(6),$$

since the higher value can get the higher ranking by the construction of this scheme.

The ranking of the students on the subvectors, z_1^2, z_1^3, and z_1^4 depends upon the ranking of the problems, which means that the student can get higher rank if he takes higher scores in the higher ranking problems. Similarly, the ranking of the problems on the subvectors, z_2^2, z_2^3, and z_2^4 depend upon the ranking of the students, which means that the problem can get higher rank if it takes higher points in the higher ranking students.

Now each vector in (2.4) consists of two subvectors, namely, z_1^n and z_2^n, so that we can express the mixed ranking, students and problems. It has very interesting result that the problem $z(5)$ got the highest rank over all evaluation in the vector $A^4 y$, which tells us that the result of the mixed ranking is the following;

$$z(5) > z(1) > z(2) > z(3) > z(4) > z(6).$$

This result may tell us that the teacher who made the examination could give the students the most difficult problem and the easiest problem in the sense of the mixed ranking.

Step-3

Since the matrix A is irreducible, we may apply Perron-Frobenius Theorem to find out the spectral radius λ corresponding to the unique eigenvector r whose elements are all positive. The power method makes it easy task to find out the unique eigenvector. Let the initial vector

$$y = [\ 1,\ 1,\ 1,\ 1,\ 1,\ 1\ ]^T$$

we apply the power method, then it follows that

$$\lim_{n\to\infty} \frac{A^n y}{||A^n y||} = r.$$

Another word, the vector r is the limit of the ranking vector at each step n.

Definition 3.2 *The vector r is said to be the ranking vector, if a matrix A is irreducible and the vector r is the one corresponding to the spectral radius λ of A.*

Now we apply the power method on A and find out the spectral radius λ and its corresponding ranking vector r, that is,

$$\lambda = 14.3197$$

and

$$r = [\ 0.476380,\ 0.469897,\ 0.465724,\ 0.332259,\ 0.365585,\ 0.302178\ ]^T.$$

Therefore the final ranking among the students who made same score on the examination is that $z(1)$ is the first, $z(2)$ is the second, and $z(3)$ is the third. Similarly the final ranking of the problems in the examination is that $z(5)$ is the first, $z(4)$ is the second, and $z(6)$ is the third.

4 An equally weighted irreducible matrix and its influence on the Ranking

We have set that $a[i,j] = 0.1 \quad (1 \le i,j \le 3 \text{ or } 4 \le i,j \le 6, \text{ where } i \ne j)$ in (2.2) of the last section, since there are some interactions between $a(i)$ and $a(j)$ $(i \ne j)$ and we considered that each weight of the interactions is same.

In this section, we will set $a(i,j) = \alpha$ for all i,j and study the influence of the α on the creation of the ranking. Note that we still assume that each weight of the interaction is same. We define the irreducible matrix B as follows;

$$B = \begin{pmatrix} 0 & \alpha & \alpha & 8 & 7 & 5 \\ \alpha & 0 & \alpha & 7 & 6 & 7 \\ \alpha & \alpha & 0 & 5 & 6 & 9 \\ 2 & 3 & 5 & 0 & \alpha & \alpha \\ 3 & 4 & 4 & \alpha & 0 & \alpha \\ 5 & 3 & 1 & \alpha & \alpha & 0 \end{pmatrix}.$$

Letting the initial vector

$$y = [\ 1,\ 1,\ 1,\ 1,\ 1,\ 1\ ]^T,$$

we compute B^1y, B^2y and B^3y then,

$$B^1y = [\ 20+2\alpha,\ 20+2\alpha,\ 20+2\alpha,\ 10+2\alpha,\ 11+2\alpha,\ 9+2\alpha\ ]^T$$

a(i)	a(4)	a(5)	a(6)	total
a(1)	8	7	5	20
a(2)	7	5	8	20
a(3)	5	8	7	20
mean	6.7	6.7	6.7	—

Table 2: each student's score for each problem and the mean.

$$B^2y = \begin{bmatrix} 202 + 80\alpha + 4\alpha^2 \\ 199 + 80\alpha + 4\alpha^2 \\ 197 + 80\alpha + 4\alpha^2 \\ 200 + 40\alpha + 4\alpha^2 \\ 220 + 41\alpha + 4\alpha^2 \\ 180 + 39\alpha + 4\alpha^2 \end{bmatrix}$$

$$B^3y = \begin{bmatrix} 4040 + 1198\alpha + 240\alpha^2 + 8\alpha^3 \\ 3980 + 1198\alpha + 240\alpha^2 + 8\alpha^3 \\ 3940 + 1198\alpha + 240\alpha^2 + 8\alpha^3 \\ 1986 + 1200\alpha + 120\alpha^2 + 8\alpha^3 \\ 2190 + 1260\alpha + 123\alpha^2 + 8\alpha^3 \\ 1804 + 1140\alpha + 117\alpha^2 + 8\alpha^3 \end{bmatrix}$$

As long as $\alpha > 0$ and α is not so large number comparing with the score which is an element in the matrix, the students' ranking of B^2y is equivalent to that of A^2y and so is the problems' rankng. This property will be kept, when n goes to infinity. Even if we use either the matrix A or the matrix B, we will be able to derive the same ranking on both students' and problems' ranking.

We make use of the power method to find out the ranking vector, which we suppose that the same examination are given infinitely many times to the same students under the same problems, moreover, the students will make the same score every time. We are actually simulating the limit of the ranking on the students who made the same score, even though the only one examination has been given. It is possible that the ranking will be fixed after the certain number of iterations. After passing $n = 2$, we can obtain the same ranking in this examination.

5 The case that an irreducible matrix can not rank the students who made the same score on an examination

In this section we will discuss an example that the previous method does not work for ranking the students who made the same score on an examination.

Referring Table 2, we will construct the irreducible matrix C as we constructed the matrix A in the section 2. The matrix C will be;

$$C = \left(\begin{array}{ccc|ccc} 0 & 0.1 & 0.1 & 8 & 7 & 5 \\ 0.1 & 0 & 0.1 & 7 & 5 & 8 \\ 0.1 & 0.1 & 0 & 5 & 8 & 7 \\ \hline 2 & 3 & 5 & 0 & 0.1 & 0.1 \\ 3 & 5 & 2 & 0.1 & 0 & 0.1 \\ 5 & 2 & 3 & 0.1 & 0.1 & 0 \end{array} \right).$$

Now let the initial vector be

$$y = [\ 1,\ 1,\ 1,\ 1,\ 1,\ 1\ ]^T,$$

and suppose that the eigenvalue μ and it's corresponding eigenvector u, then by applying the power method we will have the following μ and u;

$$\mu = 14.3421$$

and

$$u = [\ 0.471405,\ 0.471405,\ 0.471405,\ 0.333333,\ 0.333333,\ 0.333333\ ]^T.$$

Each row sum of the submatrices in C is same, moreover the means of the students' score are all same, that is, each row sum of the submatrix on the upper block is same so is that of the submatrix on the lower block, and therefore we have such a u for finding ranking vector. This implies that we cannot make an appropriate ranking both on the students and problems in the examination.

6 Conclusion and another possible application of this method

The purpose of this ranking method is to rank participants that have same points in a competition without making an extra competition such as a playing off. In an examination such as the above , the more we have students who make the same score, the more we do not have the case, since we will expect that each student may be able to make a different score on each problem and so make a different mean.

This method is applicable not only to an examination but also any other competitions or contests such as golf, football, baseball, and an prediction of all kind of bettings as well.

REFERENCES

1. Abraham Berman and Robert J.Plemmons,Nonnegative Matrices in Mathematical Sciences, Academic Press, New York, 1979

2. James P.Keener, The Perron-Frobenius Theorem and The Ranking of Football,SIAM Review, 35(1993),No.1, pp80 - pp93.

3. James M. Ortega, Numerical Analysis, A Second Course, Academic Press, New York, 1972

4. K.Oshima, I.Hofuku, The Ranking Procedure on an Examination by means of the Ranking Vector, Transactions of the Japan Society for Industrial and Applied Mathematics,Vol. 6 No.1(1996), pp133 - pp146

7th Int. Coll. on Differential Equations, pp. 321-328
D. Bainov (Ed)

CHAOTIC MIXING OF VISCOUS FLUIDS: A MARANGONI-CONVECTION-DRIVEN EXAMPLE

AUGUST PALANQUES-MESTRE and GERARD ALBÀ-SOLER

Department of Applied Mathematics and Analysis, University of Barcelona, Gran Via 585, 08007-Barcelona, Spain

ABSTRACT

In recent years, the mixing of viscous fluids has been linked to 'chaos' theory. We prove that the perturbed flow around a bubble that moves subject to the combined action of buoyancy and a downward vertical temperature gradient is chaotic. Furthermore, we give analytical and numerical evalutations of the phase space transport that takes place in the vicinity of the bubble.

Keywords: lobe dynamics, primary and secondary intersection points, transition numbers, rotary numbers, Melnikov function, secondary Melnikov functions, the "falling" bubble.

1. INTRODUCTION

Twelve years ago, the pioneering work of Aref, Ottino and other authors linked the mixing of viscous fluids to 'chaos' theory. In fact, fluid mixing provides one of the few settings where chaotic processes can be identified in terms of directly observable physical phenomena. If the velocity field of a viscous fluid flow is time periodic, the study of the motion of the fluid is reduced to the study of a Poincar map. Since the phase space of this map is actually the physical space occupied by the fluid, the kinematical mixing is given by the transport phenomena in the Poincar map. The velocity field itself is not chaotic. Chaotic advection becomes apparent by following the path of the fluid particles. Thus, the kinematical mixing of viscous (or very viscous) fluids relies on what is called 'Lagrangian' turbulence.

The underlying structure which determines the transport properties is the treillis, i.e. the heteroclinic (or homoclinic) tangle created by the intersections of stable and unstable manifolds of the hyperbolic fixed points. The transport that takes place during a period of the Eulerian velocity field is directly related to the primary intersecion of the manifolds, and therefore to the computation of the Melnikov function (or the adiabatic Melnikov function, for very large periods). Secondary Melnikov functions help us to detect secondary intersections and to go beyond one period.

Among the paradoxes of hydrodynamics, Birkhoff mentioned the paradox of a "falling" bubble moving in a viscous fluid toward a heat source under the action of thermocapillary forces. In this paper it is proved that, for some range of values of the Bond number ($3/2 < \mathbf{Bo} < 9/2$), small perturbations of either the gravitational force or the temperature gradient lead to chaotic advection in the vicinity of the bubble. The Melnikov function and the second Melnikov function are computed. Then, analytical evaluations of the lobe dynamics and the efficiency of the mixing can be given.

The paper is organized as follows. In Section 2, basic facts about the kinematical mixing of viscous fluids are reviewed. Section 3 is devoted to the introduction of the basic analytic tools in studying the trellises formed by stable and unstable manifolds of two-dimensional maps. In Section 4, we review the flow around the "falling" bubble, we prove that the perturbed flow is chaotic, and we make use of the tools introduced in the previous section to give an analytic evaluation of the lobe dynamics. In Section 5, we briefly describe how to obtain a numerical evaluation of the lobe dynamics.

2. KINEMATICAL MIXING OF VISCOUS FLUIDS

Let $v(x,t)$, $x \in \mathbb{R}^2$ or $\mathbb{R}^3$, be an Eulerian velocity field of a fluid (a solution of the incompressible Navier-Stokes or Stokes equations, for instance) that is T-periodic in time. The motion of the fluid particles in the flow is given by:

$$\frac{dx}{dt} = v(x, t/T) \tag{2.1}$$

The study of this system can be reduced to the study of a two- or three-dimensional Poincar map. The phase space of the Poincar map is actually the physical space occupied by the fluid. Therefore, the dynamics of the Poincar map are, in fact, the dynamics of the fluid motion. Studying this map is equivalent to sampling fluid particle trajectories at time intervals equal to the period (a sort of strobing process). The kinematical mixing of the fluid, i.e. the mixing that stems from the convective nature of the flow (and not from molecular diffusion) is given by the transport phenomena in the Poincar map. If we think fluid mixing as the efficient stretching and folding of material lines and surfaces we have to notice that stretching and folding is the fingerprint of 'chaos'. Chaotic processes can be identified in terms of directly observable physical phenomena and will lead to mixing. Notice that the velocity field $v(x,t)$ is not chaotic. In fact, most examples of mixing of fluids related to 'chaotic' dynamical systems occur in the creeping flow regime (Stokes equations).

Despite the importance of the mixing of viscous fluids in nature and industry, the mixing process had been poorly understood before the merging of the kinematics of fluid mechanics and the theory of dynamical systems. For a long time, the kinematic origin of the mixing phenomenon was overlooked and chaotic advection was confined to high Reynolds number flows or even to inviscid flows (conventional, or 'Eulerian', turbulence). In his pioneering work [1], Aref studied a 'blinking vortex' model using Poincar sections, revealing that there is nearly globally chaotic mixing for a certain range of blinking periods. Since then, many other papers have been published on this topic. Most examples of 'chaotic mixing' are given in two-dimensional flows. Some of them are somewhat idealized and admit analytical treatment. Other examples can be simulated experimentally; emphasis is, then, placed on the comparison of numerical and experimental results. Reference [2] is a good review of the subject. Reference [3] is a quite recent survey of the 'state of the art'.

Two facts have to be noticed about the connection between fluid mixing and chaotic motions. On one hand, 'chaos is good' (in order to obtain an efficient mixing). On the other hand, asymptotics and long-time behavior have little relevance to the problem; the mixing is related with finite-time and transient phenomena (e.g., for a mixing experiment one may be interested in 20-40 perturbation periods).

3. TRANSPORT IN TWO-DIMENSIONAL MAPS

Given a two-dimensional area- and orientation-preserving diffeomorphism F (for instance, F is a Poincar map of a time-dependent two-dimensional fluid flow), the phase space can be divided into disjoint regions using pieces of the stable and unstable manifolds of hyperbolic fixed points. These pieces have to meet either in the hyperbolic fixed points or in primary intersection points (pips). (An intersection point of a stable and an unstable manifold is a pip if the pieces of the manifolds joining the point and the hyperbolic points do not intersect). The transport between these regions (the advection, in the fluid flow case) is associated with the intersections of the stable and unstable manifolds and the subsequent tangling of the manifolds. The first attempt to describe and classify the trellises formed by the tangling of the manifolds is due to R.W. Easton [4].

In the integrable case (when stable and unstable manifolds coincide), we have complete barriers, across which no phase space can be transported. Breaking integrability entails the destruction of these barriers; phase space is transported across what was once a complete barrier. The pieces of the stable and unstable manifolds form partial barriers; the transport across these barriers is described in terms of images of turnstile lobes, i.e. the lobes that the manifolds form between adjacent pips. Rom-Kedar and Wiggins have given formulae that compute this transport in terms of the intersections of the turnstile lobes and their images [5].

Here, we will make four assumptions (that define the setting that is found later on):

(i) The map F has two hyperbolic fixed points p_I and p_{II}.

(ii) One branch of the stable manifold of p_{II} transversally intersects one branch of the unstable manifold of p_I; and one branch of the stable manifold of p_I coincides with one branch of the unstable manifold of p_{II}.

(iii) The other branches of the stable and unstable manifolds of p_I and p_{II} extend to infinity, possessing no homoclinic orbits.

(iv) Given a pip q_0, the segments of the manifolds connecting q_0 to $F^{-1}q_0$ intersect each other only in a pip r_0.

Actually, one can assume (iv) with no loss of generality. Assumptions (i), (ii) and (iii) imply that the phase space is divided in a bounded region and four unbounded regions. Complete barriers separate three of the unbounded regions; from now, these regions will be omitted. The fourth unbounded region is separated from the bounded region by a partial barrier. This partial barrier is given, for instance, by $W^u[p_I, q_0]$ (the piece of the unstable manifold that connect p_I to a pip q_0) and $W^s[q_0, p_{II}]$ (the piece of the stable manifold that connect q_0 to p_{II}). A turnstile lobe D_0 is formed by the piece of the stable manifold that connects $F^{-1}q_0$ to r_0 ($S_0^D = W^s[F^{-1}q_0, r_0]$) and the piece of the unstable manifold that connects these two points ($U_0^D = W^u[F^{-1}q_0, r_0]$). And a second lobe E_0 is formed by the pieces of the manifolds that connect r_0 to q_0 ($S_0^E = W^s[r_0, q_0]$ and $U_0^E = W^u[r_0, q_0]$). All the other turnstile lobes are images or pre-images of D_0 or E_0.

Under F, the D_0 lobe is mapped from the bounded region to the unbounded one, while the E_0 lobe is mapped from the unbounded region to the bounded one. Assumption (iii) (called open map hypothesis) implies that there is no mechanism for re-entering to the bounded region, i.e. the phase space contained in the D_0 does not re-enter to the bounded region in subsequent iterations of the map F. In other words, there is no intersection between the E_0 lobe and the images of the D_0 lobe. On the other hand, by a straightforward area-preservation argument, part of the phase space that is entering the bounded region must be detrained in subsequent iterations of F. There are i, $i \geq 0$, such that $F^i E_0 \cap D_0 \neq \emptyset$.

The escape rate, i.e. the phase space that escapes from the bounded region at the nth iteration, will be denoted by c_n. If $\mu(L)$ denotes the area of the L lobe, it is straightforward to prove that:

$$c_n = \mu(D_0) - \sum_{j=0}^{n-1} \mu(F^j E_0 \cap D_0) \tag{3.1}$$

The phase space that stays in the bounded region S after the nth iteration is given by $R_n = \mu(S) - \sum_{j=1}^{n} c_n$. Therefore, the area of the invariant set in S is given by $\lim_{n\to\infty} R_n$.

Notice that there are i, $i \geq 0$, such that $F^i E_0 \cap D_0 \neq \emptyset$ because there are i, $i \geq 0$, such that $F^i U_0^E \cap S_0^D \neq \emptyset$. (Since $F^i S_0^E \subset W^s[q_0, p_{II}]$ and $U_0^D \subset W^u[p_I, q_0]$, for any i, $i \geq 0$, $F^i S_0^E \cap U_0^D = \emptyset$). Let $s_0 \in F^i U_0^E \cap S_0^D$. The pieces $W^u[p_I, s_0]$ and $W^s[s_0, p_{II}]$ have intersections other than the point s_0 (for example, the point q_0). The lobe dynamics is related to the appearance of secondary intersection points (sips). In fact, under the assumptions made, for any secondary intersection point s there are $m \in Z$ and $j \geq 0$ such that $F^m s \in F^j U_0^E \cap S_0^D$. j is called the transition number of the point s. Given a treillis, the structure index l is given by the minimal value of i such that $F^i E_0 \cap D_0 \neq \emptyset$. Therefore, a treillis has structure index l_0 if and only if all the sips have transition number $j \geq l_0$. Given a secondary intersection point s, the rotary number of s counts the number of times the orbit of s encircles the invariant set in S. For example, $s_0 \in F^j U_0^E \cap S_0^D$ is a k-rotary sip of transition number j if the orbit $F^i s_0$ encircles around the invariant set $k-1$ times as i increases from $-j-1$ to 0.

Suppose now that F is the Poincar map of a $1\frac{1}{2}$-degree-of-freedom Hamiltonian system. Let H be the unperturbed time-independent hamiltonian and let H_1 be a small periodic perturbation (of order ϵ).

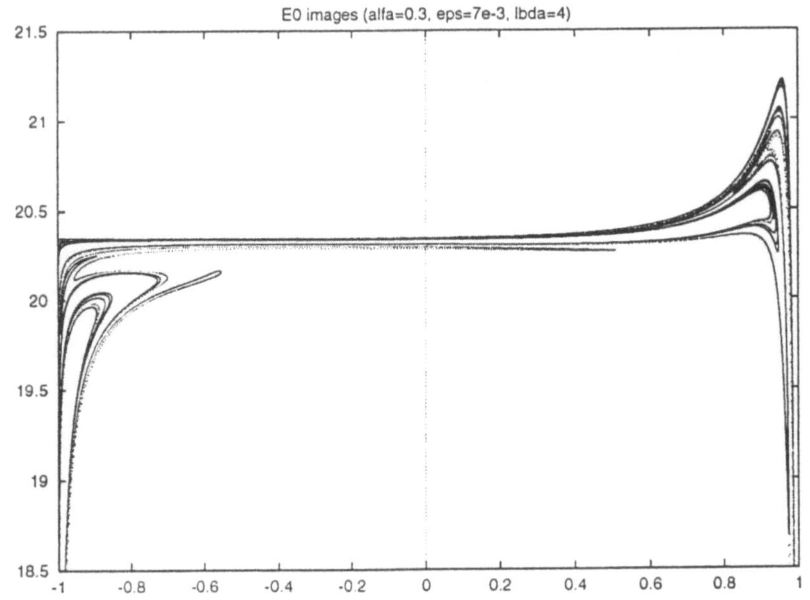

Figure 1.

A perturbative evaluation for the splitting of the separatrix is given by the Melnikov function:

$$M_1(t_0, \phi_0) = \int_{-\infty}^{+\infty} dt \; \{H, H_1\}(q(t-t_0), p(t-t_0), t + \frac{\phi_0}{2\pi}T) \qquad (3.2)$$

Simple zeros of $M_1(t_0, \phi_0)$ correspond to transverse primary heteroclinic intersections, i.e. pips. Furthermore, the area of any turnstile lobe is given by the integration of M_1 between the corresponding pips.

Recently, Rom-Kedar [6] introduced secondary Melnikov functions to obtain a perturbative evaluation of the secondary intersection points and the related parameters (transition and rotary numbers). Assume that $H = 0$ in the unperturbed separatrix and choose ϕ_0 such that $M_1(0, \varphi_0) = 0$. Let $P(h)$ the period of the unperturbed periodic orbit of energy h. (Notice that the unperturbed bounded region is filled by peridic orbits). The second Melnikov function is defined by:

$$M_2(t_0) = M_1(t_0, \phi_0) + M_1(t_0 + P(M_1(t_0)), \phi_0) \quad t_0 \in (0, T) \qquad (3.3)$$

Simple zeros of $M_2(t_0)$ correspond to transverse 2-rotary sips. The transition number of a such a sip is given by:

$$j(t_0) = \left[\frac{t_0 + P(M_1(t_0))}{T}\right] \qquad (3.4)$$

(where $[x]$ denotes the integer part of x). Higher order Melnilkov functions can be constructed for evaluating k-rotary sips ($k > 2$). See reference [6] for more details.

4. ANALYTIC EVALUATION OF THE LOBE DYNAMICS

Consider a bubble that moves subject to the combined action of buoyancy and a downward vertical temperature gradient. (This problem was first discussed in detail by Young, Goldstein and Block [7].) In the vicinity of the bubble, the temperature variation along the surface of the bubble results in an opposite variation of the interfacial tension. As a consequence, the fluid is dragged by the bubble in the direction of the colder pole. Far from the bubble, buoyancy drags the fluid in the opposite direction. If we neglect convective transport terms in the energy and momentum equations, and neglect deformations of the bubble from the spherical shape, the unique solution for the flow in the bubble reference frame is axially symmetrical and given by:

$$v_r = [1 - (1 + \frac{\lambda}{2})r^{-1} + \frac{\lambda}{2}r^{-3}]\cos\theta \qquad v_\theta = -[1 - \frac{1}{2}(1 + \frac{\lambda}{2})r^{-1} - \frac{\lambda}{4}r^{-3}]\sin\theta \qquad (4.1)$$

The topology of the velocity field depends on the value of λ, and this parameter is given by the Bond number: $\lambda = 6(2\mathbf{Bo} - 3)^{-1}$. If $\lambda > 1$, i.e. $3 < 2\mathbf{Bo} < 9$, there are two saddle points heteroclinically connected. (Buoyancy exceeds thermocapillary forces; the bubble moves upward but thermocapillary forces are strong enough to create a return flow in the vicinity of the bubble.)

Small periodic perturbations of either the gravitational force or the temperature gradient lead to a quasi-static flow provided that the time scale of the perturbation (i.e. the period T) is much larger than the time-scale of the diffusion of momentum and the time-scale of the diffusion of energy. Then, time enters only as a periodic parameter that modifies λ; thus, the perturbation of the velocity field is given by:

$$v'_r = \frac{\epsilon}{2}[-r^{-1} + r^{-3}]\cos\theta \sin\frac{2\pi t}{T} \qquad v'_\theta = \frac{\epsilon}{4}[r^{-1} + r^{-3}]\sin\theta \sin\frac{2\pi t}{T} \qquad (4.2)$$

where ϵ is a small parameter.

Define new variables $q = r^3$, $p = \cos\theta$. The phase space regions $p > 1$ and $p < -1$ are unphysical. The region $q < 1$ corresponds to the bubble. And the region $q > 1$ corresponds to the fluid around the bubble. The area of any domain in this region multiplied by a factor $2\pi/3$ gives the volume of the related domain in the vicinity of the bubble. In the new variables, the unperturbed velocity field is given by the Hamiltonian $H(q,p) = f(p)g(q)$ where:

$$f(p) = 1 - p^2 \qquad g(q) = \frac{3}{2}[-q^{2/3} + (1 + \frac{\lambda}{2})q^{1/3} - \frac{\lambda}{2}q^{-1/3}] \tag{4.3}$$

We will focus on the hyperbolic points $(\mp 1, q_0)$, where $q_0 = [(\lambda/4)(1+\sqrt{1+8/\lambda})]^3$. The heteroclinic connection is given by $\{(q_0, \tanh(K_1(t-t_0))), t \in \mathbb{R}\}$ where $K_1 = -g'(q_0) = 2^{-1}q_0^{-4/3}[q_0 - \lambda] > 0$. This separatrix divides the physical phase space in two parts: a bounded region $(1 < q < q_0)$ and an unbouded region $(q > q_0)$. The bounded region is filled by periodic orbits of energy $h > 0$. The period of a given inner orbit is:

$$P(h) = \left|\frac{dA(h)}{dh}\right| = \int_{q_-(h)}^{q_+(h)} dq\; g(q)^{-1}\left(1 - \frac{h}{g(q)}\right)^{-1/2} \tag{4.4}$$

where $A(h)$ is the phase space area encircled by the orbit and $q_+(h)$, $q_-(h)$ are such that $g(q_\pm(h)) = h$. For small values of h one can expand the integral as a series:

$$P(h) = \gamma \ln(\frac{K}{h}) + O(h \ln h) + O(h) + \ldots \tag{4.5}$$

where:

$$\gamma = \gamma(\lambda) = \frac{2}{q_0^{1/3} - 1}\left[\frac{1}{1 - q_1^{1/3}} + \frac{q_0}{q_0^{1/3} - q_1^{1/3}}\right] \tag{4.6}$$

and $q_1 = [(\lambda/4)(1 - \sqrt{1+8/\lambda})]^3$.

The separatrix is splitted by the perturbation $H_1(p,q,t) = \epsilon f(p) g_1(q)\sin(2\pi t/T)$ where $g_1(q) = (3/4)[q^{1/3} - q^{-1/3}]$. A perturbative evaluation of the distance between the perturbed manifolds in the Poincar section is given by the Melnikov integral:

$$M_1(t_0, \phi_0) = \int_{-\infty}^{+\infty} dt\; \{H, H_1\}(q_0, \tanh(K_1(t-t_0)), t + \frac{\phi_0}{2\pi}T) \tag{4.7}$$

The computation of this integral is straightforward and leads to:

$$M_1(t_0, \phi_0) = \epsilon\frac{4\pi K_2}{K_1 T}\frac{\pi^2}{K_1 T}\left[\sinh\left(\frac{\pi^2}{K_1 T}\right)\right]^{-1}\cos(\frac{2\pi t_0}{T} + \phi_0) \tag{4.8}$$

where $K_2 = g_1(q_0) > 0$. As expected, the Melnikov function has simple zeros, proving that perturbed manifolds intersect transversally. In other words, the flow around the perturbed bubble is chaotic. In the following, we choose $\phi_0 = -\pi/2$ (we fix a Poincar section). Actually, the zeros of $M_1(t_0, -\pi/2)$ correspond to primary intersection points. Since there are two zeros along a period of the perturbation, it follows that there is only one pip between another pip and its image by the Poincar map. Therefore, the segments of stable and unstable manifolds that connect a pip and its image form two turnstile lobes.

The integral of the Melnikov function between two adjacent pips gives the area of the corresponding lobe. In this case, all the lobes have the same area:

$$A_0 = \epsilon\frac{4K_2}{K_1}\frac{\pi^2}{K_1 T}\left[\sinh\left(\frac{\pi^2}{K_1 T}\right)\right]^{-1} \tag{4.9}$$

As long as there is no secondary intersection of transition number $j = 0$, this area measures the extent of the kinematical mixing that takes place, during a period of the perturbation, in the vicinity of the bubble.

In order to go beyond one period and obtain perturbative information about the lobe dynamics, one must compute second (and higher) Melnikov functions. It is straightforward to prove that:

$$M_2(t_0) = 2\epsilon\frac{4\pi K_2}{K_1 T}\frac{\pi^2}{K_1 T}\left[\sinh\left(\frac{\pi^2}{K_1 T}\right)\right]^{-1}\sin(\frac{2\pi t_0}{T}+\frac{\pi P(M_1(t_0))}{T})\cos(\frac{\pi P(M_1(t_0))}{T}) \quad (4.10)$$

Therefore, one obtains two families of 2-rotary sips:

$$\frac{2\pi t_0}{T}+\frac{\pi P(M_1(t_0))}{T} = (j+1)\pi \qquad j \geq 0 \quad (4.11)$$

$$\frac{\pi P(M_1(t_0))}{T} = (2j+1)\frac{\pi}{2} \qquad j \geq 0 \quad (4.12)$$

The index j is, in fact, the transition number of the sip. Making use of expressions (4.8) and (4.5), it is quite easy to construct a bifurcation diagram that describes the changes in the lobe dynamics as the strength of the perturbation is increased. (The strenght of the perturbation is given by the 'small' parameter ϵ.)

Define:

$$\epsilon_I^j = \sqrt{1+\frac{T^2}{\gamma^2\pi^2}\frac{KK_1T}{4\pi K_2}\left[\frac{\pi^2}{K_1T}\right]^{-1}\sinh\left(\frac{\pi^2}{K_1T}\right)\exp[-\frac{1}{\gamma}(jT+T-2t_1)]} \quad (4.13)$$

$$\epsilon_{II}^j = \frac{KK_1T}{4\pi K_2}\left[\frac{\pi^2}{K_1T}\right]^{-1}\sinh\left(\frac{\pi^2}{K_1T}\right)\exp[-\frac{1}{\gamma}(jT+\frac{T}{2})] \quad (4.14)$$

where $t_1 \in [0, T/4]$ is such that $\tan(2\pi t_1/T) = \gamma\pi/T$. And notice that:

$$\ldots < \epsilon_{II}^{j+1} < \epsilon_I^j < \epsilon_{II}^j < \epsilon_I^{j-1} < \ldots \quad (4.15)$$

If $\epsilon < \epsilon_I^j$ there is no secondary point of transition number $\leq j$. If $\epsilon_I^j < \epsilon < \epsilon_{II}^j$ there are two transversal 2-rotary sips of transition number j (and no sip of transition number $< j$). In this case, the $F^j E_0$ lobe 'intersects' the D_0 lobe. If $\epsilon_{II}^j < \epsilon < \epsilon_I^{j-1}$ there are four transversal 2-rotary sips of transition number j (and no sip of transition number $< j$). In this case, the $F^j E_0$ lobe 'crosses' the D_0 lobe. When $\epsilon = \epsilon_I^j$ or $\epsilon = \epsilon_{II}^j$ for some j, secondary heteroclinic tangencies occur.

It is obvious that if $\epsilon < \epsilon_I^j$ the structure index of the treillis is $\geq j+1$. Therefore, after time $(j+1)T$ (i.e. after $j+1$ iterations of the related Poincar map), the phase space area which escapes from the bounded region is:

$$A_j = \sum_{i=1}^{j+1}\mu(D_0) = (j+1)\mu(D_0) \quad (4.16)$$

This area gives a lower bound for the phase space area involved in the chaotic advection when $\epsilon < \epsilon_I^j$.

5. NUMERICAL EVALUATION OF THE LOBE DYNAMICS

We numerically compute the escape rates c_n from the formula (3.1). we first find the heteroclinic points defining lobes E_0,D_0 as zeros of function $H(z,w) = F^{k_1}(f(z)) - F^{-k_2}(g(w))$, where $f(z)$, $g(w)$ are local representations of the invariant manifolds W^u, W^s, that define the mixing boundary. k_1, k_2 are fixed positive integers once we have an initial guess (this can be obtained from a representation of the manifolds using $f(z)$, $g(w)$). Then the boundaries of the E_0 lobe are approximated by polygonal lines of images $F^{k_1}(f(z_i))$, $z_i \in (z_{q_0}, z_{r_0})$ and $F^{-k_2}(g(w_i))$, $w_i \in (w_{r_0}, w_{q_0})$. In the same way we approximate D_0 boundary by a polygon.

Intersections of the $F^j E_0$'s with D_0 are obtained from a polygon clipping algorithm. Then its areas are computed using a square-subdivision recursive algorithm. Both need a test to whether accept or not a point is interior to a polygon. We decide that a point is inside the polygon iff a random line though it cuts the polygon in an odd number of edges.

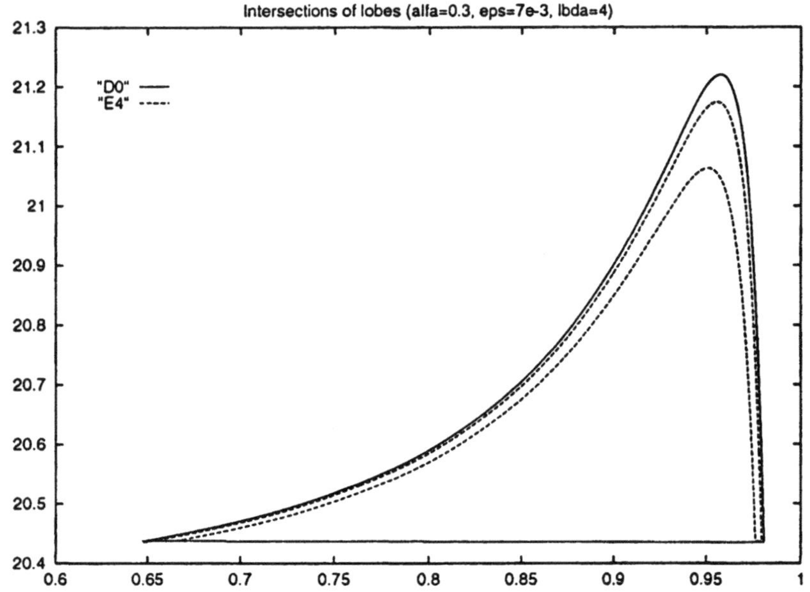

Figure 2.

REFERENCES

1. H. Aref, J. Fluid. Mech. 143, 1-21 (1984).
2. J.M. Ottino, The kinematics of mixing: stretching, chaos, and transport, Cambridge University Press (1989).
3. H. Aref and M.S. El Naschie (Eds.), Chaos Applied to Fluid Mixing, Pergamon (1995).
4. R.W. Easton, Trans. Am. Math. Soc. 294, 719-732 (1986).
5. V. Rom-Kedar and S. Wiggins, Arch. Rat. Mech. Anal. 109, 239-298 (1990).
6. V. Rom-Kedar, Chaos 5, 385-401 (1995).
7. N.O. Young, J.S. Goldstein and M.J. Block, J. Fluid Mech. 6, 350-356 (1959).

7th Int. Coll. on Differential Equations, pp. 329-338
D. Bainov (Ed)

Numerical solutions to the minimum-time problem for linear second-order conflict-controlled systems

V.S.Patsko, V.L.Turova

Institute of Mathematics and Mechanics, Ekaterinburg, Russia

Abstract

The time-optimal problem for linear differential games in the plane is considered. An algorithm for constructing level sets of the value function is proposed. Numerical examples are presented.

Keywords: *Differential games; time-optimal control; value function.*

Introduction

In this paper, we consider differential games [1–3] with the linear dynamics and geometrical bounds on controls

$$\dot{x} = Ax + u + v, \quad x \in R^2, \; u \in P, \; v \in Q. \tag{1}$$

Here P and Q are convex closed polygons in the plane. The terminal set M (a convex polygon in the plane) is given. The first player who governs the control parameter u seeks to minimize the time of attaining M from some initial point m, the aim of the second player governing the control vector v is opposite. So, the payoff of the game is the time of attaining the set M. The permissible controls are feedback controls.

We are interested in finding $W(\theta, M)$, $\theta > 0$. Each of them is the set of all initial states x_0 such that the first player guarantees the transition of the state vector to M by the time θ. The set $W(\theta, M)$ is the level set (the Lebesgue set) of the value function of the minimum-time game problem.

In terms of [2, 3], the set $W(\theta, M)$ is also called t-section of the maximal u-stable bridge (the cross-section with the hyperplane $t = \theta$).

The paper is devoted to the numerical construction of $W(\theta, M)$.

In many cases, the sets $W(\theta, M)$ can be found only numerically (even for examples with very simple dynamics, for instance, $\dot{x}_1 = x_2 + v$, $\dot{x}_2 = u$, $\quad |u| \leq 1$, $|v| \leq 1$).

The application of backward procedures is the typical way for solving control and differential game problems. General ideas of backward procedures were considered in papers of R.Bellman, R.Isaacs, W.Fleming, L.S.Pontryagin, and B.N.Pshenichny.

The most developed results [4–8] related to algorithmic implementations of the backward constructions to differential games were obtained for the case where a linear system should be brought to a convex terminal set M at a given time moment, and we are interested in finding the set of all states from which this transfer can be done. In this case, the application of the backward procedure gives t-sections of the maximal u-stable bridge. The algorithms use the property: the convexity of the target set implies the convexity of

t-sections of the maximal stable bridge. This makes the problem easier and enables to apply numerical methods to some important practical problems [9, 10].

The above mentioned feature is not inherent to differential games with the nonfixed time of termination: as a rule, t-sections of maximal stable bridges are not convex. Numerical methods for solving problems with the nonfixed time of termination and nonconvex problems with the fixed terminal time are studied in papers of V.N.Ushakov and his collaborators [11, 12]. Recently, numerical methods for constructing value functions and their level sets based on the notion of viscosity solutions of Hamilton-Jacobi (Bellman-Isaacs) equations were developed [13, 14].

The algorithm described below is based on the ideas of the algorithms proposed in [4] for linear differential games with the fixed time of termination.

STATEMENT OF THE PROBLEM

We now define the set $W(\theta, M)$ more precisely [2, 3]. Let $\mathcal{U}$ be the set of all positional strategies U of the first player. Namely, this is the set of all functions defined on $[0,\theta] \times R^2$ and taking values in P. Let σ be an arbitrary partition of the segment $[0,\theta]$ formed by the points $0 = t_1 < t_2 < ... < t_n = \theta$, let $d(\sigma)$ be its diameter, and let $v(\cdot)$ be a measurable function of time with values in Q. Let $y(\cdot\,; \sigma, x_0, U, v(\cdot))$ denotes the Euler spline emanating from the point x_0. We denote by $W(\theta, M)$ the set of all points $x_0 \in R^2$ for each of which there exist a strategy $U \in \mathcal{U}$ and a mapping $\varepsilon \to \delta(\varepsilon)$ from R_+ to R_+ such that for any $\varepsilon > 0$, any σ with the diameter $d(\sigma) \leq \delta(\varepsilon)$, and any function $v(\cdot)$ with values in Q there exists a time $t \in [0,\theta]$ at which $y(t; \sigma, x_0, U, v(\cdot))$ belongs to the ε-neighborhood of M.

Such a definition is equivalent to other well-known definitions [3, 13, 14] of the solvability set $W(\theta, M)$ of the time-optimal game problem. We give this definition because it shows properties of the optimal guaranteeing strategy of the first player in terms of the bundle of motions generated by various controls of the second player.

THE IDEA OF THE ALGORITHM

The set $W(\theta, M)$ is formed via a step-by-step backward procedure giving a sequence of embedded sets

$$W(\Delta, M) \subset W(2\Delta, M) \subset W(3\Delta, M) \subset ... \subset W(i\Delta, M) \subset ... \subset W(\theta, M). \qquad (2)$$

Here Δ is the step of the backward procedure. Each set $W(i\Delta, M)$ consists of all initial points such that the first player brings system (1) into the set $W((i-1)\Delta, M)$ within the time duration Δ (we put $W(0, M) = M$).

Before doing the first step of the backward procedure, we find a usable part Γ_0 of the boundary of M. In accordance with [1], the usable part is a curve or several curves on the boundary of M attainable for trajectories of system (1) from points lying in the exterior of M close to the boundary of M. The usable part is defined by the following formula

$$\Gamma_0 = \mathrm{cl}\{x \in \partial M : \min_{u \in P} \max_{v \in Q} \langle \ell, Ax + u + v \rangle < 0, \ \forall \ell \in K_x\}.$$

Here K_x is the cone of outward normals to the set M at x. Since the target set is convex, each curve of the usable part is locally convex in the following sense: the normal to the curve at a point x keeps its rotation in only direction when x moves along the curve.

Let us introduce the term "front". We put $F_0 = \Gamma_0$. The front F_i is the set of all points on the boundary of $W(i\Delta, M)$ for which the minimum guaranteeing time of the achievement of $W((i-1)\Delta, M)$ is equal to Δ. For other points of the boundary of $W(i\Delta, M)$ the

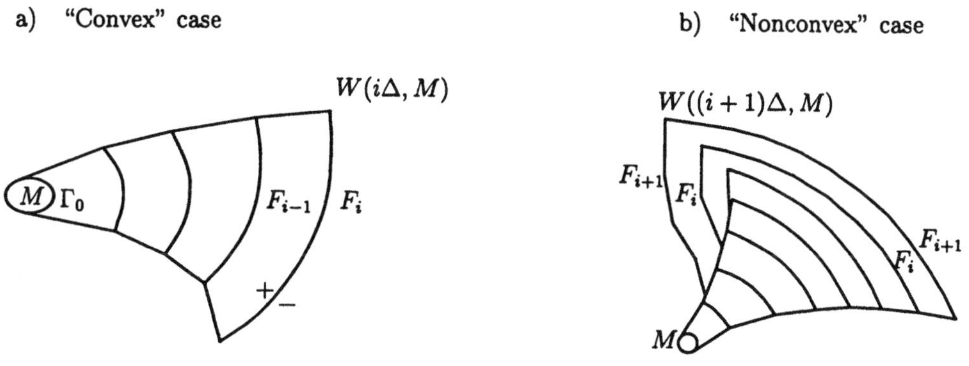

Figure 1. Construction of the sets $W(i\Delta, M)$.

optimal time is less than Δ. The line $\partial W(i\Delta, M) \setminus F_i$ possesses the properties of barriers [1]. The front F_i is designed using the previous front F_{i-1}.

Due to the linearity of system (1), the fronts F_1, $F_2, \ldots,$ $F_i, \ldots$ inherit (Figure 1a) the property of the local convexity of Γ_0 at the initial stage of constructions, and this property is kept until the current front F_i does not meet the set $W((i-1)\Delta, M)$. Straight lines connecting the endpoints of F_i with the corresponding endpoints of F_{i-1} form the continuation of the barrier lines. The boundary of $W(i\Delta, M)$ is formed by the front F_i, the above mentioned continuations of the barrier lines, and the line $\partial W((i-1)\Delta, M) \setminus F_{i-1}$. The property of the local convexity of fronts enable us to employ (with some small modifications) procedures which were developed for the construction of t-sections of maximal stable bridges in the case of linear differential games with the convex target set and the fixed time of termination [4].

If the next front F_i meets the already constructed set $W((i-1)\Delta, M)$, we say that the front collides with $W((i-1)\Delta, M)$. The situation of "collision" means that the current front meets the barrier part of the boundary of $W((i-1)\Delta, M)$ or the part $\partial M \setminus \Gamma_0$ of the boundary of M. To construct the next front F_{i+1}, we should take into account that F_i and the boundary of $W((i-1)\Delta, M)$ have the nonconvex conjunction (Figure 1b). Due to the properties of the plane, the complement of $W(i\Delta, M)$ is locally convex near the conjunction point. So, assuming that the second player seeks to bring the system to the complement, and the first player has the opposite objective, we can use the ideas of the "convex" algorithms from [4]. After combining the curve which is obtained from the locally convex part of the front with the one obtained from the nonconvex conjunction, we get a new front F_{i+1} that may be not locally convex.

So, the algorithm consists of the following operations: 1) finding the usable part on the boundary of the target set; 2) constructing the next front using the previous front; 3) testing the intersections of the current front with the barrier part of the already constructed set and the boundary of M. In the case the intersection is detected, further computations are being carried out taking into account the arising nonconvex conjunction and possible splitting the front into several parts.

Classification of singular lines

The analysis of singular lines where the value function and optimal trajectories have peculiarities is of great importance in differential games. The algorithm proposed can be provided with some simple diagnostics for finding and classification of singular lines and, in particular, for finding the most complicated among them equivocal [1, 15] lines.

In order to explain the possibility of the algorithmic classification of singular lines, we give some more detailed description of local constructions carried out in the process of finding the next front. The sets P, Q are assumed to be segments.

The front F_i is stored as an ordered collection of points. We consider each front as a polygonal line consisting of links defined by pairs of neighboring points. The normals to the links are directed to the front's outward side which is said to be negative (Figure 1). The opposite side of the front is called positive. We call an apex of the polygonal line to be the point of the local convexity if the angle between the positive sides of the neighboring links is less than π. We call this apex the point of the local concavity if the above mentioned angle is greater than π or equal to π.

Each link ab of the front F_i is assigned the segment with endpoints α, β which are defined by the formulas

$$\alpha = a - \Delta(A\,a + u^\circ + v^\circ), \quad \beta = b - \Delta(A\,b + u^\circ + v^\circ).$$

Here, u° and v° are extremal controls obtained from the following relations

$$u^\circ = arg \min_{u \in P} \langle \ell_{[ab]}, u \rangle, \quad v^\circ = arg \max_{v \in Q} \langle \ell_{[ab]}, v \rangle, \tag{3}$$

where $\ell_{[ab]}$ is the normal to the link ab. We say that the segment $\alpha\beta$ is generated by the segment ab with the use of extremal trajectories emanating from ab with the controls u°, v°. If the extremal controls of the players are the same for two neighboring links ab, bc, then the constructions done according to the above rule give four points, and two of these points coincide. So, we obtain two adjoint segments $\alpha\beta$ and $\beta\gamma$ which are generated by the links ab and bc. If the extremal controls defined by two neighboring normals do not coincide at least for one of the players, then we obtain two non-joint segments $\alpha\beta_1$ and $\beta_2\gamma$. In this case, some additional constructions should be done.

Let an apex b at which two neighboring links of the front F_i join be the point of the local convexity. The following cases are possible.

A. The extremal control of the second player is the same for the normals $\ell_{[ab]}$ and $\ell_{[bc]}$; the extremal control of the first player for $\ell_{[ab]}$ differs from that one for $\ell_{[bc]}$. In this case, two points β_1 and β_2 are associated with the point b. The segment with the endpoints β_1, β_2 is called the insertion due to P (Figure 2*A*).

B. The extremal control of the first player is the same for the normals $\ell_{[ab]}$ and $\ell_{[bc]}$; the extremal control of the second player for $\ell_{[ab]}$ differs from that one for $\ell_{[bc]}$. The segments $\alpha\beta_1$, $\beta_2\gamma$ generated by ab, bc intersect each other in a way shown in Figure 2*B*. Let ξ be the intersection point. We delete the parts $\beta_1\xi$, $\beta_2\xi$. There are two extremal trajectories that arrive at the point ξ. The first trajectory starts from the segment ab, the second one comes from bc. Assuming the motions go forward in time, we obtain that they disperse at the point ξ with respect to the segment $b\xi$.

C. The extremal controls of the first and the second players for $\ell_{[ab]}$ differ from those for $\ell_{[bc]}$. There are normals ℓ_P and ℓ_Q to P and Q that lie between $\ell_{[ab]}$, $\ell_{[bc]}$. Suppose that ℓ_Q lies between ℓ_P and $\ell_{[ab]}$. In this case, we add a segment $\beta_1\beta_3$ to the segment $\alpha\beta_1$ (Figures $2C_1$, $2C_2$). The endpoints β_1, β_3 are obtained from the point b using two different extremal controls of the first player and the extremal control of the second player corresponding to the normal $\ell_{[ab]}$. We call this additional segment the insertion due to P. We intersect the polygonal line $\alpha\beta_1\beta_3$ with the segment $\beta_2\gamma$. The intersection point ξ belongs either to $\beta_1\beta_3$ or $\alpha\beta_1$. We delete the parts $\beta_3\xi$, $\beta_2\xi$. If the point ξ belongs to $\alpha\beta_1$ (Figure $2C_1$), then two

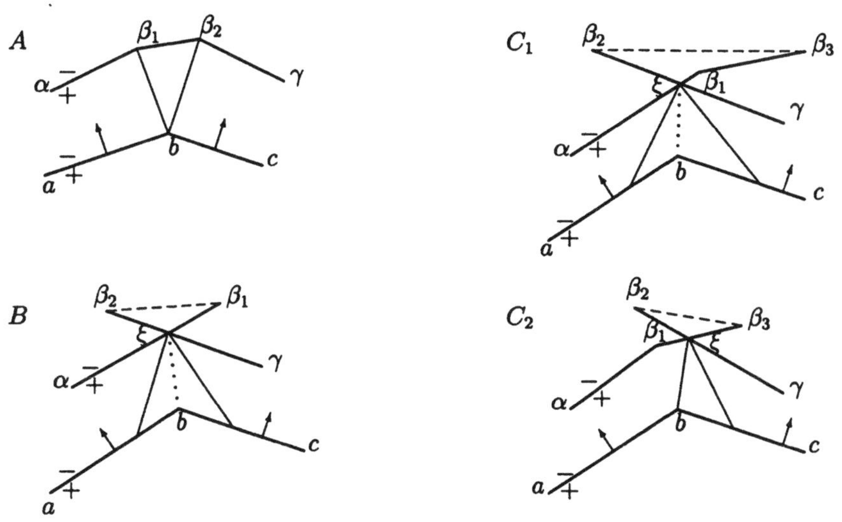

Figure 2. Constructions in the case of local convexity.

retrograde trajectories arrive at this point: one of them starts from ab, the other starts from bc. If ξ belongs to the insertion due to P (Figure $2C_2$), then one trajectory comes to ξ from the segment bc and the other trajectory comes from the point b. The latter is not extremal because it is obtained with some not extremal control of the first player ensuring the transfer to the point ξ and the extremal control of the second player corresponding to $\ell_{[ab]}$. Considering trajectories in the forward time, we obtain two trajectories emanating from ξ: one of them is extremal and the other one is not extremal.

If the normal ℓ_Q lies between $\ell_{[ab]}$ and ℓ_P, the difference is that the insertion due to P adjoins the segment $\beta_2\gamma$ generated by bc.

Practically, the algorithm we use is more complicated than the local constructions described above. Namely, we intersect polygonal lines consisting of a large number of links. Links of fronts can get smaller. Nevertheless, the algorithm does not require to decrease the step Δ in accordance with the length of links.

The classification of singular lines is done as follows.

In the case A, we have two points β_1, β_2 of the front F_{i+1} which give the insertion due to P. One of these two points (any of them) is called a switch point of the first player.

In the case B, the intersection point ξ is called a dispersal point of the second player. This name characterizes the fact that two optimal forward time trajectories obtained with different controls of the second player emanate from such a singular point. Each of these trajectories is the extremal trajectory; one of them goes to the region where the optimal control of the second player takes the first of two extremal values; the other trajectory goes to the region where the optimal control of the second player takes the second extremal value.

The intersection point ξ which appears in the case C is the dispersal point of the second player if it does not belong to the insertion due to P. Two optimal trajectories which are both extremal emanate from this point. The intersection point is called an equivocal point of the second player if it belongs to the insertion due to P. In this case, two trajectories emanate from such a point: the first one is extremal and the second one is not extremal.

It depends on the behavior of the second player which of these two trajectories is realized.

Some similar classification of singular points can be done in the case of the local concavity.

As a result, we obtain a collection of singular lines (switch lines, dispersal lines, equivocal lines) after finishing computations.

EXAMPLES

In this section, numerical examples of computing the sequence $\{W(i\Delta, M)\}$ are given. The step Δ was equal to 0.05 in all cases. The terminal set M was a small regular octagon with the center at the origin in the examples 1, 3, and with the center at the point $(-2.2, 0.4)$ in the example 2.

1. The well-known example of the time-optimal problem for oscillating systems in the theory of optimal control [16] has the form

$$\begin{aligned} \dot{x}_1 &= x_2 \\ \dot{x}_2 &= -x_1 + u, \quad |u| \leq 1. \end{aligned} \tag{4}$$

The sets $W(\tau, M)$, $\tau = 2\Delta k$, $k = \overline{1, 80}$, are depicted in Figure 3. Note, that $W(\tau, M)$ is convex for any τ.

2. Now, consider the following differential game:

$$\begin{aligned} \dot{x}_1 &= x_2 + u_1 + v_1 \\ \dot{x}_2 &= -x_1 + u_2 + v_2 \end{aligned} \tag{5}$$

$$u = (u_1, u_2)' \in P, \quad v = (v_1, v_2)' \in Q$$

whose dynamics is similar to (4). The set P is the vertical segment with the endpoints $(0, -2.5)$, $(0, 2.5)$, and Q is the segment with the apexes $(-5, 1.5)$, $(-1, -1.5)$. In Figure 4a, the sets $W(\tau, M)$, $\tau = i\Delta$, $i = \overline{1, 132}$ are depicted. The computations are carried out up to $\tau = 6.6$. At $\tau = 6.6$, the front collides with the terminal set M and is divided into two parts. Further constructions are being done independently for these two parts. The computations for the upper part are continued till $\tau = 11.6$ in Figure 4b, and we fill up the gap G. The front which corresponds to the maximal $\tau = 11.6$ lies approximately in the middle of G. In Figure 4b, only two fronts constructed from the lower part are shown. The accumulation of fronts generates black regions in Figure 4, which means very fast changing the value function (though it is continuous).

3. Figure 5 corresponds to the following system

$$\begin{aligned} \dot{x}_1 &= 0.35x_1 + x_2 + v \\ \dot{x}_2 &= -0.8x_1 + u \end{aligned} \tag{6}$$

$$-2 \leq u \leq 1.5, \quad -6.1 \leq v \leq -4.$$

The level sets $W(\tau, M)$ are computed for $\tau = i\Delta$, $i = \overline{1, 189}$. Up to $\tau = 5.7$, the front moves between two barriers emanating from the set M. The left barrier terminates at $\tau = 5.7$. For $\tau > 5.7$, the front begins to go around this barrier so that one of its endpoints slides along the outward side of the barrier. At $\tau = 8.15$, the front collides with some early part of the left barrier from outside. For $\tau > 8.15$, the left and right endpoints of the front move towards each other along the left barrier. The computation is finished at $\tau = 9.45$.

In this example, the set filled up with the fronts computed for $\tau \leq 9.45$ is the maximal set where the optimal guaranteeing time is less than infinity. The first player can not

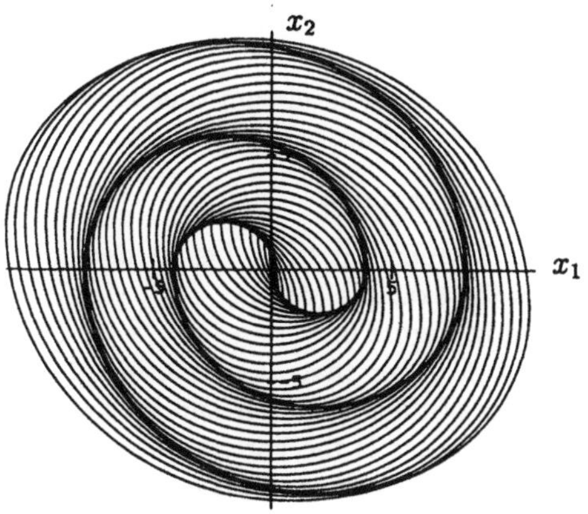

Figure 3. Canonical control problem (4).

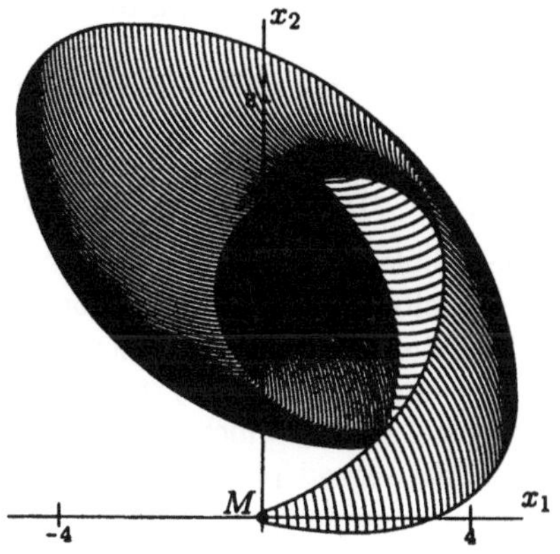

Figure 5. Level sets of value function in problem (6).

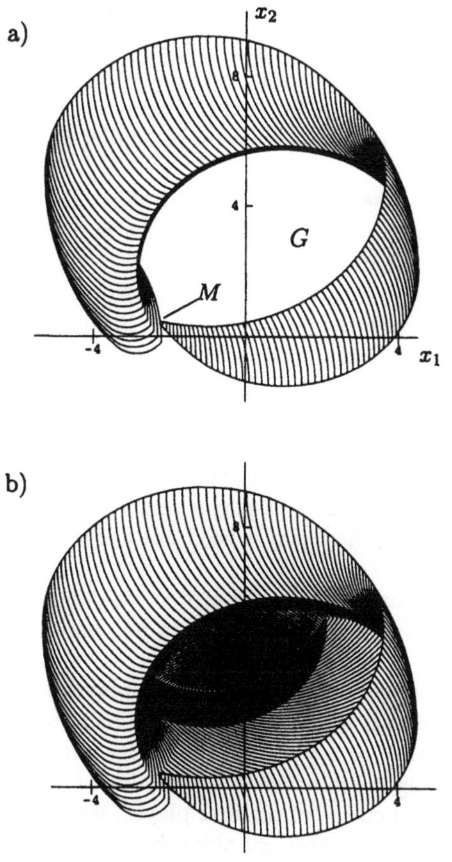

Figure 4. Level sets of value function in problem (5). Filling the gap G.

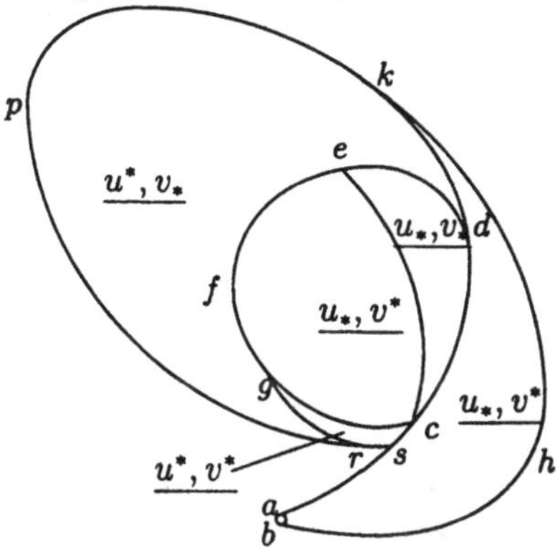

Figure 6. Structure of first and second player's optimal controls.

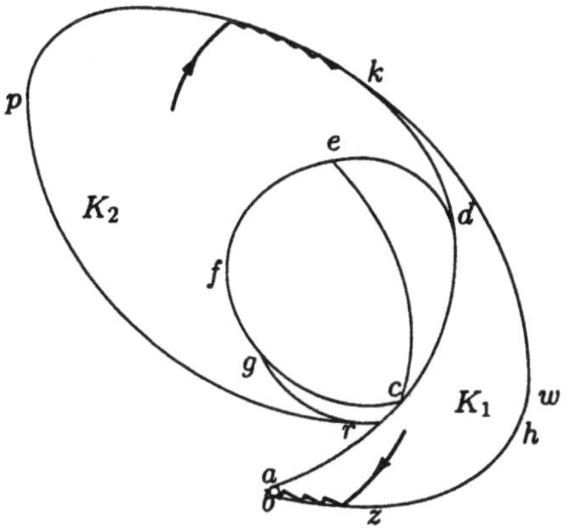

Figure 7. Sliding the motions along barrier lines.

guarantee the transferring to M from initial points lying outside this set within a finite time.

In [17], a number of numerical examples computed for the minimum-time second-order problem with different eigenvalues of the matrix A are given.

Optimal feedback control and optimal trajectories

The problem of finding optimal controls is rather independent task. Let us demonstrate the possibility of constructing optimal controls in the example (6).

The singular lines for the game (6) are depicted in Figure 6.

The barrier line $acdef$ terminates at the point f. After that, it is continued by the equivocal line fg which splits into the switch line gc of the first player and the switch line gr of the second player at the point g. The curve $bhkprs$ is the barrier, the curve dk is the equivocal line, and the curve ec is the switch line of the second player. The singular lines listed above divide the set where the problem has a solution into subsets so that the optimal controls of the players take constant values in the interiors of the subsets. These constant values are equal to the minimal and maximal admissible values of the control parameters: $u_* = -2$, $u^* = 1.5$, $v_* = -6.1$, $v^* = -4$. On the boundaries and near the boundaries of the subsets, the optimal controls are defined in a special manner.

The feedback control we specified is optimal in the interiors of the subsets (Figure 6). If the boundary of some subset contains a part of a barrier line, it can require the first and the second player to control in a special manner near this line to provide the time of termination to be close to the value of the game. The value function is discontinuous on barriers. The side of a barrier faced to the set where the value function is smaller is said to be positive. The opposite side is called negative.

Let us consider, for example, the set bounded by the curves $ascdk$ and $abhk$. Denote it by K_1 (Figure 7). In the interior of K_1, the optimal controls are $u = u_*$, $v = v^*$. The boundary of K_1 involves the arcs acd and bhk of the barriers.

The barrier acd is the trajectory of the system with $u = u_*$, $v = v_*$. The control u_* of the first player prevents trajectories from penetrating from the positive side to the negative side for all admissible controls of the second player. The control v_* of the second player prevents trajectories from penetrating from the negative side to the positive one for all admissible controls of the first player. According to the terminology of R.Isaacs [1], the controls $u = u_*$, $v = v_*$ ensure the property of semipermeability. Since the protective control of the first player coincides with the control value which is optimal in the interior of K_1, the trajectory generated by the optimal control of the first player can not come to the arc acd from the interior of K_1. So, in this case, the first player need not operate in a special manner near this arc.

The arc bhk is divided by the point w into two parts so that the protective control is $u = u^*$ for the arc bw and $u = u_*$ for wk. Near wk, the first player does not need any special way of control.

If a trajectory comes to the curve bw (or it lies very close to bw), the first player must change his control $u = u_*$ for $u = u^*$. After the changing, the trajectory can go back to the interior of K_1, then the first player switches his control to $u = u_*$, the trajectory comes to the curve bw again, and so on. As a result, a sliding mode which is schematically shown in Figure 7 occurs. Decreasing the step of the control choice of the first player, we obtain a trajectory which slides along the curve bw. Limit trajectories go along bw towards the point b. The "slowest" trajectory among them delivers the value of the game.

The arc bw is divided by the point z into two parts. For the part bz, the slowest trajectory goes with $v = v_*$, whereas the control u depends on the current point on this curve and takes values from the interval (u_*, u^*). The motion along the arc zw is feasible for $v = v^*$ only. This value is protective for the second player. In this case, the optimal limit trajectory is obtained with $u = u^*$, $v = v^*$.

Similarly, one can consider the behavior of the first player near the barrier $kprs$ which lies on the boundary of the set K_2 contoured by the curve $kprgfedk$. Here we also have a part of the barrier which lie near the point k and for which the protective control of the first player does not coincide with the optimal control $u = u^*$ inside K_2. After arriving at this part, the first player must change his control for $u = u_*$. As a result, we obtain a sliding motion near the barrier towards the point k. The optimal limit trajectory goes exactly along the barrier. For the barrier fed, the protective control of the first player is $u = u^*$. So, the first player need not handle in a special manner near fed.

Since the second player must prevent trajectories from penetrating from the negative side of a barrier to the positive one, it can require from the second player a special way to form his control when trajectories approach the negative side of the barrier.

In Figure 8, optimal limit trajectories emanating from the initial point $x_0 \in K_3$ are shown. At the initial stage, the optimal trajectory goes with the controls u_*, v^* and it is unique. After approaching the equivocal line fg, the optimal trajectory splits into two trajectories: one goes along the equivocal line fg, the other goes to the region K_2 where $u = u^*$, $v = v_*$ are optimal. The equivocal line ends at the point f, so, there exists a trajectory going along this line up to the point f and leaving this line at the point f. Three trajectories leaving fg are shown in Figure 8. Each of them approaches the barrier pk, then goes along pk and arrives at the point k. After that, they go along the equivocal line kd, and they can bifurcate at each point of kd. After the bifurcation, one of the trajectories continues to go along the equivocal line, the other leaves this line. In Figure 8, two trajectories leaving kd are shown. Moving in the interior of K_1, where $u = u_*$, $v = v^*$ are optimal, these trajectories approach the barrier bw and go along bw.

All trajectories described are optimal. The dark region in Figure 9 is filled with the optimal trajectories starting from the point x_0. The splitting occurs on the equivocal lines fg and kd.

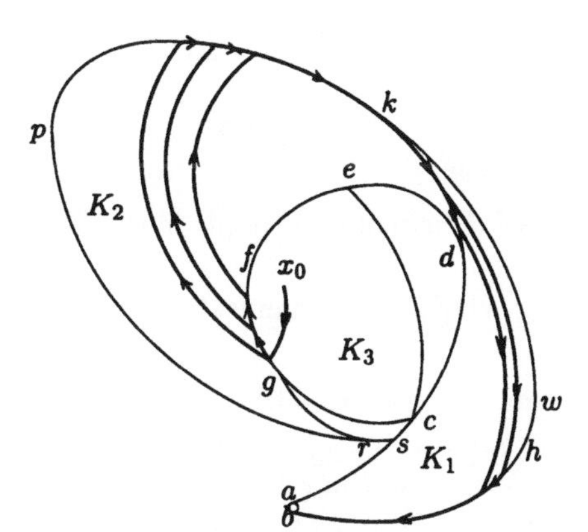

Figure 8. Splitting the optimal trajectories on equivocal lines.

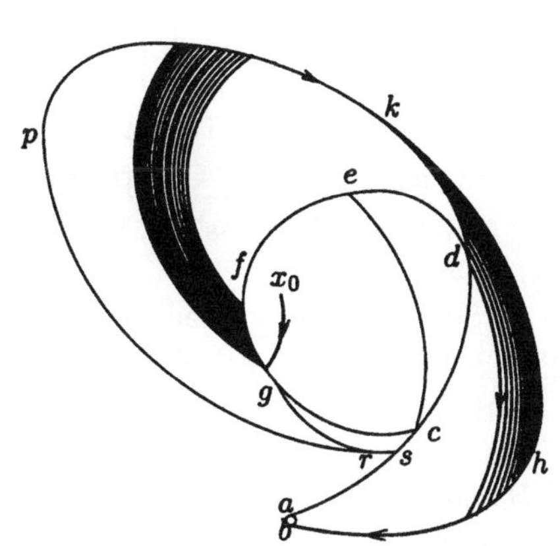

Figure 9. The set filling by optimal trajectories from $x_0 \in K_3$.

Summarizing, the principle of the construction of optimal trajectories is the following. When constructing optimal trajectories outside singular lines, we take into account that optimal controls are determined from the relations (3) with $\ell_{[ab]}$ replaced by the normal to the front at the current point. When specifying trajectories which go along singular lines, we take into account the behavior of trajectories on each of singular lines.

Conclusion

Effective algorithms can be created for solving differential games in the plane. In this paper, a general scheme of the algorithm for linear time-optimal differential games is considered. Possibility of the classification of singular lines in the computation process is demonstrated. Examples of computing level sets of the value function are given. Also, examples that serve for the explanation of the behavior of optimal trajectories are presented.

References

1. R.Isaacs, Differential Games, John Wiley, New York (1965).

2. N.N.Krasovskii and A.I.Subbotin, Positional differential games, Nauka, Moscow (1974).

3. N.N.Krasovskii and A.I.Subbotin, Game-theoretical control problems, Springer-Verlag, New York (1988).

4. A.I.Subbotin and V.S.Patsko (Eds.), Algorithms and programs for solving linear differential games , Inst. of Math. and Mech., Ural Scientific Center, Sverdlovsk (1984).

5. M.A.Zarkh and V.S.Patsko, Engrg. Cybernetics, 26(4), 92–99 (1988).

6. A.M.Taras'yev, A.A.Uspenskii, and V.N.Ushakov, in: Control in Dynamic Systems, A.I.Subbotin and V.N.Ushakov, (Eds.), Ural Scientific Center, Sverdlovsk (1990) pp. 93–100.

7. N.D.Botkin and E.A.Ryazantseva, in: Proceedings of the Institute of Mathematics and Mechanics, 2, Ju.S.Osipov, (Ed.), Ekaterinburg (1992) pp. 128–134.

8. M.A.Zarkh and A.G.Ivanov, in: Proceedings of the Institute of Mathematics and Mechanics, 2, Ju.S.Osipov, (Ed.), Ekaterinburg (1992) pp. 140–155.

9. V.S.Patsko, N.D.Botkin, V.M.Kein, V.L.Turova, and M.A.Zarkh, JOTA, 83(2), 237–267 (1994).

10. B.N.Sokolov and V.L.Turova, Mechanics of Solids, 23(5), 13–21 (1988).

11. V.N.Ushakov, Engrg. Cybernetics, 18(4), 16–23 (1981).

12. V.A.Ivanov, A.M.Taras'yev, V.N.Ushakov, and A.P.Khripunov, J. Appl. Math. Mech., 57(3), 419–425 (1993).

13. A.I.Subbotin, Generalized solutions of first-order PDEs: the dynamical optimization perspective, Birkhäuser, Boston (1995).

14. M.Bardi and M.Falcone, SIAM J. Control Optim., 28, 950–965 (1990).

15. A.A.Melikyan, in: Lecture Notes in Control and Inform. Sciences, 156, R.P.Hamalainen and H.K.Ehtamo, (Eds.), Springer-Verlag, Berlin (1991) pp. 81–90.

16. L.S.Pontryagin, V.G.Boltjanskii, R.V.Gamkrelidze, and E.F.Mischenko, The Mathematical Theory of Optimal Processes , Interscience, New York (1962).

17. V.S.Patsko and V.L.Turova, Numerical solution of two-dimensional differential games, Preprint, Inst. of Math. and Mech., Ural Branch of Russian Academy of Sciences, Ekaterinburg (1995).

7th Int. Coll. on Differential Equations, pp. 339-346
D. Bainov (Ed)

NONLINEAR THERMOELASTICITY AND SYMMETRY

REINHARD RACKE

Faculty of Mathematics, University of Constance
P.O.Box 5560, 78434 Konstanz, Germany

Dedicated to Prof. Dr. Dr.h.c Rolf Leis on the occasion of his 65th birthday in 1996

ABSTRACT

We report on initial boundary value problems arising in nonlinear thermoelasticity. For these hyperbolic-parabolic coupled systems, the situation for spatially one-dimensional systems is well understood, i.e. a lot is known about the existence and uniqueness of solutions. We shall summarize some results on existence and on the asymptotic behavior of solutions.

In three or two space dimensions the picture is far from being complete. Besides local well-posedness results, there are only few results on global solutions to the nonlinear system. Also the asymptotic behavior is not yet completely understood, even for the linearized system. If there is symmetry present, as for example in the radially symmetrical case, positive answers to the existence problem for small data and to the question of asymptotic behavior can be given. We report on results for homogenous, initially isotropic thermoelasticity, where the rotation of the displacement vanishes, as it is the case under radial symmetry. Then a global existence theorem for small data can be proved and the asymptotic behavior as time tends to infinity can be described.

KEYWORDS

Global solution, asymptotic behavior, radial symmetry

1 INTRODUCTION

The equations of thermoelasticity describe elastic and heat-conducting media, given in terms of a hyperbolic-parabolic initial boundary value problem in one, two or three space dimensions. One of the central questions is to find out whether the hyperbolic part arising from elasticity or the parabolic part arising from heat conduction predominates. It turns out that this depends on the space dimension, a fact that can also be regarded as a consequence of symmetry available in one dimension.

The situation for spatially one-dimensional models is more or less well understood. In Section 2 we shall summarize results on the existence of smooth, small solutions to the nonlinear systems as well as on the asymptotic behavior of solutions as time goes to infinity.

In three as well as in two space dimensions the picture is much more complicated and only few results are known on the asymptotic behavior of solutions to the linearized systems in terms of decay rates as time tends to infinity, or the existence of global small solutions resp. blow-up results for the nonlinear system; only the local well-posedness is well understood in most general cases, see [1–18]. Even for the linearized equations, the knowledge on precise decay rates is restricted to symmetrical situations like all of $\mathbb{R}^3$ or radially symmetrical situations as we shall see in Section 3, where recent results on global existence for small data and on the asymptotic behavior of solutions for certain symmetrical situations will be presented.

2 ONE-DIMENSIONAL MODELS

For spatially one-dimensional models, there are many papers on the linearized equations and on the asymptotic behavior of the correponding solutions, also on the global existence of small smooth solutions to the nonlinear system together with the description of the asymptotic behavior of solutions, and on possible blow-up phenomena for large data, respectively. This holds both for bounded and for unbounded domains and for different boundary conditions, see [2,3,11,15,19–45].

Consider for example the following nonlinear model, which assumes a homogeneous medium for simplicity.

The reference configuration is represented by the interval $\Omega := (0,\ell)$, ℓ being a fixed positive number. The equations for the displacement $u = u(t,x)$ from the reference configuration and the temperature difference $\theta = \theta(t,x) = T_a(t,x) - \tau_0$, where T_a is the absolute temperature and τ_0 is the constant reference temperature, read as follows:

$$u_{tt} - S(u_x,\theta)_x = f_1 \ in\ [0,\infty)\times\Omega, \tag{2.1}$$

$$(\theta+\tau_0)N(u_x,\theta)_t - Q(u_x,\theta_x,\theta)_x = f_2 \ in\ [0,\infty)\times\Omega. \tag{2.2}$$

Here S is the Piola–Kirchhoff stress tensor, N is the specific entropy and Q is the heat flux. S, N and Q are assumed to be real-valued functions satisfying

$$\frac{\partial S}{\partial u_x}(0,0) = \alpha > 0, \quad \frac{\partial N}{\partial \theta}(0,0) = \delta > 0, \quad -\frac{\partial S}{\partial \theta}(0,0) = \frac{\partial N}{\partial u_x}(0,0) = \gamma \neq 0,$$

$$\frac{1}{\tau_0}\frac{\partial Q}{\partial \theta_x}(0,0,0) = \beta > 0, \quad \frac{\partial Q}{\partial u_x}(0,0,0) = \frac{\partial Q}{\partial \theta}(0,0,0) = 0.$$

Initial resp. boundary conditions are given by

$$u(0,x) = u_0(x), \quad u_t(0,x) = u_1(x), \quad \theta(0,x) = \theta_0(x) \ in\ \Omega, \tag{2.3}$$

$$u(t,0) = u(t,1) = \theta(t,0) = \theta(t,1) = 0 \quad \text{for} \quad t \geq 0 \tag{2.4}$$

([rigidly clamped, constant temperature] at the boundary).

Let

$$u_2 := u_{tt}(t=0), \quad \theta_1 := \theta_t(t=0)$$

be given formally through the differential equation, explicitly in terms of the initial data u_0, u_1, θ_0. Assuming $Q = Q(\theta_x)$ for simplicity, we have the following global existence result, which demonstrates what kind of solutions are discussed in this framework, and what is the typical asymptotic behavior as time tends to infinity.

Theorem 2.1 *Assume* $u_i \in W^{3-i,2}$, $i = 0,1,2$, $\theta_j \in W^{3-j,2}$, $j = 0,1$ *and* $|\theta_0(x)| < \tau_0$ *in* $\overline{\Omega}$, $f_1, f_2 \in C^2([0,\infty), L^2) \cap C^1([0,\infty), W^{1,2})$. *Let* $\lambda(t) := \|\overline{D}^1(f_1,f_2)(t,\cdot)\|_{L^2}^2 + \|\partial_t^2(f_1,f_2)(t,\cdot)\|_{L^2}^2 + \|\partial_t\partial_x(f_1,f_2)(t,\cdot)\|_{L^2}^2$. *Then there exists* $\delta > 0$ *such that if*

$$\|(u_0,u_1,\theta_0,\theta_1)\|_{W^{2,2}}^2 + \|u_2\|_{W^{1,2}}^2 + \sup_{t\geq 0}\lambda(t) \leq \delta$$

then (2.1)–(2.4) *has a unique global solution*

$$u \in \bigcap_{j=0}^{3} C^j([0,\infty), W^{3-j,2}), \quad \theta \in \bigcap_{j=0}^{1} C^j([0,\infty), W^{3-j,2}).$$

Moreover, there are constants $c_1, c_2 > 0$ *such that for* $t \geq 0$

$$\|\overline{D}^3 u(t)\|_{L^2}^2 + \|\overline{D}^2\theta(t)\|_{L^2}^2 + \|\theta_{txx}(t)\|_{L^2}^2 + \|\theta_{xxx}(t)\|_{L^2}^2$$

$$\leq c_1 e^{-c_2 t}\Big(\|(u_0,u_1,\theta_0,\theta_1)\|_{W^{2,2}}^2 + \|u_2\|_{W^{1,2}}^2 + \int_0^t e^{c_2 r}\lambda(r)dr\Big).$$

Here, $\overline{D}^m$ denotes all derivatives up to order m, and the spaces $W^{m,p} = W^{m,p}(\Omega)$ denote the usual Sobolev spaces, $m \in \mathbb{N}_0$, $1 \leq p \leq \infty$.

The proof of Theorem 2.1 from our work with Shibata and Zheng [41] derives appropriate a priori estimates to extend a locally (in time) existing solution to a global one. The necessary estimates involve sophisticated estimates of boundary terms arising through partial integrations within the classical energy method; they go back to and extend the linear case studied by Muñoz Rivera [35].

Exploiting the exponential stability, we obtain the existence of global, small periodic solutions:

Corollary 2.2 *Suppose that* f_1 *and* f_2 *are periodic with respect to* t *with period* $\omega > 0$ *arbitrary:*

$$f_k(t+\omega,\cdot) = f_k(t,\cdot), \quad k = 1,2, \quad \textit{for all } t.$$

Then there exists a small constant $\varepsilon_1 > 0$ *such that if*

$$\sup_{t\geq 0}\lambda(t) < \varepsilon_1,$$

then the problem (2.1)–(2.3) *admits a unique solution*

$$u \in \bigcap_{j=0}^{3} C^j([0,\infty), W^{3-j,2}), \quad \theta \in \bigcap_{j=0}^{1} C^j([0,\infty), W^{3-j,2}),$$

$$\theta \in C^2([0,\infty), L^2),$$

which is periodic with period ω.

The investigations of nonlinear systems started with the paper of Slemrod [43], also for a bounded interval, but for mixed Dirichlet-Neumann type boundary conditions. In the meantime, many papers appeared (see the references mentioned above), which deal with

- other reasonable boundary conditions like thermally insulated for the heat conduction part for example,
- unbounded domains like all of $\mathbb{R}$ or the half-line,
- time-independent forces,
- inhomogeneous media,
- blow-up results for large data,
- propagation of singularities.

We remark that the assumption of a purely longitudinal motion can be interpreted as a symmetry assumption. The medium is isotropic and all vector fields are gradient fields. This will also be the case in the next section, where the assumption that the rotation of the displacement vanishes identically will lead to a global existence theorem.

3 TWO- OR THREE-DIMENSIONAL MODELS

In three space dimensions, local well-posedness in rather general anisotropc, inhomogeneous media is given, but the global existence of small solutions or the description of the asymptotic behavior of solutions even to the linearized system is known only in some special cases. The Cauchy problem, where the medium is initially isotropic and fills all of $\mathbb{R}^3$, is one example. Here the symmetry of $\mathbb{R}^3$ is important, which allows the classical decomposition of vector fields into divergence- and curl-free components, respectively. Another example will be described below, where the rotation curl u, u being the displacement vector, vanishes identically, e.g. in radially symmetrical situations.

In order to demonstrate the complexity of the possible behavior of solutions, we remind the following facts about the linearized systems in a homogenous, isotropic situation under Dirichlet type boundary conditions in a bounded domain in $\mathbb{R}^3$:

In a cube, the solution tends to zero, but the rate of decay may be arbitrarily slow. In a ball, there are oscillating solutions that do not decay at all, but if the data are radially symmetrical, then the solution tends to zero exponentially. Of course, the geometry of the domain plays an important rôle.

In the sequel we shall report on recent results for symmetrical situations where the rotation of the displacement vector vanishes identically; the results are taken from our joint work with Jiang and Muñoz Rivera [9].

We consider the equations of linear thermoelasticity for a homogeneous, isotropic medium with bounded reference configuration $\Omega \subset \mathbb{R}^n$, $n = 2$ or 3, having a smooth boundary $\partial\Omega$:

$$u_{tt} - \mu\Delta u - (\mu+\lambda)\nabla \operatorname{div} u + \beta\nabla\theta = 0, \quad (3.1)$$

$$c\theta_t - \kappa\Delta\theta + \beta \operatorname{div} u_t = 0, \quad (3.2)$$

where $u(x,t) = (u_1, \cdots, u_n)$, $x \in \Omega$, $t \geq 0$, is the displacement, $\theta = \theta(x,t)$ is the temperature deviation, μ and λ are the Lamé moduli satisfying $\mu > 0$ and $2\mu + n\lambda > 0$, $c, \kappa > 0$ and $\beta \neq 0$ are given constants depending on the material properties.
Additionally, one has initial conditions

$$u(x,0) = u^0(x), \quad u_t(x,0) = u^1(x), \quad \theta(x,0) = \theta^0(x), \quad x \in \Omega, \tag{3.3}$$

and Dirichlet boundary conditions

$$u(\cdot,t)_{|\partial\Omega} = 0, \quad \theta(\cdot,t)_{|\partial\Omega} = 0, \qquad t \geq 0. \tag{3.4}$$

Using the energy method, one can prove

Theorem 3.3 *Let (u,θ) be the solution to (3.1)–(3.4) and assume*

$$\operatorname{rot} u = 0 \ \text{in} \ \Omega \times [0,\infty).$$

Then there are constants $\Gamma \geq 1$ and $\gamma > 0$ independent of the initial data and of t such that

$$E(t) + \int_0^t e^{\gamma s} \|\nabla\theta_t(s)\|_{L^2}^2 ds \leq \Gamma E(0),$$

where

$$E(t) := e^{\gamma t} \left\{ \sum_{k=0}^{2} \|\partial_t^k u(t)\|_{W^{2-k,2}}^2 + \|\theta_t(t)\|_{L^2}^2 + \|\theta(t)\|_{W^{2,2}}^2 \right\}.$$

If u is a radially symmetrical vector field, then there exists a function $\phi : \mathbb{R}_0^+ \longrightarrow \mathbb{R}$ with $u(x) = x\phi(r)$, hence the rotation of u vanishes. Therefore, we immeditately have

Corollary 3.4 *Theorem 3.1 applies to radially symmetrical situations.*

Finally, we also obtain a global existence result for small data for the nonlinear system, yielding the exponential decay of solutions; this also applies to radially symmetrical situations.

Without loss of generality, we restrict ourselves to three space dimensions. Then the equations of nonlinear thermoelasticity for a homogeneous medium with unit reference density in $\Omega \subset \mathbb{R}^3$ read

$$\frac{\partial^2 u_i}{\partial t^2} = C_{i\alpha j\beta}(\nabla u, \theta)\frac{\partial^2 u_j}{\partial x_\alpha \partial x_\beta} + \tilde{C}_{i\alpha}(\nabla u, \theta)\frac{\partial \theta}{\partial x_\alpha}, \qquad i = 1,2,3, \tag{3.5}$$

$$a(\nabla u, \theta)\theta_t = \frac{1}{b(\theta)} \operatorname{div} q(\nabla u, \theta, \nabla\theta) + \tilde{C}_{i\alpha}(\nabla u, \theta)\frac{\partial^2 u_i}{\partial x_\alpha \partial t}. \tag{3.6}$$

Here (and throughout this section) the Einstein summation convention is used; $u = (u_1, u_2, u_3)^T$, $q = (q_1, q_2, q_3)^T$, $a \geq a_0 > 0$, b is a C^∞-function such that $b(\theta) = \theta + \tau_0$ for $|\theta| \leq \tau_0/2$ and $0 < b_1 \leq b(\theta) \leq b_2 < \infty$, b_1, b_2 constants, $-\infty < \theta < \infty$, $\tau_0 > 0$ the reference temperature. (3.1), (3.2) are derived for small values of $|\theta|$, i.e. for $|\theta| \leq \tau_0/2$, which is a posteriori justified by the smallness of the solutions obtained later.

We consider (3.3), (3.4) as the initial and boundary conditions for (3.5), (3.6).

We assume that the medium is initially isotropic:

$$\begin{aligned} C_{i\alpha j\beta}(0,0) &= \lambda\delta_{i\alpha}\delta_{j\beta}+\mu(\delta_{ij}\delta_{\alpha\beta}+\delta_{\alpha j}\delta_{i\beta}),\quad \tilde{C}_{i\alpha}(0,0)=-\beta\delta_{i\alpha},\quad \frac{\partial q_i(0,0,0)}{\partial(\partial\theta/\partial x_j)}=k\delta_{ij},\\ a(0,0) &= c, \end{aligned}$$

and that

$$C_{i\alpha j\beta}(\nabla u,\theta)=C_{j\beta i\alpha}(\nabla u,\theta),\quad \frac{\partial q_i(\nabla u,\theta,\nabla\theta)}{\partial(\partial\theta/\partial x_j)}=\frac{\partial q_j(\nabla u,\theta,\nabla\theta)}{\partial(\partial\theta/\partial x_i)},$$

$$\frac{\partial q_i(0,0,0)}{\partial\theta}=0,\quad \frac{\partial q_i(0,0,0)}{\partial(\partial u_\alpha/\partial x_\beta)}=0,\quad 1\le i,j,\alpha,\beta\le 3.$$

Using (3.3), we can write (3.1), (3.2) in the form:

$$u_{tt}-\mu\Delta u-(\mu+\lambda)\nabla\operatorname{div}u+\beta\nabla\theta = f(\nabla u,\theta,\nabla^2u,\nabla\theta), \tag{3.7}$$

$$c\theta_t-\kappa\Delta\theta+\beta\operatorname{div}u_t = g(\nabla u,\theta,\nabla^2u,\nabla^2\theta,\nabla u_t), \tag{3.8}$$

where $\kappa := k/\tau_0>0$, $f=(f_1,f_2,f_3)^T$, and

$$f_i := (C_{i\alpha j\beta}(\nabla u,\theta)-C_{i\alpha j\beta}(0,0))\frac{\partial^2u_j}{\partial x_\alpha\partial x_\beta}+\left(\tilde{C}_{i\alpha}(\nabla u,\theta)-\tilde{C}_{i\alpha}(0,0)\right)\frac{\partial\theta}{\partial x_\alpha},$$

$$i=1,2,3,$$

$$\begin{aligned} g &:= c\left(\frac{1}{a(\nabla u,\theta)b(\theta)}\frac{\partial q_i(\nabla u,\theta,\nabla\theta)}{\partial(\partial\theta/\partial x_j)}-\frac{1}{a(0,0)b(0)}\frac{\partial q_i(0,0,0)}{\partial(\partial\theta/\partial x_j)}\right)\frac{\partial^2\theta}{\partial x_i\partial x_j}\\ &+c\left(\frac{\tilde{C}_{i\alpha}(\nabla u,\theta)}{a(\nabla u,\theta)}-\frac{\tilde{C}_{i\alpha}(0,0)}{a(0,0)}\right)\frac{\partial^2u_i}{\partial x_\alpha\partial t}+\frac{c}{a(\nabla u,\theta)b(\theta)}\frac{\partial q_i(\nabla u,\theta,\nabla\theta)}{\partial\theta}\frac{\partial\theta}{\partial x_i}\\ &+\frac{c}{a(\nabla u,\theta)b(\theta)}\frac{\partial q_i(\nabla u,\theta,\nabla\theta)}{\partial(\partial u_\alpha/\partial x_\beta)}\frac{\partial^2u_\alpha}{\partial x_i\partial x_\beta}. \end{aligned}$$

Assuming in addition that for $(\tilde{u},\tilde{\theta})$ with $\operatorname{rot}\tilde{u}=0$

$$C_{i\alpha j\beta}(\nabla\tilde{u},\tilde{\theta})\frac{\partial^2\tilde{u}_j}{\partial x_\alpha\partial x_\beta}\equiv A_{ij}(\nabla\tilde{u},\tilde{\theta})\Delta\tilde{u}_j,\quad i=1,2,3,$$

we obtain, using the energy method again,

Theorem 3.5 *Let (u,θ) be the local solution of (3.7), (3.8), (3.3), (3.4) in some time interval $[0,T]$. Assume that $\operatorname{rot}u=0$ for $(x,t)\in\Omega\times(0,T)$. Then there is a constant $\varepsilon>0$ such that if*

$$\sum_{j=0}^{4}\|u_j\|^2_{W^{4-j,2}}+\sum_{j=0}^{2}\|\theta_j\|^2_{W^{4-j,2}}+\|\theta_3\|^2_{L^2}\le\varepsilon^2,$$

then $T=\infty$. Moreover, $\|u(t)\|_{W^{4,2}}$, $\|\theta(t)\|_{W^{4,2}}$ decay to zero exponentially as $t\to\infty$.

In particular, this applies to radially symmetrical situations.

We give an example for nonlinearities in the radially symmetrical case. Let

$$C_{i\alpha j\beta}(\nabla u,\theta):=\tilde{\lambda}(\operatorname{div}u,\theta)\delta_{i\alpha}\delta_{j\beta}+(\delta_{ij}\delta_{\alpha\beta}+\delta_{\alpha j}\delta_{i\beta})\tilde{\mu},\quad \tilde{C}_{i\alpha}(\nabla u,\theta):=-\tilde{\beta}(\operatorname{div}u,\theta)\delta_{i\alpha},$$

$$q(\nabla u,\theta,\nabla\theta):=\tilde{\kappa}(\theta)\nabla\theta,\quad a(\nabla u,\theta):=\tilde{a}(\operatorname{div}u,\theta),$$

where $\tilde{\lambda}$, $\tilde{\beta}$, $\tilde{\kappa}$ and $\tilde{a}$ are smooth functions of $\operatorname{div} u$ and θ, $\tilde{\mu}$ is a positive constant. Let $\tilde{\lambda}$, $\tilde{\mu}$, $\tilde{\beta}$, $\tilde{\kappa}$, $\tilde{a}$ satisfy

$$3\tilde{\lambda}(0,0) + 2\tilde{\mu} > 0, \quad \tilde{\beta}(0,0) \neq 0, \quad \tilde{\kappa}(0) > 0, \quad \tilde{a}(0,0) > 0,$$
$$-\tilde{\beta}_w(w,\theta) = \tilde{\lambda}_\theta(w,\theta), \quad -\tilde{\beta}_\theta(w,\theta) = \tilde{a}_w(w,\theta), \qquad w,\ \theta \in \mathbb{R}.$$

These conditions also guarantee the existence of a Helmholtz potential $\Psi = \Psi(\nabla u, \theta)$ (cf. [39]) satisfying

$$C_{i\alpha j\beta} = \frac{\partial^2 \Psi}{\partial(\partial u_i/\partial x_\alpha)\partial(\partial u_j/\partial x_\beta)}, \quad \tilde{C}_{i\alpha} = \frac{\partial^2 \Psi}{\partial(\partial u_i/\partial x_\alpha)\partial\theta}, \quad a = -\frac{\partial^2 \Psi}{\partial \theta^2}.$$

ACKNOWLEDGEMENT

The author gratefully acknowledges partial support by the Deutsche Forschungsgemeinschaft and by the organizers of the Colloquium, where this lecture has been given.

REFERENCES

1. A. Chrzęszczyk, A., Arch. Mech.,39, 605–617 (1987).
2. C.M. Dafermos, Arch. Rational Mech. Anal., 29, 241–271 (1968).
3. C.M. Dafermos, Lec. Notes Math., 503, 295–306 (1976).
4. W. Dan, Math. Meth. Appl. Sci., 18, 1053–1082 (1995).
5. G. Dassios, and M. Grillakis, Arch. Rational Mech. Anal., 87, 49–91 (1984).
6. J. Gawinecki, Ann. Pol. Math., 2, 135–145 (1991).
7. D.B. Henry, Publ. Secc. Mat., Univ. Autòn. Barc., 31, 29–84 (1987).
8. S. Jiang, SFB 256 Preprint, University of Bonn, 351 (1994).
9. S. Jiang, J.E. Muñoz Rivera and R. Racke, Quart. Appl. Math., (to appear).
10. S. Jiang, S., R. Racke, Math. Meth. Appl. Sci., 12, 315–339 (1990).
11. R. Leis, Math. Meth. Appl. Sci., 2, 379–396 (1980).
12. R. Leis, Math. Meth. Appl. Sci., 3, 312–317 (1981).
13. T. Mukoyama, Tsukuba J. Math., 13, 363–386 (1989).
14. G. Ponce and R. Racke, J. Differential Equations 87, 70–83 (1990).
15. R. Racke, Proc. Roy. Soc. Edinburgh, 107A, 289–298 (1987).
16. R. Racke, Math. Z., 203, 649–682 (1990).
17. R. Racke, Math. Meth. Appl. Sci., 12, 267–273 (1990).
18. R. Racke, Asymptotic Analysis, 3, 105–132 (1990).
19. J.A. Burns, Z. Liu and S. Zheng, J. Math. Anal. Appl., 179, 574–591 (1993).
20. C.M. Dafermos and L. Hsiao, Quart. Appl. Math., 44, 463–474 (1986).

21. E. Feireisl, Comment. Math. Univ. Carolinae, 31, 243–255 (1990).

22. S.W. Hansen, J. Math. Anal. Appl., 167 429–442 (1992).

23. D.B. Henry, A. Perissinitto Jr. and O. Lopes, Nonlinear Analysis, T.M.A., 21, 65–75 (1993).

24. W.J. Hrusa and S.A. Messaoudi, Arch. Rational Mech. Anal., 111, 135–151 (1990).

25. W.J. Hrusa and M.A. Tarabek, Quart. Appl. Math., 47, 631–644 (1989).

26. S. Jiang, Proc. Roy. Soc. Edinburgh, 115A, 257–274 (1990).

27. S. Jiang, SFB 256 Preprint, University of Bonn, 138, (1990).

28. S. Jiang, Nonlinear Analysis, T.M.A., 19, 107–121 (1992).

29. S. Jiang, Nonlinear Analysis, T.M.A., 20, 1245–1256 (1993).

30. S. Kawashima, Systems of a hyperbolic-parabolic composite type, with applications to the equations of magnetohydrodynamics. Thesis, Kyoto University (1983).

31. S. Kawashima and M. Okada, Proc. Jap. Acad., Ser. A, 53, 384–387 (1982).

32. S. Kawashima and Y. Shibata, Czech. Math. J., 45, 39–67 (1995).

33. J.U. Kim, SIAM J. Math. Anal., 23, 889–899 (1992).

34. Z. Liu and S. Zheng, Quart. Appl. Math., 51, 535–545 (1993).

35. J.E. Muñoz Rivera, Funkcial. Ekvac., 35, 19–30 (1992).

36. J.E. Muñoz Rivera, SIAM J. Math. Anal. 24, 1–17 (1993).

37. J.E. Muñoz Rivera, Applic. Anal., 53, 55–65 (1994).

38. J.E. Muñoz Rivera and R. Racke, SIAM J. Math. Anal., 26, 1547–1563 (1995).

39. R. Racke, Lectures on nonlinear evolution equations. Initial value problems. Aspects of Mathematics E19. Friedr. Vieweg & Sohn, Braunschweig/Wiesbaden (1992).

40. R. Racke and Y. Shibata, Arch. Rational Mech. Anal. 116, 1–34 (1991).

41. R. Racke, Y. Shibata and S. Zheng, Quart. Appl. Math., 51, 751–763 (1993).

42. Y. Shibata, Banach Center Publications, 27, 457–480 (1992).

43. M. Slemrod, Arch. Rational Mech. Anal., 76, 97–133 (1981).

44. S. Zheng, Sci. Sinica, Ser. A, 27, 1274–1286 (1984).

45. S. Zheng, Sci. Sinica, Ser. A, 30, 1133–1149 (1987).

7th Int. Coll. on Differential Equations, pp. 347-354
D. Bainov (Ed)

Saddle point Methods in Spatial Models of Biological Systems.

John Radcliffe and Linda Rass.

School of Mathematical Sciences,

Queen Mary and Westfield College,

Mile End Road, London E1 4NS, England.

Abstract. A saddle point method is described which can be used to obtain the speed of propagation in spatial multitype biological models. The technique is applicable to a wide range of models both in discrete and continuous time.

We show how it can be used to find the speed of first spread of infection in epidemic models, of new alleles in genetics models, and new strategies in game theoretic models. It can also be used to obtain the speed of the forward tail of the distribution of furthest spread in branching process models.

Key Words: Spatial Spread, Saddle-point Method, Epidemics, Branching Processes, Genetics, Evolutionary Games.

1. INTRODUCTION.

There are many models describing spatial spread in mathematical biology where the spatial aspect is incorporated into the model by means of contact distributions. The particular areas which concern us are multi-type epidemics, contact branching processes, genetics and evolutionary games.

A method of obtaining the speed of first spread in such systems is described. This we call the saddle point method. The saddle point method is applied to the approximate equations which are valid in the forward front of the model. It can be used in discrete and continuous time. It has been applied [1,2,3,4] to find the speed of first spread of infection in epidemics and contact branching processes; and of new genes and strategies in genetics and evolutionary game theory [5,6].

Exact results exist [7] for the speed of propagation of infection in n-type epidemics in which a return to the susceptible state is not allowed. A survey of these exact results is given in [8]. Exact results have also been proved for the spread of gene in a discrete time model with $n = 2$ genes [9,10,11,12,13,14]. The only exact results for more complex models in genetics are those of [15,16] for a sex linked gene. The genetic results for $n = 2$ have been extended [5] to cover the spread of a strategy in evolutionary game theory.

In all these models where we have an exact result, the answer obtained by the saddle point method agrees with this. However the saddle point method gives results for situations in which exact results are not available, e.g. multi-type epidemics in which a return to the susceptible state is allowed [3]. It can also be applied in much more general situations far less tractable to exact analysis, e.g. in genetics and evolutionary game theory models involving n alleles/genotypes or strategies.

The method was first developed in continuous time and applied to non-reducible n-type epidemics [1], then extended to cover reducible epidemics [2]. The saddle point method for discrete time systems is given in [6], where it is also applied to various multi-type models. These include models in which the migration is dependent on the genotype, and bimatrix games.

The purpose of this paper is to give a survey of the saddle point method. The saddle point method for continuous and discrete time is described briefly. For simplicity we restrict attention to spatial spread in R. We then illustrate how the method may be applied to a wide variety of situations. Many more variations and more complex models can be written down for which the linearised equations in the forward front fit into the format described in sections 2 or 5. The method described in this paper can be used to obtain the speed of first spread for such models.

2. THE SADDLE POINT METHOD FOR CONTINUOUS TIME MODELS.

The saddle point method is applied to a system of integro-differential equations. For epidemic models these describe the spatial spread of infection introduced into a heterogeneous population of susceptible individuals; each type of constant density in R. For contact branching process models they describe the spread of the tail of the distribution of furthest spread of types when a specific type at position zero initiates the process at time zero. We only consider non-reducible models.

In the forward front these equations may be approximated by a system of linear differential equations of the form

$$\frac{\partial y_i(s,t)}{\partial t} = \sum_{j=1}^{n} a_{ij} \int_R y_j(s-r,t) p_{ij}(r) dr - \mu_i y_i(s,t), \quad (i = 1, ..., n). \tag{2.1}$$

For epidemic models $y_i(s,t)$ measures the proportion of type i individuals at position s who are infectives at time t. For the contact branching process $y_i(s,t) = P(U(t) > s|$ one type i individual at position s at time 0). Here $U(t)$ is the position of furthest spread to the right of any type.

Laplace transform equations (2.1) with respect to s. Let $L_i(\lambda,t) = \int_{-\infty}^{\infty} e^{\lambda s} y_i(s,t) ds$, $\{\mathbf{L}(\lambda,t)\}_i = L_i(\lambda,t)$ and $P_{ij}(\lambda) = \int_{-\infty}^{\infty} e^{\lambda r} p_{ij}(r) dr$. Hence

$$\frac{\partial \mathbf{L}(\lambda,t)}{\partial t} = (\mathbf{A}(\lambda) - \mu \mathbf{I})\mathbf{L}(\theta,t),$$

where $\mu = \max(\mu_1, ..., \mu_n)$, $\{\mathbf{A}(\lambda)\}_{ij} = a_{ij} P_{ij}(\lambda) + \delta_{ij}(\mu - \mu_i)$, and δ_{ij} is the Kronecker delta.

Thus $\mathbf{L}(\lambda, t) = e^{(\mathbf{A}(\lambda)-\mu \mathbf{I})t}\mathbf{L}(\lambda, 0)$. The Laplace transforms are inverted using the expression $y_j(r,t) = \frac{1}{2\pi i}\int_{c(t)-i\infty}^{c(t)+i\infty} e^{-r\lambda}\{\mathbf{L}(\lambda,t)\}_j d\lambda$ for a suitable choice of c(t).

For a given small $\eta > 0$ we define s for given t by $\eta = \int_{r\geq s} y_j(r,t)dr$. For this value of s, $\lambda = c(t)$ is chosen to be the saddle point of $Re(g(\lambda))$, where $g(\lambda) = (\rho(\mathbf{A}(\lambda))t - \lambda s)$ and $\rho(\mathbf{A}(\lambda))$ is the Perron-Frobenius root of $\mathbf{A}(\lambda)$. The speed of spread of the forward front is then $\lim_{t\to\infty} s/t$.

When $\rho(\mathbf{A}(0)) > \mu$ the saddle-point method gives the speed of spread of the forward front for all types as

$$c = \max\left(0, \inf_{\lambda\in(0,\Delta_V)} f(\lambda)\right) \text{ where } f(\lambda) = \frac{\rho(\mathbf{A}(\lambda)) - \mu}{\lambda}. \tag{2.2}$$

Note that Δ_V is the minimum of the abscissae of convergence of the $P_{ij}(\lambda)$ in the positive half of the complex plane. When $\rho(\mathbf{A}(0)) \leq \mu$ it suggests the speed of spread is 0.

3. AN EPIDEMIC MODEL.

In this section we consider an $S \to I \to R$ epidemic. Consider n types of individuals, each type having uniform density in $\mathbf{R}$, and consisting of susceptible, infectious and removed individuals. Denote the proportions of susceptible, infectious and removed individuals of type i at position s and time t by $x_i(s,t)$, $y_i(s,t)$ and $z_i(s,t)$ respectively; so that $x_i(s,t) + y_i(s,t) + z_i(s,t) = 1$. The density of type i individuals is σ_i. Let λ_{ij} be the rate of infection of susceptible individuals of type i by infectious individuals of type j. The contact distribution representing the distance r over which infection occurs has density $p_{ij}(r)$. The removal rate for infectious individuals of type i is μ_i.

The epidemic occuring over all real t is described by the equations

$$\begin{aligned}
\frac{\partial x_i(s,t)}{\partial t} &= -x_i(s,t)\sum_{j=1}^{n}\lambda_{ij}\sigma_j\int_{-\infty}^{\infty} p_{ij}(r)y_j(s-r,t)dr,\\
\frac{\partial y_i(s,t)}{\partial t} &= x_i(s,t)\sum_{j=1}^{n}\lambda_{ij}\sigma_j\int_{-\infty}^{\infty} p_{ij}(r)y_j(s-r,t)dr - \mu_i y_i(s,t),\\
\frac{\partial z_i(s,t)}{\partial t} &= \mu_i y_i(s,t), \quad (i = 1,...,n).
\end{aligned} \tag{3.1}$$

In the extreme forward region, where there is little infection, we may approximate $x_i(s,t)$ by one, so that the $y_i(s,t)$ approximately satisfy the equations

$$\frac{\partial y_i(s,t)}{\partial t} = \sum_{j=1}^{n}\lambda_{ij}\sigma_j\int_{-\infty}^{\infty} p_{ij}(r)y_j(s-r,t)dr - \mu_i y_i(s,t), \quad (i = 1,...,n). \tag{3.2}$$

These are just equations (2.1) with $a_{ij} = \sigma_j\lambda_{ij}$. It follows that the speed of first spread of infection for all types is given by equation (2.2).

A similar result has been obtained for the $S \to I \to S$ epidemic [2], and the saddle point method has also been applied to more complex epidemic models [3].

4. THE CONTACT BIRTH-DEATH PROCESSES.

The saddle point method has been applied to various contact branching processes in both discrete and continuous time, [1,3,4]. In this section we consider a continuous time contact birth-death process.

An n-type contact birth-death process in R commences with a single type i individual at position $s = 0$ at time $t = 0$. The probability that a type i individual gives birth to an offspring in a time interval $(t, t+\delta t)$ is $\alpha_i \delta t + o(\delta t)$. The probability that this offspring is of type j is q_{ij}, where $\sum_j q_{ij} = 1$. Let $\lambda_{ij} = \alpha_i q_{ij}$. The position s of such a new individual relative to the parent has contact distribution $p_{ij}(s)$. The probability that an individual of type i dies in $(t, t+\delta t)$ is $\mu_i \delta t + o(\delta t)$.

Consider the spread to the right. Let U(t) be the position of the furthest individual from 0 at time t in the given direction, and let $y_i(s,t) = P[U(t) > s|$ one type i individual at position 0 at time $t = 0]$.

The usual conditional arguments lead to the following equations for $y_i(s,t)$:

$$\frac{\partial y_i(s,t)}{\partial t} = (1 - y_i(s,t)) \sum_{j=1}^{n} \lambda_{ij} \int_{-\infty}^{\infty} y_j(s-r,t) p_{ij}(r) dr - \mu_i y_i(s,t), \quad (i = 1, ..., n). \quad (4.1)$$

Note that the equations hold for all s, and that $y_i(s,0) = 1$ for $s < 0$ and $y_i(s,0) = 0$ for $s \geq 0$.

The approximate equations which hold in the forward tail are

$$\frac{\partial y_i(s,t)}{\partial t} = \sum_{j=1}^{n} \lambda_{ij} \int_{-\infty}^{\infty} p_{ij}(r) y_j(s-r,t) dr - \mu_i y_i(s,t), \quad (i = 1, ..., n). \quad (4.2)$$

These are just equations (2.1) with $a_{ij} = \lambda_{ij}$. It follows that the speed of first spread of infection for all types is given by equation (2.2).

5. THE SADDLE-POINT METHOD FOR DISCRETE TIME MODELS.

The saddle-point results for discrete time models were proved in [6]. In this section we describe the method briefly, so that it can be applied in sections 6,7 and 8 to certain spatial models in genetics and evolutionary games. The models considered in this paper involve certain types, and $x_i^{(m)}(s)$ denotes the proportion of type i in the mth generation at position s. The type may represent an individual playing a particular strategy, or in genetics models it represents a particular allele (genotype) when the migration density is independent of (dependent on) the genotype. We suppose that initially the population consists of certain types, $i = 1, ..., n_1$, e.g. individuals playing strategies in game theory or alleles (genotypes) in genetics. Other types $n_1 + 1, ..., n$ are introduced into a bounded region B and the speed of first spread of these types is considered. Far from the region B, the linearised equations in the models considered in this paper are of the form

$$x_i^{(m+1)}(\mathbf{s}) = \sum_{j=n_1+1}^{n} \gamma_{ij} \int_R p_{ij}(r) x_j^{(m)}(s-r)dr \tag{5.1}$$

for $i = n_1 + 1, ..., n$, where the γ_{ij} will depend upon the particular application.

Consider the speed of spread of the forward front for type $(n_1 + i)$. Take a fixed η, which is small and positive. Then define $s(m)$ so that $\int_{s(m)}^{\infty} x_{n_1+i}^{(m)}(u)du = \eta$. The speed of first spread of a type $(n_1 + i)$ individual is defined as $c = \lim_{n\to\infty} \frac{s(m)}{m}$.

Let $\{L^{(m)}(\lambda)\}_i = \int_{-\infty}^{\infty} e^{\lambda s} x_i^{(m)}(s)ds$ and $P_{ij}(\lambda) = \int_{-\infty}^{\infty} e^{\lambda s} p_{ij}(s)ds$. Taking the Laplace transform of each of equations (5.1) we obtain

$$\mathbf{L}^{(m+1)}(\lambda) = \mathbf{A}(\lambda)\mathbf{L}^{(m)}(\lambda),$$

where $\{\mathbf{A}(\lambda)\}_{ij} = \gamma_{n_1+i,n_1+j} P_{n_1+i,n_1+j}(\lambda)$.
Then

$$\mathbf{L}^{(m)}(\lambda) = (\mathbf{A}(\lambda))^m \mathbf{L}^{(0)}(\lambda). \tag{5.2}$$

Inverting equation (5.2) for a suitable choice of $\theta(m)$ to obtain $x_{n_1+i}^{(m)}(r)$, and integrating over $r \geq s(m)$, we obtain

$$\eta = \frac{1}{2\pi i} \int_{\theta(m)-i\infty}^{\theta(m)+i\infty} \lambda^{-1} e^{-\lambda s(m)} \{(\mathbf{A}(\lambda))^m \mathbf{L}^{(0)}(\lambda)\}_{n_1+i} d\lambda.$$

Let $g(\lambda) = -\lambda s(m) + m a(\lambda)$ where $a(\lambda) = \log(\rho(\mathbf{A}(\lambda)))$ and $\rho(\mathbf{A}(\lambda))$ is the Perron-Frobenius root of $\mathbf{A}(\lambda)$. It is shown in [6] that $Re(g(\lambda))$ has a saddle-point on the real line at $\lambda = \theta$ where θ satisfys the relation $a'(\theta) = s(m)/m$. The $\theta(m)$ used in the integration is taken as this saddle-point. It is shown in [6] that the speed c is given by

$$c = \max\left(0, \inf_{\lambda>0} \frac{\log(\rho(\mathbf{A}(\lambda)))}{\lambda}\right). \tag{5.3}$$

6. THE SPREAD OF A STRATEGY.

Consider a population in which each individual can play one of the strategies $S_1, ...S_{n-1}$. A new strategy S_n is introduced into the population in which the the proportions of individuals playing the strategies $S_1, ..., S_{n-1}$ are in stable equilibrium, with a proportion η_i of individuals playing strategy S_i, $i = 1, ...n-1$, at all s. Let $x_n^{(m)}(s)$ denote the proportion of individuals playing the strategy S_n at position s in the mth generation.

Then

$$x_n^{(m+1)}(s) = \int_{-\infty}^{\infty} p(s-r) g_n(\mathbf{x}^{(m)}(r))dr, \tag{6.1}$$

where

$$g_n(\mathbf{x}^{(m)}(r)) = \frac{x_n^{(m)}(r)\{\mathbf{W}\mathbf{x}^{(m)}(r)\}_n}{(\mathbf{x}^{(m)}(r))'\mathbf{W}\mathbf{x}^{(m)}(r)}.$$

Here $\{\mathbf{x}^{(m)}(r)\}_n = x_n^{(m)}(r)$.

In this evolutionary game theory model $\mathbf{W}$ represents a payoff matrix. It is assumed that $w_{ij} > 0$ for all i, j. However, $\mathbf{W}$ need not be symmetric. Let the matrix $\mathbf{W}$ be partitioned into

$$\mathbf{W} = \begin{pmatrix} \mathbf{W}^* & \mathbf{b} \\ \mathbf{c}' & d \end{pmatrix}.$$

It is shown in [5] that the linearised equation in the forward front is

$$x_n^{(m+1)}(s) = \gamma \int_{-\infty}^{\infty} p(s-r) x_n^{(m)}(r) dr, \qquad (6.2)$$

where $\gamma = \dfrac{\mathbf{c}'\eta^*}{\eta^{*\prime}\mathbf{W}^*\eta^*}$, $\eta^{*\prime} = (\eta_1, ..., \eta_{n-1})$ and $\eta_i = \{\mathbf{W}^{*-1}\mathbf{1}\}_i / \mathbf{1}'\mathbf{W}^{*-1}\mathbf{1} > 0$ for $i = 1, ..., n-1$.

The saddle point method is used to find the speed of spread c of the forward front. It is shown that

$$c = \max\left(0, \inf_{\lambda > 0}\left(\frac{\log P(\lambda) + \log \gamma}{\lambda}\right)\right).$$

7. THE SPREAD OF AN ALLELE.

A genetics model, analogous to the model of section 6, can be considered with alleles $A_1, ..., A_n$, in which $\mathbf{W}$ is a fitness matrix which is symmetric. This leads in an identical manner to the same expression c for the speed of first spread of a new allele A_n. If $n = 2$ and we consider the speed of first spread of an allele A_2 through a population which initially consists entirely of allele A_1, then $\gamma = w_{21}/w_{11}$, and the result agrees with that obtained by the exact method.

We now consider a two allele model in which the fitness matrices are different for males and females. This is a spatial version of the model described in [17]. Suppose there are two alleles A_1 and A_2. Let $x^{(m)}(r)$ and $y^{(m)}(r)$ denote the proportions of the alleles A_1 in the male and female populations respectively at position r in the mth generation. Suppose the fitness matrices for males and females are the 2×2 matrices $\mathbf{W} = (w_{ij})$ and $\mathbf{V} = (v_{ij})$ respectively.

The spatial model is described by the equations

$$\begin{aligned} x^{(m+1)}(r) &= \int_{-\infty}^{\infty} p(r-s) g_M(x^{(m)}(s), y^{(m)}(s)) ds, \\ y^{(m+1)}(r) &= \int_{-\infty}^{\infty} q(r-s) g_F(x^{(m)}(s), y^{(m)}(s)) ds, \end{aligned} \qquad (7.1)$$

where

$$\begin{aligned} g_M(x,y) &= \frac{w_{11}xy + \frac{1}{2}w_{12}\{x(1-y) + y(1-x)\}}{w_{11}xy + w_{12}\{x(1-y) + y(1-x)\} + w_{22}(1-x)(1-y)}, \\ g_F(x,y) &= \frac{v_{11}xy + \frac{1}{2}v_{12}\{x(1-y) + y(1-x)\}}{v_{11}xy + v_{12}\{x(1-y) + y(1-x)\} + v_{22}(1-x)(1-y)}. \end{aligned}$$

Consider the propagation of allele A_1 through a population in which initially only allele A_2 is present. In the forward front, the approximate equations are

$$\begin{aligned} x^{(m+1)}(r) &= \frac{w_{12}}{2w_{22}} \int_{-\infty}^{\infty} p(r-s)\left(x^{(m)}(s)+y^{(m)}(s)\right) ds, \\ y^{(m+1)}(r) &= \frac{v_{12}}{2v_{22}} \int_{-\infty}^{\infty} q(r-s)\left(x^{(m)}(s)+y^{(m)}(s)\right) ds. \end{aligned} \tag{7.2}$$

For this model the matrix $A(\lambda)$ is given by

$$\mathbf{A}(\lambda) = \frac{1}{2}\begin{pmatrix} \dfrac{w_{12}P(\lambda)}{w_{22}} & \dfrac{w_{12}P(\lambda)}{w_{22}} \\ \dfrac{v_{12}Q(\lambda)}{v_{22}} & \dfrac{v_{12}Q(\lambda)}{v_{22}} \end{pmatrix}$$

and hence $\rho(\mathbf{A}(\lambda)) = \frac{1}{2}\left(\frac{w_{12}P(\lambda)}{w_{22}} + \frac{v_{12}Q(\lambda)}{v_{22}}\right)$. It follows from the general result of section 2 that

$$c = \max\left(0, \inf_{\lambda>0}\left(\frac{\log(w_{12}P(\lambda)/w_{22} + v_{12}Q(\lambda)/v_{22}) - \log 2}{\lambda}\right)\right).$$

8. A MULTIPLE ALLELE MODEL WITH SEX-LINKED LOCUS.

A two allele model with sex-linked locus is considered in [15,16]. Let $\{\mathbf{y}\}_i = y_i^{(m)}(s)$ and $\{\mathbf{z}\}_i = z_i^{(m)}(s)$ be the proportions of the allele A_i in the female and male outputs at location s in generation m prior to mating. The male is haploid for the X chromosome therefore only has n genetic types, the ith being A_i. The female is diploid, so has $n(n+1)/2$ genotypes A_iA_j for $1 \le i \le j \le n$. The survival fitness of the male A_i genotype is $v_i = \{\mathbf{v}\}_i$ and of the female A_iA_j genotype is $w_{ij} = \{\mathbf{W}\}_{ij}$. The equations for the model are

$$\begin{aligned} y_i^{(m+1)}(s) &= \int_{\infty}^{\infty} \frac{y_i^{(m)}(r)\{\mathbf{W}\mathbf{z}^{(m)}(r)\}_i + z_i^{(m)}(r)\{\mathbf{W}\mathbf{y}^{(m)}(r)\}_i}{(2(\mathbf{y}^{(m)}(r))'\mathbf{W}\mathbf{z}^{(m)}(r))} p(s-r)dr, \\ z_i^{(m+1)}(s) &= \int_{-\infty}^{\infty} \frac{v_i y_i^{(m)}(r)}{(\mathbf{v}'\mathbf{y}^{(m)}(r))} q(s-r)dr, \quad (i=1,...,n). \end{aligned} \tag{8.1}$$

Consider the spread of an allele A_n when introduced into a bounded region B of a population containing $(n-1)$ alleles in stable equilibrium, there being proportions η_i (ζ_i) for females (males) of allele A_i for $i = 1, ..., n-1$ at all positions s. Here

$$\begin{aligned} \{\eta\}_i &= \frac{\{(\mathbf{W}^* diag(\mathbf{v}^*) + diag(\mathbf{v}^*)\mathbf{W}^*)^{-1}\mathbf{1}\}_i}{(\mathbf{1}'(\mathbf{W}^* diag(\mathbf{v}^*) + diag(\mathbf{v}^*)\mathbf{W}^*)^{-1})\mathbf{1})} > 0, \\ \{\zeta\}_i &= \frac{\{(\mathbf{W}^* + diag(\mathbf{v}^*)\mathbf{W}^*(diag(\mathbf{v}^*)^{-1})^{-1}\mathbf{1}\}_i}{(\mathbf{1}'(\mathbf{W}^* + diag(\mathbf{v}^*)\mathbf{W}^* diag(\mathbf{v}^*)^{-1})^{-1}\mathbf{1})} > 0, \end{aligned}$$

for $i = 1, ..., n-1$, where $\mathbf{W}^*$ and $\mathbf{v}^*$ are the parts of the fitness matrices relating to the first $(n-1)$ alleles. Define $\eta_n = 0$ and $\zeta_n = 0$. Let $\{\eta\}_i = \eta_i$ and $\{\zeta\}_i = \zeta_i$. Then the zeroth generation has $\mathbf{y}^{(0)}(s) \equiv \eta$ and $\mathbf{z}^{(0)}(s) \equiv \zeta$ for $s \notin B$.

The approximate equations which are valid in this forward region are

$$\begin{aligned} y_n^{(m+1)}(s) &= \int_{-\infty}^{\infty} \frac{y_n^{(m)}(r)\{\mathbf{W}\zeta\}_n + z_n^{(m)}(r)\{\mathbf{W}\eta\}_n}{(2\eta'\mathbf{W}\zeta)} p(s-r)dr, \\ z_n^{(m+1)}(s) &= \int_{-\infty}^{\infty} \frac{v_n y_n^{(m)}(r)}{(\mathbf{v}'\eta)} q(s-r)dr. \end{aligned} \tag{8.2}$$

Note that equations (8.2) are just a special case of equations (2.1) with n replaced by $2n$, $n_1 = 2n-2$, and $x_{2n-1}^{(m)}(s) = y_n^{(m)}(s)$ and $x_{2n}^{(m)}(s) = z_n^{(m)}(s)$. Also $p_{i,2n-1}(r) = p(r)$, $p_{i,2n}(r) = q(r)$ for $i = (2n-1), 2n$ and $\gamma_{2n-1,2n-1} = \{\mathbf{W}\zeta\}_n/(2\eta'\mathbf{W}\zeta)$, $\gamma_{2n-1,2n} = \{\mathbf{W}\eta\}_n/(2\eta'\mathbf{W}\zeta)$, $\gamma_{2n,2n-1} = v_n/(\mathbf{v}'\eta)$ and $\gamma_{2n,2n} = 0$. Hence we obtain the result that the speed of first spread of allele A_n amongst both males and females is c, where

$$c = \max\left(0, \inf_{\lambda>0}\left(\frac{\log(\gamma_{2n-1,2n-1}(P(\lambda)/2)) + \log\left(1+\sqrt{1+4\beta Q(\lambda)/P(\lambda)}\right)}{\lambda}\right)\right),$$

where $\beta = \gamma_{2n-1,2n}\gamma_{2n,2n-1}/\gamma_{2n-1,2n-1}^2$ and $P(\lambda)$ and $Q(\lambda)$ are the 2-sided Laplace transforms of the migration distributions $p(s)$ and $q(s)$.

This result corresponds to the result obtained in [15,16] for the case $n = 2$.

REFERENCES

[1] J.Radcliffe and L.Rass, Rocky Mountain J. Mathematics, 14 No 3 599-617 (1984).
[2] J.Radcliffe and L.Rass, Rocky Mountain J. Mathematics, 23 No 2 725-752 (1993).
[3] J.Radcliffe and L.Rass, Mathematical Population Dynamics: Analysis of Hetrogeneity, Wuerz Publishing Ltd., (1995) pp 147-170.
[4] J.Radcliffe and L.Rass, Proceedings of First World Conference on Branching Processes Springer Lecture Notes in Statistics No. 99, (1995) pp.169-179.
[5] J.Radcliffe and L.Rass, Proceedings of the 4th International Conference on Mathematical Population Dynamics, in press.
[6] J.Radcliffe and L.Rass, Math. Biosciences, in press.
[7] J.Radcliffe and L.Rass, J. Math. Biology, 23, 341-359 (1986).
[8] L.Rass and J.Radcliffe, Seventh International Conference on Differential Equations.
[9] H.F.Weinberger, Lecture Notes in Mathematics Vol 648, Springer, New York (1978) pp.47-98.
[10] H.F.Weinberger, SIAM J. Math. Anal., 13, 353-396 (1982).
[11] P.Cregan and R.Lui, J. Math. Biology, 20, 59-68 (1984).
[12] R.Lui, SIAM J. Math. Anal., 13 No 6, 913-937 (1982).
[13] R.Lui, SIAM J. Math. Anal., 13 No 6, 938-953 (1982).
[14] R.Lui, J. Math. Biol., 16, 199-220 (1983).
[15] R.Lui, Math. Biosciences, 93, 269-295 (1989).
[16] R.Lui, Math. Biosciences, 93, 297-312 (1989).
[17] W.J.Ewens, Mathematical Population Genetics, Springer-Verlag, New York (1979).

7th Int. Coll. on Differential Equations, pp. 355-362
D. Bainov (Ed)

The asymptotic spatial behaviour of a class of epidemic models.

Linda Rass and John Radcliffe.

School of Mathematical Sciences,

Queen Mary and Westfield College,

Mile End Road, London E1 4NS, England.

Abstract. A survey is given of the exact results concerning the spread of infection in continuous space contact models describing a class of multi-type epidemics. In addition an upper bound is obtained for the spatial final size and a simpler proof given for the pandemic theorem.

Key Words: Contact spatial models, Epidemics, Asymptotic Behaviour.

1. INTRODUCTION.

There are many examples of epidemics involving several types. These include host-vector diseases such as malaria and yellow fever, more general two type epidemics with carriers/non-carriers and vaccinated/non-vaccinated, and multitype epidemics such as rabies.

In this paper we restrict our attention to epidemics where individuals once infected cannot return to the susceptible state. The timescale of the epidemic is taken to be small relative to the lifetime of individuals, so that births into the system may be neglected. An epidemic is termed non-reducible when any type of individual may, possibly through a sequence of infections, infect any other type; otherwise it is said to be reducible.

There now exists a fairly comprehensive theory for the spread of infection in continuous space contact models of both non-reducible and reducible n type epidemics. The purpose of this paper is to present a survey of these results, with an emphasis on the non-reducible case. In addition an upper bound is obtained for the spatial final size and a simpler method of proof of the pandemic theorem is given.

The one type models considered were of two kinds; one with constant infection and removal rates, termed $S \rightarrow I \rightarrow R$, (see [1], [2], [3]); the other with varying infectivity (see [4], [5], [6]). The first model can be shown to lead to the examination of the same system of equations as the second kind of model, with the function describing the infectivity at a specified time since infection taking a particular form. General results for multitype models were obtained in a series of papers by the present authors ([7], [8], [9], [10], [11]).

Consider the spatial spread of a deterministic epidemic in a homogeneous population.

One can think of the spread of infection as a wave travelling through the population. Section 2 discusses the speeds at which wave solutions can exist and the uniqueness of such a solution at a given speed. There is a general criterion for the existence of wave solutions at some speeds (the criterion for the occurence of a major epidemic). When this criterion is met, there is a minimum speed at which wave solutions exist.

The next question concerns the speed of propagation of the epidemic. The initial infection may be taken to be confined to some compact region. Although the speed of spread of infection will initially depend on the quantity and configuration of the initial infection, intuition suggests that it will eventually settle down to a constant speed. We need to define precisely what is meant by this speed. The essential idea behind the definition of the asymptotic speed of propagation is easily described. Suppose this speed is c^*. Then if you run faster than c^* you will leave the epidemic behind; whereas if you run slower than c^* then you will be surrounded by the epidemic. A more precise formulation is given in section 3. Consider rotationally symmetric contact distributions. When no major epidemic occurs, no matter how slowly you run in any given direction, you will always leave the epidemic behind. In the non-reducible case, when the criterion for a major epidemic is met, the asymptotic speed of propagation c^* is in fact the minimum speed c_0 for which wave solutions exist. Although these results are still presumably true in the non-rotationally symmetric case, a proof is still lacking (see [12]).

Provided the criterion for a major epidemic is met, the proportion eventually suffering the epidemic at each spatial position for each type can be shown to have a lower bound. This is called the pandemic theorem. The bound is linked to the bound obtained in the corresponding non-spatial model. This is discussed in section 4.

A further problem concerns convergence to a wave form. When the conditions for a major epidemic are met, consider running at speed c_0 in the epidemic on the line, and looking along its length at the proportions of infectives. It seems intuitively reasonable that the proportions would converge to the unique wave solution associated with the minimum speed c_0. Although similar results have been proved for genetic models ([13], [14]) and branching-diffusions ([15]), this result has not yet been proved for spatial epidemic models, and remains an outstanding problem.

2. WAVE SOLUTIONS.

Consider n types of individuals, type i having uniform density σ_i in $\mathbf{R}^N$. Each type consists of susceptible and infectious individuals. Denote the proportions of susceptible individuals of type i at position $\mathbf{s}$ and time t by $x_i(\mathbf{s},t)$. Also let $I_i(\mathbf{s},t,\tau)d\tau$ be the corresponding proportion who were infected in the time interval $(t-\tau-d\tau, t-\tau)$. The rate of infection of susceptibles of type i by infectious individuals of type j who were infected time τ ago is $\lambda_{ij}(\tau)$ and $p_{ij}(\mathbf{r})$ is the contact distribution representing the distance $\mathbf{r}$ over which infection occurs.

The epidemic occuring over all $\mathbf{s} \in \mathbf{R}^N$ and time $t \in \mathbf{R}$ is described by the equations

$$\begin{aligned}
\frac{\partial x_i(\mathbf{s},t)}{\partial t} &= -x_i(\mathbf{s},t)\sum_{j=1}^{n}\sigma_j\int_{-\infty}^{\infty}\int_0^{\infty} I_j(\mathbf{s}-\mathbf{r},t,\tau)p_{ij}(\mathbf{r})\lambda_{ij}(\tau)d\tau d\mathbf{r},\\
I_i(\mathbf{s},t,0) &= -\frac{\partial x_i(\mathbf{s},t)}{\partial t}, \qquad (2.1)\\
I_i(\mathbf{s},t,\tau) &= I_i(\mathbf{s},t-\tau,0), \quad (i=1,\dots,n).
\end{aligned}$$

Consider equations (2.1). We look for wave solutions travelling with speed $c > 0$ in a direction with direction cosines ξ. Let $p^*_{ij}(s)$ be the projection of the contact distribution in this direction. Writing $x_i(\mathbf{s},t) = x_i(u)$, where $u = \xi'\mathbf{s} - ct$ we obtain the equations

$$x_i'(u) = x_i(u)\sum_{j=1}^{n}\int_{-\infty}^{\infty} x_j'(u+t)v_{ij}(t)dt, \qquad (2.2)$$

where

$$v_{ij}(t) = c^{-1}\sigma_j\int_0^{\infty} p^*_{ij}(s-t)\lambda_{ij}(s/c)ds. \quad (i=1,\dots,n). \qquad (2.3)$$

All individuals are specified to be susceptible before the wave, i.e. $x_i(\infty) = 1$ for $i = 1,\dots,n$. The $p^*_{ij}(s)$ are also restricted to be exponentially dominated in the tail.

Note from equations (2.1) that $x_i(u)$ is monotone increasing. Integrating equations (2.2) we obtain an equivalent system of equations,

$$w_i(u) = \sum_{j=1}^{n}\int_{-\infty}^{\infty}(1-e^{-w_j(u+t)})v_{ij}(t)dt, \quad (i=1,\dots,n), \qquad (2.4)$$

where $w_i(u) = -\log x_i(u)$, with $w_i(u)$ monotone decreasing and $w_i(\infty) = 0$.

The wave solutions for the $S \to I \to R$ epidemic also leads to equation (2.4) with $\lambda_{ij}(t) = \lambda_{ij}e^{-\mu_j t}$. Results for our model with varying infectivity are therefore also valid for this model with constant infection and removal rates λ_{ij} and μ_i.

Define $\{\mathbf{\Gamma}\}_{ij} = \sigma_j\int_0^{\infty}\lambda_{ij}(\tau)d\tau$. Restrict $\mathbf{\Gamma}$ to be non-reducible. It is easily established that no non-trivial wave solution is possible at any speed if $\rho(\mathbf{\Gamma}) \le 1$. Take the limit as $u \to -\infty$ in (2.4). The limit $l_i = \lim_{u\to-\infty}(1-x_i(u))$ exists since $x_i(u)$ is monotone and bounded. Then

$$\mathbf{m} - \mathbf{\Gamma}\mathbf{l} = \mathbf{0},$$

where $\{\mathbf{l}\}_i = l_i$ and $\{\mathbf{m}\}_i = -\log(1-l_i)$. Multiply by the left eigenvector of $\mathbf{\Gamma}$ corresponding to the Perron-Frobenius root $\rho(\mathbf{\Gamma})$. It immediately follows that $\mathbf{l} = \mathbf{0}$, which corresponds to the trivial solution $w_i(u) = 0$ for all u and i.

When $\rho(\mathbf{\Gamma}) > 1$, define $W_i(\lambda) = \int_{-\infty}^{\infty} e^{\lambda u} w_i(u)du$, $V_{ij}(\lambda) = \int_{-\infty}^{\infty} e^{-\lambda u} v_{ij}(u)du$ and $T_i(\lambda) = \int_{-\infty}^{\infty} e^{\lambda u}(w_i(u) - (1 - e^{-w_i(u)}))du$. Let $\{\mathbf{W}(\lambda)\}_i = W_i(\lambda)$, $\{\mathbf{T}(\lambda)\}_i = T_i(\lambda)$ and $\{\mathbf{V}(\lambda)\}_{ij} = V_{ij}(\lambda)$. It can be shown that these all exist for $0 < Re\lambda < \delta$, for some $\delta > 0$. Taking the Laplace transform of equation (2.4) we obtain

$$(\mathbf{V}(\lambda) - \mathbf{I})\mathbf{W}(\lambda) = \mathbf{V}(\lambda)\mathbf{T}(\lambda).$$

Let $K_c(\lambda) = \rho(\mathbf{V}(\lambda))$ and $\mathbf{u}'(\lambda)$ be the corresponding left eigenvector. Hence

$$(K_c(\lambda) - 1)\mathbf{u}'(\lambda)\mathbf{W}(\lambda) = \mathbf{u}'(\lambda)\mathbf{V}(\lambda)\mathbf{T}(\lambda).$$

Note that $K_c(0) = \rho(\mathbf{\Gamma}) > 1$. Also within the region of convergence of the transforms the right hand side stays positive. Let Δ_V be the minimum of the abscissae of convergence of the $\{\mathbf{V}(\lambda)\}_{ij}$ in the positive half of the complex plane.

The behaviour of $K_c(\lambda)$ determines whether a non-trivial wave solution exists, with $\mathbf{W}(\lambda)$ becoming infinite when $K_c(\lambda)$ first equals 1. No wave solution is possible if $K_c(\lambda) > 1$ for all $0 \leq \lambda \leq \Delta_V$. The function $K_c(\lambda)$ can be shown to be convex in λ and monotone decreasing in c for each λ. Hence there is a c_0 such that $K_c(\lambda) = 1$ has a positive root for each positive speed $c \geq c_0$, and no root for each positive speed $c < c_0$. In fact wave solutions can be shown to exist at all positive speeds $c \geq c_0$. No wave solutions exist at any positive speed $c < c_0$.

The results on wave solutions from [7] are summarised in the following theorem

Theorem 2.1. *(Wave solutions for $\mathbf{\Gamma}$ non-reducible.)*

If $\rho(\mathbf{\Gamma}) \leq 1$, no wave solution exists at any speed $c > 0$. When $\rho(\mathbf{\Gamma}) > 1$, there exists a real positive finite c such that $K_c(\lambda) < 1$ for some $\lambda \in (0, \Delta_V)$. In this case let $c_0 = \inf\,(c \in R_+ : K_c(\lambda) < 1$ *for some* $\lambda \in (0, \Delta_V))$.

(i) If $c_0 = 0$, non-trivial wave solutions exist for all speeds $c > 0$.

(ii) If $c_0 > 0$, non-trivial wave solutions exist for all speeds $c \geq c_0$. No non-trivial wave solution exists for c such that $0 < c < c_0$.

(iii) For each fixed $c > c_0$, the non-trivial wave solution at speed c is unique modulo translation.

(iv) If $c_0 > 0$ and $K_{c_0}(\Delta_V) \neq 1$, then the non-trivial wave solution at speed c_0 is unique modulo translation.

Corresponding results for the reducible case are given in [10].

3. THE SPEED OF PROPAGATION.

Now suppose that instead of the epidemic occuring over all time, it is initiated at time $t = 0$ by the introduction of infective individuals from outside into a population consisting entirely of susceptible individuals. We use the same definitions as in section 2. Additionally $\epsilon_j(\mathbf{s},\tau)d\tau d\mathbf{s}$ represents the number of individuals from outside who are in the N dimensional interval $(\mathbf{s}, \mathbf{s}+d\mathbf{s})$ and were themselves infected in the time interval $(-\tau - d\tau, -\tau)$. The infection rate of a type j susceptible by a type i individual from outside who was infected time τ ago is $\lambda^*_{ij}(\tau)$ and $p^*_{ij}(\mathbf{r})$ is the corresponding contact distribution. The epidemic is described by the equations

$$\begin{aligned} \frac{\partial x_i(\mathbf{s},t)}{\partial t} &= -x_i(\mathbf{s},t)\left[\sum_{j=1}^{n}\sigma_j\int_{R^N}\int_0^t I_j(\mathbf{s}-\mathbf{r},t,\tau)p_{ij}(\mathbf{r})\lambda_{ij}(\tau)d\tau d\mathbf{r} + h_i(\mathbf{s},t)\right], \\ I_i(\mathbf{s},t,0) &= -\frac{\partial x_i(\mathbf{s},t)}{\partial t}, \\ I_i(\mathbf{s},t,\tau) &= I_i(\mathbf{s},t-\tau,0), \quad (i=1,...,n), \end{aligned} \tag{3.1}$$

where

$$h_i(\mathbf{s},t) = \sum_{j=1}^{k}\int_{R^N}\int_0^\infty \varepsilon_j(\mathbf{s}-\mathbf{r},\tau)p^*_{ij}(\mathbf{r})\lambda^*_{ij}(t+\tau)d\tau d\mathbf{r}.$$

The initial conditions are $x_i(\mathbf{s},0) = 1$ for $i = 1,...,n$.

Again we may consider an equivalent system of equations,

$$w_i(\mathbf{s},t) = \sum_{j=1}^{n}\int_{R^N}\int_0^t (1-\exp\{-w_j(\mathbf{s}-\mathbf{r},t-\tau)\})p_{ij}(\mathbf{r})\gamma_{ij}(\tau)d\tau d\mathbf{r} + H_i(\mathbf{s},t), \quad (i=1,...,n),$$

where $w_i(\mathbf{s},t) = -\log x_i(\mathbf{s},t)$, $\gamma_{ij}(\tau) = \sigma_j\lambda_{ij}(\tau)$ and $H_i(\mathbf{s},t) = \int_0^t h_i(\mathbf{s},u)du$. As before $\{\mathbf{\Gamma}\}_{ij} = \int_0^\infty \gamma_{ij}(\tau)d\tau$.

The proof concerning the speed of propagation requires the contact distributions to be radially symmetric. Take $\mathbf{\Gamma}$ to be non-reducible. When $\mathbf{\Gamma}$ is finite with $\rho(\mathbf{\Gamma}) > 1$, then the speed c^* is shown to be c_0, the minimum speed for which a wave solution exists. Note that $c_0 > 0$. When $\rho(\mathbf{\Gamma}) \leq 1$ the speed c^* is shown to be zero. If $\mathbf{\Gamma}$ has an infinite element, then $c^* = \infty$. If c_0 is defined as in theorem 2.1 for all $\mathbf{\Gamma}$, then in all cases $c^* = c_0$.

The asymptotic speed of propagation amongst type i individuals, as defined in [16], is c^* if, for any c_1 and c_2 with $0 < c_1 < c^* < c_2$,
(i) $(1 - x_i(\mathbf{s},t))$ tends uniformly to zero in the region $|\mathbf{s}| \geq c_2 t$;
(ii) $(1-x_i(\mathbf{s},t))$ is bounded away from zero uniformly in the region $|\mathbf{s}| \leq c_1 t$ for t sufficiently large. This may be rewritten in terms of $w_i(\mathbf{s},t)$.

The main results from [9] are summarised in the following theorem:

Theorem 3.1. *(The asymptotic speed of propagation for $\mathbf{\Gamma}$ non-reducible.)*

(i) For any $c_2 > 0$ such that $K_{c_2}(\lambda) < 1$ for some $\lambda \in (0, \Delta_V)$), i.e. for any $c_2 > c_0$, $\lim_{t\to\infty}\sup\{w_i(\mathbf{s},t) : |\mathbf{s}| \geq c_2 t\} = 0$ for $i = 1,...,n$.

(ii) If $\rho(\Gamma) > 1$, then for any $c_1 \in (0, c_0)$, there exist positive constants d_i and constants T_i sufficiently large such that $\min\{w_i(\mathbf{s}, t) : |\mathbf{s}| \leq c_1 t\} \geq d_i$ for all $t > T_i$ and $i = 1, ..., n$.

The first part is relatively simple to prove. If $\rho(\Gamma) \leq 1$, then $c_0 = 0$. For any $c > 0$, $K_c(0) \leq 1$ and $K_c'(0) < 0$, so that there exists a $\lambda > 0$ with $K_c(\lambda) < 1$. If $\rho(\Gamma) > 1$, for any $c > c_0$ there is also a $\lambda > 0$ with $K_c(\lambda) < 1$. Take $c_0 < c < c_2$ and choose $\lambda > 0$ with $K_c(\lambda) < 1$. Equation (5.1) is rewritten in terms of $u_i(\mathbf{s}, t) = w_i(\mathbf{s}, t)e^{\lambda(\{\mathbf{s}\}_1 - ct)}$. A (unique) solution $u_i(\mathbf{s}, t)$ is constructed iteratively, for which $u_i = \sup_{\mathbf{s},t} u_i(\mathbf{s}, t)$ is bounded above (by B say). Hence

$$w_i(\mathbf{s}, t) \leq B \exp(\lambda(ct - |\mathbf{s}|)).$$

Therefore $\sup\{w_i(\mathbf{s}, t) : |\mathbf{s}| \geq c_2 t\} \leq Be^{-\lambda t(c_2 - c))} \to 0$ as $t \to \infty$.

The second part of the proof is more complex, so only a brief indication of the method of proof is given; for precise details see [9].

The result that $K_c(\lambda) > 1$ for all λ for any $c_1 < c < c_0$ implies an equivalent result holds for R and T large with the contact distributions $p_{ij}(\mathbf{r})$ truncated for $\mathbf{r} > R$ and the infection rates $\lambda_{ij}(\tau)$ truncated for $\tau > T$. This is used to identify an operator E_T and construct a subsolution $\psi(\mathbf{s}, t)$ with $w_i(\mathbf{s}, t) \geq E_T(w_i(\mathbf{s}, t))$, for $t \geq T$, and $E_T(\psi(\mathbf{s}, t)) \geq \psi(\mathbf{s}, t)$, with the second inequallity strict for $\mathbf{s} \in \operatorname{supp}\psi$.

The subsolution $\psi(\mathbf{s}, t)$ is constructed, for some positive Q and M and for any $\sigma > 0$, so that $\psi(\mathbf{s}, t) \leq \sigma M$ with the equallity holding for $|\mathbf{s}| \leq Q + ct$ for $t \geq T$. It can be shown that there exists a t_0 such that $\inf_{|\mathbf{s}| \leq R, t \geq t_0} w_i(\mathbf{s}, t) > 0$. Hence, for a suitably chosen σ, the subsolution can be wedged below $w_i(\mathbf{s}, t)$ to give $w_i(\mathbf{s}, t_0 + t) \geq \psi(\mathbf{s}, t)$ for all $0 \leq t \leq T$.

A comparison principle based on the operator E_T enables the result to be extended so that $w_i(\mathbf{s}, t_0 + t) \geq \psi(\mathbf{s}, t)$ for all $t \geq 0$. Since the constructed $\psi(\mathbf{s}, t) = \sigma M$ for $|\mathbf{s}| \leq Q + ct$ and $t \geq T$, we obtain $w_i(\mathbf{s}, t) \geq \sigma M$ for $|\mathbf{s}| \leq Q + c(t - t_0)$ and $t \geq t_0 + T$.

As $c_1 < c$ this implies $w_i(\mathbf{s}, t) \geq \sigma M$ for $|\mathbf{s}| \leq c_1 t$, for t sufficiently large. This establishes part (ii) of the proof for the speed of propagation.

Note that if at least one (projected) contact distribution is not exponentially dominated in the tail, then for any $c > 0$ the subsolution can be constructed, so that no matter how fast you run you will always be overtaken by the epidemic. The speed of propagation is therefore said to be infinite in this case.

The speed of propagation in the reducible case is obtained in [11].

4. THE PANDEMIC THEOREM.

This is a spatial version of the final size for a non-spatial epidemic. A survey of the non-spatial results is given in [17]. The spatial results require radial symmetry of the contact distributions. Results can be proved directly in 1 and 2 dimensions as in

[8]. Alternatively results pertaining to part (ii) of the velocity of propagation can be strengthened to sharpen the lower bounds d_i (see [9]). This can be interpreted in terms of $v_i(\mathbf{s},t) = 1 - x_i(\mathbf{s},t)$.

Note that $x_i(\mathbf{s},t)$ is monotone increasing in t and is bounded. Hence $v_i(\mathbf{s}) = 1 - \lim_{t\to\infty} x_i(\mathbf{s},t)$ exists and measures the proportion of individuals at point $\mathbf{s}$ who eventually suffer the epidemic. Also $a_i(\mathbf{s}) = \lim_{t\to\infty} H_i(\mathbf{s},t)$ exists. The pandemic theorem gives a lower bound for the final size $v_i(\mathbf{s})$ when $\rho(\mathbf{\Gamma}) > 1$.

Theorem 4.1. *(The pandemic theorem for $\mathbf{\Gamma}$ non-reducible).*

Let $\mathbf{\Gamma}$ be non-reducible with $\rho(\mathbf{\Gamma}) > 1$ and the $p_{ij}(r)$ exponentially dominated in the forward tail for all i,j. Then for all $\mathbf{s}$ and $i = 1,...,n$, $v_i(\mathbf{s}) \geq \eta_i(\mathbf{0})$, where $\eta_i(\mathbf{a})$ is defined as follows:

(i) If $\rho(\mathbf{\Gamma})$ is finite, then $y_i = \eta_i(\mathbf{a})$, $(i = 1,...,n)$, is the unique positive solution to

$$-\log(1-y_i) = \sum_{j=1}^{n} \gamma_{ij} y_j + \{\mathbf{a}\}_i \quad (i = 1,...,n), \tag{4.1}$$

(ii) If $\mathbf{\Gamma}$ has an infinite element in every row, then $\eta_i(\mathbf{a}) = 1$ for $(i = 1,...,n)$.

(iii) Let $\mathbf{\Gamma}$ is partioned (by permutation of the indices) into

$$\mathbf{\Gamma} = \begin{pmatrix} \mathbf{\Gamma}_{11} & \mathbf{\Gamma}_{12} \\ \mathbf{\Gamma}_{21} & \mathbf{\Gamma}_{22} \end{pmatrix}$$

where $(\mathbf{\Gamma}_{11}\mathbf{\Gamma}_{12})$ has no infinite element, and $(\mathbf{\Gamma}_{21}\mathbf{\Gamma}_{22})$ has at least one infinite element in each row. Let $\eta(\mathbf{a})$ be the n-vector with corresponding partition $\eta(\mathbf{a})' = (\eta_1'(\mathbf{a}), \eta_2'(\mathbf{a}))$. Then $\eta_2(\mathbf{a}) = \mathbf{1}$, and $\mathbf{y} = \eta_1(\mathbf{a})$ is the unique positive solution to

$$\mathbf{u} = \mathbf{\Gamma}_{11}\mathbf{y} + \mathbf{\Gamma}_{12}\mathbf{1} + \mathbf{a}_1,$$

where $\{\mathbf{u}\}_i = -\log(1 - \{\mathbf{y}\}_i)$.

Note that if $\rho(\mathbf{\Gamma}) \leq 1$ and $\mathbf{a} = \mathbf{0}$ we define $\eta(\mathbf{0}) = \mathbf{0}$.

A brief indication of the proof of the pandemic theorem will be given for case (i), the other cases present no additional difficulty. Letting * indicate convolution, the final size equations for case (i) are

$$-\log(1 - v_i(\mathbf{s})) = \sum_{j=1}^{n} \gamma_{ij} p_{ij} * v_j(\mathbf{s}) + a_i(\mathbf{s}).$$

Let $v_i = \inf_{\mathbf{s}} v_i(\mathbf{s})$. Then

$$-\log(1 - v_i) \geq \sum_{j=1}^{n} \gamma_{ij} v_j.$$

From theorem 3.1 of [17] this implies that either $v_i = 0$ for all i, or $v_i \geq \eta_i(\mathbf{0})$ for all i.

The case $v_i = 0$ for all i can be excluded using section 3. There is a lower bound for $w_i(\mathbf{s},t)$ and hence for $v_i(\mathbf{s},t)$. For any $c_1 \in (0,c_0)$, there exist T and D_i such that $\inf\{v_i(\mathbf{s},t) : |\mathbf{s}| \le c_1 t\} \ge D_i$ for $t \ge T$. Hence $v_i(\mathbf{s}) \ge D_i$, and therefore $v_i > 0$ all i. Thus $v_i \ge \eta_i(\mathbf{0})$ for all i, and so $v_i(\mathbf{s}) \ge \eta_i(\mathbf{0})$ for all i and $\mathbf{s}$. This completes the proof.

We can also establish an upper bound and consider what happens if the amount of initial infection tends to zero. Again we just consider the case when $\mathbf{\Gamma}$ has all finite entries. Define $b_i = \sup_{\mathbf{s}} a_i(\mathbf{s})$, $u_i = \sup_{\mathbf{s}} v_i(\mathbf{s})$ and $\{\mathbf{b}\}_i = b_i$. Then $-\log(1-u_i) \le \sum_j^n \gamma_{ij}u_j + b_i$.

Let $\alpha = \max_i(u_i/\eta_i(\mathbf{b}))$. Choose i so that the maximum is achieved. For this i,

$$-\log(1-\alpha\eta_i(\mathbf{b})) = -\log(1-u_i) \le \sum_j^n \gamma_{ij}\alpha\eta_j(\mathbf{b}) + b_i = \alpha[-\log(1-\eta_i(\mathbf{b}))] + b_i(1-\alpha).$$

Hence

$$\left[\frac{-\log(1-\alpha\eta_i(\mathbf{b}))}{\alpha\eta_i(\mathbf{b})}\right] - \left[\frac{-\log(1-\eta_i(\mathbf{b}))}{\eta_i(\mathbf{b})}\right] \le \frac{b_i(1-\alpha)}{\alpha\eta_i(\mathbf{b})}. \tag{4.2}$$

Suppose $\alpha > 1$. Since $-\log(1-x)/x$ is a strictly monotone increasing function of x for $x > 0$, the left hand side of the inequality (4.2) is positive. But the right hand side of the inequality cannot be positive. Hence a contradiction is obtained. Therefore $\alpha \le 1$. This gives an upper bound for the spatial final size, since $v_i(\mathbf{s}) \le u_i \le \eta_i(\mathbf{b})$ for all $\mathbf{s}, i$.

Hence we obtain the bounds $\eta_i(\mathbf{0}) \le v_i(\mathbf{s}) \le \eta_i(\mathbf{b})$ for the spatial final size. If the total amount of initial infection tends to zero, then $\mathbf{b} \to \mathbf{0}$, so $\eta_i(\mathbf{b}) \to \eta_i(\mathbf{0})$ (using theorem 3.1, part (v), of [17]). Hence $v_i(\mathbf{s}) \to \eta_i(\mathbf{0})$ for all $i, \mathbf{s}$; the same limit as in the non-spatial case.

Note that the upper bound is still valid if $\rho(\mathbf{\Gamma}) \le 1$. From theorem 3.1, part (v), of [17], $\eta_i(\mathbf{b}) \to 0$ when $\mathbf{b} \to \mathbf{0}$. Hence as the total amount of initial infection tends to zero, $v_i(\mathbf{s}) \to 0$ for all $\mathbf{s}$ and i; an equivalent result to the non-spatial case.

The pandemic theorem for the reducible case is given in [11].

REFERENCES

[1] C. Atkinson and G.E.H. Reuter, Math. Proc. Camb. Phil. Soc., **80**, 315-330 (1976).
[2] A.D. Barbour, Math. Proc. Camb. Phil. Soc., **82**, 127-130 (1977).
[3] K.J. Brown and J. Carr, Math. Proc. Camb. Phil. Soc., **81**, 431-433 (1977).
[4] O. Diekmann, J. of Differential Equations, **33**, 58-73 (1978).
[5] O. Diekmann and H.G. Kaper, Nonlin. Anal., Th. and Appl. 2, **6**, 721-737 (1978).
[6] H. R. Thieme, J. Reine. Angew. Math., **306**, 94-121 (1979).
[7] J. Radcliffe and L. Rass, J. Math. Biol., **17**, 45-66 (1983).
[8] J. Radcliffe and L. Rass, J. Math. Biol., **19**, 309-327 (1984).
[9] J. Radcliffe and L. Rass, J. Math. Biol., **23**, 341-359 (1986).
[10] J. Radcliffe and L. Rass, Lect. Notes in Pure and Applied Math., **131**, 93-114 (1991).
[11] J. Radcliffe and L. Rass, Rocky Mountain J. Math., **26**, (1996), in press.
[12] F. Van den Bosch, J.A.J. Metz, O. Diekmann, J. Math. Biol., **28**, 529-565 (1990).
[13] R. Lui, SIAM J. Math. Anal., **13**, 913-937 (1982).
[14] R. Lui, SIAM J. Math. Anal., **13**, 938-953 (1982).
[15] M. Bramson, Mem. Am. Math. Soc., **44**, (1983).
[16] D.G. Aronson and H.F. Weinberger, Adv. Math., **30**, 33-76 (1978).
[17] J. Radcliffe and L. Rass, Sixth International Coll. on Diff. Equations, in press.

7th Int. Coll. on Differential Equations, pp. 363-370
D. Bainov (Ed)

A Generalized Diffusion Equation

JOHN SCHMEELK
Department of Mathematical Sciences, Box 2014, Virginia Commonwealth University, Richmond VA 23284-2024, USA

Abstract. A generalized space, $\Gamma^{s,\mu,\mathcal{M}}$, and its dual are introduced where μ is a sequence of positive numbers, $\mathcal{M}$ is a space of continuous functions and $s \geq 1$. A generalized Laplacian operator, Δ, is introduced applied to functionals in $\Gamma^{s,\mu,\mathcal{M}}$. Together with a classical time derivative the generalized diffusion equation is shown to have a local solution in a space, $\Gamma^{s',\mu,\mathcal{M}}$ where $s' \geq s$.

Keywords: Generalized diffusion equation, distributions, Beurling ultradistributions, operational methods

1. INTRODUCTION

The classical diffusion equation, $u_t = ku_{xx}$, together with its fundamental source solution, $u(x,t) = (4\pi kt)^{-1/2}e^{-x^2/4kt}$, $t > 0$, $-\infty < x < \infty$, has been investigated and results reported on in many classical as well as many recent research papers. However if we select a distributional milieu and consider the fundamental source solution as a sequence in t, holding k as a parameter that we obtain a fundamental sequence for the Dirac delta functional. More specifically if we focus on the space of tempered distributions we would have

$$\lim_{t\to 0}(4\pi k_0 t)^{-1/2}e^{-x^2/4k_0 t} = \delta(t)$$

where the lim is understood to be in the distributional sense,

$$\lim_{t\to 0}\left\langle (4\pi k_0 t)^{-1/2}e^{-x^2/4k_0 t}, \phi(t)\right\rangle = \langle \delta(t), \phi(t)\rangle,$$

for every rapid descent test function, $\phi(t) \in S$. It is in this sense together with a generalized Laplacian operator [20, 23] that our generalized diffusion equation will be studied.

The controls placed on the spaces implemented in this paper will be in the spirit of Komatsu and Beurling's ultradistributions [11, 12, 13]. We will give a brief review of our nonlinear functional spaces. Section 3 will develop a generalized Laplacian operator placed in our generalized space $\Gamma^{s,\mu,\mathcal{M}}$. This operator was introduced in reference [20] where a specific sequence, $m_q = q!^{1/p}$, $p \geq 1$ was selected and condition (M3) to be presented in section 2 was relaxed to condition (M3)$'$ namely $\sqrt[q]{m_q} \to \infty$. We now give a local solution to our problem for any sequence satisfying the original Komatsu condition (M3).

Section 4 will sketch an existence proof for our generalized diffusion equation. This proof is an outgrowth of techniques implemented in other existence type theorems found in references [20–24].

2. THE SPACES $\Gamma^{\mu,\mathcal{M}}$ AND THEIR DUALS, $(\Gamma^{\mu,\mathcal{M}})'$

Throughout the paper we will assume that we have a monotonically increasing sequence of positive numbers, $\mu = (m_q)_{q\in\mathbb{N}_0}$ satisfying the conditions, $(\mathrm{M}_1) - (\mathrm{M}_3)$ from reference [11]. Specifically these conditions are

$$(\mathrm{M}_1) \qquad m_q^2 \le m_{q-1}m_{q+1}, \quad q = 1,2,\dots ;$$

$$(\mathrm{M}_2) \qquad \exists A,\ H \in \mathbb{R}_+ \text{ such that } m_{q+1} \le A \cdot H^q \cdot m_q, \quad q = 0,1,2,\dots ;$$

$$(\mathrm{M}_3) \qquad \sum_{q=1}^{\infty} \frac{m_{q-1}}{m_q} \ne \infty.$$

It is convenient to take $m_0 = 1$. Moreover we will also assume that we have a sequence of continuous even functions on $\mathbb{R}$ denoted $\mathcal{M} = (m_p(\cdot))_{p\in\mathbb{N}_0}$ satisfying

$$1 = m_0(t) \le m_1(t) \le \dots , \quad t \in \mathbb{R}. \tag{2.1}$$

We require the usual conditions, (P), (M), and (N) from reference [17] be placed on the functions in $\mathcal{M}$. An infinitely differentiable function, ϕ on $\mathbb{R}^q$, $q \in \mathbb{N}$ will be in the test space $K_p\{m_p\}$ iff for every $p \in \mathbb{N}_0$ the norms,

$$\|\phi\|_p = \sup\left\{|\phi^j(t)|m_p(\|t\|): \quad t \in \mathbb{R}^q, \quad |j| \le p\right\} \tag{2.2}$$

are finite. The notation, $|j|$ signifies $\sum_{i=1}^{\infty} j_i$, $j_i \in \mathbb{N}$ $(1 \le i \le q)$ and $\|t\|$ denotes the usual Euclidean norm in $\mathbb{R}^q$.

One of the more important examples are when the sequence of functions, $\mathcal{M}$, are selected to be

$$m_p(\|t\|) = \left(1 + \|t\|^2\right)^{p/2}, \quad p \in \mathbb{N}_0$$

which gives us the Schwartz space of rapid descent test functions and the dual space, $K_q'\{m_p\}$ becomes the Schwartz space of tempered distributions [28].

We now continue with our constructions. We denote by $\|\cdot\|_{-p}$, $p \in \mathbb{N}_0$ the following norm on $K_q'\{m_p\}$,

$$\|x\|_{-p} = \sup\left\{|\langle x,\phi\rangle|: \quad \phi \in K_q\{m_p\}, \quad \|\phi\|_p \le 1\right\}. \tag{2.3}$$

Next we introduce multilinear functionals, a_q, where

$$a_q: \quad K'_q\{m_p\} \times \ldots \times K'_q\{m_p\} \mapsto C, \quad q \in \mathbb{N}$$

and by definition $a_0 \in \mathbb{C}$. We designate

$$\|a_q\|_p = \sup\left\{|a_q[x, \ldots, x]| : \quad x \in K'_q\{m_p\}, \quad \|x\|_{-p} \leq 1\right\} \tag{2.4}$$

and denote the column vector as

$$\Phi = \begin{bmatrix} a_0 \\ \vdots \\ a_q \\ \vdots \end{bmatrix} \doteq [a_0, a_1, \ldots, a_q, \ldots,]^t .$$

A functional, Φ, will be in the space $\Gamma^{s,\mu,\mathcal{M}}$, $s > 1$ iff the norms,

$$|||\Phi|||^{(p)}_{s,\mu,m} = \sup_{q \in \mathbb{N}_0} \frac{\|a_q\|_p \cdot m_q}{s^q}, \tag{2.5}$$

are finite for every $p \in \mathbb{N}_0$. Clearly for $s' > s > 1$, we have $\Gamma^{s,\mu,\mathcal{M}} \subset \Gamma^{s',\mu,\mathcal{M}}$. We can define our generalized space $\Gamma^{\mu,\mathcal{M}}$ to be union of spaces $\Gamma^{s,\mu,\mathcal{M}}$, i.e.

$$\Gamma^{\mu,\mathcal{M}} = \bigcup_{s>1} \Gamma^{s,\mu,\mathcal{M}}, \tag{2.6}$$

or an inductive limit space. In our sequel we can also use the representation given in reference [21] whereby

$$\Phi_{ker} = [\phi_0, \phi_1, \ldots, \phi_q, \ldots]^t \tag{2.7}$$

where

$$\phi_q(t_1, \ldots, t_q) \doteq a_q\left[\delta(x_1 - t_1), \ldots, \delta(x_q - t_q)\right]. \tag{2.8}$$

As in reference [22] one can prove

Theorem 2.1. *Each function, ϕ_q, given in expression (2.8) is infinitely differentiable on $\mathbb{R}^q$ and belongs to the space, $K_q\{m_p\}$.*

Theorem 2.2. *The row vector, F, where*

$$F \leftrightarrow [F_0, F_1, \ldots, F_q, \ldots]$$

can be observed as an element in the space, $\left(\Gamma^{\mu,\mathcal{M}}\right)'$, iff for each $s > 1$, $\Phi \in \Gamma^{s,\mu,\mathcal{M}}$, m sufficiently large and $p > m$, it holds,

$$\sum_{q=0}^{\infty} \|F_q\|_{-m} \frac{s^q}{m_q} |||\Phi|||^{(p)}_{s,\mu,m} < \infty.$$

3. THE GENERALIZED LAPLACIAN OPERATOR

We consider an infinite-dimensional Laplacian operator, Δ, defined on our generalized space, $\Gamma^{\mu,\mathcal{M}}$. The details surrounding its construction can be found in references [20, 22, 25].

Definition 3.1. The infinite-dimensional Laplacian, Δ, is defined as follows:

$$\Delta : \Gamma^{\mu,\mathcal{M}} \mapsto \Gamma^{\mu,\mathcal{M}}$$

$$\Delta : [a_0, a_1, \ldots, a_q, \ldots]^t \mapsto$$

$$\left[1 \cdot 2 \int_{\mathbf{R}^n} a_2[\delta_\tau, \delta_\tau]\, d\tau, \ldots, (q+1)(q+2) \int_{\mathbf{R}^n} a_{q+2}[\delta_\tau, \delta_\tau, \ldots]\, d\tau, \ldots\right]^t.$$

Lemma 3.1. *Given a sequence, μ, where $q+1 \leq L\dfrac{m_{q+1}}{m_q}$ for all $q \in \mathbb{N}$ and some $L > 0$. Then the operator defined as*

$$\Gamma^{\mu,\mathcal{M}} \mapsto \Gamma^{\mu,\mathcal{M}}\, [a_0, \ldots, a_q, \ldots]^t \mapsto [1 \cdot 2a_2, \ldots, (q+1)(q+2)a_{q+2}, \ldots]^t$$

is well defined.

Proof. Select a $\Phi \mapsto [a_0, \ldots, a_q, \ldots]^t$ in $\Gamma^{\mu,\mathcal{M}}$. Then there exists $s > 1$ such that $\Phi \in \Gamma^{s,\mu,\mathcal{M}}$. Select $s' > s$ and consider

$$\left|\left|\left|\left[1 \cdot 2a_2, \ldots, (q+1)(q+2)a_{q+2}, \ldots\right]^t\right|\right|\right|^{(p)}_{s',\mu,m} = \sup\left\{\frac{m_q}{(s')^q}(q+1)(q+2)\|a_{q+2}\|_p : q \in \mathbb{N}_0\right\}$$

$$\leq (s')^2 \sup_{q\in\mathbb{N}_0}\left\{\frac{m_{q+2}\|a_{q+2}\|_p}{(s)^{q+2}}\right\} \sup\left\{\left(\frac{s}{s'}\right)^{q+2} \frac{m_q}{m_{q+2}} L \cdot \frac{m_{q+2}}{m_{q+1}} \cdot L\frac{m_{q+1}}{m_q} : q \in \mathbb{N}_0\right\},$$

so we obtain

$$\left|\left|\left|\left[1 \cdot 2a_2, \ldots, (q+1)(q+2)a_{q+2}, \ldots\right]^t\right|\right|\right|^{(p)}_{s',\mu,m} \leq C_p (s')^2 L^2 \sup\left\{\left(\frac{s}{s'}\right)^q : \quad q \in \mathbb{N}_0\right\} < \infty. \quad \square$$

Theorem 3.1. *Given a sequence, μ where $q+1 \leq L\dfrac{m_{q+1}}{m_q}$ for all $q \in \mathbb{N}$ and some $L > 0$. Then the generalized Laplacian Δ, given in Definition 3.1 maps $\Gamma^{\mu,\mathcal{M}}$ into itself and moreover this mapping is continuous.*

Proof. We consider $\Delta : \Gamma^{s,\mu,\mathcal{M}} \mapsto \Gamma^{s',\mu,\mathcal{M}}$ where $s' > s$. Then for sufficiently large p we have

$$|||\Delta\Phi|||^{(p)}_{s',\mu,m} = \sup_q \frac{(q+1)(q+2)\left\|\displaystyle\int_{\mathbf{R}^n} a_{q+2}[\delta_\tau, \delta_\tau, \ldots]d\tau\right\|_p m_q}{(s')^q}$$

$$= \sup_q \frac{(q+1)(q+2)\left\|\displaystyle\int_{\mathbf{R}^n} \frac{a_{q+2}[(1+\tau^2)\delta_\tau, (1+\tau^2)\delta_\tau, \ldots]}{(1+\tau^2)^2} d\tau\right\|_p m_q}{(s')^q}$$

$$\leq \sup_q \frac{(q+1)(q+2)\left\|a_{q+2}\right\|_p \pi m_q}{(s')^q}$$

$$= \sup_q \frac{(q+1)(q+2)\left\|a_{q+2}\right\|_p m_{q+2}}{m_{q+2}} \frac{s^{q+2}}{s^{q+2}} \pi \frac{m_q}{(s')^q}$$

$$\leq \sup_q \frac{\left\|a_{q+2}\right\|_p m_{q+2}}{(s)^{q+2}} \sup_q \frac{(q+1)(q+2)}{m_{q+2}} \left(\frac{s}{s'}\right)^q s^2 \pi m_q$$

$$\leq \left|\left|\left|\Phi\right|\right|\right|^{(p)}_{s,\mu,\mathcal{M}} K \sup_q \left[\left(\frac{s}{s'}\right)^q \frac{m_q}{m_{q+2}} \cdot L \cdot \frac{m_{q+2}}{m_{q+1}} \cdot L \cdot \frac{m_{q+1}}{m_q}\right]$$

$$\leq C_p K \left\{\sup_q \left(\frac{s}{s'}\right)^q\right\} < \infty. \quad \square$$

4. INVERSE TRANSFORMATIONS

Let $u(t)$ be a $\Gamma^{s,\mu,\mathcal{M}}$ valued function where $t \in [\alpha, \beta] \subset \mathbb{R}$. $u(t)$ is strongly continuous at $t \in [\alpha, \beta]$ whenever $\varepsilon > 0$, $p \in \mathbb{N}_+$ there exists a $\delta_{t_0,\varepsilon,p} > 0$ if $|t - t_0| \leq \delta_{t_0,\varepsilon,p}$ implies $\left|\left|\left|u(t) - u(t_0)\right|\right|\right|^{(p)}_{s,\mu,\mathcal{M}} < \varepsilon$. Likewise $u(t)$ is strongly differentiable whenever $\exists\, \gamma(t_0) \in \Gamma^{s,\mu,\mathcal{M}}$ such that $\varepsilon > 0$, $p \in \mathbb{N}_+$ $\exists \delta_{t_0,\varepsilon,p} > 0$ such that $0 < |t - t_0| < \delta_{t_0,\varepsilon,p}$ then

$$\left\|\frac{u(t) - u(t_0)}{t - t_0} - \gamma(t_0)\right\|^{(p)}_{s,\mu,\mathcal{M}} < \varepsilon.$$

The vector space of continuous parametric functionals which are strongly differentiable ℓ times is denoted $C^\ell\left([\alpha, \beta]; \Gamma^{s,\mu,\mathcal{M}}\right)$.

Theorem 4.1. *Given:*

(i) *a continuous parametric functional* $u(t) \in C^0\left([\alpha, \beta]; \Gamma^{s,\mu,\mathcal{M}}\right)$;

(ii) *a scalar valued function of the form* $B(\xi) = \sum_{i=0}^{\infty} b_i(\xi)^i$ *with real,* b_i, *and nonzero radius of convergence,* $|\xi| < \rho$;

(iii) *a polynomial transformation of the form,* $P\left(Q^t_{t_0}\Delta\right) = Q^t_{t_0} P_1(\Delta) + \ldots + \left(Q^t_{t_0}\right)^\ell P_\ell(\Delta)$, *where* P_λ $(1 \leq \lambda \leq \ell)$ *are real-valued polynomials of degree* n_λ, $(1 \leq \lambda \leq \ell)$ *and where we formally replace the independent variable with our infinite dimensional Laplacian operator, (i.e.* $P_\lambda(\Delta) = d^\lambda_{n_\lambda}(\Delta)^{n_\lambda} + \ldots + d_0$, $d^\lambda_i \in \mathbb{R}$, $0 \leq i \leq n_\lambda$, $(1 \leq \lambda \leq \ell)$);

(iv) *a sequence,* μ *where* $q + 1 \leq L\dfrac{m_{q+1}}{m_q}$ *for all* q.

Then the series,

$$B\left(P\left(Q^t_{t_0}\Delta\right)\right) = \sum_{i=0}^{\infty} b_i \left(P\left(Q^t_{t_0}\Delta\right)\right)^i,$$

obtained by substituting $P\left(Q^t_{t_0}\Delta\right)$ for ξ into $B(\xi)$ and formally expanding it into a series of $Q^t_{t_0}\Delta$ monomials is applied to $u(t)$. This gives another parametric generalized functional, $\nu(t_0) \in \Gamma^{s',\mu,\mathcal{M}}$ continuous for $t \in [t_0, t] \subset [\alpha, \beta]$ where $s' \geq (\ell+1)s$.

Proof. A generalization of the proof in reference [20] implementing the techniques found in reference [21]. □

5. EXISTENCE FOR A SOLUTION TO A GENERALIZED DIFFUSION EQUATION

We define the generalized diffusion equation to be

$$\frac{\partial^\ell u(\tau)(x)}{\partial \tau^\ell} = \frac{\partial^{\ell-1}}{\partial \tau^{\ell-1}} P_1(\Delta)u(\tau)(x) + \ldots + P_\ell(\Delta)u(\tau)(x) \tag{5.1}$$

where $\tau \in [0,T] \subset \mathbb{R}$; $x \in K'\{m_p\}$ and $u(\tau)(x) \in \Gamma^{s,\mu,\mathcal{M}}$, with the initial conditions,

$$u(\tau)(x)\Big|_{\tau=\tau_0} \in \Gamma^{s,\mu,\mathcal{M}}, \ldots, \frac{\partial^{\ell-1}}{\partial \tau^{\ell-1}} u(\tau)(x)\Big|_{\tau=\tau_0} \in \Gamma^{s,\mu,\mathcal{M}}. \tag{5.2}$$

We can apply the transformations, $\left(Q^t_{t_0}\right)^\ell$ to (5.1) injecting the initial conditions (5.2) whenever necessary. This process gives us a transformation of the form

$$\left[I - \left(Q^t_{t_0}\right) P_1(\Delta) - \ldots - \left(Q^t_{t_0}\right) P_\ell(\Delta)\right],$$

which will have a local solution to problem (5.1)–(5.2) whenever the requisite hypothesis of Theorem 4.1 are satisfied.

REFERENCES

[1] F. A. Berezin and M. A. Shubin, *The Schrödinger Equations,* Kluwer Academic Publishers, Norwell, MA, 1991.

[2] F. A. Berezin, *The Method of Second Quantization,* Academic Press, New York and London, 1966.

[3] N. N. Bogoliubov and D. V. Shirkov, *Introduction to the Theory of Quantized Fields,* John Wiley & Sons, New York, 1976.

[4] F. Cholewinski, Generalized Fock spaces and associated operators, *SIAM Journal of Mathematical Analysis,* **15** (1984), No. 1, 177–202.

[5] V. A. Fock, *Fundamentals of Quantum Mechanics,* MIR Publishers, Moscow, 1978.

[6] A. Friedman, *Generalized Functions and Partial Differential Equations,* Prentice Hall Inc., New Jersey, 1963.

[7] K. O. Friedrichs and H. N. Shapiro, *Integration of Functional,* Lecture Notes, New York University, Courant Institute of Mathematical Sciences, 1957.

[8] I. M. Gelfand and G. E. Shilov, *Generalized Functions,* Vol. II (Trans. Morris D. Friedman, Amiel Fernstein, Christian P. Pellzer), Academic Press, New York, 1960.

[9] I. M. Gelfand and N. Y. Vilenkin, *Generalized Functions,* Vol. IV (Trans. Amiel Feinstein), Academic Press, New York, 1964.

[10] D. Kastler, *Introduction a L'Electrodynamique Quantique,* Dunod, Paris, 1961.

[11] H. Komatsu, Ultradistributions, I, Structural theorems and a characterization, *Journal of the Faculty of Science, Tokyo, Section 1A Mathematics,* **20** (1973), 25–105.

[12] P. Kristensen, I. Mejlbo and E. T. Poulsen, Tempered distribution in infinitely many dimensions I, Canonical field operators, *Communications Mathematical Physics,* **1** (1965), 175–214.

[13] P. Kristensen, I. Mejlbo and E. T. Poulsen, Tempered distribution in infinitely many dimensions II, Displacement operators, *Math. Scand.,* **14** (1964), 129–150.

[14] P. Kristensen, I. Mejlbo and E. T. Poulsen, Tempered distribution in infinitely many dimensions III, Linear transformations of field operators, *Communications Mathematical Physics,* **6** (1967), 29–48.

[15] T. P. G. Liverman, *Generalized Functions and Direct Operational Methods,* Prentice Hall, Englewood Cliffs, New Jersey, 1964.

[16] S. Pilipovic, Structural theorems for periodic ultradistributions, *Proceedings of the American Mathematical Society,* **98,** (1986), No. 2, 261–266.

[17] S. Pilipovic and A. Takaci, Space H'(M) and convolutors, *Proc. of the Moscow Conference on Generalized Functions,* Moscow, 415–427, 1981.

[18] C. M. Roumieu, Sur quelques extensions de la notion de distributions, *Ann. Scient. E. Norm. Sup.,* **77** (1960), 41–121.

[19] J. Rzewuski, On entire functionals in quantum field theory, *Report on Mathematical Physics,* **1** (1970), No. 1, 1–27.

[20] J. Schmeelk, An infinite dimensional Laplacian operator, *J. Differential Equations,* **36** (1980), No. 1, 74–88.

[21] J. Schmeelk and A. Takaci, Ultra annihilation and creation operators, *Portugaliae Mathematica,* **49** (1992), No. 3, 263–279.

[22] J. Schmeelk, Infinite dimensional Fock spaces and associated creation and annihilation operators, *J. of Math. Analysis and Appl.,* **134** (1988), 432–461.

[23] J. Schmeelk, Infinite dimensional Fock spaces and Schrödinger equation solutions, *Applicable Analysis,* **40** (1991), 233–250.

[24] J. Schmeelk, Application of test surfunctions, *Applicable Analysis,* **17** (1984), 169–185.

[25] J. Schmeelk, Infinite dimensional parametric distributions, *Applicable Analysis,* **24** (1987), No. 4, 291–319.

[26] F. Treves, *Ovcyannikov Theorem and Hyperdifferential Operators,* Notes de Mathematica, No. 46, Institute de Mathematica Pura e Applicada, Brasil, 1968.

[27] V. V. Zarinov, Compact families of locally convex topological vector spaces, Frechet–Schwartz and dual Frechet–Schwartz spaces, *Russian Math. Surveys,* **34** (1979), No. 4, 105–143.

[28] A. H. Zemanian, *Distribution Theory and Transform Analysis,* McGraw-Hill, New York, 1965.

7th Int. Coll. on Differential Equations, pp. 371-379
D. Bainov (Ed)

Analysis of a Degenerate Hopf bifurcation in a PID controlled CSTR

Stefano Serra* and Cristina Tablino Possio**

*Dipartimento di Informatica, Università di Pisa, C.so Italia, 40, 56100 Pisa, Italy.
E-mail: Serra@morse.dm.unipi.it
**Dipartimento di Matematica, Università di Milano, via Saldini, 50, 20133 Milano, Italy.
E-mail: Tablino@vmimat.mat.unimi.it

Abstract The analytical analysis of a nongeneric Gavrilov-Guckenheimer bifurcation unfolding is performed in the case of a typical PID controlled chemical reactor model. Nongeneric Type I Hopf degeneracies are always present in PID controlled dynamic systems, in general giving rise to complex nonlinear dynamic phenomena.
Key words: Dynamic systems, bifurcation theory, degenerate Hopf bifurcations, chemical reactors.

1 INTRODUCTION

In the last decades nonlinear dynamics has begun one of the most promising fields in a great variety of applied sciences, starting from the characterisation of related complex phenomena, through the analysis of their dependence on system parameters, arriving till their exploitation for practical purposes, as in the case of *Controlling chaos.*

Chemically reacting systems surely result a wide class of realistic problems that may show surprisingly complicated patterns of dynamic behaviour, even if simple mathematical models are considered in order to describe involved phenomena. In particular, the Continuous Stirred Tank Reactor (CSTR) results the typical reference system among lumped dynamic systems in the field of chemical reactor modelling. In fact, owing to both the simplicity and strong nonlinearity of the corresponding mathematical model, this reactor can be successfully considered as a prototype for chemical process design, in which possible continuous variations in space can be approximated by a proper series of subsequent CSTRs. This is the main reason of the great attention devoted to such a subject in chemical engineering literature: CSTRs with complex reactions [13, 12, 6, 23, 4, 19], with a traditional controller [10, 21, 24, 26, 27, 28], with external capacitance [22, 20], with forcing [16, 14, 3, 17, 29].

It is worthwhile outlining that chemical reactors are often coupled with less or more sophisticated types of controllers in order to improve system performances, e.g. in order to make shorter the transient time before convergence to an asymptotically stable invariant state or to make asymptotically stable an asymptotically unstable invariant state, showing optimal properties with respect to the applicative point of view. Starting from these considerations, Proportional-Integral (PI) and Proportional-Integral-Derivative (PID) controllers result a simple, not expensive and, in general, successful tool in chemical reactor processes despite their linear nature. Nevertheless, they are just the direct responsibles for the appearance of a specif type of high codimesion bifurcations, in the neighbourhood of which complex dynamic phenomena (simultaneous periodic solutions, isolas of periodic orbits, quasi-periodic oscillations, period doubling cascades leading to chaos) may originate according to the statements of Normal Form Theory with respect to their parametric unfolding [7, 1, 9, 2]. In fact, some attention has been paid in chemical engineering literature on this subject; mention can be made to some papers [22, 6, 19, 20, 26, 27, 28, 24] approaching the analysis by different ways.

So, in this perspective, the most effective method to foresee and to plainly understand the existing patterns of nonlinear dynamics is ensured by Bifurcation Theory: this powerful theory allows the detection of all distinct structurally stable modes of behaviour that can be obtained by varying system parameters on their whole admissible domain, the main tool being generic procedures in order to separate the parameter space into regions showing system solutions with different qualitative structure. So, in such, at first sight, paradoxical way the definition and the analysis of robust phenomena, required from an applicative point of view, is strictly linked to the location of bifurcation boundaries in the system parameter space. More surprising, the spectral properties of the Jacobian matrix, evaluated at the considered invariant state, i.e. the first order dynamic approximation, give first strong information about new appearing structurally stable modes according to Normal form theory. For instance the so called Type I Hopf degeneracies are defined according to the number and type of null real part eigenvalues of the Jacobian matrix evaluated at the equilibrium point and they correspond to the case where the first hypothesis of the Hopf bifurcation theorem is not fulfilled [18].

The paper is organised as follows. In section 2 the PID controller law is defined in the case of a generic $n-$dimensional time continuous dynamic system and its application to the CSTR is considered. In section 3 some general bifurcational properties of PID controlled time continuous dynamic systems are outlined, together with the non-generic nature of the involved Type I Hopf degeneracies. Section 4 is devoted to the analysis of the Gavrilov-Guckenheimer bifurcation unfolding for the PID controlled CSTR: owing to its specific non-generic nature careful asymptotical expansions with respect to the Jacobian eigenvalues and to the unfolding parameters are required. Lastly special attention is paid to the analysis of the asymptotical stability properties of the observed family of invariant tori, also with respect to the derivative control parameter k_D values.

2 PID CONTROLLED DYNAMIC SYSTEMS

Let us consider a $n-$ dimensional uncontrolled time continuous dynamic system

$$d\mathbf{x}/dt = \mathbf{f}(\mathbf{x}, u) \quad \mathbf{x} \in \mathbb{R}^n, u \in \mathbb{R} \tag{2.1}$$

such that $\mathbf{x_s} = \mathbf{0}$ is an equilibrium point for a zero value of the parameter u. At each time t, a PID controller moves the parameter u from its reference value $u = 0$ according to the actual displacement of a phase variable x^*, chosen in the set $\{x_1, \dots, x_n\}$, with respect to its reference value $x_s^* = 0$ as to the following law

$$u = -k_P x^* - k_I z - k_D f^*, \quad z = \int_0^t x^* dt \tag{2.2}$$

where f^* is the component of the function $\mathbf{f}$ corresponding to the phase variable x^*. By differentiating the z relation in Eq. (2.2) with respect to the time, the following $(n+1)$–dimensional PID controlled dynamic system is obtained

$$\begin{cases} d\mathbf{x}/dt &= \mathbf{f}(\mathbf{x}, -k_P x^* - k_I z - k_D f^*) \\ dz/dt &= x^*. \end{cases} \tag{2.3}$$

In this paper a PID controlled CSTR is considered as practical example in the general class of PID controlled time continuous dynamic systems. This reactor consists of a vessel, placed in a coolant bath at constant temperature y_e; the vessel is stirred sufficiently rapidly so that, the new entering flow element leaves the global contents of the vessel spatially homogeneous at any time. Under this assumption the considered reacting system is correctly described by means of two ordinary differential equations, respectively representing the material and the energy conservation laws. More precisely, we are considering a single ideal non-isothermal CSTR where a generic 1st order reaction takes place in the liquid phase, the reaction rate following the Arrhenius law $r = a(1-x)\exp(-b/y)$, where x is the extent of conversion and y is the reactor temperature. Now, the PID controller law tries to vanish the reactor temperature displacement $y - y_{set}$ by varying the displacement $y_0 - y_{0set}$ of the reactor inlet temperature, assumed as manipulable parameter u according to Eq. (2.2): y_{0set} is the inlet reactor temperature value for which y_{set} results the reactor temperature at the equilibrium point for the uncontrolled reactor. In such a way, by considering a fully dimensionless notation [21], the following $3D$ autonomous ODEs system is obtained

$$\begin{cases} dx/dt &= -x + a(1-x)\exp(-b/y) \\ dy/dt &= [y_{0set} - k_P(y-y_{set}) - z - y - c(y-y_e) + a(1-x)\exp(-b/y)]/(1+k_D) \\ dz/dt &= k_I(y - y_{set}). \end{cases} \tag{2.4}$$

Notice that the equilibrium point $(x, y, z) = (x_s, y_s, 0)$, without any dependence on control parameters, is simply the extension of the equilibrium point $(x, y) = (x_s, y_s)$ of the uncontrolled reactor, whose $2D$ model is obtained for $(k_I, k_P, k_D) = (0, 0, 0)$, $z = 0$.

Such a model offers several interesting phenomena related to nonlinear dynamics Theory, as already appreciated in [21] where Feigenbaum universality property of the observed period doubling scenario was proved. Moreover, these complex phenomena can obtain a more thorough explanation when framed in the rich global bifurcation structure proper of this low dimensional multiparameter dynamic system [26, 28].

3 BIFURCATIONAL PROPERTIES OF PID CONTROLLED DYNAMIC SYSTEMS

PID controlled dynamic systems result a strongly interesting class of dynamic systems since some general properties can be stated about their codimension 1 bifurcational behaviour. More precisely, if the equilibrium point $\mathbf{x_s}$ for the uncontrolled system at $u = 0$ does not show a fold bifurcation or a Hopf bifurcation, the following two statements can be proved [5]. Clearly, this is just an academic case since from an applicative point of view it is unnatural to control an equilibrium point under bifurcation conditions.

Now, the former statement concerns the system bifurcational behaviour along the hyperplane $k_I = 0$; in any case, it is worthwhile outlining that control parameters k_I, k_P, k_D are the most natural ones with respect to a bifurcational analysis since the equilibrium point of both the uncontrolled and controlled system does not show dependence from them.

Proposition 3.1 [5] *A PI(D) controlled dynamic system always possesses a fold bifurcation boundary (one null eigenvalue) located at the hyperplane $k_I = 0$. In addition, this results the unique possible fold bifurcation boundary under the quoted hypotheses.*

Nevertheless, this fold bifurcation boundary is non-generic since at $k_I = 0$ there is a break-down of the system dimensionality, coming back to the original value showed in the P controlled case or in the uncontrolled one. However, despite this break, the crossing of such a boundary always involves a change in the global stability properties of physically significant equilibria. The latter statement establishes a systematical relation existing between P and PI(D) controlled dynamic systems. Once again under the quoted weak hypotheses, the following holds:

Theorem 3.1 [5] *Let Γ be the Hopf bifurcation locus in the plane (k_I, k_P) for a generic PI(D) controlled dynamic system and let k_P^* be the fold bifurcation value for the P controlled case. The locus Γ begins at $(k_I, k_P) = (0, k_P^*)$ with a null frequency ω and approaches infinity continuously, as ω approaches infinity, without intersecting itself. The locus also passes through all existing points $(k_I, k_P) = (0, k_P^{**})$, where k_P^{**} are Hopf bifurcation values for the P controlled case.*

By joining these two statements, the following one clearly holds:

Proposition 3.2 *For a generic PI(D) controlled dynamic system the fold and Hopf bifurcation boundaries always intersect at higher codimension bifurcation loci, in the simplest case:*

- *the locus $(0, k_P^*(k_D))$ resulting in a Bogdanov-Tackens (BT) bifurcation (double null eigenvalue),*
- *all loci $(0, k_P^{**}(k_D))$ resulting in a Gavrilov-Guckenheimer (GG) bifurcation (one null eigenvalue plus one imaginary pair).*

In such a way, a generic PI(D) controlled dynamic system a priori locates high codimension Hopf bifurcation degeneracies related to the failure of the first hypothesis in the Hopf bifurcation theorem [18], i.e. when more than one null eigenvalue and/or one imaginary pair of eigenvalues simultaneously appear, so revealing themselves to be the source of complex dynamic phenomena. Refer to the relevant literature [1, 9, 30, 2] for the complete classification of Type I Hopf degeneracies and the related existing polynomial normal form unfoldings, to which any dynamic system can be reduced, if necessary with the preliminary aid of center manifold projection [11]. Since these bifurcation boundaries can be though as "organising" loci, in which neighbourhood families of invariant tori or period doubling sequences leading to chaos can appear, the theoretical and practical interest is clearly understood. Nevertheless, owing to the non-generic nature of the involved fold bifurcation, all these bifurcation loci show a corresponding non-generic nature, so requiring careful asymptotical expansions for their unfolding analysis, as exemplified in the following section in the case of the GG bifurcation curve shown by the PID controlled CSTR.

4 THE ANALYSIS OF GG BIFURCATION IN THE PID CONTROLLED CSTR

The analysis of GG bifurcation in the PID controlled CSTR can be performed by means of several subsequent steps. First, it is required to carefully locate the GG bifurcation in the control parameter space (k_I, k_P, k_D). Notice that in the case of the considered dynamic system the related defining conditions are easily obtained with respect to the algebraic invariants of the Jacobian J_s evaluated at the equilibrium point as

$$tr(J_s(k_I, k_P, k_D)) = det(J_s(k_I, k_P, k_D)) = 0, \quad m_2(J_s(k_I, k_P, k_D)) > 0. \tag{4.1}$$

This task is further simplified since these algebraic invariants are linear with respect to the control parameters, so that the obtained GG locus is the half-straight line

$$k_{P,GG}(k_D) = -d(1 + k_D) + e, \quad k_{I,GG} = 0, \quad m_2(J_s(k_{I,GG}, k_{P,GG}, k_D)) > 0. \tag{4.2}$$

Here and hereafter refer to [25] for any detail concerning analytical calculations and the whole analytical procedure. Now, in the second step the reduction to the corresponding normal form system is performed (the center manifold projection is unnecessary in the present case) so that two subsequent transformations are considered: first the Jacobian matrix is put in its canonical Jordan form L by the usual linear transformation, so obtaining $du/dt = L\mathbf{u} + \mathbf{g}(\mathbf{u}, \mu, \nu, \sigma), \mathbf{u} \in \mathbb{C}^3$ and after a nonlinear polynomial transformation $\mathbf{u} = \mathbf{v} + \mathbf{P}(\mathbf{v})$ is applied in order to reduce nonlinear contributions of $\mathbf{g}$ to the so called resonant terms, exploiting the essential nonlinear features of the system at the GG bifurcation [8]. In such a way the GG bifurcation normal form unfolding results in cylindrical co-ordinates with respect to parameters $\mu = k_I$, $\nu = k_P - k_{P,GG}$ and $\sigma = k_D > -1$ as

$$\begin{cases} dr/dt = \alpha(\mu,\nu,\sigma)r + a_1(\mu,\nu,\sigma)rz + a_2(\mu,\nu,\sigma)r^3 + a_3(\mu,\nu,\sigma)rz^2 + O(4) \\ dz/dt = \lambda(\mu,\nu,\sigma)z + b_1(\mu,\nu,\sigma)r^2 + b_2(\mu,\nu,\sigma)z^2 + b_3(\mu,\nu,\sigma)r^2z + b_4(\mu,\nu,\sigma)z^3 + O(4) \\ d\Theta/dt = \omega_0 + O(1), \end{cases} \tag{4.3}$$

$\alpha(\mu, \nu, \sigma)$ and $\lambda(\mu, \nu, \sigma)$ respectively being the real part of conjugate complex eigenvalues and the real eigenvalue vanishing at $(\mu, \nu, \sigma) = (0, 0, \sigma)$ $\forall \sigma$. This normal form results slightly different from both the Gavrilov and Guckenheimer normal form: the present choice is due to the natural parametrisation with respect to Jacobian eigenvalues so that the break-down of the system dimensionality at $k_I = 0$ is directly represented. Lastly, the analysis of the obtained normal form unfolding has to be carried out in order to outline all complex dynamic phenomena that may occur in the neighbourhood of the present bifurcation.

Since the system (4.3) truncated at cubic terms is rotationally symmetric about the z axis, its analysis can be performed, with a significant reduction in complexity, by analysing only the planar system in the phase space (r, z) for which three different equilibria exist:

$$\begin{array}{ll} (i) & r = 0, z = 0; \\ (ii) & r = 0, z = -(\lambda(\mu,\nu,\sigma) + O(2))/b_2(\mu,\nu,\sigma); \\ (iii) & b_1(\mu,\nu,\sigma)r^2 = \chi(\mu,\nu,\sigma), \quad z = -(\alpha(\mu,\nu,\sigma) + O(2))/a_1(\mu,\nu,\sigma), \\ & \chi(\mu,\nu,\sigma) = -(\lambda(\mu,\nu,\sigma) + b_2(\mu,\nu,\sigma)z)z. \end{array} \tag{4.4}$$

It is worthwhile outlining that the most interesting equilibrium point is the third one because it represents a periodic solution for $\chi(\mu, \nu, \sigma)/b_1(\mu, \nu, \sigma) > 0$ with respect to the $3D$ system, while the other ones simply result steady state solutions $(r = 0)$.

Before to deal with possible bifurcations of the equilibrium point (*iii*) a fundamental remark must be done, it being the basic motivation of the subsequent considered procedure. By direct calculations it is proved [25] that at the GG bifurcation $\forall\sigma\ b_i(0,0,\sigma)=0$, $b_{i,\nu}(0,0,\sigma)=0$, $i=1,\dots,4$, so that the truncated system dimensionality decreases by one. This pathological feature is not so surprising if one remembers that the same decreasing of dimensionality arises for the original dynamic system (2.3) along the plane $\mu = k_I = 0$; this involves the existence of special curves along which no convergence of (ii) and (iii) to the trivial solution (i) occurs for vanishing μ and ν. So, one of the strongest properties of the quoted system is exactly preserved by nonlinear transformations involved during the reduction to the GG normal form. In addition, it is quite natural to think that this will result a natural property also in the case of generic PID controlled dynamic systems. Nevertheless, just because of such nongeneric nature of GG bifurcations, the planar system (4.3) requires to consider (at least) the first order terms of coefficients b_i, λ, α with respect to the parameters μ, ν so that the following asymptotic expansion formula for the periodic branch (iii) is obtained:

$$(iii)\ z=-(\alpha_\mu\mu+\alpha_\nu\nu)/a_1+O(2),\ \ r^2=\lambda_\mu(\alpha_\mu\mu^2+\alpha_\nu\nu\mu+O(3))/(a_1(b_{1,\mu}\mu+O(2))) \tag{4.5}$$

where the coefficients α_μ, α_ν, λ_μ, $b_{1,\mu}$, a_1 all result different from 0 [25]. Here and hereafter the explicit dependency of all coefficients on μ, ν and σ is neglected if unnecessary.

The next relevant step consists on the detection and definition of possible bifurcations of the periodic branch (4.5), in order to investigate the potential richness of nonlinear patterns of dynamic behaviour that may be present in the full dimension dynamic system.

For instance, the occurrence of a Hopf bifurcation along the branch (*iii*) of the planar system means the appearance of a limit cycle centred in such equilibrium point; but, with respect to the $3D$ system, the rising of a limit cycle is then observed around each point of the curve (*iii*) $r^2 = \chi(\mu,\nu,\sigma)/b_1(\mu,\nu,\sigma) > 0$ so that the birth of a family of invariant tori is simply detected by means of two subsequent Hopf bifurcations: the former is the GG Hopf degeneracy for the system (2.3) and the latter the codimension 1 Hopf bifurcation for the planar system (4.3), respectively defining the two different frequencies of the quasiperiodic motion. Starting from these considerations, first we will give the defining conditions of the latter Hopf bifurcation in terms of the parameters (μ,ν) and of the coefficients in the normal form; after the asymptotical stability properties of the (possible) family of invariant tori will be investigated.

By considering the first two equations of system (4.3) truncated at the fourth order and the algebraic invariants of the corresponding Jacobian matrix evaluated at (iii) and by using the well known conditions for Hopf bifurcation in a $2D$ system ($tr\,J_{(iii)}(\mu,\nu,\sigma)=0$, $det\,J_{(iii)}(\mu,\nu,\sigma)>0$) we obtain that the toroidal branch is defined as

$$\nu_{Tor}(\mu,\sigma) = ((k_P)_{Tor}-(k_P)_{GG})(\mu,\sigma) = -(a_1b_{1,\mu}+3a_2\alpha_\mu)\mu/(3a_2\alpha_\nu)+O(\mu^2) \tag{4.6}$$

and this linear dependence of $\nu_{Tor}(\mu,\sigma)$ completely avoids possible pathological situations for $\mu\to 0$ as previously remarked, since $r^2_{(iii)}(\mu,\nu_{Tor}(\mu,\sigma)) = O(\mu)$ collapses into the trivial equilibrium point $(r,z)=(0,0)$. Nevertheless, owing to the vanishing of b_i normal form coefficients at the GG bifurcation, some additional conditions must be stated:

• the condition $det\,J_{(iii)}>0$ requires that $(\mu,\nu_{Tor}(\mu,\sigma))$ belongs (for small values of μ) to the halfspace

$$r^2_{(iii)}\mu(3a_2\lambda_\mu-2a_1b_{1,\mu})>0, \tag{4.7}$$

• the analysis makes sense if and only if $r^2_{(iii)}>0$, so $(\mu,\nu_{Tor}(\mu,\sigma))$ should also verify the inequality

$$\lambda_\mu(\alpha_\mu\mu+\alpha_\nu\nu_{Tor}(\mu,\sigma))/(b_{1,\mu}a_1)>0; \tag{4.8}$$

therefore the locus $(\mu, \nu_{Tor}(\mu, \sigma))$ must belong, for $\mu > 0$ or for $\mu < 0$, to the intersection of both the (4.7) and (4.8) halfspaces to make possible the occurrence of a family of invariant tori with secondary frequency $\Omega(\mu_{Tor}(\nu, \sigma), \nu, \sigma)$. Both conditions (4.7) and (4.8) will be verified in the case of the considered set of system parameter values [27], together with the well known transversality condition

$$A_2 = (\partial A(\mu, \nu)/\partial \mu)|_{\nu=\nu_{Tor}(\mu,\sigma)} = \lambda_\mu(1 + 3a_2\alpha_\mu/(a_1 b_{1,\mu}))/2 + O(1) \neq 0, \tag{4.9}$$

(for (μ, ν) not long a curve tangent to the $\mu = 0$ plane) so that Type II Hopf bifurcation degeneracies [7] are avoided.

Now, we are concerning with the characterisation of local and global asymptotical stability properties of the family of invariant tori detected via the Hopf bifurcation of the equilibrium point (iii) of the planar system (4.3): the standard procedure [11] has been considered in order to transform the system into its corresponding Hopf bifurcation normal form in polar co-ordinates $u = \rho e^{i\psi}$

$$\begin{cases} d\rho/dt &= \rho(A_2\eta + c_1(0, \sigma)\rho^2 + O(3)) \\ d\psi/dt &= \Omega_0(1 + O(1)). \end{cases} \tag{4.10}$$

Special attention is paid to the coefficient $c_1(\eta, \sigma) = Re(K_1(\eta, \sigma))$, $\eta = \mu - \mu_{Tor}(\nu, \sigma)$ evaluated at $\eta = 0$, that defines the Hopf bifurcation supercritical or subcritical nature and, therefore, the local asymptotical stability properties of the family of invariant tori (refer to [25] for any detail in calculations). By following the standard procedure this relation holds

$$Re(K_1(\eta, \sigma))_{|\eta=0} = \left[\left(G_{2,0}G_{1,1} - 2|G_{1,1}|^2 - |G_{0,2}|^2/3\right) i/(2\Omega_0) + G_{2,1}/2\right]_{|\eta=0}, \tag{4.11}$$

where in the present case the secondary frequency results $\Omega_0(\mu_{Tor}(\nu, \sigma), \nu, \sigma) = \sqrt{3a_2\lambda - 2a_1b_1 \cdot r_{(iii)}}$ and $G_{i,j}$ functions are linear combinations of the second and third order derivatives of the planar system (4.3) evaluated at (iii) along the branch (4.6), according to the linear transformation that put $J_{(iii)}(\mu, \nu_{tor}(\mu, \nu))$ in its canonical Jordan form.

The theoretic calculation showed that $c_1(0, \sigma) = \hat{c}_1(\sigma)(1 + O(\nu))$ where $\hat{c}_1(\sigma)$ is an absolute constant value with respect to μ and ν. The numerical evaluation of the involved quantities, for the reference set of system parameter values and for $\mu \to 0$ along the toroidal branch, totally agrees with the quoted theoretical expectation. In addition, the theoretically calculated branch h^1_* exhibits a perfect agreement with the corresponding branch h^1 (at least in the neighbourhood of the analysed bifurcation), previously numerically calculated with the help of LOCBIF package [15], but without obtaining any information about asymptotical stability properties. Just with respect to the local asymptotical stability properties of such toroidal branch different results are obtained according to the value of the control parameter $\sigma = k_D$:

- Neutral saddle bifurcation (two real opposite eigenvalues) for $-1 < k_D < -2.6 \cdot 10^{-3}$,
- subcritical Hopf bifurcation for $-2.6 \cdot 10^{-3} < k_D < 6.3881 \cdot 10^{-2}$,
- supercritical Hopf bifurcation $6.3881 \cdot 10^{-2} < k_D < (k_D)_{G1} = 1.076 \cdot 10^{-1}$,

where $(k_D)_{G1}$ is the derivative control parameter value corresponding to the G1 bifurcation (triple null eigenvalue), i.e. the terminal point of the GG bifurcation half-straight line in the parameter space (k_I, k_P, k_D). In addition, the third real eigenvalue is always negative in the related interval of k_D values, so that the global asymptotical stability properties of the family of invariant tori are uniquely defined by the nature of the quoted planar Hopf bifurcation. Notice also that $\Omega_0(\nu, \sigma)$ (the second frequency of the present motion) is $O(1)$ with respect to ν, while the first frequency ω_0 is $O(0)$ with respect to ν, i.e. a constant value: so that the quasi-periodic solution of the full $3D$ system shows a fast frequency modulated by a slow one.

The validity of the considered procedure is guaranteed for k_D values far enough from the $k_{D,G1}$ value, since at this bifurcation point there is a break-down in the continuity of the canonical Jordan form and completely different asymptotical expansions hold for the Jacobian J_s eigenvalues.

REFERENCES

[1] V.I.Arnold,Geometrical methods in the theory of Ordinary Differential Equations, Springer, New York (1983).

[2] V.I.Arnold, V.S.Afrajmovich, Yu.S.Il'yashenko and L.P.Sil'nikov, in Dynamical Systems - Encyclopaedia of Mathematical Sciences, Vol. 5, V. I. Arnold (Ed.), Springer, New York (1994).

[3] G.T.Caneba and M.J.Crossey, Chem. Engng. Comm., 51, 1-18 (1987).

[4] P.C.H.Chan, D.G.Retzloff, R.Mohamed, B.Berdouzi, C.Chicone and D.Offin, Chem. Engng. Comm., 57, 105-138 (1987).

[5] H.-C.Chang and L.-H.Chen, Chem. Engng. Sci., 39, 1127-1142 (1984).

[6] W.W.Farr and R.Aris, Can. Math. Soc. Conf. Proc., 8, 397-418 (1987).

[7] M.Golubitsky and W.F.Langford, J. Differential Eqns., 41, 375-415 (1981).

[8] J.Guckenheimer, SIAM J. Math. Anal., 15, 1-49 (1984).

[9] J.Guckenheimer and P.Holmes, Nonlinear oscillations, Dynamical Systems and Bifurcations of Vector Fields, Springer, New York (1983).

[10] S.Hasegawa, N.Watanaba and M.Matsubara, Chem. Engng. Comm., 30, 35-53 (1984).

[11] B.D.Hassard, N.D.Kazarinoff and N.D.Wan, Theory and Applications of Hopf bifurcations, Cambridge University Press, Cambridge (1981).

[12] D.V.Jorgensen and R.Aris, Chem. Engng. Sci., 38, 44-53 (1983).

[13] C.Kahlert, O.E.Roessler, A.Varma, Modelling Chemical Reaction Systems, Springer, Heidelberg (1981).

[14] I.G.Kevrekidis, R.Aris and L.D.Schmidt, Chem. Engn. Sci., 41, 1549-1560 (1986).

[15] A.I.Knibik, Y.A.Kuznetsov, V.V.Levitin and E.V.Nikolaev, Physica D, 62, 360-371 (1993).

[16] J.C.Mankin and J.L.Hudson, Chem. Engn. Sci., 39, 1807-1814 (1984).

[17] M.A.McKarnin, L.D.Schmidt and R.Aris, Chem. Engn. Sci., 43, 2833-2844 (1988).

[18] J.E.Marsden and M.McCraken, The Hopf bifurcation and its applications, Springer, New York (1986).

[19] J.Moiola, M.C.Colantonio, A.Desages and J.Romagnoli, Chem. Engn. Sci., 45, 297-305 (1990).

[20] J.Moiola, A.Desages and J.Romagnoli, Chem. Engn. Sci., 46, 1475-1490 (1991).

[21] L.Pellegrini and G.Biardi, Computers Chem. Engng., 14, 1237-1247 (1990).

[22] J.B.Planeaux and K.F.Jensen, Chem. Engng. Sci., 41, 1497-1523 (1986).

[23] D.G.Retzloff, P.C.H. Chan, C.Chicone, D.Offin and R.Mohamed, Physica D 25, 131-154 (1987).

[24] S.Serra and C.Tablino Possio, TR n. 15 Dip. di Matematica, University of Calabria, Italy (1995).

[25] S.Serra and C.Tablino Possio, TR n. 5 Dip. di "Energetica - S. Stecco" , University of Florence, Italy (1996).

[26] C.Tablino Possio and L.Pellegrini, TR n. 22 Dip. of Matematica, University of Milano, Italy (1995).

[27] C.Tablino Possio and L.Pellegrini, in press on Int. J. Bif. and Chaos (1996).

[28] C.Tablino Possio and L.Pellegrini, Proceedings of the 2nd European Nonlinear Oscillation Conference (Prague, September 9-13, 1996).

[29] M.A.Taylor and I.G.Kevrekidis, Chem. Engng. Sci., 48, 2129-2149 (1993).

[30] P.Yu and K.Huseyin, IEEE Trans. autom. Control., 33, 28-41 (1988).

7th Int. Coll. on Differential Equations, pp. 381-388
D. Bainov (Ed)

ON THE ARITHMETIC PROPERTIES OF THE MODULAR FUNCTION INDUCED FROM THE HYPERGEOMETRIC DIFFERENTIAL EQUATION OF TYPE (3, 6)

Hironori SHIGA
Faculty of Science, Chiba University,
Yayoi-cho 1-33, Inage-ku, 263 Chiba, Japan

Abstract

In this paper we study the value of the modular function induced from the hypergeometric function of type $(3,6)$, and we construct certain class fields over the CM fields K of degree 8 with $\sqrt{-1} \in K$.

1 Introduction

In this note we study the arithmetic properties of a certain modular function of 4 variables, it is the inverse of the Schwarz mapping for the hypergeometric differential equation of type $(3,6)$ with special parameter $(\lambda_1, \cdots, \lambda_6) = (1/2, \cdots, 1/2)$. This modular function is studied recently by Matsumoto, Sasaki and Yoshida ([MSY] and [M]), and their investigation is based on the theory of the hypergeometric differential equation induced by Gelfand, Aomoto and others. As the result of this paper

(i) we give the necessary and sufficient condition for the value of this modular function at an algebraic point to be an algebraic number,

(ii) we construct the Shimura class field for the number field of degree 8 of the form $K = L(\sqrt{-1})$ with a totally real quartic field L.

2 The modular function of type $(3,6)$

In this section we relate the general properties of the modular function of type $(3,6)$ after [MSY] and [M].

Let $\mathcal{F}$ be the family of algebraic surfaces obtained as the double covering over $\mathbf{P}^2$ ramifying along 6 lines in general position. Always we consider the minimal nonsingular model of these surfaces. As well known they become $K3$ surfaces. Our modular function will be constructed as the inverse of the period mapping for the family $\mathcal{F}$.

Now the element of $\mathcal{F}$ is given in the form

$$S: \quad z^2 - \prod_{i=1}^{6}(x_{1i} + x_{2i}s + x_{3i}t) = 0 \tag{1}$$

in affine (s,t,z)-space. Where the parameter matrix

$$X = \begin{bmatrix} x_{11} & \cdots & x_{16} \\ x_{21} & \cdots & x_{26} \\ x_{31} & \cdots & x_{36} \end{bmatrix}$$

is supposed to satisfy the condition: (C) any 3×3 minor

$$d_{ijk} = \det \begin{bmatrix} x_{1i} & x_{1j} & x_{1k} \\ x_{2i} & x_{2j} & x_{2k} \\ x_{3i} & x_{3j} & x_{3k} \end{bmatrix}$$

becomes never zero for $\{i,j,k\} \subset \{1,2,3,4,5,6\}$. Put M be the totality of such 3×6 matrices. Suppose 2 elements (x_{ij}) and $x'_{ij})$ of M. If we have the relation

$$(x'_{ij}) = T(x_{ij}) \begin{bmatrix} d_1 & & & 0 \\ & \cdot & & \\ & & \cdot & \\ 0 & & & d_6 \end{bmatrix}$$

with $d_i \in \mathbf{C}^*$ $(i = 1, \cdots, 6)$ and a certain element T in $GL(3, \mathbf{C})$, we get two isomorphic surfaces. So the quatient space

$$X = GL(3, \mathbf{C}) \backslash M / (\mathbf{C})^6$$

is the parameter space of our family of surfaces. In this way we can regard the family $\mathcal{F}$ as the fibre space over X. Moreover we can find the element of M of the form

$$\begin{bmatrix} 1 & 0 & 0 & 1 & 1 & 1 \\ 0 & 1 & 0 & 1 & x'_1 & x'_2 \\ 0 & 0 & 1 & 1 & x'_3 & x'_4 \end{bmatrix}$$

in an arbitrary orbit $GL(3, \mathbf{C})(x_{ij})((\mathbf{C}^*)^6)$. Hence we know that $\mathcal{F}$ is a 4-parameter family and X is a Zariski open subset of $\mathbf{P}^4$.

Next we examine the property of the period mapping for our family $\mathcal{F}$. Because the minimal nonsingular model $\tilde{S}$ of the surface S is a $K3$ surface, there exists unique holomorphic 2-form on $\tilde{S}$ up to a constant factor, and the 2nd homology group $H_2(S, \mathbf{Z})$ is isomorphic to the free $\mathbf{Z}$-module of rank 22. This holomorphic 2-form is given by $(1/z) ds \wedge dt$. On our surface S we can find 16 independent algebraic elements of $H_2(S, \mathbf{Z})$, these are the 15 exceptional curve comming from the intersection points of the ramification loci and the hyperplane section. If we consider the quatient module of H_2 by this submodule, we obtain a free $\mathbf{Z}$-module of rank 6. We can visualize the generator system of this quatient module as the following. Here we assume the initial element $x_0 = (x_{ij})$ of M is a real matrix and the last columm of x_0 is given by ${}^t(1,0,0)$, and we denote the corresponding surface by S_0 for the moment. Set D_{ijk} be the triangle bounded by three lines ℓ_i, ℓ_j, ℓ_k, where ℓ_i is the line defined by the i-th columm of x_0:

$$\ell_i : \quad x_{1i} + x_{2i}s + x_{3i}t = 0.$$

Let $C_1, \cdots, C_6$ be these triangles:

$$C_1 = D_{123}, C_2 = D_{125}, C_3 = D_{135}, C_4 = D_{145}, C_5 = D_{234}, C_6 = D_{345}.$$

They constitute the required generator system on S_0.

If we consider any arc on X starting from x_0, we can make the continuation of the system $\{C_1, \cdots, C_6\}$ using the topological trivialization of the fibre space $(\mathcal{F}, X)$. Therefore we can construct an multivalued analytic mapping $\Phi : X \longrightarrow \mathbf{P}^5$ defined by

$$\Phi * x = (x_{ij}) \mapsto [\zeta_1, \cdots, \zeta_6] = [\int_{C_1} \frac{ds \wedge dt}{z}, \cdots, \int_{C_6} \frac{ds \wedge dt}{z}].$$

This is our period mapping. Based on this construction we can state the properties of Φ. The precise exposition is given in [M] and [MSY].

1). Let $\mathcal{D}$ be the image of X under the mapping Φ, and we perform the coordinate change of $\mathbf{P}^5$ given by the matrix:

$$P = \begin{pmatrix} 1 & 0 & 0 & 0 & 0 & 0 \\ 0 & -1 & 0 & 0 & 0 & 0 \\ 0 & 0 & -1 & 0 & 0 & 0 \\ 0 & 0 & 0 & -1 & 0 & 0 \\ 0 & 0 & 0 & 0 & \frac{-1}{1-i} & \frac{-i}{1-i} \\ 0 & 0 & 0 & 0 & \frac{-1}{1+i} & \frac{i}{1+i} \end{pmatrix}.$$

We set ${}^t[z_1, \cdots, z_6] = P\,{}^t[\zeta_1, \cdots, \zeta_6]$. By this transformation the image is mapped to the type IV domain

$$\mathbf{D}^+ = \{z = [z_1, \cdots, z_6] \in \mathbf{P}^5 : {}^t zAz = 0, {}^t\overline{z}Az > 0, \Im(\frac{z_3}{z_1}) > 0\}$$

, where A indicates the 6×6 matrix given by

$$\begin{pmatrix} 0 & 1 \\ 1 & 0 \end{pmatrix} \oplus \begin{pmatrix} 0 & 1 \\ 1 & 0 \end{pmatrix} \oplus \begin{pmatrix} -1 & 0 \\ 0 & -1 \end{pmatrix}.$$

The domain $\mathbf{D}^+$ is biholomorphically equivalent with

$$\mathbf{H}_2 = \{W \in M(2, \mathbf{C}) : \frac{W - {}^t\overline{W}}{2\sqrt{-1}} > 0\}.$$

The equivalence mapping from $\mathbf{H}_2$ to $\mathcal{D}$ is given by

$$x = (x_{ij})_{1 \le i,j \le 2} \quad \mapsto \quad {}^t[d_{34}, d_{12}, d_{14}, d_{23} \cdot d_{13}, d_{24}],$$

where d_{ij} indicates $\det \begin{pmatrix} \tilde{x}_{i1} & \tilde{x}_{j1} \\ \tilde{x}_{i2} & \tilde{x}_{j2} \end{pmatrix}$ for the extended matrix

$$\tilde{x} = (\tilde{x}_{ij}) = \begin{pmatrix} 1 & 0 & x_{11} & x_{12} \\ 0 & 1 & x_{21} & x_{22} \end{pmatrix}.$$

2). Set

$$G = Isom^+(A, \mathbf{Z}) = \{g \in PSL(6, \mathbf{Z}) : {}^t gAg = A, g(\mathbf{D}^+) = \mathbf{D}^+\},$$

and set

$$\Gamma = \{\gamma \in M(4, \mathbf{Z}\sqrt{-1}) : {}^t \overline{\gamma} J \gamma = J\},$$

with $J = \begin{pmatrix} O & E_2 \\ -E_2 & O \end{pmatrix}$. These 2 groups G and Γ act on $\mathbf{D}^+$ and $\mathbf{H}_2$, respectively, and they are isomorphic via the above equivalence. The fundamental group $\pi_1(X)$ induces the (projective) monodromy transformation group saying $Mono$ acting on $\mathbf{D}^+$ and it is naturally a subgroup of G. In fact we have

$$Mono = G(2) = \{g \in G : g \equiv E_6 \pmod 2\} \rtimes < \tau >,$$

where τ indicates the involution on $\mathbf{H}_2$ induced from $W \mapsto {}^t W$.

3). We have the following

Proposition 2.1 *Let us consider the mapping*

$$\rho : \mathbf{H}_2 \longrightarrow \mathcal{S}_4$$

given by

$$\rho(W) = \frac{1}{2}\begin{pmatrix} W +{}^t W & i(W -{}^t W) \\ -i(W -{}^t W) & W +{}^t W \end{pmatrix},$$

where $\mathcal{S}$ indicates the Siegel upper half space. The mapping ρ induces the modular embedding of $\mathbf{H}_2/\Gamma$ into $\mathcal{S}_4/SP(4,\mathbf{Z})$. Namely

$$\rho(\mathbf{H}_2)/Sp(4,\mathbf{Z}) \mid_{\rho(\mathbf{H}_2)} \cong \mathbf{H}_2/\Gamma,$$

where

$$\rho(\mathbf{H}_2) \mid_{\rho(\mathbf{H}_2)} = \{\gamma \in SP(4,\mathbf{Z}) : \gamma(\rho(\mathbf{H}_2)) = \rho(\mathbf{H}_2)\}.$$

Remark 2.1 The image $\rho(\mathbf{H}_2)$ is a 4-dimensional linear subvariety in $\mathcal{S}$.

4). Now we employ a new coordinate system to point out the surface in (1) instead of the old coordinate $(x_{\ell m}) \in X$. Let us note there are 10 divisions of the set $\{1,2,3,4,5,6\}$ in two equal parts. They induces 10 different products of minors:

$$\begin{array}{ll} Z_1 = d_{123}d_{456} & Z_2 = -d_{124}d_{356} \\ Z_3 = -d_{125}d_{346} & Z_4 = -d_{135}d_{246} \\ Z_7 = d_{136}d_{245} & Z_8 = -d_{145}d_{236} \\ Z_9 = -d_{146}d_{235} & Z_{10} = d_{156}d_{234}. \end{array}$$

These values determine a point on $\mathbf{P}^9$, and it is easy to check the point is independent of the choice of the representative $(x_{\ell m}$ in X. We denote this mapping by q, and we denote

$$Y = q(X) = \{[Z_1, \cdots, Z_{10}] \in \mathbf{P}^9 : (x_{\ell m}) \in X\}.$$

Naturally Y is biholomorphically equivalent to X via the mapping q, and we obtain $\mathbf{P}^4$ as the compactification $\overline{Y}$ of Y.

K. Matsumoto ([M]) has introduced the follow ing theta function (theta constants) to describe the modular function for the family $\mathcal{F}$. According to Freitag we define the theta function on $\mathbf{H}_2$:

$$\Theta\begin{bmatrix} a \\ b \end{bmatrix}(W) = \sum_{n \in \mathbf{Z}[i]^2} \exp(\pi i)[(n+a)^* W (n+a) + 2\Re(b^* n)],$$

where $W \in \mathbf{H}_2$, $a, b \in (\mathbf{Z}[i]/2)^2$.

These theta functions are related with the ordinary Riemann theta :

$$\Theta\begin{bmatrix} a \\ b \end{bmatrix}(W) = \vartheta\begin{bmatrix} \Re a & \Im a \\ \Re b & \Im b \end{bmatrix}(\rho(W)).$$

We make the correspondence between the special 10 theta characteristics and 10 divisions of the set $\{1, 2, 3, 4, 5, 6\}$ into two equal parts:

$$\frac{1}{1+i}\begin{bmatrix} 0 & 0 \\ 0 & 0 \end{bmatrix} \longleftrightarrow (135)(246) \qquad \frac{1}{1+i}\begin{bmatrix} 0 & 0 \\ 0 & 1 \end{bmatrix} \longleftrightarrow (136)(245)$$

$$\frac{1}{1+i}\begin{bmatrix} 0 & 0 \\ 1 & 0 \end{bmatrix} \longleftrightarrow (146)(235) \qquad \frac{1}{1+i}\begin{bmatrix} 0 & 0 \\ 1 & 1 \end{bmatrix} \longleftrightarrow (145)(236)$$

$$\frac{1}{1+i}\begin{bmatrix} 0 & 1 \\ 0 & 0 \end{bmatrix} \longleftrightarrow (134)(256) \qquad \frac{1}{1+i}\begin{bmatrix} 0 & 1 \\ 1 & 0 \end{bmatrix} \longleftrightarrow (156)(234)$$

$$\frac{1}{1+i}\begin{bmatrix} 1 & 0 \\ 0 & 0 \end{bmatrix} \longleftrightarrow (125)(346) \qquad \frac{1}{1+i}\begin{bmatrix} 1 & 0 \\ 0 & 1 \end{bmatrix} \longleftrightarrow (126)(345)$$

$$\frac{1}{1+i}\begin{bmatrix} 1 & 1 \\ 0 & 0 \end{bmatrix} \longleftrightarrow (124)(356) \qquad \frac{1}{1+i}\begin{bmatrix} 1 & 1 \\ 1 & 1 \end{bmatrix} \longleftrightarrow (123)(456)$$

By $\theta(ijk)(lmn)$ we denote the theta function with the corresponding charasteristic.

Using the above notations we can state the explicit representation of the modular function for $\mathcal{F}$ due to K. Matsumoto[M].

Theorem 2.1 *Theorem M The mapping from* $\mathbf{H}_2$ *to* $\overline{Y}$*:*

$$\lambda : [d_{ijk}d_{lmn}]_{ijk} = [\theta(ijk)(lmn)]_{ijk}$$

is the inverse of the period mapping Φ *(via the equivalence by the mapping* q*). Moreover* λ *determines an biholomorphic equivalence from the Satake compactification of* $\mathbf{H}_2/M$ *ono to the compactification* $\mathbf{P}^4$ *of* Y.

3 The arithmetic properties

The symmetric group S_6 acts on X as the permutation group of 6 columns. Hence S_6 acts on Y also.

Remark 3.1 The quatient space $X/S_6 \cong Y/S_6$ is the moduli space for our family of $K3$ surfaces with double covering structure (namely, as the space of polarized $K3$ surfaces with the polarization induced from the ramification divisor). So the inverse mapping $J : \mathbf{H}_2 \longrightarrow Y/S_6$ of the reduced period mapping

$$\Phi' : Y/S_6 \longrightarrow \mathbf{H}_2$$

gives the modular function (mapping) for our family $\mathcal{F}$. Moreover, the corresponding monodromy group is the full modular group Γ (see [M]).

We denote by $R = \mathbf{C}[Z_1, \cdots, Z_{10}]$ the graded polynomial ring over $\mathbf{C}$ generated by 10 linear functions on Y. The group S_6 acts on this graded ring also, and we denote the induced invariant subring by

$$R^{S_6} = \{P(Z_1, \cdots, Z_{10}) \in R : P^\sigma = P \quad \forall \sigma \in S_6\}$$

Proposition 3.1 *The invariant subring* R^{S_6} *is isomorphic with the polynomial ring*

$$\mathbf{C}[g_2, g_3, g_4, g_5, g_6]$$

with 5 generators. The generators are given by

$$g_k = \sum_{i=1}^{10} Z_i^k \quad (k = 2, 3, 4, 5, 6)$$

Proof. The group S_6 acts on the subspace Y of $\mathbf{P}^9$. To describe the action of S_6 on Y we choose the coordinate $[Z_1, Z_2, Z_3, Z_6, Z_7]$ for Y. This defines the projective coordinate of $\mathbf{P}^4$ (the compactification of Y). Going back to the definition of $\{Z_i\}$ we know the rests are given by

$$\begin{aligned} Z_4 &= -Z_1 - Z_2 + Z_3 & Z_5 &= -Z_1 + Z_6 - Z_7 \\ Z_8 &= -Z_2 - Z_3 - Z_7 & Z_5 &= Z_1 - Z_3 - Z_6 \\ Z_{10} &= -Z_2 - Z_6 + Z_7 \end{aligned}$$

Let us calculate the action on $\mathbf{P}^4$ of the generators $(1i)$ $(i = 2, 3, 4, 5, 6)$ of S_6. The permutation (12) induces

$$[Z_1, Z_2, Z_3, Z_6, Z_7] \longmapsto [-Z_1, -Z_2, -Z_3, -Z_9, -Z_8].$$

So it induces the matrix representation

$$\rho_{12} = \begin{pmatrix} -1 & 0 & 0 & 0 & 0 \\ 0 & -1 & 0 & 0 & 0 \\ 0 & 0 & -1 & 0 & 0 \\ -1 & 0 & 1 & 1 & 0 \\ 0 & -1 & 1 & 0 & 1 \end{pmatrix}$$

with respect to the basis $[Z_1, Z_2, Z_3, Z_6, Z_7]$ By the same way we obtain the matrices corresponding to $(1, i)$ $(i = 3, 4, 5, 6)$. Based on this calculation we obtain the table of characteristic polynomials. AS the consequence we can obtain the Poincaré polynomial $F(t)$ for our representation:

$$F(t) = \frac{1}{\sharp S_6} \sum_{\gamma \in S_6} \frac{1}{P_\gamma(t)} = \frac{t^{15}}{(t^2-1)(t^3-1)(t^4-1)(t^5-1)(t^6-1)}$$

where P_γ indicates the characteristic polynomial of the representation $\rho_{|gamma}$ of $\gamma \in S_6$. If we compare this result with Molien's theorem we know the invariant subring R^{S_6}, that is the polynomial ring generated by 5 generators with degree 2,3,4,5 and 6, respectively. We can examine directly that $g_k(Z_1, \cdots, Z_{10}) = \sum_{i=1}^{10} Z_i^k$ is invariant under the action of the generator system $(12), \cdots, (16)$ of S_6. Hence these polynomials are the generators of our invariant subring.

Q.E.D.

Now we define our modular mapping J from $\mathbf{H}_2$ to $\overline{Y}/S_6$ by

$$J(W) = (\frac{g_4}{g_2^2}, \frac{g_6}{g_2^3}, \frac{g_3^2}{g_2^3}, \frac{g_5^2}{g_2^5}).$$

Then we have the following:

Theorem 3.1 *1). Let W be an algebraic point on $\mathbf{H}_2$. Then $J(W)$ is an algebraic point if and only if $\rho(W) \in S_4$ is a CM-point.*

2). Let $W = \begin{pmatrix} w_{11} & w_{12} \\ w_{21} & w_{22} \end{pmatrix}$ be an algebraic point on $\mathbf{H}_2$, and set $\rho(W) = (\omega_{ij})$. Suppose $\rho(W)$ is a simple CM-point, then 2 fields $\mathbf{Q}(w_{ij})$ and $\mathbf{Q}(w_{kl})$ obtained by the adjunction of the coordinate are the same, and this number field takes the form $L(\sqrt{-1})$ with a certain totally real quartic field L.

Remark 3.2 We call a point $\Omega \in \mathcal{S}$ is a CM point when the corresponding abelian variety with the principal polarization is of CM type, and we call a simple CM point if it is a simple abelian variety of CM type.

Remark 3.3 Let $K = L(\sqrt{-1})$ be a CM field with a certain qurtic field L. If we consider the pairing with the primitive CM type $\Phi = (\sigma_1, \sigma_2, \sigma_3, \sigma_4)$, where σ_i indicates the embedding of K into $\mathbf{C}$, then the corresponding abelian variety A is simple CM. We can find its period matrix Ω on $\rho(\mathbf{H}_2)$, and we have

$$\mathbf{Q}(\omega_{ij}) = K, \qquad \mathrm{End}_0(A) \cong K.$$

Proof of the theorem. 1). We assume always W is algebraic, and it is equivalent to say $\Omega = \rho(W)$ is algebraic. It is assured in [S] that we have the following: $A(\Omega)$ is defined over $\overline{\mathbf{Q}}$ if and only if $\Omega = \rho(W)$ is a CM point. By the theorem of Igusa A_Ω is realized as the intersection of some quadratic surfaces with coefficients in ϑ constants $\vartheta\begin{bmatrix} \varepsilon \\ \varepsilon' \end{bmatrix}(0, \Omega)$ $(2\varepsilon, 2\varepsilon' \in \{0, 1\})$. So A_Ω is defined over $\overline{\mathbf{Q}}$ if and only if all the ratios of theta constants are algebraic numbers. We call this property "relative algebraic".

By (1) it is equivalent to say 10 our thetas $\Theta\begin{bmatrix} a \\ b \end{bmatrix}(W)$ are relatively algebraic. According to Matsumoto theta map theorem it is nothing but the point $Y(W)$ is algebraic.

2). In the previous paper [S],[S-W] it is shown that: if A_Ω is a simple CM abelian variety, then we have $\mathbf{Q}(\tau_{ij}) = \mathrm{End}_0(A_\Omega)$.

Let Λ be the lattice in $\mathbf{C}^4$ generated by ${}^t(\tau_{1i}, \cdots, \tau_{4i})$ $(i = 1, 2, 3, 4)$. Then we find that Λ is invariant under the transformation by $T = \begin{pmatrix} O & E_2 \\ E_2 & O \end{pmatrix}$. This transformation stands for $\sqrt{-1}$. So the adjunction field by coordinates should be the required form. And we can deduce it must coincides with $\mathbf{Q}(w_{ij})$.

Q.E.D.

Proposition 3.2 *Set $\mathcal{A}$ be the family of 4-dimensional principally polarized abelian varieties A with the following properties:*

1) $\mathbf{Q}(\sqrt{-1}) \subset \mathrm{End}_0(A)$

2) the action of $\sqrt{-1}$ on the tangent space has eigen values 1,1,-1,-1.

Then $\mathcal{A}$ is parametrized by $\rho(\mathbf{H}_2)/\Sigma$ as the subspace of $\mathcal{S}_4/Sp(4, \mathbf{Z})$.

Proof. The action of $\sqrt{-1}$ on the tangent space has two eigen spaces of dimention $r = 2, s = 2$. By the argument in [W-S] we know that the parameter space is a type VI domain with the dimension rs. On the otherhand $\rho(\mathbf{H}_2)/\Sigma$ is actually contained in this parameter space. If we consider the compactification of these 2 spaces, they must coincide.

Q.E.D.

Remark 3.4 Let $K = L(\sqrt{-1})$ be an arbitrary CM field of the type we are discussing. If we choose a primitive CM type Φ of this CM field, the pair (K, Φ) determines an abelian variety A with $K = \mathrm{End}_0(A)$. By the above proposition we can find a point $\rho(W)$ in $\rho(\mathbf{H}_2)$ corresponding to this abelian variety.

Theorem 3.2 *Let W be a point on $\mathbf{H}_2$. Suppose $\Omega = \rho(W)$ be a simple CM point, and we denote the field of the coordinate adjunction by $K = L(\sqrt{-1}) = \mathbf{Q}(w_{11}, \cdots, w_{22})$, and let Φ be the CM type of K corresponding to the abelian variety A_Ω. Then the field*

$$K\left(\frac{g_4}{g_2^2}(W), \frac{g_6}{g_2^3}(W), \frac{g_3^2}{g_2^3}(W), \frac{g_5^2}{g_2^5}(W)\right) \qquad (*)$$

gives the Shimura class field for the reflex K' of K. Especially, if $K/\mathbf{Q}$ is a galois extension, the above adjunction field is the Shimura class field of K itself.

Proof. The fundamental theorem of complex multiplication by Shimura-Taniyama says that the field of moduli of (A_Ω, Φ) gives the Shimura class field.The field $(*)$ is the field of moduli. So we get the above conclusion.

Q.E.D.

Remark 3.5 We have 4 cyclotomic fields $\mathbf{Q}(\zeta)$ of degree 8. Our theorem is valid for these cases. But they all have class number 1, so the corresponding values of our modular function do not go out from the original field $\mathbf{Q}(\zeta)$

REFERENCES

[A] K. Aomoto, On the structure of integrals of power products of linear functions, Sci. Papers, Coll. Gen. Education, Univ. Tokyo, 27(1977), 49-61.

[G] I.M. Gelfand, General theory of hypergeometric functions, Soviet Math. Dokl. 33(1986), 573-577.

[H1] R.P. Holzapfel, Ball model for Hilbert's twlfth problem, Max Plank Inst. Bonn, (1990), bd.95.

[H2] R.P. Holzapfel, On algebraic values of the Picard modular function, Spetial Differential Equations, Proc. Taniguchi workshop 1991, 48-61.

[M] K. Matsumoto, Theta functions on the bounded symmetric domain of type $I_{2,2}$ and the period map of a 4-parameter family of $K3$ surfaces, Math. Annalen, 295(1993), 383-409.

[MSY] Matsumoto, Sasaki, Yoshida, The monodromy of the period map of a 4-parameter family of $K3$ surfaces and the hypergeometric function of type (3,6), Int. J. Math., vol.3(1992), 1-164.

[S-T] Shimura.G and Y. Taniyama, Complex multiplication of abelian varieties and its application to number theory, Publ. Math. Soc. Japan, 1961.

[Shim] G. Shimura, Automorphic forms and the periods of abelian varieties, J. Math. Soc. Japan, 31(1979), 561-592.

[S1] H. Shiga, On the representation of Picard modular function by θ constants I-II, Pub. RIMS Kyoto Univ., 24(1988), 311-360.

[S2] H. Shiga, On the transcendency of the values of the modular function at algebraic points, S.M.F., Astérisques 209(1992), 293-305.

[S-W] Shiga.H and J. Wolfart, Criteria for complex multiplication and transcendence properties of automorphic fuctions, J. reine angew. Math., Bd 463(1995), 1-25.

[W] G.Wüstholz, Algebraische Punkte auf analytischen Untergruppen algebraischer Gruppen, Ann. Math.,(1989), 501-517.

7th Int. Coll. on Differential Equations, pp. 389-394
D. Bainov (Ed)

Caloric morphisms and a generalization of the Appell transformation

KATSUNORI SHIMOMURA

Department of Mathematical Sciences, Ibaraki University Mito, Ibaraki 310, Japan

Abstract. We give a systematic way to construct caloric morphisms between different dimensional Euclidean spaces.
Keywords: heat equation, caloric morphism, Appell transformation

1. Introduction

For non-negative integer k, $\mathbf{R}^{k+1}$ denotes the $k+1$-dimensional Euclidean space with coordinates $(t, x) = (x_0, x_1, \ldots, x_k)$. Let $D \subset \mathbf{R}^{m+1}$, $E \subset \mathbf{R}^{n+1}$ be domains, where m, n are positive integers.

Consider a mapping $f(t,x) = (f_0(t,x), f_1(t,x), \ldots, f_n(t,x)) : D \to E$ which preserves solution of heat equation with weight function φ. We denote by $(t,x) = (t, x_1, \ldots, x_m)$, $(\tau, y) = (\tau, y_1, \ldots, y_n)$ the points of $\mathbf{R}^{m+1}$, $\mathbf{R}^{n+1}$ respectively. We shall use the following:

$$\nabla := \left(\frac{\partial}{\partial x_1}, \ldots, \frac{\partial}{\partial x_m}\right), \quad \Delta := \sum_{j=1}^{m} \frac{\partial^2}{\partial x_j^2}, \qquad H := \frac{\partial}{\partial t} - \Delta.$$

A continuous function h is said to be caloric if h satisfies the heat equation

$$Hh = 0.$$

in the sense of distribution. Since H is a hypoelliptic operator, every caloric function is of class C^∞.

Definition 1. A continuous mapping $f : D \to E$ is said to be a caloric morphism, if there exists a positive continuous function φ on D, we call this function a weight for f, such that for every caloric function $h(\tau, y)$ on E, $\varphi(t,x)(h \circ f)(t,x)$ is also a caloric function on D.

Sometimes, we also call the pair (f, φ) a caloric morphism.

In this paper, we shall give a systematic way to construct caloric morphisms between different dimensional Euclidean spaces. It is an application of the characterization of the caloric morphism, which was given in [1]. Applying the construction to the Appell transformations, we can obtain a generalization of the Appell transformations.

2. Characterization of caloric morphism

First we recall the characterization of the caloric morphism (see [1]).

Theorem 1. *Let $f : D \to E$ be a C^∞-mapping. Then the following statements are equivalent:*

1. *f is a caloric morphism.*
2. *There exists a positive C^∞-function φ on D such that for any caloric polynomial $P(\tau, y)$ of degree ≤ 4, $\varphi(t,x)(P \circ f)(t,x)$ is a caloric function on D.*
3. *There exists a positive caloric function φ on D such that:*

   (i) $\varphi H f_j = 2\nabla\varphi\cdot\nabla f_j, \quad 1 \leq j \leq n$, *where $\cdot$ denotes the Euclidean inner product in $\mathbf{R}^m$.*

   (ii) $\nabla f_i\cdot\nabla f_j = 0, \quad 1 \leq i,j \leq n,\ i \neq j,$

   (iii) $|\nabla f_i| = |\nabla f_j|, \quad 1 \leq i,j \leq n,$

   (iv) $\nabla f_0 = 0$, *(f_0 depends only on t)*

   (v) $\dfrac{df_0}{dt} = |\nabla f_j|^2$, *($\dfrac{df_0}{dt} \geq 0$ and $|\nabla f_j|^2$ is independent of x).*
4. *There exist a positive caloric function φ on D and a function $\lambda(t) \geq 0$ on D such that*

$$H\left(\varphi(t,x)(u \circ f)(t,x)\right) = \lambda^2(t)\varphi(t,x)(Hu \circ f)(t,x)$$

   holds for every C^2 function u on E.

By the condition 3, φ is a C^∞-function and f is a C^∞-mapping.

Corollary 1. *If $n > m$, there are no non-constant caloric morphisms.*

proof. By (ii) and (iii) of Theorem 1 (3), $\nabla f_1, \ldots, \nabla f_n$ are n orthogonal vectors in $\mathbf{R}^m$ with same length. Since $n > m$, $\nabla f_1 = \cdots = \nabla f_n = 0$ for all (t, x) and then by (v), $\frac{df_0}{dt} = 0$ for all t.

In the case $m = n$, as shown in [3], f must be a composition of the following three type of transformations.

Example 1 (Appell Transformation)

$$f(t, x) = \Big(-\frac{1}{t}, \frac{x}{t} \Big), \quad \varphi(t, x) = \frac{1}{(4\pi |t|)^{n/2}} e^{-\frac{|x|^2}{4t}}$$

is a caloric morphism from $\mathbf{R}^{n+1} \setminus \{t = 0\}$ *to* $\mathbf{R}^{n+1} \setminus \{\tau = 0\}$.

Example 2 (Translation)

$$f(t, x) = (t + c, x + ta + b), \quad \varphi(t, x) = e^{\frac{1}{4}|a|^2 t + \frac{1}{2} ax}$$

where a, $b \in \mathbf{R}^n$, $c \in \mathbf{R}$, *is a caloric morphism from* $\mathbf{R}^{n+1}$ *to* $\mathbf{R}^{n+1}$.

Example 3 (Similarity in x)

$$f(t, x) = (\lambda^2 t, \lambda O x), \quad \varphi(t, x) = 1,$$

where $\lambda > 0$, $O \in O(n)$, $a \in \mathbf{R}$, *is a caloric morphism from* $\mathbf{R}^{n+1}$ *to* $\mathbf{R}^{n+1}$.

So, we assume $m > n$ in the following.

Example 4 (Symmetrization in $\mathbf{R}^m$ with respect to a codimension 2 subspace)

Let $m \geq 4$, $n = m - 2$, $D = \{(t, x); t \neq 0,\ |x'| > 0\}$ *and* $E = \{(\tau, y); \tau \neq 0,\ y_1 > 0\}$ *where* $x' = (x_1, x_2, x_3, 0, \ldots, 0)$ *for* $x = (x_1, \ldots, x_m)$. *Then*

$$\begin{cases} f_0(t, x) = -t^{-1}, \\ f_1(t, x) = t^{-1}|x'|, \\ f_j(t, x) = t^{-1} x_{j+2}, \quad (2 \leq j \leq n), \end{cases} \qquad \varphi(t, x) = |x'|^{-1} |t|^{-\frac{m-2}{2}} \exp\Big(-\frac{|x|^2}{4t} \Big),$$

is a caloric morphism from D *onto* E.

The following theorem shows that we can superpose the caloric morphisms if they have the same target space. So, we can construct the new caloric morphisms by the superposition.

Theorem 2. *Let $n, m_1, \ldots, m_k$ be positive integers, k be an open interval and D_j be domains in $\mathbf{R}^{m_j}$ $(1 \le j \le k)$. From k caloric morphisms $(f^j, \varphi^j) : I \times D_j \subset \mathbf{R}^{m_j+1} \to \mathbf{R}^{n+1}$ $(1 \le j \le k)$, we can construct a caloric morphism $(f, \varphi) : I \times D_1 \times \cdots \times D_k \to \mathbf{R}^{n+1}$ by*

$$\begin{aligned} f_i(t, x^1, \ldots, x^k) &= f_i^1(t, x^1) + f_i^2(t, x^2) + \cdots + f_i^k(t, x^k), \\ \varphi(t, x^1, \ldots, x^k) &= \varphi^1(t, x^1)\varphi^2(t, x^2) \cdots \varphi^k(t, x^k), \end{aligned}$$

here we use the subscript i to denote the i-th coordinate $(0 \le i \le n)$ of $\mathbf{R}^{n+1}$, and $x^j = (x_1^j, \ldots, x_{m_j}^j)$ denotes the coordinate of $\mathbf{R}^{m_j}$.

Proof. Since each (f^j, φ^j) is a caloric morphism, Theorem 1 implies for $1 \le j \le k$,

$$\varphi^j H^j f_i^j = 2\nabla^j \varphi^j \cdot \nabla^j f_i^j, \quad \nabla^j f_i^j \cdot \nabla^j f_l^j = \delta_{il} \frac{df_0^j}{dt}, \quad H^j \varphi^j = 0, \quad 1 \le i, l \le n,$$

where H^j and ∇^j denote the operators in $\mathbf{R}^{m_j+1}$. Then together with

$$\begin{aligned} \nabla f_i &= (\nabla^1 f_i^1, \nabla^2 f_i^2, \ldots, \nabla^k f_i^k), \\ \nabla \varphi &= \varphi\Big(\frac{\nabla^1 \varphi^1}{\varphi^1}, \frac{\nabla^2 \varphi^2}{\varphi^2}, \ldots, \frac{\nabla^k \varphi^k}{\varphi^k}\Big), \\ H f_i &= H^1 f^1 + H^2 f^2 + \cdots + H^k f^k, \end{aligned}$$

we have

$$\begin{aligned} 2\nabla\varphi \cdot \nabla f_i &= \varphi\Big(\frac{2\nabla^1 \varphi^1 \cdot \nabla^1 f_i^1}{\varphi^1}, \frac{2\nabla^2 \varphi^2 \cdot \nabla^2 f_i^2}{\varphi^2}, \ldots, \frac{2\nabla^k \varphi^k \cdot \nabla^k f_i^k}{\varphi^k}\Big) \\ &= \varphi(H^1 f_i^1 + H^2 f_i^2 + \cdots + H^k f_i^k) \\ &= \varphi H f_i, \end{aligned}$$

and

$$\begin{aligned} \nabla f_i \cdot \nabla f_l &= \nabla^1 f_i^1 \cdot \nabla^1 f_l^1 + \nabla^2 f_i^2 \cdot \nabla^2 f_l^2 + \cdots + \nabla^k f_i^k \cdot \nabla^k f_l^k \\ &= \delta_{il}\Big(\frac{df_0^1}{dt} + \frac{df_0^2}{dt} + \cdots + \frac{df_0^k}{dt}\Big) \\ &= \delta_{il}\frac{df_0}{dt}. \end{aligned}$$

Finally we obtain $H\varphi = 0$ because

$$\begin{aligned} \frac{\partial \varphi}{\partial t} &= \varphi\Big(\frac{1}{\varphi^1}\frac{\partial \varphi^1}{\partial t} + \frac{1}{\varphi^2}\frac{\partial \varphi^2}{\partial t} + \cdots + \frac{1}{\varphi^k}\frac{\partial \varphi^k}{\partial t}\Big), \\ \Delta\varphi &= \varphi\Big(\frac{\Delta^1 \varphi^1}{\varphi^1} + \frac{\Delta^2 \varphi^2}{\varphi^2} + \cdots + \frac{\Delta^k \varphi^k}{\varphi^k}\Big). \end{aligned}$$

Thus f is a caloric morphism by Theorem 1.

3. Generalization of Appell transformation

As a consequence of Theorem 2, we can construct the generalized Appell transformations by superposing the Appell transformations.

Let $n = m_1 = m_2 =, \cdots, = m_k$ and

$$f^j(t, x^j) = \left(\frac{\alpha_j t + \beta_j}{\gamma_j t + \delta_j}, \frac{x^j + tc^j + d^j}{\gamma_j t + \delta_j}\right), \quad j = 1, \ldots, k,$$

where $c^j = (c_1^j, \ldots, c_n^j), d^j = (d_1^j, \ldots, d_n^j) \in \mathbf{R}^n$, $\begin{pmatrix} \alpha_j & \beta_j \\ \gamma_j & \delta_j \end{pmatrix} \in SL(2, \mathbf{R})$ with $\gamma_2 \cdots \gamma_k \neq 0$. We put f and φ as

$$\begin{aligned} f_0(t) &= \sum_{j=1}^{k} \frac{\alpha_j t + \beta_j}{\gamma_j t + \delta_j}, \\ f_i(t, x) &= \sum_{j=1}^{k} \frac{x_i^j + tc_i^j + d_i^j}{\gamma_j t + \delta_j}, \quad 1 \leq i \leq n, \end{aligned}$$

if $\gamma_1 \neq 0$,

$$\varphi(t, x) = C \prod_{j=1}^{k} |\gamma_j t + \delta_j|^{-n/2} \exp\left(- \frac{|\gamma_j x^j + \gamma_j d^j - \delta_j c^j|^2}{4\gamma_j(\gamma_j t + \delta_j)}\right),$$

if $\gamma_1 = 0$,

$$\varphi(t, x) = C \exp\left(\frac{|c^1|^2 t}{4} + \frac{c^1 \cdot x^1}{2}\right) \prod_{j=2}^{k} |\gamma_j t + \delta_j|^{-n/2} \exp\left(- \frac{|\gamma_j x^j + \gamma_j d^j - \delta_j c^j|^2}{4\gamma_j(\gamma_j t + \delta_j)}\right),$$

where $C > 0$, $x = (x^1, \ldots, x^k)$ and $x^j \in \mathbf{R}^n$ $(1 \leq j \leq k)$. Then (f, φ) is a caloric morphism from $\mathbf{R}^{ln+1} \setminus \{\prod_{j=1}^{k}(\gamma_j t + \delta_j) = 0\}$ to $\mathbf{R}^{n+1}$.

In the case $m < 3n$, we can prove that if f has the form

$$f_i(t, x) = \sum_{j=1}^{m} a_{ij}(t) x_j + b_i(t), \quad 1 \leq i \leq n,$$

then (f, φ) is a composition of a generalized Appell transformation and a projection in a suitable coordinates of $\mathbf{R}^m$.

References

[1] K. Shimomura, The caloric morphisms on the Euclidean spaces preprint.

[2] B. Fuglede, Harmonic morphisms between Riemannian manifolds, Ann. Inst. Fourier, Grenoble, **28** 2, 1978, 107–144.

[3] H. Leutwiler, On the Appell transformation, Potential Theory, J.Král, J.Lukeš, I.Netuka, J.Veselý ed., Plenum, New York, 1988.

7th Int. Coll. on Differential Equations, pp. 395-403
D. Bainov (Ed)

Boundary Behaviors of Some Families of Special Functions

H.M. SRIVASTAVA

Department of Mathematics and Statistics, University of Victoria
Victoria, British Columbia V8W 3P4, Canada
E-Mail: HMSRI@UVVM.UVIC.CA

Abstract: In many areas of applications using various families of special functions such as hypergeometric functions of one, two, and more variables, one often finds the need for the asymptotic behaviors of these functions near the boundaries of the regions of convergence of the series defining them. With this point in view, we present here a systematic investigation of the various results depicting the asymptotic behaviors of certain classes of hypergeometric series in one, two, and more variables near the boundaries of their regions of convergence. A number of directions for further studies on these topics will also be considered.

1991 Mathematics Subject Classification. Primary 33C05, 33C20, 33C65; Secondary 34B30, 35B40, 35C20.

Key Words and Phrases. Asymptotic expansions, special functions, hypergeometric functions, Pochhammer symbol, Gauss's summation theorem, Appell functions, Kampé de Fériet functions, Lauricella functions, multiple hypergeometric series.

1. INTRODUCTION AND PRELIMINARIES

In terms of the Pochhammer symbol (or shifted factorial) $(\lambda)_n$ given by

$$(\lambda)_n := \frac{\Gamma(\lambda+n)}{\Gamma(\lambda)} = \begin{cases} 1 & (n=0) \\ \lambda(\lambda+1)\cdots(\lambda+n-1) & (n \in \mathbb{N} := \{1,2,3,\cdots\}), \end{cases} \tag{1.1}$$

a generalized hypergeometric function ${}_pF_q(z)$ (with p numerator parameters $a_1,\cdots,a_p$ and q denominator parameters $b_1,\cdots,b_q$) is defined by

$${}_pF_q[(a_p);\,(b_q);\,z] = {}_pF_q\left[\begin{matrix} a_1,\cdots,a_p; \\ b_1,\cdots,b_q; \end{matrix}\; z\right] := \sum_{n=0}^{\infty} \frac{\prod_{j=1}^{p}(a_j)_n}{\prod_{j=1}^{q}(b_j)_n}\,\frac{z^n}{n!} \tag{1.2}$$

$$(p \leq q \text{ and } z \in \mathbb{C}; \quad p = q+1 \text{ and } z \in \mathcal{U} := \{z : z \in \mathbb{C} \text{ and } |z| < 1\}),$$

provided that no zeros appear in the denominator; here, for convenience, (a_p) abbreviates the array of p parameters $a_1, \cdots, a_p$, with similar interpretations for (b_q), *et cetera*. Setting

$$\omega = \sum_{j=1}^{p} b_j - \sum_{j=1}^{p+1} a_j \,, \tag{1.3}$$

a generalized hypergeometric series ${}_{p+1}F_p(z)$ converges absolutely also on the boundary $\partial\mathcal{U}$ when $\Re(\omega) > 0$. Moreover, when ω is an integer, we say that the generalized hypergeometric function ${}_{p+1}F_p(z)$ is ω-balanced.

The asymptotic behaviors of various classes of special functions such as hypergeometric functions of one, two, and more variables near the boundaries of the regions of convergence of the series defining them are useful in several areas of applications involving these functions. Numerous results on asymptotic behaviors of this type have been given in the literature (*cf.*, *e.g.*, [15] to [18], [24], and [29]); some of these results were used also in solving certain boundary value problems involving the Euler-Darboux equation and in the study of various operators of fractional calculus (*cf.* [14] and [32]). The object of this paper is to present a systematic investigation of the various results exhibiting the asymptotic behaviors of certain generalized hypergeometric series in one and more variables near the boundaries of their regions of convergence.

For the sake of ready reference whenever it is needed in this paper, we begin by recalling the asymptotic behaviors of the hypergeometric function ${}_{p+1}F_p(z)$ near $z = 1$ when $p \in \mathbb{N}$. Indeed, under various sets of parametric constraints, the asymptotic behaviors of the Gaussian hypergeometric function ${}_2F_1(z)$ near $z = 1$ can be derived from the following well-known analytic continuation formula [10, p. 108]

$$\begin{aligned} {}_2F_1(a,b;c;z) &= A_1 \quad {}_2F_1(a,b;\, a+b-c+1;\, 1-z) \\ &+ A_2\,(1-z)^{c-a-b} \quad {}_2F_1(c-a, c-b;\, c-a-b+1;\, 1-z) \\ &(|\arg(1-z)| \leq \pi - \epsilon \quad (0 < \epsilon < \pi))\,, \end{aligned} \tag{1.4}$$

where

$$A_1 := \frac{\Gamma(c)\Gamma(c-a-b)}{\Gamma(c-a)\Gamma(c-b)} \quad \text{and} \quad A_2 := \frac{\Gamma(c)\Gamma(a+b-c)}{\Gamma(a)\Gamma(b)}.$$

When $z = 1$ and $\Re(c-a-b) > 0$, (1.4) immediately yields the familiar Gauss summation theorem (*cf.*, *e.g.*, [33, p. 282]):

$$\begin{gathered} {}_2F_1(a,b,;\; c;\; 1) = A_1 = \frac{\Gamma(c)\Gamma(c-a-b)}{\Gamma(c-a)\Gamma(c-b)} \\ (\Re(c-a-b) > 0;\;\; c \neq 0, -1, -2, \cdots)\,. \end{gathered} \tag{1.5}$$

In case, however, $\Re(c-a-b) \leq 0$, the value ${}_2F_1(a,b;\ c;\ 1)$ does not exist, and we have the order estimates:

$$ {}_2F_1(a,b;\ c;\ 1-\rho) = \begin{cases} O(\log\rho) & (\rho \to 0+;\ \Re(c-a-b)=0) \\ O\left(\rho^{\Re(c-a-b)}\right) & (\rho \to 0+;\ \Re(c-a-b)<0)\,. \end{cases} \tag{1.6} $$

An asymptotic formula exhibiting the behavior of a *zero-balanced* Clausenian hypergeometric function ${}_3F_2(z)$ near $z=1$ was stated (*without proof*) by Ramanujan [13]. Many different proofs of Ramanujan's formula have appeared in the literature (see, for example, Saigo [16], Evans and Stanton [9], and Bühring [3]). In particular, Bühring [3] also generalized Ramanujan's formula to hold true for an m-balanced hypergeometric ${}_3F_2$ series which, in the ${}_2F_1$ case, would correspond to the familiar analytic continuation formula (1.4) when $c = a+b+m \quad (m \in \mathbb{N})$.

2. The Zero-Balanced ${}_pF_{p-1}(z)$ Series

While referring to the complexity in the proof of Ramanujan's formula by Evans and Stanton [9], Evans [8, p. 553] posed the problem of finding analogues of Ramanujan's formula for ${}_4F_3$ and other *higher-order* hypergeometric series. In an attempt to solve this problem of Evans [8], Saigo and Srivastava [26] established a formula exhibiting the behavior of a zero-balanced hypergeometric series ${}_pF_{p-1}(z)$ near the boundary point $z=1$ of its region of convergence. A markedly different approach than theirs to such problems involving ${}_4F_3$ and other *higher-order* hypergeometric series was made earlier by Nørlund [12] and, more recently, by Marichev and Kalla [11].

In order to recall the work of Saigo and Srivastava [24], we begin by considering the Kampé de Fériet series (*cf.* [1, p. 150]; see also [31, p. 27, Equation 1.3(28) *et seq.*]):

$$ \Omega(x,y) = F^{0:p-1;2}_{1:p-3;0}\left[\begin{matrix} \text{———} : & (\alpha_{p-1});\ \beta_{p-2}-\alpha_p, \beta_{p-1}-\alpha_p; \\ \beta_{p-2}+\beta_{p-1}-\alpha_p : & (\beta_{p-3});\ \text{———}; \end{matrix}\ x,y\right] \tag{2.1} $$

$$ (p \geq 3;\quad \max\{|x|,\ |y|\} < 1)\,. $$

We write $\Omega(x,y)$ as an infinite series of powers of y, and separate the first term. On the other hand, upon writing $\Omega(x,y)$ as an infinite series of powers of x, we apply the Gaussian summation theorem (1.5) in order to express $\Omega(x,1)$ in terms of ${}_pF_{p-1}(x)$.

Equating these two expressions for $\Omega(x,1)$, and setting $x = 1-\rho$, we obtain the reduction formula:

$$
\begin{aligned}
{}_pF_{p-1}\left[\begin{matrix}(\alpha_p);\\(\beta_{p-1});\end{matrix}\;1-\rho\right] &= \frac{\Gamma(\beta_{p-2})\Gamma(\beta_{p-1})}{\Gamma(\alpha_p)\Gamma(\beta_{p-2}+\beta_{p-1}-\alpha_p)} \\
&\quad\cdot {}_{p-1}F_{p-2}\left[\begin{matrix}(\alpha_{p-1});\\(\beta_{p-3}),\beta_{p-2}+\beta_{p-1}-\alpha_p;\end{matrix}\;1-\rho\right] \\
&\quad+\frac{(\beta_{p-2}-\alpha_p)(\beta_{p-1}-\alpha_p)\Gamma(\beta_{p-2})\Gamma(\beta_{p-1})}{\Gamma(\alpha_p)\Gamma(\beta_{p-2}+\beta_{p-1}-\alpha_p+1)} \\
&\quad\cdot F^{0:p-1;3}_{1:p-3;1}\left[\begin{matrix}\text{———} : & (\alpha_{p-1}); & \beta_{p-2}-\alpha_p+1,\beta_{p-1}-\alpha_p+1,1;\\ \beta_{p-2}+\beta_{p-1}-\alpha_p+1: & (\beta_{p-3}); & 2;\end{matrix}\;1-\rho,1\right] \qquad (2.2)\\
&\qquad (p\geq 3;\quad \rho>0;\quad \Re(\alpha_p)>0).
\end{aligned}
$$

It is not difficult to verify that the Kampé de Fériet series, occurring on the right-hand side of (2.2), converges when $\rho \to 0+$, provided that

$$\Re\left(\sum_{j=1}^{p-1}\beta_j-\sum_{j=1}^{p}\alpha_j\right)>-1,$$

which is indeed satisfied if (for example) the hypergeometric series ${}_pF_{p-1}$, occurring on the left-hand side of (2.2), is zero-balanced.

Now we set $p = 3$ in the reduction formula (2.2) and make use of the well-known result (*cf.*, *e.g.*, [7, p. 110, Equation 2.10(12)]; see also [15, p. 63, Equation (1)]). By making use of the Gaussian summation theorem (1.5) once again, we thus obtain the known formula which, as we mentioned in the preceding section, was considered by (among others) Ramanujan [13], Berndt [2], Saigo [16], Evans and Stanton [9], Evans [8], and Bühring [3]. In a similar manner, setting $p = 4$ in the reduction formula (2.2) and applying the aforementioned result for ${}_3F_2$, we obtain the corresponding asymptotic formula for ${}_4F_3$.

Iterating this process, we are led from (2.2) to the general formula:

$$\frac{\Gamma(\alpha_1)\cdots\Gamma(\alpha_p)}{\Gamma(\beta_1)\cdots\Gamma(\beta_{p-1})}\, {}_pF_{p-1}\left[\begin{matrix}(\alpha_p);\\(\beta_{p-1});\end{matrix}\ 1-\rho\right]$$

$$= -[2\gamma+\psi(\alpha_1)+\psi(\alpha_2)+\log\rho] + \frac{\beta_1-\alpha_3}{\alpha_1\alpha_2}\left(\sum_{j=1}^{p-1}\beta_j - \sum_{j=3}^{p}\alpha_j\right)$$

$$\cdot\, {}_4F_3\left[\begin{matrix}\beta_1-\alpha_3+1,\ \sum_{j=2}^{p-1}\beta_j-\sum_{j=3}^{p}\alpha_j+1,\ 1,\ 1;\\ \alpha_1+1,\ \alpha_2+1,\ 2;\end{matrix}\ 1\right]$$

$$+\sum_{k=3}^{p-1}\frac{(\beta_{k-1}-\alpha_{k+1})\left(\sum\limits_{j=k}^{p-1}\beta_j-\sum\limits_{j=k+1}^{p}\alpha_j\right)\prod\limits_{j=1}^{k}\Gamma(\alpha_j)}{\Gamma\left(\sum\limits_{j=k-1}^{p-1}\beta_j-\sum\limits_{j=k+1}^{p}\alpha_j+1\right)\prod\limits_{j=1}^{k-2}\Gamma(\beta_j)} \tag{2.3}$$

$$\cdot\, F^{0:\ \ k;3}_{1:k-2;1}\left[\begin{matrix}\text{———}: & (\alpha_k); & \beta_{k-1}-\alpha_{k+1}+1,\ \sum\limits_{j=k}^{p-1}\beta_j-\sum\limits_{j=k+1}^{p}\alpha_j+1,\ 1;\\ \sum\limits_{j=k-1}^{p-1}\beta_j-\sum\limits_{j=k+1}^{p}\alpha_j+1: & (\beta_{k-2}); & 2;\end{matrix}\ 1,1\right] + o(1)\quad(\rho\to 0+)$$

$$(p\geq 3;\ \alpha_1+\cdots+\alpha_p=\beta_1+\cdots+\beta_{p-1};\ \Re(\alpha_j)>0,\quad j=3,4,\cdots,p),$$

where $\psi(z) := \Gamma'(z)/\Gamma(z)$ denotes the Digamma function (that is, the logarithmic derivative of the Gamma function) and $\gamma := -\psi(1)$ is the Euler-Mascheroni constant.

Since an empty sum is interpreted to be zero, (2.3) yields the asymptotic formulas for ${}_3F_2$ and ${}_4F_3$ in its special cases when $p=3$ and $p=4$, respectively. Thus, in view of these asymptotic formulas, the general result (2.3) can be proven directly by induction on $p \in \mathbb{N}$.

3. Further Results and Observations

Since the publication of the work of Saigo and Srivastava [25], which we have detailed reasonably in the preceding section, Bühring [4] has given a number of analytic continuation formulas [*op. cit.*, Theorems 2, 3, and 4] for the generalized hypergeometric function ${}_{p+1}F_p(z)$ for $|1-z|<1$, $|\arg(1-z)|<\pi$, and $p\geq 2$, in the different cases when ω defined by (1.3) is not equal to an integer, is equal to an integer $\nu \geq 0$, or is equal to an integer $-\nu \leq 0$. In particular, in the zero-balanced case, Bühring [4, Corollary 1] gives an analytic continuation formula which would yield a representation for ${}_pF_{p-1}(1-\rho)$ (when $\rho \to 0+$) different from (2.3) above (see also Bühring [5] and [6]). Some applications of the general asymptotic formula (2.3) were considered recently by Srivastava [30].

Various hypergeometric functions of two, three, and more variables are known to satisfy systems of partial differential equations (see, for details, Srivastava and Karlsson [31]). The boundary behaviors of many of these multiple hypergeometric series have been investigated recently. Apart from the references cited in Section 1 of this presentation, one may refer to the works on this subject by Saigo *et al.* (*cf.*, *e.g.*, [19] to [22], [24], and [26] to [28]).

There indeed are numerous multiple hypergeometric series [including (for example) the Kampé de Fériet hypergeometric series in two variables, Lauricella's *third* hypergeometric series $F_C^{(n)}$ in n variables, and so on] for which a systematic study of the boundary behaviors is conspicuously missing in the mathematical literature. Additionally, one may be encouraged to turn one's attention to some of these already considered multiple hypergeometric series, but under substantially different parametric constraints. Solutions to such boundary behavior problems are potentially useful in areas where these multiple hypergeometric series emerge naturally.

Acknowledgements

The present investigation was supported, in part, by the *Natural Sciences and Engineering Research Council of Canada* under Grant OGP0007353.

References

[1] P. Appell *et* J. Kampé de Fériet, *Fonctions Hypergéométriques et Hypersphériques; Polynômes d'Hermite*, Gauthier-Villar, Paris, 1926.

[2] B.C. Berndt, Chapter 11 of Ramanujan's second notebook, *Bull. London Math. Soc.* **15**(1983), 273-320.

[3] W. Bühring, The behavior at unit argument of the hypergeometric function ${}_3F_2$, *SIAM J. Math. Anal.* **18**(1987), 1227-1234.

[4] W. Bühring, Generalized hypergeometric functions at unit argument, *Proc. Amer. Math. Soc.* **114**(1992), 145–153.

[5] W. Bühring, Transformation formulas for terminating Saalschützian hypergeometric series of unit argument, *J. Appl. Math. Stochast. Anal.* **8**(1995), 189–194.

[6] W. Bühring, Terminating Saalschützian hypergeometric series at unit argument, *Math. Japon.* **44**(1996), 15–23.

[7] A. Erdélyi, W. Magnus, F. Oberhettinger, and F.G. Tricomi, *Higher Transcendental Functions*, Vol. I, McGraw-Hill, New York, Toronto, and London, 1953.

[8] R.J. Evans. Ramanujan's second notebook: Asymptotic expansions for hypergeometric series and related functions, in *Ramanujan Revisited* (G.E. Andrews *et al.*, Editors) (Proceedings of the Ramanujan Centenary Conference, University of Illinois, June 1-5, 1987), Academic Press, New York, 1988, pp. 537-560.

[9] R.J. Evans and D. Stanton, Asymptotic formulas for zero-balanced hypergeometric series, *SIAM J. Math. Anal.* **15**(1984), 1010-1020.

[10] W. Magnus, F. Oberhettinger, and R.P. Soni, *Formulas and Theorems for the Special Functions of Mathematical Physics*, Die Grundlehren der Mathematischen Wissenschaften in Einzeldarstellungen, Band **52**, Third enlarged edition, Springer-Verlag, New York, 1966.

[11] O.I. Marichev and S.L. Kalla, Behaviour of hypergeometric function ${}_pF_{p-1}(z)$ in the vicinity of unity, *Rev. Técn. Fac. Ingr. Univ. Zulia* **7**(1984), 1-8.

[12] N.E. Nørlund, Hypergeometric functions, *Acta Math.* **94**(1955), 189-349.

[13] S. Ramanujan, *Notebooks of Srinivasa Ramanujan*, Vol. 2, Tata Institute of Fundamental Research, Bombay, 1957.

[14] M. Saigo, A certain boundary value problem for the Euler-Darboux equation. II and III, *Math. Japon.* **25**(1980), 211-220; **26**(1981), 103-119.

[15] M. Saigo, On a property of the Appell hypergeometric function F_1, *Math. Rep. College General Ed. Kyushu Univ.* **12**(1980), 63-67.

[16] M. Saigo, On properties of the Appell hypergeometric functions F_2 and F_3 and the generalized Gauss function ${}_3F_2$, *Bull. Central Res. Inst. Fukuoka Univ.* **66**(1983), 27-32.

[17] M. Saigo, On properties of the Lauricella hypergeometric function F_D, *Bull. Central Res. Inst. Fukuoka Univ.* **104**(1988), 13-31.

[18] M. Saigo, On properties of hypergeometric functions of three variables, F_M and F_G. *Rend. Circ. Mat. Palermo* (2) **37**(1988), 449-468.

[19] M. Saigo, On behaviors of hypergeometric series of one, two, three and n variables near boundaries of their convergence regions of logarithmic case (Kyoto, 1989), *Sûrikaisekikenkyûsho Kôkyûroku* **714**(1990), 91–109.

[20] M. Saigo, The asymptotic behaviors of triple hypergeometric series, F_G, F_K, F_N and F_R near boundaries of their convergence regions (Kyoto, 1989), *Sûrikaisekikenkyûsho*

Kôkyûroku **714**(1990), 110–126.

[21] M. Saigo and N.T. Hai, The asymptotic representations of Appell series F_1 and Horn series H_3 near boundaries of their convergence regions, *Fukuoka Univ. Sci. Rep.* **21**(1991), 17–23.

[22] M. Saigo, O.I. Marichev, and N.T. Hai, Asymptotic representations of Gaussian series ${}_2F_1$, Clausenian series ${}_3F_2$ and Appell series F_2 and F_3 near boundaries of their convergence regions, *Fukuoka Univ. Sci. Rep.* **19**(1989), 83–90.

[23] M. Saigo and H.M. Srivastava, The behaviors of the Appell double hypergeometric series F_4 and certain Lauricella triple hypergeometric series near the boundaries of their convergence regions, *Fukuoka Univ. Sci. Rep.* **19**(1989), 1-10.

[24] M. Saigo and H.M. Srivastava, The asymptotic behaviour of Lauricella's second hypergeometric series $F_B^{(n)}$ in n variables near the boundaries of its convergence region, *Fukuoka Univ. Sci. Rep.* **19**(1989), 71–82.

[25] M. Saigo and H.M. Srivastava, The behavior of the zero-balanced hypergeometric series ${}_pF_{p-1}$ near the boundary of its convergence region, *Proc. Amer. Math. Soc.* **110**(1990), 71–76.

[26] M. Saigo and H.M. Srivastava, Some asymptotic formulas exhibiting the behaviors of the triple hypergeometric series F_S and F_T near the boundaries of their convergence regions, *Funkcial. Ekvac.* **34**(1991), 423–448.

[27] M. Saigo and H.M. Srivastava, The asymptotic behavior of the Srivastava hypergeometric series H_C near the boundary of its convergence region, *Kyungpook Math. J.* **32**(1992), 152–158.

[28] M. Saigo and H.M. Srivastava, A theorem on the asymptotic behavior of Lauricella's first hypergeometric series $F_A^{(n)}$ near the boundary of its convergence region (Varna, 1991), in *Complex Analysis and Generalized Functions* (I. Dimovski and V. Hristov, Editors), Bulgarian Academy of Sciences, Sofia, 1993, pp. 258–267.

[29] M. Saigo and H.M. Srivastava, The behaviour of the fourth type of Lauricella's hypergeometric series in n variables near the boundaries of its convergence region, *J. Austral. Math. Soc. Ser. A* **57**(1994), 281-294.

[30] A.K. Srivastava, Asymptotic behaviour of certain zero-balanced hypergeometric series, *Proc. Indian Acad. Sci. Math. Sci.* **106**(1996), 39–51.

[31] H.M. Srivastava and P.W. Karlsson, *Multiple Gaussian Hypergeometric Series*, Halsted Press (Ellis Horwood Limited, Chichester), John Wiley and Sons, New York, Chichester, Brisbane, and Toronto, 1985.

[32] H.M. Srivastava and M. Saigo, Multiplication of fractional calculus operators and boundary value problems involving the Euler-Darboux equation, *J. Math. Anal. Appl.* **121**(1987), 325-369.

[33] E.T. Whittaker and G.N. Watson, *A Course of Modern Analysis: An Introduction to the General Theory of Infinite Processes and of Analytic Functions; With an Account of the Principal Transcendental Functions*, Fourth edition, Cambridge University Press, Cambridge, London, and New York, 1927.

7th Int. Coll. on Differential Equations, pp. 405-407
D. Bainov (Ed)

On a Coordinate Transformation to Express an Explicit Duck Solution in the FitzHugh–Nagumo Equation

KIYOYUKI TCHIZAWA
Department of Mathematics, Musashi Institute of Technology, 1-28-1, Tamazutsumi, Setagaya-Ku, Tokyo, 158, JAPAN

Abstract. This paper shows that there is a suitable coordinate transformation which gives a local coordinate to express the explicit duck solutions in the FitzHugh-Nagumo (FHN) equation.

Keywords: FitzHugh-Nagumo equation, Constrained system, Duck solution

1. INTRODUCTION

In this paper, we construct a transformation of the coordinates for showing the existence of an explicit duck solution [2] in the FHN equation. As a result, it can be shown that the shadow of the explicit duck solutions may contain not only a pseudo singular saddle point but also a pseudo singular node point.

Generally, on the following system (1.1):

$$\begin{aligned} dx/dt &= f(x,y,z), \\ dy/dt &= g(x,y,z), \\ \varepsilon dz/dt &= h(x,y,z), \end{aligned} \tag{1.1}$$

where ε is infinitesimally small, we need the following two conditions to represent the explicit duck solution [1]:

$$\text{(i)} \quad h(\mathbf{0}) \cong h_y(\mathbf{0}) \cong h_z(\mathbf{0}) \cong f(\mathbf{0}) \cong 0,$$

$$\text{(ii)} \quad g(\mathbf{0}) \not\cong 0, \quad h_x(\mathbf{0}) \not\cong 0, \quad h_{zz}(\mathbf{0}) \not\cong 0,$$

where $\mathbf{0} = (0,0,0)$ is a pseudo singular point.

In the FHN equation, under some conditions, exchanging the variables successively, the following new FHN system (1.2) is obtained [2]:

$$\begin{aligned} dX/dt &= c(Z - \gamma X), \\ dY/dt &= 1, \\ \varepsilon dZ/dt &= -\rho(Z) - X + Y, \quad \rho(Z) = Z(Z-1)(Z-a), \end{aligned} \tag{1.2}$$

where $0 < a < 1/2$, c and γ are any constants and ε is infinitesimal. Note that this system does not satisfy the above conditions by itself.

2. A TRANSFORMATION OF THE COORDINATES

In this section, we provide a transformation of the coordinates to express an explicit duck solution in the FHN equation.

Theorem. *By a suitable coordinate transformation, the system* (1.2) *is transformed to satisfy the conditions* (i) *and* (ii).

Proof. First, let change the variables X, Y and Z so that the origine is a pseudo singular point $P_0(x_0, y_0, z_0)$ to the variables u, v and w:

$$u = X - x_0, \quad v = Y - y_0, \quad w = Z - z_0. \tag{2.1}$$

Secondarily, let construct a linear transformation of the coordinate which gives a local coordinate to represent an explicit duck solution. In the coordinate, the system satisfies the conditions (i) and (ii). It is important to recognize the fact: every component of the matrix associated with the transformation may contain a non-standard number. As we consider all of its combinatorial cases, it becomes clear that the following transformation (2.2):

$$\begin{pmatrix} x \\ y \\ z \end{pmatrix} = \begin{pmatrix} 1 & -1 & 0 \\ \alpha & \beta & 0 \\ 0 & 0 & 1 \end{pmatrix} \begin{pmatrix} u \\ v \\ w \end{pmatrix}, \qquad (\alpha,\ \beta \text{ are any constants}), \tag{2.2}$$

only remains a possibility to accomplish our purpose. For the simplicity, we choose $\alpha = -1$, $\beta = 2$. In the system (1.2), by exchanging the coordinates through this transformation and form (1.2) and (2.1), we get the system (2.3):

$$\begin{aligned} du/dt &= c\Big\{(w + z_0) - \gamma(u + x_0)\Big\} = c(w - \gamma u) + c(z_0 - \gamma x_0), \\ dv/dt &= 1, \\ \varepsilon dw/dt &= -\rho(w + z_0) - u + v - x_0 + y_0. \end{aligned} \tag{2.3}$$

Furthermore, from (2.2) and (2.3), the following equations are established:

$$\begin{aligned} dx/dt &= du/dt - dv/dt \\ &= c(w - \gamma u) + c(z_0 - \gamma x_0) - 1 \\ &= cz - c\gamma(2x + y) + c(z_0 - \gamma x_0) - 1 \equiv \overline{f}(x, y, z), \end{aligned} \tag{2.4}$$

$$dy/dt = -du/dt + 2dv/dt \tag{2.5}$$

$$= -c(w - \gamma u) - c(z_0 - \gamma x_0) + 2$$
$$= -cz + c\gamma(2x + y) - c(z_0 - \gamma x_0) + 2 \equiv \overline{g}(x, y, z),$$

$$\text{(2.6)} \qquad \varepsilon dz/dt = \varepsilon dw/dt$$
$$= -\rho(w + z_0) - u + v - x_0 + y_0$$
$$= -\rho(z + z_0) - (2x + y) + (x + y) - x_0 + y_0$$
$$= -\rho(z + z_0) - x - x_0 + y_0 \equiv \overline{h}(x, y, z).$$

As the origine is one of the two pseudo singular points, $\overline{h}(\mathbf{0}) = 0$, $\overline{h}_z(\mathbf{0}) = 0$ and $\overline{h}_x(\mathbf{0})\overline{f}(\mathbf{0}) + \overline{h}_y(\mathbf{0})\overline{g}(\mathbf{0}) = 0$ are satisfied. See [2]. Therefore, the following equations (2.7) and (2.8) are established:

$$\text{(2.7)} \qquad -\rho(z_0) - x_0 + y_0 = 0,$$
$$\text{(2.8)} \qquad -c(z_0 - \gamma x_0) + 1 = 0.$$

Substituting (2.7) and (2.8) to (2.4), (2.5) and (2.6), the following system (2.9) is obtained:

$$\text{(2.9)} \qquad \begin{aligned} dx/dt &= -c\gamma(2x + y) + cz, \\ dy/dt &= c\gamma(2x + y) - cz + 1, \\ \varepsilon dz/dt &= -\rho(z + z_0) - x + \rho(z_0). \end{aligned}$$

It satisfies the conditions (i) and (ii) obviously. □

Remark. The above theorem, from the result of E. Benoit [1], ensures the existence of the explicit duck solutions. Moreover the shadow of them may be admissible to contain not only a pseudo singular saddle point but also a pseudo singular node point in the FHN equation.

REFERENCES

[1] E. Benoit, Canards et enlacements, *Math. Pub. I.H.E.S.*, (1990), 63–91.

[2] N. Kakiuchi and K. Tchizawa, On the existence of a duck solution and delay in the FitzHugh–Nagumo equation, *Proceedings of the 7-th International Colloquium on Differential Equations,* Plovdiv, Bulgaria, Aug. 18–23, (1996).

7th Int. Coll. on Differential Equations, pp. 409-412
D. Bainov (Ed)

On the Completeness of the Eigenfunctions of Elliptic Operators

Takesi Tokita
Faculty of Education, University of Fukui
Fukui 910, Japan

Abstract

We show that the eigenfuctions of elliptic operators make a complete system in $L^2(\Omega)$. S.Agmon obtained important results on the general elliptic operators with the boundary conditions. We obtain results on the problems to which Agmon's method cannot apply. Key lemmas are 1/2-estimates which L.Hörmander obtained by using the theory of pseudo-differential operators. We can have the coercive inequality for the boundary value problems introducing a new variable if the boundary conditions cover the elliptic operators. Our problems are not the case.

1. Introduction

The purpose of this paper is to show that the set of eigenfunctions and generalized eigenfunctions of the following elliptic oprators is complete in $L^2(\Omega)$. The space $L^2(\Omega)$ consists of the second power integrable functions defined in the domain Ω in R^2 with the smooth boundary $\partial\Omega$.

$$A(D)u(x) - \lambda u(x) = (D_1^2 + D_2^2)u(x) - \lambda u(x) = f(x) \quad \text{in } \Omega, \tag{1}$$

$$B(x, D)u(x) = b_1(x)D_1u(x) + b_2(x)D_2u(x) + u(x) = 0 \quad \text{on } \partial\Omega \tag{2}$$

where we write $i = \sqrt{-1}$, $D_j = i^{-1}\partial/\partial x_j$ $j = 1,2$ and λ for a complex number with the absolute value $|\lambda|$ and the argument $arg\lambda = \theta$. The domain Ω is simply connected and contained in the right half plane; $\Omega \subset \{(x_1, x_2) \in R^2; x_1 > 0\}$. The boundary $\partial\Omega$ contains the segment $I_0 = \{(0, x_2); -5\delta \le x_2 \le 5\delta\}$ where δ=1/8. The functions $b_j(x)$ $j = 1,2$ which are smooth on $\partial\Omega$ $b_j(x) = 0$ $j = 1,2$ $x \in \partial\Omega\backslash I_0$ and $b_1(0, x_2) = 1$, $b_2(0, x_2) = \sqrt{-4 + ix_2}$ for $-\delta < x_2 < \delta$, specially $b_2(0,0) = 2i$. We denote $\overline{\Omega} = \Omega \cup \partial\Omega$ and $\| u \|^2 = \int_\Omega | u |^2 dx$. Now we describe the main theorem.

Theorem 1 *If $u(x) \in C^2(\overline{\Omega})$ which is 2 times continuously differentiable in $\overline{\Omega}$ satisfies the boundary conditon (2) we can find positive constants Λ and C such that.*

$$\| u \| \le C |\lambda|^{-1} \| (A(D) - \lambda)u \| \tag{3}$$

for $\Lambda \le |\lambda|$ and $arg\lambda = \pm\pi/300$,

$$\| u \| \le C |\lambda|^{-3/4} \| (A(D) - \lambda)u \| \tag{4}$$

for $\Lambda \le |\lambda|$ and $arg\lambda = -\pi$.

Remark: taking $\pm\pi/300$ for convenience any sufficiently small number may be taken.

2. Estimates of the solutions of the boundary value problems

We need the following Theorem 2 to derive the Theorem 1. By introducing the third real variable x_3 we define $\tilde{A}(D;\theta)$ as follows

$$\tilde{A}(D;\theta)v = (D_1^2 + D_2^2)v - e^{i\theta}D_3^2 v = f \quad \text{in } \Gamma \tag{5}$$

where $\Gamma = \Omega \times \mathrm{R} = \{x = (x_1, x_2, x_3); (x_1, x_2) \in \Omega, x_3 \in \mathrm{R}\}$.

Theorem 2 *There exists a positive constant C such that the following estimates hold if $v(x) \in C^2(\overline{\Omega} \times R)$ satisfies the boundary condition (2) on $\partial\Gamma$ and the function $v(x)$ is identically zero outside $\overline{\Omega} \times (-1, 1)$*

$$\| v \|_{2,\Gamma} \le C\,(\| \tilde{A}(D;\theta)v \|_{\Gamma} + \| v \|_{\Gamma}) \tag{6}$$

in the case of $\theta = \pm\pi/300$,

$$\| v \|_{3/2,\Gamma} \le C\,(\| \tilde{A}(D;\theta)v \|_{\Gamma} + \| v \|_{\Gamma}) \tag{7}$$

in the case of $\theta = -\pi$.

The inequality (6) implies (3). See Agmon [1]. Also inequality (7) implies (4). See Tokita [8]. Substituting $u(x)\exp ix_3\sqrt{|\lambda|}$ into the formulas (6) and (7) we can derive (3) and (4). We write

$$\| v \|_{m,\Gamma}^2 = \sum_{|\alpha|\le m} \| D^\alpha v \|_{\Gamma}^2 \quad m = 0, 1, 2, \cdots$$

where α stands for the multi-index whose length $\alpha_1 + \alpha_2 + \alpha_3$ is denoted by $|\alpha|$. The function space $\mathrm{H}^m(\Gamma)$ is the completion of $\mathrm{C}^m(\Gamma)$ under the norm above. The norm $\| * \|_{j/2,\Gamma}$ $j = 1, 3$ is definded by means of interpolation from $\| * \|_{m,\Gamma}$ $m = 0, 1, 2$ (see Lions and Magenes [5]). Let $V(x)$ be an extension of $v(x)$ where the support of $V(x)$ is contained in a neighborhood of Γ and $V = v$ in Γ. Then the following inequality is well known. There exists a positive constant C such that

$$\| V \|_{3/2,R^3} \le C(\| v \|_{3/2,\Gamma})$$

where

$$\| V \|_{3/2,R^3}^2 = (2\pi)^{-3} \int_{R^3} (1 + \xi_1^2 + \xi_2^2 + \xi_3^2)^{3/2} |FV|^2 d\xi,$$

$$FV(\xi) = \int_{R^3} V(x) \exp i(x_1\xi_1 + x_2\xi_2 + x_3\xi_3) dx.$$

The formula (6) is the coercive inequality. See [2],[6],[7]. In the case of $\theta = -\pi$ the boundary condition (2) does not the elliptic operator (5). We derive the inequality (7) using 1/2-estimates which L. Hörmander[4] obtained from the theory of pseudo-differential operator.

3. Completeness of eigenfunctions

The estimates (3) show that the solution of the boundary value problems (1), (2) is uniqe in the case of $\lambda = \lambda_1$ being $arg\lambda_1 = \pi/300$ and $\Lambda \le |\lambda_1|$. The adjoint boundary value problems of (1) and (2) have also similar estimates. See Shimakura [7]. These imply that there exists the inverse T of the operator $A(D) - \lambda_1$ which is a one-to-one mapping

$L^2(\Omega) \longrightarrow \{u \in H^2(\Omega); B(x,D)u = 0 \text{ on } \partial\Omega\}$. Besides T is a compact operator in $L^2(\Omega)$. Writing the resolvent of T by $R(\lambda\,;T) = (\lambda I - T)^{-1}$ we have

$$R(\lambda^{-1};T) = \lambda I - \lambda^2 R(\lambda\,;A) \tag{8}$$

The $L^2(\Omega)$-valued function $R(\lambda\,;T)f$ is analytic except for $\lambda=0$ and λ being eigenvalues $\{\mu_k\}_{k=1,2,\cdots}$ of T. We have the Laurent expansion in the neighborhood of $\lambda = \mu_k$

$$R(\lambda;T)f = \frac{\phi_{k,1}}{(\lambda-\mu_k)^{j_k}} + \frac{\phi_{k,2}}{(\lambda-\mu_k)^{j_k-1}} + \cdots + \frac{\phi_{k,j_k}}{(\lambda-\mu_k)} + g(\lambda,\ T, f). \tag{9}$$

Here the functions $\{\phi_{k,j}\}_{j=1,2,\cdots,j_k}$ satisfy $(T-\mu_k I)^j \phi_{k,j} = 0\; j = 1,2,\cdots,j_k$. The following estimates of the resolvent of T with respect to λ are well known. See Dunford-Scwartz[3] and [1].

Theorem 3 *There exists a sequence* $\{\rho_k\}_{k=1,2,\cdots}$ $\rho_k \to \infty, k \to \infty$ *for any positive number* ϵ *such that*

$$\| R(\lambda^{-1};T) \| \leq \exp|\lambda|^{1+\epsilon} \tag{10}$$

for $|\lambda| = \rho_k,\ k = 1,2,\cdots$.

We omit the proof. We only remark that we have the following

$$TL^2(\Omega) \subset H^2(\Omega), \quad \Omega \subset R^2.$$

Namely the operator T has 2 regularity up and works in 2 dimensional space.

The inequalities (3) and (4) imply that

$$\| R(\lambda;A) \| = O(|\lambda|^{-1}), \quad |\lambda| \to \infty \tag{11}$$

for $arg\lambda = \pm\pi/300$,

$$\| R(\lambda;A) \| = O(|\lambda|^{-3/4}), \quad |\lambda| \to \infty \tag{12}$$

for $arg\lambda = -\pi$.

We define $sp(T)$ to be the closure in $L^2(\Omega)$ of the space spanned by all generalized eigenfunctions of T. The proof of the completeness, that is $sp(T) = L^2(\Omega)$, is by contradiction. Then we can find $f^* \in L^2(\Omega)$ such that f^* is orthogonal to $sp(T)$. Setting $F(\lambda) = (f^*, R(\lambda^{-1};T)f)$, $F(\lambda)$ is an entire function from the definition and (9). We need the well known Phrgmén-Lindelöf's theorem. Applying the theorem to $F(\lambda)$ in the regions with the angles into which the complex plane is divided by the rays $arg\lambda = \pm\pi/300, -\pi$ we obtain from (8),(11),(12) and Theorem 3

$$F(\lambda) = O(|\lambda|^{5/4}), \quad |\lambda| \to \infty. \tag{13}$$

On the other hand the Taylor development in the neighborhood of $\lambda = 0$ is

$$F(\lambda) = \lambda(f^*, f) + \lambda^2(f^*, Tf) + \cdots \tag{14}$$

From (13) and (14) we have $(f^*, Tf) = 0$. Since f is arbitary while the image of T is dense in $L^2(\Omega)$ we have $f^* = 0$. This shows that $sp(T) = L^2(\Omega)$.

REFERENCES

[1] S. Agmon, On the eigenfunctions and the eigenvalues of general elliptic boundary value problems. Comm. Pure Appl. Math., 15(1962) 119-147

[2] S. Agmon, Lectures on elliptic boundary value problems. Van Nostrand, 1965.

[3] Danford, N., and Schwartz, J.T., Linear Operators, part II, section XI. 1963

[4] L. Hörmander, Pseudo-differential operators and non-elliptic boundary problems. Ann. of Math., 83(1966), 129-209.

[5] J.L. Lions and E. Magenes,Non homogeneous boundary value problems and applications I. Springer-Verlag, Berlin, 1972. (Translated from French.)

[6] S. Mizohata, The theory of partial differential equation. Cambridge University Press, 1973. (Translated from Japanese.)

[7] N. Shimakura, Partial differential operators of elliptic type. American Mathematical Society. 1992 (Translated from Japanese.)

[8] T. Tokita, On the behavior of the solutions of $(\lambda - \Delta)u = f$ as $\lambda \to \infty$. Memoirs of the Faculty Education Fukui University Series II, 46(1995)

7th Int. Coll. on Differential Equations, pp. 413-420
D. Bainov and V. Covachev (Eds)

AN ALGORITHM FOR HIGH ACCURACY PRODUCT IN REDUNDANT REPRESENTATION

Kumiko TSUJI
Kyushu Teikyo Junior College,
4-3-124 Shin-Katsutachi-Machi Ohmuta 836, Japan

Abstract

An algorithm of multiplication with high accuracy product is the same with the parallel algorithm of multiplication for signed digit numbers, replacing base r by large base β^t and a signed digit by a signed fraction of t digits from -1 position to -t position (which is abbreviated as FSD). The terms ,each of which is product of FSD in multiplicand and that in multiplier, are added in parallel in each rank. The sum in each rank is separated into a minimal redundant FSD and transfer FSD which is added to minimal redundant FSD in upper rank by 1. It is possible to obtain variation(increment or decrement) of FSD in product with respect to the variation of multiplicand (multiplier) in parallel in ranks. Conversely, the unknown variation of FSD in multiplicand is determined by the variations of the sums in 1st $\sim$ 3rd ranks for ranks suffering the perturbations due to variation in multiplicand. But it is shown that the unknown variation is not determined by the variations of FSD's in any other ranks in product.

Keywords: round-off error, Fortran, multiplication, redundant representation system

1 Introduction

In [2], an algorithm for sum of floating-point numbers is established by transforming floating-point numbers into maximal redundant representations with large base in order to treat the number independently in parts. In [1], the multiplication in redundant representation number system with large base is used to obtain high accuracy product.

High accuracy product has many digits. It is necessary to obtain the digits separating into several parts. If a error arises in some part by input multiplicand or multiplier error, it is necessary to correct this part only lefting the accurate digits. If more high accuracy product is necessary, it is useful to left upper digits as it is and to supplement the digits in lower positions. In the present algorithm, the numbers are added independently in part grouping the digits. A group of t digits is taken as a "maximal redundant FSD" and r is taken as β^t. t is the environment parameter of mantissa length for a floating-point number. Since t digits can be stored into a floating-point number, a group of t digits fits the FORTRAN program.

Definition 1.1 (minimal redundant FSD, maximal redundant FSD) *Let α be a fraction number in a set which constitutes of numbers initiating from 0 terminating $1-\beta^{-t}$ with step size β^{-t}. Let $D(\alpha)$ be a set defined as $D(\alpha) = \{x : x = -\alpha, \alpha, (\beta^{-t})\}$. A fraction in set $D(1/2)$ is called a "minimal redundant FSD." A fraction in set $D(1-\beta^{-t})$ is called a "maximal redundant FSD."*

Let m be the minimum exponent and M be the maximum exponent in a floating-point system. Let M_0 be maximum exponent such that M_0 is divisible by t and $M \geq M_0 + t$.

$$(1) \qquad v = \sum_{k_1}^{k_2} v(k)\beta^{M_0-tk}, \ (0 \leq k_1, k_2 \leq N) \ where \ N = \lfloor (M_0 - m + 1)/t \rfloor - 1.$$

Definition 1.2 (Maximal redundant representation, minimal redundant representation, rank, rank number) *The $\beta^t - ary$ power series (1) is called a "maximal redundant representation", if $v(k)$ is maximal redundant FSD i.e. $v(k) \in D(1-\beta^{-t})$. The (1) is called "mimimal redundant representation", if $v(k)$ is a minimal redundant FSD i.e. $v(k) \in D(1/2)$. β^{M_0-tk} is called a "rank". k is called "rank number."*

Definition 1.3 (CSB, difference FSD) *Let x be a floating-point number such that $|x| \leq 1$. A carry signed bit is defined as $B(x) = sign(x)$ if $|x| \geq 1/2 + \beta^{-t}; B(x) = 0$ if $|x| \leq 1/2$. "carry signed bit" is abbreviated as CSB.*

Let p and q be $\beta^t - ary$ maximal redundant representation of the form $p = \sum_{i=i_1}^{i_2} p(i)\beta^{M_0-ti}; q = \sum_{j=j_1}^{j_2} q(j)\beta^{M_0-tj}$. Here $i_1 + j_1 \geq M_0/t$ and $i_2 + j_2 \leq \lfloor (2M_0/t - m + 1)/t \rfloor - 2$. Let v be the product of p and $q : v = \sum_{i=i_1}^{i_2} \sum_{j=j_1}^{j_2} p(i)q(j)\beta^{M_0-t(i+j)}$. The set of digits $p(i)q(j)$ in positions from -1 to $-2t$ are separated into two sets y_0 and $y_1 : y_0$ constitutes of the digits in positions from -1 to $-t$ and y_1 the digits in positions from $-t$ to $-2t$.

The character of the algorithm is that the fraction signed digits y_0 and $y_1\beta^t$ with same rank are added in parallel. The digits y_0 or $y_1\beta^t$ with same rank is summed independently. y_0 is added by the following algorithm (a). (a) Let the FSD's in pre-partial sum be stored into two variables $s(k-1)$ and $w(k)$. Let $w(k)$ store the difference FSD expressed in a minimal redundant FSD. Let $s(k-1)$ store the sum of CSB's. FSD y_0 is rounded by CSB operator and is separated into CSB $J1$ and difference FSD $w_3 : J1 = B(y_0); w_3 = y_0 - B(y_0)$. w_3 is added to pre-difference FSD $w(k) : w_2 = w(k) \oplus w_3$. The sum w_2 is rounded by CSB operator and the difference value is put into $w(k) : J2 = B(w_2); w(k) = w_2 - J2$. $w(k)$ is a difference digit in $D(1/2)$. The sum $J1 + J2$ is transfered to $s(k-1)$ and added to the pre-sum $s(k-1)$. The sum $s(k-1) + (J1 + J2)\beta^{-t}$ is stored into variable $s(k-1)$ and the post-sum of CSB's is obtained. $s(k-1)$ is the sum of CSB's in $D(1/2 - \beta^{-t})$ and is called a transfer digit. (b) y_1 is added to $s(k)$ and $w(k+1)$ in pre-partial sum by procedure (a) with respect to y_1 for rank β^{k+1}.

Finally, the total sum v is stored in $s(k-1)$, $w(k)$ $s(k)$ and $w(k+1)$ with $k = i+j-M_0/t$ for $i = i_1, i_2$ and $j = j_1, j_2$. $s(k)$ and $w(k)$ are added for $k = i_1 + j_1 - 1 - M_0/t, \ldots, i_2 + j_2 + 1 - M_0/t$ and stored into $v(k)$. Since $s(k) \in D(1/2 - \beta^{-t})$ and $w(k) \in D(1/2)$, $v(k) \in D(1 - \beta^{-t})$. The total sum v is obtained in FSD's $v(k)$. v is a maximal redundant representaion.

In the present paper, it is shown that the varition $\delta_s(k)$ of transfer FSD $s(k)$ due to $\delta_{x\{k+1\}}$ is either 0 or $sign(\delta_{x\{k+1\}})$ assuming $|\delta_{x\{k+1\}}| < 1$. Here $x\{k+1\}$ is the sum of the terms at rank number $k+1$. It is essential to obtain the variation of $s(k)$ due to variation of multiplicand,in order to obtained $\delta_v(k)$. But $\delta_s(k)$ is not determined which of the two if $|w(k+1)+\delta_{x\{k+1\}}|=1/2$. If $|w(k+1)+\delta_{x\{k+1\}}|\neq 1/2$ then $\delta_s(k)$ is determined. The variation $\delta_{p(n_0+i_1)}$ of $p(n_0+i_1)$ affects $v(k(l))$ for $l=n_0-1,\ldots,n_0+1+j_2-j_1$. The variations $\delta_{v(k(l))}$ due to $\delta_{p(n_0+i_1)}$ is obtained for those l after determining variations $\delta_s(k(l))$ of transfer FSD's for $l=n_0-1,\ldots,n_0+j_2-j_1$ by the sum $\delta_{x\{k(l+1)\}}+w(k(l+1))$. Here $k(l)=l+i_1+j_1-M_0/t$. In [3], the floating-point solution is obtained from the simle equation with floating-point multiplication. Let rank k whose FSD is perturbed, be known as $k=n_0+i_1$. Using the method to obtain the solution in [3], the unknown variation $\delta_p(n_0+i_1)$ of FSD in multiplicand is determined by the variations of FSD's with $1st \sim 3rd$ ranks and variation of transfer FSD transfered from 4 th rank i.e. by $\delta_v(k(n_0-1))$, $\delta_v(k(n_0))$ and $\delta_v(k(n_0+1))$ and $\delta_s(k(n_0+1))$.

2 Product of two maximal redundant representation

Let two maximal redundant representations be $p=\sum_{i=i_1}^{i_2}p(i)\beta^{M_0-ti}$; $q=\sum_{j=j_1}^{j_2}q(j)\beta^{M_0-tj}$, where $1\le j_2-j_1\le i_2-i_1$; $p(i),q(j)\in D(1-\beta^{-t})$. Put $i_2'=i_2-i_1$; $j_2'=j_2-j_1$; $\tilde{j}=j-j_1$; $n=i-i_1$. The product is given as $pq=\sum_{n=0}^{i_2'}\sum_{\tilde{j}=0}^{j_2'}p(n+i_1)q(\tilde{j}+j_1)\beta^{2M_0-t(n+\tilde{j}+i_1+j_1)}$ The term $p(n+i_1)q(\tilde{j}+j_1)\beta^{2M_0-t(n+\tilde{j}+i_1+j_1)}$ is separated into two terms $p(n+i_1)q(\tilde{j}+j_1)<-1,-t))\beta^{2M_0-t(n+\tilde{j}+i_1+j_1)}$ and $p(n+i_1)q(\tilde{j}+j_1)((-t-1,-2t>\beta^{2M_0+1-t(n+\tilde{j}+i_1+j_1)}$. Here $p(n+i_1)q(\tilde{j}+j_1)<-1,-t))$ denote $p(n+i_1)q(\tilde{j}+j_1)$ on positions from -1 to -t and $p(n+i_1)q(\tilde{j}+j_1)((-t-1,-2t>$ denote from $-t-1$ to $-2t$. The sum in the terms of product with same rank is given as follows.

Proposition 2.1 *Put $l=n+\tilde{j}$. The rank number in product is given as $k(l)=l+i_1+j_1-M_0/t$. The product pq is given as*

$$pq=\Big(\sum_{l=0}^{j_2'}+\sum_{l=j_2'+1}^{i_2'}+\sum_{l=i_2'+1}^{i_2'+j_2'+1}\Big)x\{k(l)\}\beta^{M_0-tk(l)} \tag{2}$$

Here $x\{k(l)\}$ denote the sum of the terms with rank $M_0-tk(l)$ and $x\{k(l)\}$ are given as

$$x\{k(0)\}=(p(i_1)q(j_1))<-1,-t));x\{k(l)\}=\sum_{n=0}^{l}U(n,l)+\sum_{n=0}^{l-1}L(n,l)\ if\ l=1,j_2'$$

$$x\{k(l)\}=\sum_{n=l-j_2'}^{l}U(n,l)+\sum_{n=l-1-j_2'}^{l-1}L(n,l)\ if\ l=j_2'+1,i_2'$$

$$x\{k(l)\}=\sum_{n=l-j_2'}^{i_2'}U(n,l)+\sum_{n=l-1-j_2'}^{i_2'}L(n,l)\ \ if\ l=i_2'+1,i_2'+j_2'$$

$$x\{k(i_2'+j_2'+1)\}=(p(i_2)q(j_2)\ )((-t-1,-2t>\beta^t.$$

Here $U(n,l)=(p(n+i_1)q(l-n+j_1)\ )<-1,-t))$; $L(n,l)=(p(n+i_1)q(l-1-n+j_1)\ )((-t-1,-2t>\beta^t$.

Proof. The set of $(n,\tilde{j})$ is reduced to the set of (n,l) as follows:

$$\{0\le n+\tilde{j}\le j_2',0\le n\le j_2',0\le\tilde{j}\le j_2'\}$$

$$= \{0 \le l \le j_2', 0 \le n \le j_2', l - j_2' \le n \le l\} = \{0 \le l \le j_2', 0 \le n \le l\};$$

$$\{j_2' + 1 \le n + \tilde{j} \le i_2', 0 \le n \le i_2', 0 \le \tilde{j} \le j_2'\}$$

$$= \{j_2' + 1 \le l \le i_2', 0 \le n \le i_2', l - j_2' \le n \le l\} = \{j_2' + 1 \le l \le i_2', l - j_2' \le n \le l\};$$

$$\{i_2' + 1 \le n + \tilde{j} \le i_2' + j_2', 0 \le n \le i_2', 0 \le \tilde{j} \le j_2'\}$$

$$= \{i_2' + 1 \le l \le i_2' + j_2', 0 \le n \le i_2', l - j_2' \le n \le l\} = \{i_2' + 1 \le l \le i_2' + j_2', l - j_2' \le n \le i_2'\}.$$

Product is rewritten as

$$pq = \{\sum_{l=0}^{j_2'} \sum_{n=0}^{l} + \sum_{l=j_2'+1}^{i_2'} \sum_{n=l-j_2'}^{l} + \sum_{l=i_2'+1}^{i_2'+j_2'} \sum_{n=l-j_2'}^{i_2'} \} p(n + i_1) q(l - n + j_1) \beta^{M_0 - tk(l)}.$$

3 Variation of product of two FSD's on restricted positions

Let P, Q, and $P + \delta P$ be FSD's. The variation of digits whose range of positions is restricted is considered after the product PQ varies due to variations δ_P in P. The variation restricted on $< -1, -t))$ i.e. from -1 position to $-t$ position is denoted as $\delta_{y_0}(P, Q, \delta_P)$:

$$\delta_{y_0}(P, Q, \delta_P) = \{(P + \delta_P)Q\} < -1, -t)) - (PQ) < -1, -t)).$$

The variation restricted on $((-t - 1, -2t >$ is denoted as $\delta_{y_1}(P, Q, \delta_P)$:

$$\delta_{y_1}(P, Q, \delta_P)\beta^{-t} = \{(P + \delta_P)Q\}((-t - 1, -2t > -(PQ)((-t - 1, -2t > .$$

Proposition 3.1 *Put* $U_P = (PQ) < -1, -t) + (\delta_P Q) < -1, -t));$

$$L_o = (PQ)((-t - 1, -2t > +(\delta_P Q)((-t - 1, -2t >;$$

$$S = sign\{U_P + L_o((-t))\}\beta^{-t}; C = -S + L_o((-t)) + \{S + L_o((-t - 1, -2t >\}((-t))$$

Then $\delta_{y_0}(P, Q, \delta_P) = (\delta_P Q) < -1, -t)) + C; \delta_{y_1}(P, Q, \delta_P)\beta^{-t} = (\delta_P Q)((-t - 1, -2t > -C$

Proof. We have $(P + \delta_P)Q = U_P + L_o$. If $(U_P + L_o((-t)) \;)L_o((-t - 1, -2t > \; > 0$, $U_P + L_o((-t)) = 0$ or $L_o((-t - 1, -2t >= 0$ then there is no borrow between $U_P + L_o((-t))$ and $L_o((-t - 1, -2t >$, and $C = L((-t)) : \{(P + \delta_P)Q\} < -1, -t)) = U_P + L_o((-t)) = U_P + C$. If $\{U_P + L_o((-t))\}L_o((-t - 1, -2t >< 0$, then there is a borrow between $U_P + L_o((-t))$ and $L_o((-t - 1, -2t >$, and $C = -S + L_o((-t)) : \{(P + \delta_P)Q\}((-t - 1, -2t >$
$= S + L_o((-t - 1, -2t >= L_o + S - L_o((-t)) = L_o - C.$ Q.E.D.

4 Variation of product due to multiplicand

Put $U_p = (p(n_0 + i_1)q(l - n_0 + j_1)) < -1, -t) + (\delta_p(n_0 + i_1)q(l - n_0 + j_1) \;) < -1, -t));$

$$L_o = (p(n_0 + i_1)q(l - n_0 + j_1))((-t - 1, -2t > +(\delta_p(n_0 + i_1)q(l - n_0 + j_1))((-t - 1, -2t > .$$

$$S = sign\{U_p + L_o((-t))\}\beta^{-t}; C_p(l) = -S + L_o((-t)) + \{S + L_o((-t - 1, -2t > \}((-t)).$$

$$\delta_{y_0}^{P}(k(l)) = \delta_{y_0}(p(n_0 + i_1), q(l - n_0 + j_1), \delta_p(n_0 + i_1) \;)$$

$$= \{\delta_p(n_0 + i_1)q(l - n_0 + j_1)\} < -1, -t)) + C_p(l);$$

$$\delta_{y_1}^{P}(k(l)) = \delta_{y_1}(p(n_0 + i_1), q(l - n_0 + j_1), \delta_p(n_0 + i_1) \;)$$

$$= \{\delta_p(n_0 + i_1)q(l - n_0 + j_1)\}((-t - 1, -2t > \beta^t - C_p(l)\beta^t.$$

The variation of product is given independently for each rank.

Theorem 4.1 *Let* $p(n_0+i_1)$ *change to* $p(n_0+i_1)+\delta_p(n_0+i_1)$, $(0 \le n_0 \le i_2')$. *The variation of* $x\{k(l)\}$ *is given for each* l *as* $\delta^p_{x\{k(n_0)\}} = \delta^p_{y_0}(k(n_0))$; $\delta^p_{x\{k(n_0+j_2'+1)\}} = \delta^p_{y_1}(k(n_0+j_2'))$;

$$\delta^p_{x\{k(l)\}} = \delta^p_{y_0}(k(l)) + \delta^p_{y_1}(k(l-1)) \ for \ n_0+1 \le l \le n_0+j_2'.$$

Here $k(l) = l + i_1 + j_1 - M_0/t$.

5 Example of product in redundant representation

For $p = \sum_{k=9}^{14} p(k)\beta^{M_0-tk}$; $q = \sum_{k=9}^{14} q(k)\beta^{M_0-tk}$, the product is expressed in redundant representation: $pq = \sum_{k=9}^{19} v(k)\beta^{M_0-tk}$. Let variation of $p(14)$ is given as $\delta_p(14) = -0.6(-3)$. Let $v_\delta(k)$ be the perturbed value of $v(k)$ by $\delta_p(14) = -0.6(-3)$. $v_\delta(k) = v(k)$ for $k = 1, \dots 12$, theoretically. FSD's $v(k)$ and $v_\delta(k)$ are expressed in the following table for $k = 9, 12$.

Table 1. FSD's $v(k)$ and $v_\delta(k)$ for $k = 9, 12$

k	p(k)	q(k)	$v(k) = v_\delta$	$s(k) = s_\delta(k)$	$w(k) = w_\delta(k)$
9	-0.200000(-5)	-0.200000(-5)	+0.300000(-5)	-0.100000(-5)	+0.400000(-5)
10	+0.5A830B(0)	+0.60CBC8(0)	-0.546474(0)	+0.200000(-5)	-0.546476(0)
11	-0.713305(0)	-0.2FB8DE(0)	-0.39BB4A(0)	-0.000000(-64)	-0.39BB4A(0)
12	-0.78E3D8(0)	-0.733A37(0)	+0.1F84AA(0)	+0.100000(-5)	+0.1F84A9(0)

FSD's $v_\delta(k)$ are different from FSD's $v(k)$ for $k = 13, \dots, 19$ theoretically. The following table shows the FSD's $v(k)$ for $k = 13, \dots, 19$.

Table 2. FSD's $v(k)$ for $k = 13, \dots, 19$

k	p(k)	q(k)	$v(k)$	$s(k)$	$w(k)$
13	-0.739A89(0)	+0.54C939(0)	+0.84BCF0(-1)	+0.100000(-5)	+0.84BCE0(-1)
14	+0.700000(-3)	-0.522900(0)	+0.7A12F5(0)	-0.000000(-64)	+0.7A12F5(0)
15	$M_0 = 60$		+ 0.721C30(0)	- 0.000000(-64)	+ 0.721C30(0)
16	$i_1 = 9, i_2 = 14, i_2' = 5$		+ 0.6D24F5(0)	- 0.000000 (-64)	+ 0.6D24F5(0)
17	$j_1 = 9, j_2 = 14, j_2' = 5$		+ 0.3439D4(0)	+ 0.100000(-5)	+ 0.3439D3(0)
18	$n_0 = 5, k(5) = 13$		+ 0.227DC1(0)	- 0.000000(-64)	+ 0.227DC1(0)
19	$n_0 + j_2' + 1 = 11, k(11) = 19$		- 0.1F0000(0)		- 0.1F0000(0)

The following table shows the perturbed FSD's $v_\delta(k)$ for $k = 13, \dots, 19$ due to perturbed $(p+\delta_p)(14) = +0.100000(-3)$ with variation $\delta_p(14) = -0.6(-3)$. The variations $\delta^p_{y_0}(k)$ and $\delta^p_{y_1}(k-1)$ of elements with same rank number in product are shown in the following table. Since varition $\delta^p_{y_0}(13) = 0$, it holds that $v(k) = v_\delta$, $s(k) = s_\delta(k)$ and $w(k) = w_\delta$.

Table 3. FSD's $v_\delta(k)$ for $k = 13, \dots, 19$

k	$\delta^p_{y_0}(k)$	$\delta^p_{y_1}(k-1)$	$v_\delta(k)$	$s_\delta(k)$	$w_\delta(k)$
13	- 0.000000 -64		+0.84BCF0(-1)	+0.100000(-5)	+0.84BCE0(-1)
14	- 0.245000(-3)	+0.C00000(-3)	+0.7A1CB1(0)	+0.100000(-5)	+0.7A1CB0(0)
15	+ 0.11F000(-3)	+ 0.395000(0)	+0.5492B1(0)	- 0.000000(-64)	+ 0.5492B1(0)
16	+ 0.2B3000(-3)	- 0.AACC00(0)	-0.3DA457(0)	+ 0.100000(-5)	- 0.3DA458(0)
17	- 0.1FD000(-3)	+ 0.5D4A00(0)	-0.6E7E29(0)	+ 0.100000(-5)	- 0.6E7E2A(0)
18	+0.1ED000(-3)	+ 0.48AA00(0)	+0.6B29AE(0)	- 0.000000(-64)	+ 0.6B29AE(0)
19		-0.A00000(-1)	-0.290000(0)		- 0.290000(0)

Note that the under line digits show the different digits between $v(k)$ and $v_\delta(k)$, between $s(k)$ and $s_\delta(k)$ and between $w(k)$ and $w_\delta(k)$.

6 Variation in redundant representation

Let $v = \sum_{k=-1}^{k_m} v(k)\beta^{M_0-tk}$ be the product expressed in redundant representation. If $| w + \delta_{x\{k+1\}} | \neq 1/2$ then $\delta_s(k)$ is determined in the following proposition .

Proposition 6.1 *Put* $\delta = \delta_{x\{k+1\}}$; $s_\delta = s_\delta(k)$; $w_\delta = w_\delta(k+1); w = w(k+1); s = s(k)$. *Assume that* $| \delta | < 1$.

(a) If $1/2 - sign(\delta)\delta > wsign(\delta) \geq -1/2$ *then* $\delta_s = 0$.

(b) $1/2 \geq sign(\delta)w > 1/2 - \delta sign(\delta)$ *then* $\delta_s = sign(\delta)\beta^{-t}$.

Proof.Since $s_\delta\beta^t + w_\delta = x + \delta$; $s\beta^t + w = x$, we have $(s_\delta - s)\beta^t = w + \delta - w_\delta$. (a) Let $\delta > 0$. If $-1/2+\delta \leq w+\delta < 1/2$ then $-1+\delta \leq w+\delta - w_\delta < 1$ and $s_\delta - s = 0$. Let $\delta < 0$. If $-1/2 < w + \delta \leq 1/2 + \delta$ then $-1 < w + \delta - w_\delta \leq 1 + \delta < 1$ and $s_\delta - s = 0$. (b) Let $\delta > 0$. If $1/2 + \delta \geq w + \delta > 1/2$ then $1 + \delta \geq w + \delta - w_\delta > 0$ and $s_\delta - s = \beta^{-t}$. Let $\delta < 0$. If $-1/2 + \delta \leq w + \delta < -1/2$ then $-1 + \delta \leq w + \delta - w_\delta < 0$ and $s_\delta - s = -\beta^{-t}$. Q.E.D.

Let (a_l) and (b_l) denote the conditions as follows:

(a_l) $1/2 - sign(\delta_{x\{k(l)\}})\delta_{x\{k(l)\}} > wsign(\delta_{x\{k(l)\}}) \geq -1/2$

(b_l) $1/2 \geq wsign(\delta_{x\{k(l)\}}) > 1/2 - \delta_{x\{k(l)\}}sign(\delta_{x\{k(l)\}})$.

$\delta_s(k(l-1))$ and $\delta_s(k(l))$ of CSB's are determined as follows in the following 4 cases.

If (1) i.e. (a_l) and (a_{l+1}) then $\delta_s(k(l-1)) = 0; \delta_s(k(l)) = 0$.

If (2) i.e. (a_l) and (b_{l+1}) then $\delta_s(k(l-1)) = 0$; $\delta_s(k(l)) = sign(\delta_{x\{k(l+1)\}})\beta^{-t}$.

If (3) i.e. (b_l) and (a_{l+1}) then $\delta_s(k(l-1)) = sign(\delta_{x\{k(l)\}})\beta^{-t}$; $\delta_s(k(l)) = 0$.

If (4) i.e. (b_l) and (b_{l+1}) then $\delta_s(k(l-1)) = sign(\delta_{x\{k(l)\}})\beta^{-t}; \delta_s(k(l)) = sign(\delta_{x\{k(l+1)\}})\beta^{-t}$.

The variation of $v(k(l))$ due to $\delta_{x\{k(l)\}}$ is obtained by the following theorem. Since the variation $\delta_{x\{k(l)\}}$ due to variation of multiplicand is obtained, the variation of $v(k(l))$ due to variation of multiplicand is also obtained in the 4 cases.

Theorem 6.1 *Assume that* $| \delta_{x\{k(l)\}} | < 1$ *for* $l = n_0, \ldots, n_0 + j_2' + 1$.
Let $| w(k(l)) + \delta_{x(k(l)} | \neq 1/2$ *for* $l = n_0, \ldots, n_0 + j_2' + 1$.

If (1), then $\delta_w(k(l)) = \delta_{x\{k(l)\}}$; $\delta_v(k(l)) = \delta_{x\{k(l)\}}$.

If (2), then $\delta_w(k(l)) = \delta_{x\{k(l)\}}$; $\delta_v(k(l)) = \delta_{x\{k(l)\}} + sign(\delta_{x\{k(l+1)\}})\beta^{-t}$.

If (3), then $\delta_w(k(l)) = \delta_{x\{k(l)\}} - sign(\delta_{x\{k(l)\}})$; $\delta_v(k(l)) = \delta_{x\{k(l)\}} - sign(\delta_{x\{k(l)\}})$.

If (4), then $\delta_w(k(l)) = \delta_{x\{k(l)\}} - sign(\delta_{x\{k(l)\}})$;
$\delta_v(k(l)) = \delta_{x\{k(l)\}} - sign(\delta_{x\{k(l)\}}) + sign(\delta_{x\{k(l+1)\}})\beta^{-t}$.

Variation $\delta_v(k)$ in redundant representation for example in section 5

k	$w(k) + \delta_{x\{k\}}$	case	$\delta_s(k)$	$\delta_w(k)$	$\delta_{x\{k\}}$	$\delta_v(k)$
13	/	/	0	0	0	0
14	0.7A1CB0(0)	(2)	0.100000(-5)	+ 0.9BB000(-3)	+ 0.9BB000(-3)	+0.9BC000(-3)
15	0.AB6D4F(0)	(3)	0	- 0.C6AEE1(0)	+0.39511F(0)	- 0.C6AEE1(0)
16	-0.3DA458(0)	(2)	0.100000(-5)	- 0.AAC94D(0)	- 0.AAC94D(0)	- 0.AAC94C(0)
17	0.9181D6(0)	(3)	0	- 0.A2B7FD(0)	0.5D4803(0)	- 0.A2B7FD(0)
18	0.6B29AE(0)	(1)	0	+ 0.48ABED(0)	0.48ABED(0)	+ 0.48ABED(0)
19	-0.290000(0)	(1)	0	- 0.A00000(-1)	- 0.A00000(-1)	- 0.A00000(-1)

7 Determination of a variation in a multiplicand

In the following proposition, the assumption is made such that δ_P is so small compared with P enough to satisfy

$$(A) \quad |(PQ) < -1,-t))| \geq |(\delta_P Q) < -1,-t))| + 2\beta^{-t}.$$

Put $C = -S + L_o((-t)) + \{S + L_o((-t-1,-2t>\}((-t))$ with $S = sign\{U_p + L_o((-t))\}\beta^{-t}$, $U_p = (PQ) < -1,-t)) + (\delta_P Q) < -1,-t))$, $L_o = (PQ)((-t-1,-2t> + (\delta_P Q)((-t-1,-2t>$ and $\tilde{S} = sign\{(PQ) < -1,-t))\}\beta^{-t}$. Then $C = -\tilde{S} + \{\tilde{S} + L_o\}((-t))$ is proved.

Proposition 7.1 *Assume (A). Then* 1. $S = \tilde{S}$ 2. $L_o((-t)) = 0$ *if* $\tilde{S}L_o((-t-1,-2t> < 0$ 3. $L_o((-t)) = 0$ *or* $\tilde{S}$ *if* $L_o((-t-1,-2t> = 0$ 4. $C = -\tilde{S} + \{\tilde{S} + L_o\}((-t))$

Proof. 1.$|(\delta_P Q) < -1,-t))| + 2\beta^{-t} > > (\delta_P Q) < -1,-t)) + L_o((-t))$. By (A),$S = \tilde{S}$.
2. $sign\ L_o((-t-1,-2t> = sign\ \delta_P Q((-t-1,-2t> = -\tilde{S}$. 3. If $L_o((-t)) = -\tilde{S}$ then $(PQ)((-t-1,-2t> (\delta_P Q)((-t-1,-2t> < 0$ and $L_o((-t)) = 0$ which contradicts. 4.If $\tilde{S}L_o((-t-1,-2t> < 0$, then $L_o((-t)) = 0$ and $C = -S = -\tilde{S} + \{\tilde{S} + L_o\}((-t))$. If $\tilde{S}L_o((-t-1,-2t> > 0$, then $\{\tilde{S} + L_o((-t-1,-2t>\}((-t)) = \tilde{S}$ and $C = L_o((-t)) = -\tilde{S} + \{\tilde{S} + L_o\}((-t))$. If $\tilde{S}L_o((-t-1,-2t> = 0$, then $L_o((-t-1,-2t> = 0$, $\{\tilde{S} + L_o((-t-1,-2t>\}((-t)) = \tilde{S}$. If $L_o((-t)) = \tilde{S}$ and $L_o((-t-1,-2t> = 0$, then $C = L_o((-t)) = -\tilde{S} + \{\tilde{S} + L_o\}((-t))$. If $L_o = 0$ then $C = 0 = -\tilde{S} + \{\tilde{S} + L_o\}((-t))$. Q.E.D.
Put $\tilde{\delta} = (\delta_P Q)((-t-1,-2t>; C_c = C - (\delta_P Q)((-t-1,-2t>$. $|C_c| < \beta^{-t}$ is proved.

Proposition 7.2 *Assume (A). Let cases be defined as follows.* **a** $\tilde{\delta} = 0$. **b** $\tilde{\delta} \neq 0$.
(α) $\tilde{S}\tilde{\delta} > 0$. (1)$L_o((-t)) = \tilde{S}$; (2)$L_o((-t)) = 0$.
(β) $\tilde{S}\tilde{\delta} < 0$. (1) $L_o\tilde{S} > 0$; (2) $L_o\tilde{S} < 0$; (3) $L_o = 0$.

1. *If* **a** *then* $C = 0; C_c = 0$. *If* **b**, (α) *and (1) then* $C = \tilde{S}; C_c = \tilde{S} - \tilde{\delta}$. *If* **b**, (α) *and (2) then* $C = 0; C_c = -\tilde{\delta}$. *If* **b**, (β) *and (1) then* $C = 0; C_c = -\tilde{\delta}$. *If* **b**, (β) *and (2) then* $C = -\tilde{S}; C_c = -\tilde{S} - \tilde{\delta}$. *If* **b**, (β) *and (3), then* $C = 0$; $C_c = -\tilde{\delta}$.
2. $|C_c| < \beta^{-t}$.

Proof. If **b**, (α) and (1) then $C = -\tilde{S} + 2\tilde{S} = \tilde{S}$. If (β), then $L_o((-t)) = 0$. If **b**, (β) and (1) then $L_o\tilde{S} > 0$ $\{\tilde{S} + L_o\}((-t)) = \tilde{S}; C = 0$. If **b**, (β) and (2) then $L_o\tilde{S} < 0$ $\{\tilde{S} + L_o\}((-t)) = 0; C = -\tilde{S}$. If **b**, (β) and (3), then $L_o = 0$ and $C = 0$. Q.E.D.

Let the rank number in a multiplicand for which its FSD is perturbed, is known beforehand as $n_0 + i_1$. Then the least rank number where the variation of the product arises, is $k(n_0)$. The variation in multiplicand is determined from the variation $\delta^P_{x\{k(n_0)\}} + \delta^P_{x\{k(n_0+1)\}}\beta^{-t}$
in product. Substituting $\delta_P = \delta_p(n_0 + i_1)$; $P = p(n_0 + i_1)$; $Q = q(j_1 + 1)$ into (A). (A) is denoted as (A'):

$$(A') \quad |(p(n_0 + i_1)q(j_1 + 1)) < -1,-t))| \geq |(\delta_p(n_0 + i_1)q(j_1 + 1)) < -1,-t))| + 2\beta^{-t}.$$

Since C_c includes the unknown $\delta_p(n_0 + i_1)$, C_c is denoted as $C_c(\delta_p(n_0 + i_1))$:
$C_c(\delta_p(n_0 + i_1)) = C_p(n_0) - (\delta_p(n_0 + i_1)q(j_1 + 1))((-t-1,-2t>$.

Theorem 7.1 *Assume (A'). Let* $\delta_p(n_0 + i_1)$ *be unknown variation. Let* $\delta^P_{x\{k(l)\}}$ *be given for* $l = n_0, n_0 + 1$. *Let* Q_u *be the quotient up to* $-t$ *position:*

$$Q_u = \frac{\delta^P_{x\{k(n_0)\}}\beta^t + \delta^P_{x\{k(n_0+1)\}}}{q(j_1)\beta^t + q(j_1 + 1)} < -1,-t)).$$

Let R be the remainder with quotient Q_u : $R = \delta^p_{x\{k(n_0)\}}\beta^t + \delta_{x\{k(n_0+1)\}} - Q_u(q(j_1)\beta^t + q(j_1+1))$. Then the solution $\delta_p(n_0+i_1)$ such that it takes the variation $\delta^p_{x\{k(l)\}}$, is determined by

$$\delta_p(n_0+i_1) < -1, -t >= Q_u + sign(q(j_1)R)\beta^{-t}$$
$$if \ \frac{R - C_c(Q_u + sign(q(j_1)R)\beta^{-t})}{q(j_1)\beta^t + q(j_1+1)} = sign(q(j_1)R\ )\beta^{-t};$$
$$\delta_p(n_0+i_1) < -1, -t >= Q_u \ if \ R - C_c(Q_u) = 0,$$

where $C_c(\delta_p(n_0+i_1)\ ) = -(\delta_p(n_0+i_1)q(j_1+1)\ )((-t-1, -2t > \beta^t + C_p(n_0+1)$.

Proof. Put $q_d = q(j_1)\beta^t + q(j_1+1)$; $\delta_P = \delta_p(n_0+i_1)$. $\delta^p_{x\{k(n_0)\}} = (\delta_P q(j_1)) < -1, -t)) + C_p(n_0)$; $\delta^p_{x\{k(n_0+1)\}}\beta^{-t} = (\delta_P q(j_1))((-t-1, -2t > -C_p(n_o) + (\delta_P q(j_1+1)) < -1, -t)) + C_p(n_o+1)$. $\delta^p_{x\{k(n_0)\}}\beta^t + \delta^p_{x\{k(n_0+1)\}} = \delta_P q(j_1)\beta^t + (\delta_P q(j_1+1)) < -1, -t)) + C_p(n_0+1)$. Subtracting and adding the term $C_c(\delta_P)$, $\delta^p_{x\{k(n_0)\}}\beta^t + \delta^p_{x\{k(n_0+1)\}} = \delta_P q_d + C_c(\delta_P)$. Dividing by q_d, we have

$$\delta_p(n_0+i_1) = \frac{\delta^p_{x\{k(n_0)\}}\beta^t + \delta^p_{x\{k(n_0+1)\}}}{q_d} - \frac{C_c(\delta_P\ )}{q_d}.$$

Let R be the remainder with quotient Q_u. Since $\mid R \mid < \beta^{-t} \mid q_d \mid$ and $\mid C_c(\delta_P) \mid < \beta^{-t}$,

$$\delta_P < -1, -t > -Q_u < -1, -t >= \frac{R}{q_d} < -t-1, -\infty > -\frac{C_c(\delta_P\ )}{q_d} < -t-1, -\infty > .$$

If $RC_c(\delta_P) \geq 0$, then the positions of the right hand is $< -t-1, -\infty >$ and the right hand is zero i.e. $R = C_c$. Therefore $\delta_P = Q_u$. If $RC_c(\delta_P) < 0$ then the positions of the right hand is $< -t, -\infty >$ and the right hand is $sign(Rq(j_1))\beta^{-t}$ i.e. $R - C_c(\delta_P) = sign(Rq(j_1))\beta^{-t}q_d$. Therefore $\delta_P = Q_u + \beta^{-t}sign(Rq(j_1))$. Q.E.D.

Remark If $\delta_s(k(n_0+1))$ is known, then $\delta_p(n_0+i_1)$ is determined by theorem 7.1 from the variations $\delta_v(k(n_0-1))$, $\delta_v(k(n_0))$ and $\delta_v(k(n_0+1)$ replacing $\delta^p_{x\{k(n_0)\}}\beta^t + \delta^p_{x\{k(n_0+1)\}}$ by $\delta_v(k(n_0-1))\beta^{2t} + \delta_v(k(n_0))\beta^t + \delta_v(k(n_0+1)) - \delta_s(k(n_0+1))$.

Example. Let $\delta^p_{x\{k(n_0)\}} = \delta^p_{x\{13\}} = 0 \ for \ n_0 = 5$.

$$\delta^p_{x\{14\}} = \delta^p_{y_0}(14) + \delta^p_{y_1}(13) = -0.245000(-3) + 0.C00000(-3) = 0.9BB000(-3);$$

$q_d = -1.9F3438(0)$, where $j_1 = 9$. The quotient is computed by double precision as

$$\frac{\delta^p_{x\{k(n_0+1)\}}}{q_d} \doteqdot -0.5FFDCA9C450CD3(-3).$$

$$Q_u = -0.5FF(-3); R = 0.165E43(-5); C_c(Q_u) = 0.65E438(-6).$$

$R - C_c(Q_u) = 0.FFFFF8(-6)$ and $R - C_c(Q_u)\beta^{-5} \geq \beta^{-5}$. Therefore $\delta_p(14) = Q_u = -0.5FF(-3)$ is not a solution by theorem 7.1. $\delta_p(14) = Q_u - \beta^{-t} = -0.600(-3)$ is a solution by theorem 7.1, since $C_c(\ Q_u - \beta^{-t}\ ) = -0.395000(0)$ and

$$sign(Rq(j_1))(R\beta^t - C_c(Q_u - \beta^{-t})\ ) = -1.9F343(0) = q_d.$$

References

[1] Kajiwara J. and Tsuji M, Proceeding of 5th Int.Coll. on Diff. Equ., (1995), VSP, pp.163-172.

[2] Tsuji K, Proceeding of 3rd Int. Coll. on Num. Anal. (1995), VSP,pp.181-190.

[3] Tsuji K, Proceeding of 6th Int.Coll. on Diff. Equ., (1996), VSP, pp.297-304.

7th Int. Coll. on Differential Equations, pp. 421-428
D. Bainov (Ed)

ROUND-OFF ERRORS IN FLOATING-POINT SOLUTIONS FOR CHEBYSHEV COLLOCATION POINTS

Kumiko TSUJI
Kyushu Teikyo Junior College,
4-3-124 Shin-Katsutachi-Machi Ohmuta 836, Japan

Abstract

A zero point of Chebyshev polynomial with degree N i.e. Chebyshev collocation point is computed by the Newton's iteration process $w_{n+1} = f(w_n)$. Assume that iteration points in an oscillatory state after some iterations with some period m. The round-off error of each iteration point in oscillatory state is determined by a zero of a polynomial with degree $m \times (N-1)+1$. The equation is the power series of divided differences of $f(w)$ with deviations, each deviation depends on the round-off error, the iteration point and an other iteration point on the oscillatory state. The coefficients of the power series are constituted of round-off errors of $f(w)$ on iteration points in the oscillatory state.

Keywords: round-off error, Fortran, Chebyshev polynomial, floating-point addition floating-point multiplication, Newton's method

1 Introduction

In [1], Urabe has given a relative error theory of a floating-point solution for n dimensional nonlinear equations, which is obtained by practicing the generalized Newton's iteration process $w_{n+1} = g(w_n)$ on a computer. Here the theoretical Newton's process is $x_{n+1} = f(x_n)$. He showed that the iteration points w_{i+n_0} $(i = 1, \ldots, m)$ in Newton's iteration process are in an oscillatory state taking finite values if f satisfies some Lipschitz condition and the round-off error of $f(x)$ i.e. $R(x) = f(x)-g(x)$ is small. The result in [1] that w_{i+n_0} are in oscillatory state, is used as assumption in [2]. A 2-dimensional system of linear equations has been considered and then the round-off error of each point has been shown to be a solution of two dimensional linear equations. The equation has been expressed as a power series of Lipschitz matrix. The coefficients of the power series are the round-off errors $R(x)$ at iteration points in oscillatory state.

In the present paper, the result of the theory in [2] is extended to the round-off errors of zero points of a polynomial $\sum_{n=0}^{N} A_n x^n$. Round-off error is abbreviated as RE for a zero point. Each zero point is computed by the Newton's iteration process. Assume that iteration points are in an oscillatory state after some iterations. The RE ϵ_k of each iteration point w_{k+n_0} in the oscillatory state is shown to be a zero of a polynomial with degree $m \times (N-1)+1$:

$$\epsilon_k(1-\prod_{j=1}^{m}\kappa_j(\epsilon_k)) = \sum_{i=k}^{m+k-1} R(w_{i+n_0}) \prod_{j=i+1}^{m+k-1} \kappa_j(\epsilon_k).$$

m is a period of oscillatory state. $\kappa_j(\epsilon_k)$ is a divided difference of $f(w)$ with deviation $(w_{k+n_0}+\epsilon_k)-w_{j+n_0}$. $R(w_{i+n_0})$ is the round-off error of $f(w)$ at iteration point w_{i+n_0}.

In [3], the Chebyshev polynomial $T(w)$ with degree 20 is considered. The sensitivity of the zero $cos\ (2r+1)\pi/40$ for $r=1$ is shown to be large with respect to variation in the coefficients. So the affection of RE to this zero would be large. It has been difficult to obtain zero by computing with Newton's iteration process.

In the present paper, in computing residual $T(w)$ by single precision, catastrophic cancellation arises and only one digit (which is not correct) is obtained for each w near the exact zero. The iteration $w_{n+1}=g(w_n)$ $(n=1,2,3,\ldots)$ is computed starting from the computed value $w_1 = cos(2r+1)\pi/40$ for r=1 in single precision, until w_n become oscillatory state. But stationary iteration point is obtained i.e. the period is obtained as $m=1$. The stationary iteration point is far from the exact zero. Here $g(w)=w\ominus H\otimes \tilde{T}_s(w)$. $H=1\oslash\tilde{T}'_s(w_1)\otimes 0.5$. $\tilde{T}'_s(w_1)$ is the practically computed derivative of $T(w)$ at $w=w_1$ by single precision. $\ominus$, $\otimes$ and $\oslash$ denote floating-point subtraction, multiplication and division. The RE ϵ_1 of stationary iteration point is computed with zeio of our polynomial of degree N. The approximate solution for $T(w)=0$ is obatined by $\epsilon_1+w_{1+n_0}$. The accuracy of the approximate zero $\epsilon_1+w_{1+n_0}$ is increased far more than the accuracy of the stationary iteration point w_{1+n_0}. This numerical example illustrates that the polynomial equation for RE is effective to compute the RE of zero for polynomial equation $\phi(x)=0$, even if the catastrophic cancellation arises in residual $\phi(x)$ owing to short of mantissa length in floating-point number system in computer.

2 Definitions

Let $\phi(x)$ be a polynomial such that $\phi(x)=\sum_{n=0}^{N}A_n x^n$. The generalized Newton's method is as follows for equation $\phi(x)=0$: $x_{n+1}=f(x_n)$ $(n=1,2,\ldots,\ldots)$, *where* $f(x)=x-H\phi(x)$. Here H is the practically computed value of the inverse of the Jacobian matrix $\phi_x(x_1)$ at the initial point $x=x_1$. The theoretical Newton's iteration transforms to $w_{n+1}=g(w_n)$ $(n=1,2,\ldots,\ldots)$, when the theoretical one is calculated numerically.

Definition 2.1 (Oscillatory state set with period m, Stationary iteration point, RE to oscillatory iteration point) *Let $\{w_{i+n_0}\}$ be a set of iteration points such that $w_{1+n_0}=w_{m+1+n_0}$. $\{w_{i+n_0}\}$ is called an "oscillatory state set with period m". If $m=1$, the set is constituted of one point w_{1+n_0} and w_{1+n_0} is called a "stationary iteration point." Let ϵ_k be the RE of w_{k+n_0} : $\epsilon_k=\hat{x}-w_{k+n_0}$. Here $\hat{x}$ is an exact zero such that $\phi(\hat{x})=0$. ϵ_k is called an "RE of oscillatory iteration point".*

Definition 2.2 **(Iteration round-off error,Residual)** *$\phi(x)$ for any variable x is called a "residual." $f(x)=x-H\phi(x)$ is called an "iteration function." Let w_{i+n_0} be in an oscillatory state set with period m. Let R_i be defined as $R_i=f(w_{i+n_0})-g(w_{i+n_0})$. R_i is the RE of iteration function $f(w)$ at $w=w_{i+n_0}$ which is the sum of the round-off errors of floating-point additions and multiplications. R_i is called an "iteration round-off error."*

Definition 2.3 **(Iteration divided difference)** *Let w_{i+n_0} and w_{k+n_0} be in oscillatory state set with period m. Let $\kappa_i(\epsilon_k)$ be the divided difference of iteration function $f(w)$ with* *deviation* $w_{k+n_0} + \epsilon_{k+n_0} - w_{i+n_0}$; $\kappa_i(\epsilon_k)$ *is expressed by the divided difference of* $\phi(w)$ *with* $w_{k+n_0} + \epsilon_k - w_{i+n_0}$:

$$\kappa_i(\epsilon_k) = \frac{f(w_{k+n_0}+\epsilon_k) - f(w_{i+n_0})}{w_{k+n_0}+\epsilon_k - w_{i+n_0}};\ \kappa_i(\epsilon_k) = 1 - H\frac{\phi(w_{k+n_0}+\epsilon_k) - \phi(w_{i+n_0})}{w_{k+n_0}+\epsilon_k - w_{i+n_0}}.$$

Here the divided difference of $\phi(w)$ *is expressed as a polynomial with degree* $N-1$:

$$\frac{\phi(w_{k+n_0}+\epsilon_k) - \phi(w_{i+n_0})}{w_{k+n_0}+\epsilon_k - w_{i+n_0}} = \sum_{l=1}^{N} A_l \sum_{j=0}^{l-1} (w_{k+n_0}+\epsilon_k)^j w_{i+n_0}^{l-1-j},$$

since $x^l - y^l = (x-y)\sum_{j=0}^{l-1} x^j y^{l-1-j}$. $\kappa_i(\epsilon_k)$ *is called an "iteration divided difference."*

3 Error of iteration point in oscillatory state set

The following theorem gives an equation for the RE of each point in oscillatory state set with period m.

Theorem 3.1 *Assume that:*

1. *Let* $\phi(x)$ *be the polynomial* $\phi(x) = \sum_{n=0}^{N} A_n x^n$ *with degree* N.
2. *The unique solution of* $x = f(x)$ *exists i.e.* $\hat{x} = f(\hat{x})$; $\phi(\hat{x}) = 0$.
3. *Let* $\{w_{i+n_0}\}$ *be an oscillatory state set with period* m : $w_{1+n_0} = w_{m+1+n_0}$.

Then RE ϵ_k *of oscillatory iteration point satisfies the following polynomial equation*

$$\epsilon_k\Big(1 - \prod_{i=1}^{m} \kappa_i(\epsilon_k)\Big) = \sum_{i=k}^{m+k-1} R_i \prod_{j=i+1}^{m+k-1} \kappa_j(\epsilon_k), \tag{1}$$

where $R_i = f(w_{i+n_0}) - g(w_{i+n_0})$; $\kappa_i(\epsilon_k) = 1 - H\sum_{l=1}^{N} A_l \sum_{j=0}^{l-1}(w_{k+n_0}+\epsilon_k)^j w_{i+n_0}^{l-1-j}$.

Proof. We shall prove that the RE $\epsilon_1 = \hat{x} - w_{1+n_0}$ is calculated as

$$\epsilon_1\Big(1 - \prod_{i=1}^{m} \kappa_i(\epsilon_1)\Big) = \sum_{i=1}^{m} R(w_{i+n_0}) \prod_{j=i+1}^{m} \kappa_i(\epsilon_1). \tag{2}$$

In the first stage,

$$\hat{x} - w_{2+n_0} = f(\hat{x}) - g(w_{1+n_0}) = f(\hat{x}) - f(w_{1+n_0}) + R_1 = \kappa_1(\epsilon_1)(\hat{x} - w_{1+n_0}) + R_1.$$

In order to prove by mathematical induction, assume that

$$\hat{x} - w_{m+n_0} = \prod_{i=1}^{m-1} \kappa_i(\epsilon_1)(\hat{x} - w_{1+n_0}) + \sum_{i=1}^{m-1} R_i \prod_{j=i+1}^{m-1} \kappa_j(\epsilon_1). \tag{3}$$

In m th stage,

$$\hat{x} - w_{m+1+n_0} = \kappa_m(\epsilon_1)(\hat{x} - w_{m+n_0}) + R_m. \tag{4}$$

Substituting (3) into (4), $\hat{x} - w_{m+1+n_0} = \prod_{i=1}^{m} \kappa_i(\epsilon_1)(\hat{x} - w_{1+n_0}) + \sum_{i=1}^{m} R_i \prod_{j=i+1}^{m} \kappa_j(\epsilon_1)$. Since $w_{m+1+n_0} = w_{1+n_0}$, the equation (2) is obtained putting $\epsilon_1 = \hat{x} - w_{m+1+n_0} = \hat{x} - w_{1+n_0}$. Q.E.D.

4 Catastrophic cancellation

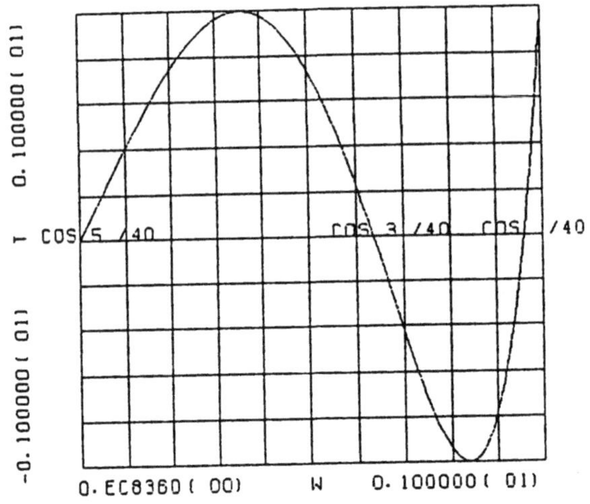

Fig.1 $cos(20cos^{-1}w)$ computed by single precision

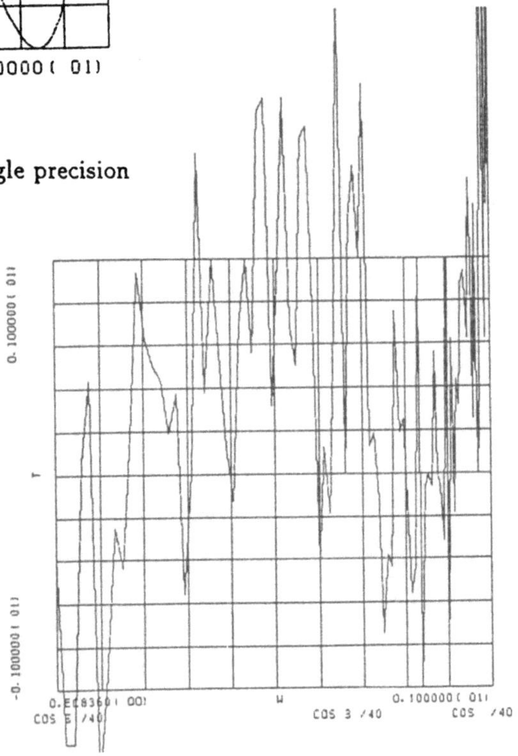

Fig.2 Chebyshev polynomial $\tilde{T}_s(w)$ computed by single precision

Chebyshev polynomial $T(w)$ with degree 20 is expressed as $cos(20cos^{-1}w)$ with cos and cos^{-1} functions. Graph of $cos(20cos^{-1}w)$ is computed by sigle precision putting $w = cos\theta$.

Since $cos(20\theta)$ is computed by intrinsic function offered by FORTRAN Language, the values are accurate in about 4 digits.

On the other hand, Chebyshev polynomial $T(w)$ is expressed also as the polynomial $T(w) = 524288w^{20} - 2621440w^{18} + 5570560w^{16} - 6553600w^{14} + 4659200w^{12} - 2050048w^{10} + 54120w^{8} - 84480w^{6} + 6600w^{4} - 200w^{2} + 1$. Polynomial $T(w)$ is computed by single precision and denoted as $\tilde{T}_s(w)$. Fig.2 shows the graph $\tilde{T}_s(w)$. Owing to the round-off errors, graph changes far from the original graph in Fig.1.The catastrophic cancellation arises in evaluation of the values of polynomial in Fig.2.

5 Stationary iteration point

Table 1 w_n and $\tilde{T}_s(w_n)$ until stationary iteration points appear.

n	w_n	$\tilde{T}_s(w_n)$
1(I)	+0.F8ED3E	+0.D00000
2	+0.FA057E	+0.800000
3	+0.FAB1F4	+.120000(1)
4	+0.FC35FE	-0.300000
5	+0.FBF551	+0.700000
6	+0.FC8C38	-0.D00000
7	+0.FB73F7	-0.900000
8	+0.FAB1F1	-0.300000
9	+0.FA7144	+0.600000
10	+0.FAF29C	+0.400000
11	+0.FB48D7	-0.100000
12	+0.FB3348	+0.C00000
13	+0.FC35F9	+0.700000
14	+0.FCCCE0	-0.E00000
15	+0.FB9F10	-0.700000
16	+0.FB0828	+0.600000
17	+0.FB8980	+0.800000
18	+0.FC35F6	-0.100000
19	+0.FC2067	+0.E00000
20	+0.FD4E36	-0.C00000
21	+0.FC4B84	+0.300000
22	+0.FC8C30	+0.500000
23	+0.FCF7FA	+0.100000
24(F)	**+0.FD0D88**	**0.000000**
25(F)	**+0.FD0D88**	**0.000000**

Fig.3 w_n and $\tilde{T}_s(w_n)$ until stationary iteration points appear.

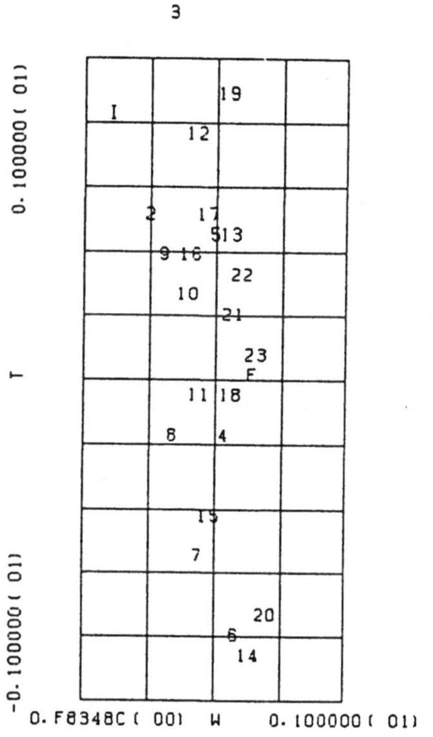

The generalized Newton's iteration process $w_{n+1} = g(w_n)$ $(n = 1, 2, \ldots, \ldots)$ where $g(w) = w \ominus H \otimes \tilde{T}_s(w)$ is computed by single precision until w_n are in an oscillatory state set with period m. But the residual is computed as zero at 24 stage i.e. $\tilde{T}_s(w_{24}) = 0$ and the stationary point is obtained. H is the practically computed value of the floating-point fraction $1.0 \oslash \tilde{T}'_s(w_1) \otimes 0.5$ and $H = -0.158ED2(-1)$. The mantissa of H is expressed by Hexadecimal expression $0.158ED2$ and the exponent is expressed by decimal integer by -1 enclosed by parenthesis. In Table 1, for Hexadecimal expression with exponent zero, (0) is omitted. $w_1 = +0.F8ED3E(0)$ is computed value of $cos\ 3\pi/40$ by single precision. $\tilde{T}'_s(w_1)$ is the computed derivative of $T(w)$ with respect to w at $w = w_1$ by single precision.

Only one digit is computed in digits $\tilde{T}_s(w_n)$ for each n and this shows that catastrophic cancellation arises in the digits. This only one digit of each $\tilde{T}_s(w_n)$ is far from $T(w_n)$ as shown in Fig.1 and Fig.2.

6 Error to stationary iteration point

Assume stationary i.e. $m = 1$. The RE ϵ_1 of the stationary iteration point is determined by the equation of a polynomial with degree N :

$$\psi(\epsilon_1) = \epsilon_1(1 - \kappa(\epsilon_1)) - R(w_{1+n_0}) = 0.$$

Here $\kappa(\epsilon_1)$ is a polynomial with degree $N - 1$:

$$\kappa(\epsilon_1) = 1 - H\sum_{l=1}^{N} A_l \sum_{j=0}^{l-1}(w_{1+n_0} + \epsilon_1)^j w_{1+n_0}^{l-1-j}.$$

ϵ_1 is computed by the Newton's method:

$$\epsilon_1^{n+1} = \epsilon_1^n - \psi(\epsilon_1^n)/\psi_{\epsilon_1}(\epsilon_1^n),\ (n = 0, 1, 2, \ldots,)$$

with initial value $\epsilon_1^0 = R(w_{1+n_0})$. Computed value ϵ_1 is

$$\epsilon_1 = -0.4204BD047218CC(-1) - 0.5C954C93\underline{7D1745}(-15).$$

Here for this ϵ_1, $\kappa(\epsilon_1)$, $\kappa'(\epsilon_1)$ $\psi_{\epsilon_1}(\epsilon_1)$, ψ_{ϵ_1} are computed by quadruple precision.

$$\kappa(\epsilon_1) = 0.ACDB15BFF9C1EB(0) + 0.095D03915E6DDA(-14);$$

$$\kappa'(\epsilon_1) = 0.7D3517C0F3C51A(1) + 0.EAFBA4D2217BCA(-13).$$

$$\psi_{\epsilon_1}(\epsilon_1) = 0.736EEBA683464D(0) + 0.79B529099EF6C7(-14);$$

$$\psi(\epsilon_1) = -0.8B157(-24).$$

For $w_{24} = +0.FD0D8$,

$$R(w_{24}) = f(w_{24}) - g(w_{24}) = -0.15710E5BDD2F86(-1) - 0.21CD86D4F4BEEF(-15).$$

$f(w_{24})$ is computed by quadruple precision. The zero $cos\frac{3\pi}{40}$ is computed by quadruple precision as follows which is correct in 28 digits and denoted by $\tilde{x}$:

$$\tilde{x} = cos\frac{3\pi}{40} = +0.F8ED3C2FB8DE73(0) + 0.3A36AB36C75229(-14)$$

Since $\epsilon_1 + w_{24} = +0.F8ED3C2FB8DE73(0) + 0.3A36AB36C\underline{82E8B}(-14)$, error of RE ϵ_1 is estimated by $\tilde{x}$ as

$$\tilde{x} - (\epsilon_1 + w_{24}) = -0.DC62(-24).$$

Therefore ϵ_1 is correct in 22 digits and incorrect in underlined digits.

7 Error using residual and derivative

Let $T(z - \delta z)$ be the residual at the approximate zero $z - \delta z$. Let $T'(z)$ be the derivative at exact zero $z = cos\ (2r + 1)\pi/40$. The RE δz is obtained fairly good by the formula

$$\delta \tilde{z} = -\frac{T(z - \delta z)}{T'(z)}$$

if the quotients $\frac{\delta z}{z - cos\ (2s+1)\pi/40}$ are small enough compared with 1:

$$\frac{\delta z}{z - cos\ (2s + 1)\pi/40} << 1\ for\ s = 0, \ldots, 19\ s \neq r.$$

The following theorem 7.1 proves this.

Theorem 7.1 *Let z be exact zero $z = cos\ \ (2r + 1)\pi/40$.*

1. *For the approximate zero $z - \delta z$,*

$$-\frac{T(z - \delta z)}{T'(z)} \sim \delta z\{1 - \sum_{s \neq r} \frac{\delta z}{z - cos\ (2s + 1)\pi/40}\}.$$

Here $T(z - \delta z) = 524288 \prod_{s=0}^{19}[z - \delta z - cos\ (2s + 1)\pi/40]$.

$$T'(z) = 524288 \prod_{s \neq r}[z - cos\ (2s + 1)\pi/40].$$

2. *If $\delta z/(z - cos(2s + 1)\pi/40) << 1$ for $s = 0, \ldots, 19$ then δz is obtained accurately by*

$$\delta z\ \sim\ -\frac{T(z - \delta z)}{T'(z)}.$$

Proof.

$$-\frac{T(z - \delta z)}{T'(z)} = \delta z \prod_{s=0}^{19}[1 - \frac{\delta z}{z - cos\ (2s + 1)\pi/40}]$$

$$\sim \delta z\{1 - \sum_{s \neq r} \frac{\delta z}{z - cos\ (2s + 1)\pi/40}$$

$$+ \sum_{s \neq r\ k \neq r} \frac{\delta z^2}{(z - cos\ (2s + 1)\pi/40)(z - cos\ (2k + 1)\pi/40)} + \ldots.\}$$

Q.E.D.

8 Ill conditioned zero

If δz and $cos\ (2r + 1)/40 - cos\ (2s + 1)/40$ are the same order for some s, then δz is not obtained accurately by the formula $\delta \tilde{z}$ in section 7. z is so called ill conditioned zero.

$cos\ 3\pi/40$ is the ill conditioned zero, since $\delta z \sim \epsilon_1 = -0.4204BD(-1)$,

$$cos\ 3\pi/40 - cos\ \pi/40 = -0.648BD4(-1);$$

$$cos\ 3\pi/40 - cos\ 5\pi/40 = 0.C69DD(-1)$$

have the same exponent -1 with δz. The residual $T(w_{24})$ and derivative $T'(\tilde{x})$ are computed by quadruple precision:

$T(w_{24}) = -0.FE9E8CBFC75108(0) - 0.51DEF31910000(-14); T'(\tilde{x}) = -0.55AC53(2)$

$\delta\tilde{z}$ is computed as

$$\delta\tilde{z} = -\frac{T(w_{24})}{T'(\tilde{x})} = -0.\underline{2F8D40518C8617(-1) - 0.D7C85AAF012480}(-15).$$

The approximate zero is

$$w_{24} + \delta\tilde{z} = 0.F\underline{A14B3FAE7379E(0) + 0.82837A550FEDB8}(-14)$$

The error of $w_{24} + \delta\tilde{z}$ is estimated by $\tilde{x}$ as

$$\tilde{x} - (w_{24} + \delta z) = -0.12777CB2E592B4(-1) - 0.84CCF1E489B8F0(-15).$$

Therefore $\delta\tilde{z}$ is incorrect from the leading digit owing that zero $\cos 3\pi/40$ is the ill conditioned zero.

9 Newton's method computing error ϵ_1

It is shown by the following theorem that the derivative $\psi_{\epsilon_1}(\epsilon_1)$ is near to 1 for stationary case i.e. $m = 1$ and ϵ_1 is obtained accurately by the Newton's method as the ϵ_1 converges to the exact value if there is some H such that $| H\phi(\hat{x}) - 1 | \leq \mu$ for small μ. However ϵ_1 is not determined by the equation $\psi(\epsilon_1) = 0$ if derivative $\phi'(\hat{x}) \neq 0$ at exact zero $\hat{x}$.

Theorem 9.1 *Put*

$$\kappa_i(\epsilon_k) = 1 - H\frac{\phi(w_{k+n_0} + \epsilon_k) - \phi(w_{i+n_0})}{w_{k+n_0} + \epsilon_k - w_{i+n_0}} \text{ and then } \kappa_i'(\epsilon_k) = \frac{-H\phi'(\hat{x}) + 1 - \kappa_i(\epsilon_k)}{\hat{x} - w_{i+n_0}}.$$

For stationary case, $\psi_{\epsilon_1}(\epsilon_1) = 1 - \kappa_1(\epsilon_1) - \epsilon_1\kappa'(\epsilon_1)$.

$$\psi_{\epsilon_1}(\epsilon_1) = H\phi'(\hat{x}) = 0 \text{ if } \phi'(\hat{x}) = 0.$$

$$| \psi_{\epsilon_1}(\epsilon_1) - 1 | < \mu \text{ if } | H\phi'(\hat{x}) - 1 | < \mu.$$

Proof. $\kappa_i'(\epsilon_k) =$

$$-H\frac{\phi'(w_{k+n_0} + \epsilon_k)}{w_{k+n_0} + \epsilon_k - w_{i+n_0}} + H\frac{\phi(w_{k+n_0} + \epsilon_k) - \phi(w_{i+n_0})}{(w_{k+n_0} + \epsilon_k - w_{i+n_0})^2} = \frac{-H\phi'(\hat{x}) + 1 - \kappa_i(\epsilon_k)}{\hat{x} - w_{i+n_0}}.$$

$\psi_{\epsilon_1}(\epsilon_1) = 1 - \kappa_1(\epsilon_1) - \epsilon_1(-H\phi'(\hat{x}) + 1 - \kappa_1(\epsilon_1))/\epsilon_1 = H\phi(\hat{x}).$ Q.E.D.

The above numerical example supports the above theorem.

$$H = -0.158ED2(-1); \phi'(\hat{x}) = -0.22AC53(2); H\phi'(\hat{x}) = 0.736EDC(0).$$

Since $| H\phi'(\hat{x}) - 1 | < 1$ is satisfied, $\psi_{\epsilon_1}(\epsilon_1) \neq 0$ and ϵ_1 has been obtained accurately by the Newton's method.

References

[1] Urabe,M.,Mem. Fac. of Sci., Kyushu Univ., Ser. A, Vol.27, No.1, (1973), pp.23-64.

[2] Tsuji,K., Proceedings 5th Int. Coll. on Diff. Equ., VSP (1995) pp. 331-340.

[3] Wilkinson,J.H.,Rounding Errors in Algebraic Processes, Notes on Applied Science No.32, National Physical Laboratory, (1963) pp. 46-47.

7th Int. Coll. on Differential Equations, pp. 429-436
D. Bainov (Ed)

Identification of Density Distribution due to One-codimensional Projected Figures

Miki Tsuji [1]

Graduate School of Mathematics, Kyushu University 33, Fukuoka, 812-81, Japan

Keywords: Density Distribution, Gauss-Weierstrass transform

Abstract

In the present paper, under the condition that direct observation is impossible in some limited region, we consider identification of unknown density distributions by analysing one-codimensional projected figures.

1 Introduction.

In the previous papers [3], [4] , and [5], we consider for $x \in \mathbf{R}^n$,

$$f_{P,K,A}(x) = K \exp\left(-\frac{(x-P)^t A^{-1}(x-P)}{4}\right) \tag{1}$$

where $K \in \mathbf{R}^+$ is a positive real number, $P = (P_1, \cdots, P_n)^t \in \mathbf{R}^n$ is a n dimensional constant vector, $A = (A_{i,j}) \in PEL_n(\mathbf{R})$ is a real Hermitian matrix and positive definite, and the multi-peak function given by the aggregation function :

$$g(x) = \sum_{l=1}^{m} f_{P^{[l]},K^{[l]},A^{[l]}}(x). \tag{2}$$

We constructed the algorithm of searching the values m and identifying $P^{[l]}$, $K^{[l]}$ and $A^{[l]}$ using the values of the Gauss-Weierstrass transformed function $G(\alpha, \beta)$ of $g(x)$, and its first derivative at several observation points.

In the present paper, under the condition that direct observation is impossible in some bounded region, we set light source plane and screen plane (Section 2) and then we consider identification of unknown density distributions by analysing one-codimensional projected figures.

---

[1]Research was supported by Grant-in-Aid for JSPS Research Fellow no.07 2309 from the Ministry of Education, Science, Sports and Culture of Japan, 1996.

2 An illustrated example of construction of shadow figure

In the present paper, under the condition that direct observation is impossible in some bounded region, firstly we set light source which distributes uniformly on one-codimensional plane and projects parallel ray with uniform intensity into this bounded region perpendicularly from this one-codimensional light source plane. We call this one-codimensional light source plane " light source plane ", simply.

Secondly we put another one-codimensional plane as the screen to obtain the shadow figure of unknown distributed density in the bounded region. Simply, we call this one-codimensional plane as the screen " screen plane".

Screen plane is parallel with the light source plane and is set so as the bounded region must be included in the zone between light source plane and screen plane.

We call this zone between light source plane and screen plane " off-limit zone ".

The shadow figure is projected on screen plane recieving the output parallel ray from off-limit zone.

We show Figure 1 as illustrated example.

Figure 1: An illustrated example in Section 2.

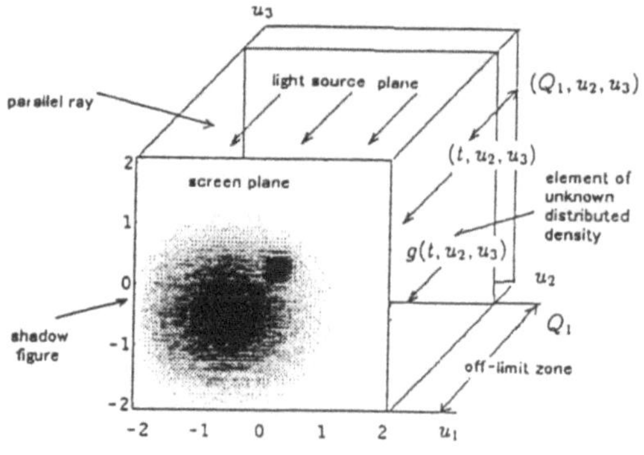

The scalar unknown density distribution is given on 3 dimensional space $x = (x_1, x_2, x_3)^t \in \mathbf{R}^3$. The dimension of light source plane is 2. If the normal vector to light source plane is given as $(1, 0, 0)^t$ and the plane passes through the point $(Q_1, 0, 0)^t$, the 2 dimensional light source plane is expressed as $x_1 = Q_1$ in x space, for a constant Q_1. Also the dimension of screen plane is 2 and since the screen plane is parallel with the light source plane, it is expressed as $x_1 = u_1$ in x space, for a constant u_1.

Off limit zone is expressed as zone between $x_1 = Q_1$ and $x_1 = u_1$ in x space.

Intensity of light being input into off-limit zone from light source plane may be decreased during of passing through off-limit zone from $x_1 = Q_1$ to $x_1 = u_1$. Thus for $v^1 = (u_2, u_3)^t$, we put $C_{Q_1,u_1}(v^1)$, such as

$$0 < C_{Q_1,u_1}(v^1) = \exp\left(-\int_{Q_1}^{u_1} g(t, v^1)dt\right) \leq 1, \tag{3}$$

as coefficient of light intensity attenuation due to unknown distributed density $g(x)$ in off-limit zone, i.e.,

$$\text{if } \int_{Q_1}^{u_1} g(t,v^1)dt \to 0, \quad \text{then } C_{Q_1,u_1}(v^1) \to 1 \quad \text{there is no shadow,}$$

$$\text{if } \int_{Q_1}^{u_1} g(t,v^1)dt \to \infty, \quad \text{then } C_{Q_1,u_1}(v^1) \to 0 \quad \text{shadow is completely dark.}$$

Light intensity at screen plane is kept as it at light source plane without decreasing for $C_{Q_1,u_1}(v^1) = 1$. On the other hand, light intensity at screen plane is decreased to 0, i.e., shadow tends to completely dark for $C_{Q_1,u_1}(v^1) \to 0$.

For $0 < \int_{Q_1}^{u_1} g(t,v^1)dt < \infty$, the darkness of shadow at v^1 on screen plane corresponds to $\int_{Q_1}^{u_1} g(t,v^1)dt$, i.e., integration of local density along light path in off-limit zone.

Shadow figure is constructed by $C_{Q_1,u_1}(v^1)$ on screen plane as the set for $-\infty < u_2 < \infty$, $-\infty < u_3 < \infty$, of dark spot at v^1 for given Q_1, u_1.

Variables of shadow figure on screen plane are u_2, u_3 for given constant Q_1, u_1, i.e., $C_{Q_1,u_1}(v^1)$ becomes a function of u_2, u_3.

Hereafter we put

$$m_{Q_1,u_1}(v^1) = -\log C_{Q_1,u_1}(v^1)$$

and handle $m_{Q_1,u_1}(v^1)$ as fundamental observed data obtained by shadow figure. On the other hand,

$$m_{Q_1,u_1}(v^1) = \int_{Q_1}^{u_1} g(t,v^1)dt \tag{4}$$

by (3).

Hence analysing shadow figure resolves itself into analysing simple form $\int_{Q_1}^{u_1} g(t,v^1)dt$.

For 3-dimensional x space, we take 3 sets of light source plane and screen plane at a right angle to each other.

For identification processing, 3 sets of shadow figure data are simultaneously used, i.e.,

$$\left[C_{Q_1,u_1}(v^1), C_{Q_2,u_2}(v^2) \text{ and } C_{Q_3,u_3}(v^3)\right],$$

or

$$\left[m_{Q_1,u_1}(v^1), m_{Q_2,u_2}(v^2) \text{ and } m_{Q_3,u_3}(v^3)\right].$$

Finally, we identify an unknown density distibution in the off limit region analysing and combining the figures of shadow on 3 screen planes, using Gauss-Weierstrass transformation.

3 Notation and Gauss-Weierstrass transform.

Hereafter, we set $n \leq 2$.

Generally, we take, for $r = 1, \cdots, n$, light source planes, where normal vector to the planes are $(0, \cdots, 0, 1_r, 0, \cdots, 0)^t$ where 1_r represents $x_r = 1$, and the planes pass through the point $(0, \cdots, 0, Q_r, 0, \cdots, 0)^t$, where Q_r represents $x_r = Q_r$, i.e., the $n-1$ dimensional light source planes as $x_r = Q_r$ in x space, for a constant Q_r.

Also we take the screen planes which are parallel with the light source planes, as $x_r = u_r$ in x space, for a constant u_r.

In this case, off limit zones are expressed as zones between $x_r = Q_r$ and $x_r = u_r$ in x space, respectively.

Let, for $v^r = (u_1, u_2, \cdots, u_{r-1}, u_{r+1}, \cdots, u_n)^t$,

$$\begin{aligned} C_{Q_r,u_r}(v^r) &= \exp\left(-\int_{Q_r}^{u_r} g(u_1, u_2, \cdots, u_{r-1}, t, u_{r+1}, \cdots, u_n)dt\right) \\ &= \exp\left(-\int_{Q_r}^{u_r} g(t, v^r)dt\right) \end{aligned} \tag{5}$$

as coefficient of light intensity attenuation due to unknown distributed density $g(x)$ in off-limit zone, which is the function with respect to v^r in the screen plane where u_r is constant, and let,

$$\begin{aligned} m_{Q_r,u_r}(v^r) &= -\log C_{Q_r,u_r}(v^r) = \int_{Q_r}^{u_r} g(u_1, u_2, \cdots, u_{r-1}, t, u_{r+1}, \cdots, u_n)dt \\ &= \int_{Q_r}^{u_r} g(t, v^r)dt \end{aligned} \tag{6}$$

Hence analysing shadow figure resolves itself into analysing simple form $\int_{Q_r}^{u_r} g(t, v^r)dt$.

We suppose that each element function of unknown distributed density in a n dimensional space is a Gaussian distribution. That is, for $x \in \mathbf{R}^n$, the distribution is expressed by

$$f_{P,K,A}(x) = K \exp\left(-\frac{(x-P)^t A^{-1}(x-P)}{4}\right) \tag{7}$$

where $K \in \mathbf{R}^+$ is a positive real number, $P = (P_1, \cdots, P_n)^t \in \mathbf{R}^n$ is a n dimensional constant vector, $A = (A_{i,j}) \in PEL_n(\mathbf{R})$ is a real Hermitian matrix and positive definite.

Then density distributions are expressed by an aggregation of Gaussian functions which is multi-peak function :

$$g(x) = \sum_{l=1}^{m} f_{P^{[l]},K^{[l]},A^{[l]}}(x). \tag{8}$$

The multi-peak function has (1) peak point at $P^{[l]}$ $(l = 1, \cdots, m)$, (2) the figure of which is characterized by the peak point position, abbreviated to *ppp* hereafter, and several parameters $K^{[l]}, A^{[l]}$ and (3) the function tends to 0 as the distance from *ppp* tends to ∞.

Our goal is identification of such an unknown density distributions $g(x)$, i.e., to identify *ppp*'s and parameters.

Define $n-1$ dimensional Gauss-Weierstrass transformed function of $m_{Q_r,u_r}(v^r)$ by

$$M_{Q_r,u_r}(\alpha, \beta) = \int_{-\infty}^{\infty} m_{Q_r,u_r}(v^r) h_{\alpha,\beta}(v^r) dv^r$$

for a function

$$h_{\alpha,\beta}(v^r) = \frac{1}{(2\sqrt{\pi})^{n-1}(\det(\alpha))^{\frac{1}{2}}} \exp\left(-\frac{(v^r-\beta)^t \alpha^{-1}(v^r-\beta)}{4}\right),$$

where $\alpha \in PEL_{n-1}(\mathbf{R})$ and $\beta \in \mathbf{R}^{n-1}$.

We construct the algorithm for identification of $g(x)$ in off-limit zone by analysing Gauss-Weierstrass transformation of its one-codimensinal projected shadow figures, i.e., by using $M_{Q_r,u_r}(\alpha, \beta)$. $\frac{\partial}{\partial \alpha_{k,l}} M_{Q_r,u_r}(\alpha, \beta)$ and $\frac{\partial}{\partial \beta_k} M_{Q_r,u_r}(\alpha, \beta)$, for all $1 \le r \le n$, at several observation points (α, β).

4 Method for identification.

We adopt 3 steps for identification.

Define

$$P^{[l]}(v^r) = (P_1^{[l]}, \cdots, P_{r-1}^{[l]}, P_{r+1}^{[l]}, \cdots, P_n^{[l]})^t. \tag{9}$$

and for $D^{[l]} = A^{[l]^{-1}}$, define

$$A^{[l]}(v^r)^{-1} = \begin{pmatrix} D^{[l]}{}_{1,1} & \cdots & D^{[l]}{}_{1,r-1} & D^{[l]}{}_{1,r+1} & \cdots & D^{[l]}{}_{1,n} \\ \vdots & \vdots & \vdots & \vdots & \vdots & \vdots \\ D^{[l]}{}_{r-1,1} & \cdots & D^{[l]}{}_{r-1,r-1} & D^{[l]}{}_{r-1,r+1} & \cdots & D^{[l]}{}_{r-1,n} \\ D^{[l]}{}_{r+1,1} & \cdots & D^{[l]}{}_{r+1,r-1} & D^{[l]}{}_{r+1,r+1} & \cdots & D^{[l]}{}_{r+1,n} \\ \vdots & \vdots & \vdots & \vdots & \vdots & \vdots \\ D^{[l]}{}_{n,1} & \cdots & D^{[l]}{}_{n,r-1} & D^{[l]}{}_{n,r+1} & \cdots & D^{[l]}{}_{n,n} \end{pmatrix} \tag{10}$$

and

$$B^{[l]}(v^r) = A^{[l]}(v^r)\left(D_{1,r}^{[l]}, \cdots, D_{r-1,r}^{[l]}, D_{r+1,r}^{[l]}, D_{n,r}^{[l]}\right)^t. \tag{11}$$

Then

$$\begin{aligned} m_{Q_r,u_r}(v^r) &= \int_{Q_r}^{u_r} \left(\sum_{l=1}^{m} f_{P^{[l]},K^{[l]},A^{[l]}}(u_1, \cdots, u_{r-1}, t, u_{r+1}, \cdots, u_n)\right)dt \\ &= \sum_{l=1}^{m} \int_{Q_r}^{u_r} \exp\left(-(t-P_r^{[l]})^2 \frac{D_{r,r}^{[l]} - B^{[l]}(v^r)^t A^{[l]}(v^r)^{-1} B^{[l]}(v^r)}{4}\right) K^{[l]} \end{aligned}$$

$$\exp\left(-\frac{\left(v^r - P^{[l]}(v^r) + (t-P_r^{[l]})B^{[l]}(v^r)\right)^t A^{[l]}(v^r)^{-1}\left(v^r - P^{[l]}(v^r) + (t-P_r^{[l]})B^{[l]}(v^r)\right)}{4}\right) dt.$$

$$\begin{aligned} &= \sum_{l=1}^{m} \int_{Q_r}^{u_r} \exp\left(-(t-P_r^{[l]})^2 \frac{D_{r,r}^{[l]} - B^{[l]}(v^r)^t A^{[l]}(v^r)^{-1} B^{[l]}(v^r)}{4}\right) \\ &\quad \times f_{P^{[l]}(v^r)-(t-P_r^{[l]})B^{[l]}(v^r),K^{[l]},A^{[l]}(v^r)}(v^r)dt. \end{aligned}$$

Note that $f_{P^{[l]}(v^r)-(t-P_r^{[l]})B^{[l]}(v^r),K^{[l]},A^{[l]}(v^r)}(v^r)$ is the function from $n-1$ dimensional space to $\mathbf{R}$.

Hence, for $K^{[l]}(v^r) = \exp\left(-(t-P_r^{[l]})^2 \frac{D_{r,r}^{[l]} - B^{[l]}(v^r)^t A^{[l]}(v^r)^{-1} B^{[l]}(v^r)}{4}\right) K^{[l]}$,

$$m_{Q_r,u_r}(v^r) = \sum_{l=1}^{m} \int_{Q_r}^{u_r} f_{P^{[l]}(v^r)-(t-P_r^{[l]})B^{[l]}(v^r),K^{[l]}(v^r),A^{[l]}(v^r)}(v^r)dt$$

and

$$\begin{aligned} M_{Q_r,u_r}(\alpha,\beta) &= \int_{-\infty}^{\infty} m_{Q_r,u_r}(v^r) h_{\alpha,\beta}(v^r) dv^r \\ &= \sum_{l=1}^{m} \int_{-\infty}^{\infty} \left[\int_{Q_r}^{u_r} f_{P^{[l]}(v^r)-(t-P_r^{[l]})B^{[l]}(v^r),K^{[l]}(v^r),A^{[l]}(v^r)}(v^r)dt\right] h_{\alpha,\beta}(v^r)dv^r \\ &= \sum_{l=1}^{m} \int_{Q_r}^{u_r} \left[\int_{-\infty}^{\infty} f_{P^{[l]}(v^r)-(t-P_r^{[l]})B^{[l]}(v^r),K^{[l]}(v^r),A^{[l]}(v^r)}(v^r) h_{\alpha,\beta}(v^r)dv^r\right] dt \end{aligned}$$

Let

$$G_{P^{[l]}(v^r),K^{[l]}(v^r),A^{[l]}(v^r)}(\alpha,\beta) = \int_{-\infty}^{\infty} f_{P^{[l]}(v^r),K^{[l]}(v^r),A^{[l]}(v^r)}(v^r)h_{\alpha,\beta}(v^r)dv^r$$

$$= \left(\frac{\det A(v^r)}{\det(\alpha + A(v^r))}\right)^{\frac{1}{2}} K^{[l]}(v^r)$$

$$\times \exp\left(-\frac{\left(\beta - P^{[l]}(v^r)\right)^t (\alpha + A^{[l]}(v^r))^{-1}\left(\beta - P^{[l]}(v^r)\right)}{4}\right)$$

Then we have

$$M_{Q_r,u_r}(\alpha,\beta) = \sum_{l=1}^{m} \int_{Q_r}^{u_r} G_{P^{[l]}(v^r)-(t-P_r^{[l]})B^{[l]}(v^r),K^{[l]}(v^r),A^{[l]}(v^r)}(\alpha,\beta)dt$$

If $B^{[l]}(v^r) = 0$, i.e., $D_{i,r}^{[l]} = 0 (i = 1,\cdots,r-1,r+1,\cdots,n)$, we have

$$M_{Q_r,u_r}(\alpha,\beta) = \sum_{l=1}^{m} \int_{Q_r}^{u_r} \left(\frac{\det A(v^r)}{\det(\alpha + A(v^r))}\right)^{\frac{1}{2}} K^{[l]}(v^r) \times \exp\left(-\frac{\left(\beta - P^{[l]}(v^r)\right)^t (\alpha + A^{[l]}(v^r))^{-1}\left(\beta - P^{[l]}(v^r)\right)}{4}\right)dt$$

$$= \int_{Q_r}^{u_r} K^{[l]}(v^r)dt \sum_{l=1}^{m}\left(\frac{\det A(v^r)}{\det(\alpha + A(v^r))}\right)^{\frac{1}{2}} \exp\left(-\frac{\left(\beta - P^{[l]}(v^r)\right)^t (\alpha + A^{[l]}(v^r))^{-1}\left(\beta - P^{[l]}(v^r)\right)}{4}\right)$$

$$= \sum_{l=1}^{m} G_{P^{[l]}(v^r),\int_{Q_r}^{u_r} K^{[l]}(v^r)dt,A^{[l]}(v^r)}(\alpha,\beta)$$

In these case, the method explained in the previous papers [5] can be applied and we obtain $P^{[l]}(v^r), \int_{Q_r}^{u_r} K^{[l]}(v^r)dt, A^{[l]}(v^r)$ for all l.

Step 1 First we take several observation points for each observation domain $\{(\alpha,\beta)\}$, for $1 \leq r \leq n$ and, as if $A^{[l]}$ had satisfied the condition $B^{[l]}(v^r) = 0$, we apply the previous method in [5] to the values $M_{Q_r,u_r}(\alpha,\beta)$, $\frac{\partial}{\partial\alpha_{k,l}}M_{Q_r,u_r}(\alpha,\beta)$ $\frac{\partial}{\partial\beta_k}M_{Q_r,u_r}(\alpha,\beta)$, expecting a little error.

We define the approximate values by $\tilde{A}^{[l']}(v^r)$, $\tilde{P}^{[l']}(v^r)$ and $\int_{Q_r}^{u_r} \tilde{K}^{[l']}(v^r)dt$.

Step 2 For each peak function, $P^{[l]}$, $K^{[l]}$, $A^{[l]}$ is approximated by selecting one set $A^{[l]}(v^r)$, $P^{[l]}(v^r)$, $\int_{Q_r}^{u_r} K^{[l]}(v^r)dt$ as appropriate combination of $\tilde{A}^{[l']}(v^r)$, $\tilde{P}^{[l']}(v^r)$, $\int_{Q_r}^{u_r} \tilde{K}^{[l']}(v^r)dt$ for all $r = 1,\cdots,n$ obtained in Step 1 and by averaging $\tilde{A}^{[l']}(v^r)$, $\tilde{P}^{[l']}(v^r)$ and calculating the approximate value of $K^{[l]}$ by the equation $K^{[l]}(v^r) = \exp\left(-(t-P_r^{[l]})^2 \frac{D_{r,r}^{[l]} - B^{[l]}(v^r)^t A^{[l]}(v^r)^{-1} B^{[l]}(v^r)}{4}\right) K^{[l]}$.

Step 3 The values P, K, A are converged by Newton method from initial values determined by Step 2.

5 Numerical example

Example 1 *Consider for example, the case when* $n = 3$, $m = 2$,

$P^{[1]} = (1.0, 1.0, 1.0)^t$, $A^{[1]} = \begin{pmatrix} 1.5 & 0.1 & 0.1 \\ 0.1 & 1.5 & 0.1 \\ 0.1 & 0.1 & 1.5 \end{pmatrix}$, $K^{[1]} = 1$ *and* $P^{[2]} = (1.1, 1.1, 1.1)^t$, $A^{[2]} = \begin{pmatrix} 1.0 & 0.1 & 0.2 \\ 0.1 & 1.0 & 0.0 \\ 0.2 & 0.0 & 1.0 \end{pmatrix}$, $K^{[2]} = 1$. *We have results in Table 1, Table 2, Table 3.*

6 Conclusion

In the present paper, under the condition that direct observation is impossible in some bounded region, firstly we set light source plane which distributes uniformly on one-codimensional plane and projects parallel ray with uniform intensity into this bounded region perpendicularly from this one-codimensional light source plane. Secondly we put another one-codimensional plane as screen plane to obtain the shadow figure of unknown distributed density in the bounded region. The shadow figure is projected on screen plane recieving the output parallel ray from off-limit zone.

We suppose that each element function of unknown distributed density in a n dimensional space is a Gaussian distribution. Then density distributions are expressed by an aggregation of Gaussian functions which is multi-peak function.

We construct the algorithm for identification of $g(x)$ in off-limit zone by analysing Gauss-Weierstrass transformation of its one-codimensinal projected shadow figures at several observation points, adopting 3 steps for identification.

ACKNOWLEDGEMENT

The author would like to express her hearty gratitude to Professor Kajiwara for his valuable suggestion.

References

[1] J.Kajiwara and H.Watanabe, Mem. Fac. Kyusyu Univ. **46** -1(1992), pp.1-30.

[2] S. Saitoh, Applicable Anal. **16**, (1983), pp. 1-6.

[3] M. Tsuji, Proceedings of the Third International Colloquium on Numerical Analysis, VSP (1995), pp.191-200.

[4] M. Tsuji and Dong Gou Zhou, Proceedings of the Second Korean-Japanese Colloquium on Finite or Infinite Dimensional Complex Analysis, (1994), pp.235-241.

[5] M. Tsuji, Proceedings of the Sixth International Colloquium on Differential Equations, pp. 305-312, VSP (1996).

Table 1: Step 1 in Example 1

r	(α,β)		$P(v^r)$, $\int_{Q_r}^{u_r} K(v^r)dt$, $A(v^r)$
$r=1$	$\beta^{[1]}=(1.0,1.0)$	$P^{[1]}(v^1)$	(1.054705642197590D+00, 1.056871078372683D+00)
	$\alpha^{[1]}=\begin{pmatrix} +0 & 0 \\ 0 & +0 \end{pmatrix}$	$A^{[1]}(v^1)$	$\begin{pmatrix} 1.496687411943486D+00 & 9.690312678377146D-02 \\ 9.690312678377146D-02 & 1.497388447701781D+00 \end{pmatrix}$
		$\int_{Q_1}^{u_1} K^{[1]}(v^1)dt$	3.110700471949995D+00
	$\beta^{[2]}=(1.0,1.0)$	$P^{[2]}(v^1)$	(1.151927817212042D+00 , 1.202431543767516D+00)
	$\alpha^{[2]}=\begin{pmatrix} 10 & 0 \\ 0 & +0 \end{pmatrix}$	$A^{[2]}(v^1)$	$\begin{pmatrix} 9.955445980967473D-01 & -8.801983351044391D-03 \\ -8.801983351044391D-03 & 9.822019459775030D-01 \end{pmatrix}$
		$\int_{Q_1}^{u_1} K^{[2]}(v^1)dt$	2.729670553101814D+00
$r=2$	$\beta^{[1]}=(1.0,1.0)$	$P^{[1]}(v^2)$	(1.054520229758465D+00, 1.054567150428310D+00)
	$\alpha^{[1]}=\begin{pmatrix} +0 & 0 \\ 0 & +0 \end{pmatrix}$	$A^{[1]}(v^2)$	$\begin{pmatrix} 1.496667656611880D+00 & 9.673222824833616D-02 \\ 9.673222824833616D-02 & 1.496590583290388D+00 \end{pmatrix}$
		$\int_{Q_2}^{u_2} K^{[1]}(v^2)dt$	3.110816113541945D+00
	$\beta^{[2]}=(1.0,1.0)$	$P^{[2]}(v^2)$	(1.152711821853530D+00, 1.099560240420430D+00)
	$\alpha^{[2]}=\begin{pmatrix} 10 & 0 \\ 0 & +0 \end{pmatrix}$	$A^{[2]}(v^2)$	$\begin{pmatrix} 9.955959112345951D-01 & 2.000225831374648D-01 \\ 2.000225831374648D-01 & 9.999889096095109D-01 \end{pmatrix}$
		$\int_{Q_2}^{u_2} K^{[2]}(v^2)dt$	2.761981822008150D+00
$r=3$	$\beta^{[1]}=(1.0,1.0)$	$P^{[1]}(v^3)$	(1.057156296584585D+00, 1.054869616572326D+00)
	$\alpha^{[1]}=\begin{pmatrix} +0 & 0 \\ 0 & +0 \end{pmatrix}$	$A^{[1]}(v^3)$	$\begin{pmatrix} 1.496439874223915D+00 & 9.681180011949775D-02 \\ 9.681180011949775D-02 & 1.495528513787162D+00 \end{pmatrix}$
		$\int_{Q_3}^{u_3} K^{[1]}(v^3)dt$	3.117565014153696D+00
	$\beta^{[2]}=(1.0,1.0)$	$P^{[2]}(v^3)$	(1.202912073661128D+00, 1.099391250141328D+00)
	$\alpha^{[2]}=\begin{pmatrix} 10 & 0 \\ 0 & +0 \end{pmatrix}$	$A^{[2]}(v^3)$	$\begin{pmatrix} 9.818617305370272D-01 & 1.000229946978174D-01 \\ 1.000229946978174D-01 & 9.996864118851789D-01 \end{pmatrix}$
		$\int_{Q_3}^{u_3} K^{[2]}(v^3)dt$	2.730710608448176D+00

Table 2: Step 2 in Example 1 (Initial values in Step 3)

	combining $P^{[1]}(v^1)$, $\int_{Q_1}^{u_1} K^{[1]}(v^1)dt$, $A^{[1]}(v^1)$; $P^{[1]}(v^2)$, $\int_{Q_2}^{u_2} K^{[1]}(v^2)dt$, $A^{[1]}(v^2)$; $P^{[1]}(v^3)$, $\int_{Q_3}^{u_3} K^{[1]}(v^3)dt$, $A^{[1]}(v^3)$ and averaging these values.
P	(1.055838263171525D+00, 1.054787629384958D+00, 1.055719114400496D+00)
A	$\begin{pmatrix} 1.496553765417897D+00 & 9.681180011949775D-02 & 9.673222824833616D-02 \\ 9.681180011949775D-02 & 1.496107962865324D+00 & 9.690312678377146D-02 \\ 9.673222824833616D-02 & 9.690312678377146D-02 & 1.496989515496084D+00 \end{pmatrix}$
K	9.854109968220490D-01
	combining $P^{[2]}(v^1)$, $\int_{Q_1}^{u_1} K^{[2]}(v^1)dt$, $A^{[2]}(v^1)$; $P^{[2]}(v^2)$, $\int_{Q_2}^{u_2} K^{[2]}(v^2)dt$, $A^{[2]}(v^2)$; $P^{[2]}(v^3)$, $\int_{Q_3}^{u_3} K^{[2]}(v^3)dt$, $A^{[2]}(v^3)$ and averaging these values.
P	(1.177811947757329D+00, 1.125659533676685D+00, 1.150995892093973D+00)
A	$\begin{pmatrix} 9.887288208858112D-01 & 1.000229946978174D-01 & 2.000225831374648D-01 \\ 1.000229946978174D-01 & 9.976155049909631D-01 & -8.801983351044391D-03 \\ 2.000225831374648D-01 & -8.801983351044391D-03 & 9.910954277935069D-01 \end{pmatrix}$
K	9.781186797611174D-01

Table 3: Step 3 in Example 1

P	(9.999999999999841D-01, 9.999999999999817D-01, 9.999999999999923D-01)
A	$\begin{pmatrix} 1.500000000000079D+00 & 1.000000000000120D-01 & 9.999999999998650D-02 \\ 1.000000000000120D-01 & 1.500000000000095D+00 & 1.000000000000322D-01 \\ 9.999999999998650D-02 & 1.000000000000322D-01 & 1.500000000000083D+00 \end{pmatrix}$
K	9.999999999997150D-01
P	(1.099999999999986D+00, 1.099999999999986D+00, 1.099999999999979D+00)
A	$\begin{pmatrix} 1.000000000000062D+00 & 9.999999999999047D-02 & 1.999999999999862D-01 \\ 9.999999999999047D-02 & 1.000000000000055D+00 & 3.311297817497393D-15 \\ 1.999999999999862D-01 & 3.311297817497393D-15 & 1.000000000000061D+00 \end{pmatrix}$
K	1.000000000000284D+00

7th Int. Coll. on Differential Equations, pp. 437-440
D. Bainov and V. Covachev (Eds)

Differential Inequalities Concerning Nonlinear Systems

CORNELIU URSESCU
Institute of Mathematics, 8 Copou blvd, 6600 Iasi, Romania

Abstract. There are given standard tangency conditions which characterize monotonicity, with regard to initial data, of solutions to differential equations of Carathéodory type.

Keywords: Carathéodory solutions, differential inequalities, tangency conditions

INTRODUCTION

In this note we bridge a gap between some nonlinear results concerning monotonic behaviour, with regard to the initial data, of solutions to differential equations. To begin with, we recall some results concerning component monotonicity. In the next section we rephrase these results for monotonicity generated by a closed set or a closed epigraph.

Let $f : \mathbb{R} \times \mathbb{R}^m \to \mathbb{R}^m$ be a Carathéodory function and consider the differential equation

$$X'(t) = f(t, X(t)).$$

Suppose, for every (τ, x), there exists a unique saturated solution X to this differential equation such that $X(\tau) = x$ and denote this solution by $F(\cdot; \tau, x)$. The function F is said to be the evolutor generated by the differential equation above. It is continuous on its domain which is open in $\mathbb{R} \times \mathbb{R} \times \mathbb{R}^m$.

Let further $g : \mathbb{R} \times \mathbb{R}^m \to \mathbb{R}^m$ be another Carathéodory function and consider the differential equation

$$Y'(t) = g(t, Y(t)).$$

Suppose, for every (τ, y), there exists a unique saturated solution Y to this differential equation such that $Y(\tau) = y$ and denote by G the corresponding evolutor. Our first main result, Theorem 3 below, originates from [1] (see also [2], [3], [4] and [5]). Here we give the statement of the originating result.

Theorem 1. *The following two conditions are equivalent:*
(a) for every $(t; \tau, x)$ *and* $(t; \tau, y)$ *(with* $t \geq \tau$ *and* $x_j \leq y_j$ *for all* j*), and for every* i,

$$F_i(t; \tau, x) \leq G_i(t; \tau, y);$$

(b) for almost every t, for every x and y (with $x_j \leq y_j$ for all j), and for every i (with $x_i = y_i$),

$$f_i(t,x) \leq g_i(t,y).$$

Let finally $h: \mathbb{R} \times \mathbb{R}^n \to \mathbb{R}^n$ be another Carathéodory function and consider the differential equation

$$Z'(t) = h(t, Z(t)).$$

Suppose, for every (τ, y), there exists a unique saturated solution Z to this differential equation such that $Z(\tau) = z$ and denote by H the corresponding evolutor. Our second main result, Theorem 3 below, originates from [6]. Here we give the statement of the originating result (sign(0)r stands for $|r|$).

Theorem 2. *Let $m = n$. The following two conditions are equivalent:*
(a) for every $(t;\tau,x)$, $(t;\tau,y)$ and $(t;\tau,z)$ (with $t \geq \tau$ and $|y_j - x_j| \leq z_j$ for all j), and for every i,

$$\left|G_i(t;\tau,y) - F_i(t;\tau,x)\right| \leq H_i(t;\tau,z);$$

(b) for almost every t, for every x, y, and z (with $|y_j - x_j| \leq z_j$ for all j), and for every i (with $|y_i - x_i| = z_i$),

$$(sign(y_i - x_i))\,(g_i(t,y) - f_i(t,x)) \leq h_i(t,z).$$

MAIN RESULTS

Theorems 1 and 2 above belong to the surface structure of a theory of which deep structure constains Theorems 3 and 4 below. There the component monotonicity is replaced by monotonicity with regard to a closed set or a closed epigraph. The core of the theory is the characterization of admissibility for differential equations by tangency conditions. For differential equations with continuous right-hand sides, the characterization has been obtained in [7]. Its proof is not much longer than the proof of the standard Cauchy-Peano theorem (cf. [8]). For differential equations with Carathéodory right-hand sides, it has been obtained in [9] (see also [10] and the attribution in [11], [12], [13]).

Let first P be a closed subset of $\mathbb{R}^m$. Next we denote by K the tangency concept devised in [14] and [15].

Theorem 3. *The following two conditions are equivalent:*
(a) for every $(t;\tau,x)$ and $(t;\tau,y)$ (with $t \geq \tau$ and $y - x \in P$),

$$G(t;\tau,y) - F(t;\tau,x) \in P;$$

(b) for almost every t and for every x and y (with $y - x \in P$),

$$g(t, y) - f(t, x) \in K_P(y - x).$$

Proof. Condition *(a)* states just the set Q of all (x, y) with $y - x \in P$ is admissible with regard to the differential system $X'(t) = f(t, X(t))$, $Y'(t) = g(t, Y(t))$. This admissibility is characterized through the tangency condition that, for almost every t and for every $(x, y) \in Q$, $(f(t, x),\ g(t, y)) \in K_Q(x, y)$. Now the conclusion follows because $K_Q(x, y)$ equals the set of all (u, v) with $v - u \in K_P(y - x)$. □

Let second P be a subset of $\mathbb{R}^n$ and let $q : \mathbb{R}^m \to \mathbb{R}^n$ be a function of which P-epigraph, $\{(u, w) : w \in q(u) + P\}$, is a closed subset of $\mathbb{R}^m \times \mathbb{R}^n$. Next we denote by D the differentiability concept devised in [16] (cf. [17]). Suppose q is D differentiable and denote the corresponding differential by D_q.

Theorem 4. *The following two conditions are equivalent:*
(a) for every $(t; \tau, x)$, $(t; \tau, y)$, and $(t; \tau, z)$ (with $t \geq \tau$ and $z \in q(y - x) + P$),

$$H(t; \tau, z) \in q\Big(G(t; \tau, y) - F(t; \tau, x)\Big) + P;$$

(b) for almost every t and for every x, y, and z (with $z \in q(y - x) + P$),

$$h(t, z) \in D_q(y - x)\Big(g(t, y) - f(t, x)\Big) + K_P(z - q(y - x)).$$

Proof. Condition *(a)* states just the set Q of all (x, y, z) with $z - q(y - x) \in P$ is admissible with regard to the differential system $X'(t) = f(t, X(t))$, $Y'(t) = g(t, Y(t))$, $Z'(t) = h(t, Z(t))$. This admissibility is characterized through the tangency condition that, for almost every t and for every $(x, y, z) \in Q$, $(f(t, x),\ g(t, y),\ h(t, z)) \in K_Q(x, y, z)$. Now the conclusion follows because $K_Q(x, y, z)$ equals the set of all (u, v, w) such that $w - D_q(y - x)(v - u) \in K_P(z - q(y - x))$. □

To close this note we consider two applications where P equals the positive orthant in $\mathbb{R}^n$. In case $m = n$ and $q_i(u) = |u_i|$, we get the result in the introduction. Let now $m = 1$, let q be a norm $\|\cdot\|$ on $\mathbb{R}^n$, and denote $D_q(u)(v)$ by $\langle u; v\rangle$, i.e., $\langle u; v\rangle = \lim_{h \downarrow 0} \dfrac{\|u + hv\| - \|u\|}{h}$. In this case, we get the equivalence of the two conditions:

(a) for every $(t; \tau, x)$, $(t; \tau, y)$, and $(t; \tau, z)$ (with $t \geq \tau$),

$$\Big\|G(t; \tau, y) - F(t; \tau, x)\Big\| \leq H(t; \tau, \|y - x\|);$$

(b) for almost every t and for every x and y,

$$\Big\langle y - x, g(t, y) - f(t, x)\Big\rangle \leq h(t, \|y - x\|).$$

This statement is justified bacause $K_P(p)$ equals either $[0, +\infty)$ or $\mathbb{R}$ provided that either $p = 0$ or $p > 0$ respectively.

REFERENCES

[1] M. Müller, Math. Z., **26**, 642 (1927).

[2] E. Kamke, Acta. Math., **58**, 82 (1932).

[3] T. Ważewski, Ann. Soc. Polon. Math., **23**, 143 (1950).

[4] C. Olech and Z. Opial, Ann. Polon. Math., **7**, 252 (1960).

[5] J. Szarski, Differential Inequalities, P.W.N., Warszawa (1967) p. 56.

[6] W. Walter, Differential- und Integral-Ungleichungen, Springer, Berlin (1964) pp. 90–107.

[7] M. Nagumo, Proc. Phys.-Math. Soc., Japan (3), **24**, 552 (1942).

[8] J. A. Yorke, Math. Systems Theory, **1**, 353 (1967).

[9] C. Ursescu, Math. Japon., **31**, 484 (1986).

[10] C. Ursescu, Preprint Series in Mathematics, University "Al. I. Cuza" Iasi, **18**, (1982).

[11] N. Pavel, Differential Equations associated to some Nonlinear Operators on Banach Spaces, (in Romanian), Editura Academiei R.S.R., Bucuresti (1977) pp. 190–203.

[12] N. Pavel, Analysis of some Nonlinear Problems in Banach Spaces and Applications, Facultatea de Matematica, Universitatea "Al. I. Cuza", Iasi (1982) pp. 323–340.

[13] N. Pavel, Differential Equations, Flow Invariance and Applications, Pitman, Boston (1984) pp. 91–98.

[14] G. Bouligand, Ann. Soc. Polon. Math., **9**, 32 (1930).

[15] F. Severi, Ann. Soc. Polon. Math., **9**, 99 (1930).

[16] F. Severi, Ann. Mat. Pura Appl. (4), **13**, 20 (1935).

[17] C. Ursescu, Atti Accad. Naz. Lincei Rend. Cl. Sci. Fis. Mat. Natur. (8), **54**, 200 (1973).

7th Int. Coll. on Differential Equations, pp. 441-448
D. Bainov (Ed)

DECAY FOR SOME SECOND ORDER NONLINEAR SYSTEMS

ENZO VITILLARO

Dipartimento di Matematica Università di Perugia
Via Vanvitelli, 1 06123 Perugia, Italy
E-mail: enzo@unipg.it

ABSTRACT. This note deals with decay properties for the solutions of a class of ordinary differential systems with time dependent restoring potential including the system

$$(|u'|^{m-2}u')' + \beta t^{\theta}|u'|^{m-2}u' + t^{\nu}|u|^{p-2}u = 0,$$

$t \in [T, \infty)$, $m > 1$, $\beta > 0$, $-1 \leq \theta < \nu/m$, $1 < p \leq m$, and the nonlinear system

$$Lu'' + T(t)u' + V(t, u') + q^2(t)Su = e(t),$$

where L and S are positive definite matrices, T is a skew-symmetric matrix continuous function, V is a quasilinear replacement of a linear resistive term $R(t)u'$ and e is continuous. We prove exponential decay to zero for the solutions when $\theta > -1$ in the first case and for the difference of two solutions, under suitable assumptions, in the second one.

KEYWORDS: exponential decay, ordinary differential equations, nonlinear networks.

1. INTRODUCTION

The aim of this note is to study the decay properties, as $t \to \infty$, for the solutions of a Lagrangian quasi-variational system of the form

$$(\nabla G(u'(t)))' + Q(t, u(t), u'(t)) + q^m(t)f(u) = 0, \qquad t \in J, \tag{1}$$

where $J = [T, \infty)$, $T > 0$, $u : J \to \mathbb{R}^N$, $N \geq 1$, G is strictly convex and m-homogeneous, $m > 1$, $Q = Q(t, u, v)$ is a continuous damping term and roughly $Q \simeq h(t)|v|^{m-2}v$, $q \in C^1(J)$, $q^m(t)f(u)$ represents an actractive force arising from a potential and roughly $f(u) \simeq |u|^{p-2}u$, $1 < p \leq m$.

Asymptotic stability for systems of a type more general than (1) has been widely studied. We refer to [1] and the bibliograpy therein and also to [2] when $Q \simeq h(t)|u|^{\alpha}|v|^{m-2}v$, $\alpha \geq 0$.

When $N = 1$, $G(v) = \frac{1}{m}|v|^m$, $Q = Q(t, v) = \delta(t)|v|^{m-2}v$, $q(t) = t^{\nu/m}$, $\nu \in \mathbb{R}$, the decay for the solutions of (1) has been studied in [3], where, when $0 < \beta \leq t\delta(t) \leq \beta_1 t^{\eta}$, $\eta > 0$ for t large enough, the authors proved algebraic decay to zero of the energy naturally associated to (1). The same authors proved in [4] that, when $m = p = 2$, $\nu = 0$ and $\delta(t) = \beta t^{\theta}$, $\beta > 0$, $-1 < \theta < 1$, exponential decay to zero of the energy occurs. This last result has been extended in [5] to the case $\nu > -2$. More recently the same type of decay has been proved in [6] when $\nu = 0$, $m > 1$, $1 < p \leq m$, $-1 < \theta < 0$ and Q is not necessarily linear even when $m = 2$. The first goal of this note is to extend the results of [6] to the case $-1 < \theta < \nu/m$, so partially extending the results of [4] and [5] to $m > 1$ and quasilinear terms Q.

When $N > 1$, $G(v) = \frac{1}{2}(Lv, v)$, $m = 2$, $f(u) = Su$, $Q = Q(v) = Tv + V(v)$ where L and S are symmetric semi-definite matrices, T is a skew symmetric matrix, V is a continous *quasi-linear replacement* (see section 4.1 below) of the linear term Rv for some symmetric semi-definite matrix R, the system (1) reduces to the homogeneous case ($e \equiv 0$) of the nonlinear system

$$Lu''(t) + Tu'(t) + V(u'(t)) + Su(t) = e(t), \tag{2}$$

where e is a continuous function. The system (2) describe a nonlinear networks of circuits in which the linear term Ru' is replaced by $V(u')$, and it was studied in [7] by Duffin, who in particular proved the exponential decay to zero of $|u_1' - u_2'|$ for each pairs u_1 and u_2 of C^1- solutions of (2) under suitable conditions (see [7, Theorem 2]).

This theorem motivated us to study the non-autonomous case in which $R = R(t)$, $T = T(t)$, $V = V(t, v)$ is a continous quasi-linear replacement of $R(t)v$ in the sense introduced in [2] (see section 4.1 below) and S is replaced by $q^2(t)S$. Then we consider the system

$$Lu''(t) + T(t)u'(t) + V(t, u'(t)) + q^2(t)Su(t) = e(t), \tag{3}$$

proving (see Theorem 3 below) that, under suitable conditions on the coefficients, for each pairs u_1 and u_2 of solutions of (3), $|u_1(t) - u_2(t)|$ and $|u_1'(t) - u_2'(t)|$ decay exponentially to zero as $t \to \infty$.

2. Formulation of the problem

We consider the system (1) under the following specific assumptions (H1–4), which are assumed to hold throughout the paper.

(H1) $G \in C^1(\mathbb{R}^N; \mathbb{R})$ and G is strictly convex and homogeneous of degree $m > 1$.

(H2) There exists $F \in C^1(\mathbb{R}^N; \mathbb{R})$ and $1 < p \le m$ such that $f(u) = \nabla_u F(u)$ and

$$pF(u) \le (f(u), u) \qquad \text{for all } u \in \mathbb{R}^N. \tag{4}$$

Moreover there exists $c_0 > 0$ such that

$$F(u) \ge c_0|u|^p \qquad \text{for all } u \in \mathbb{R}^N. \tag{5}$$

(H3) $q \in C^1(J; \mathbb{R}^+)$, $q \notin L^1(J)$.

(H4) There exist two non-negative functions σ, $\delta \in L^1_{\text{loc}}(J)$, such that for all $t \in J$, u, $v \in \mathbb{R}^N$

$$|Q(t, u, v)| \le \delta(t)|v|^{m-1}, \tag{6}$$

$$\sigma(t)|v|^m \le (Q(t, u, v), v). \tag{7}$$

Conditions (6) and (7) imply that $\sigma \leq \delta$ on J. Moreover the convexity and homogeneity of G immediately yield that for every $\lambda > 0$ and all $v \in \mathbb{R}^N$

$$G(\lambda v) = \lambda^m G(v), \qquad \nabla G(\lambda v) = \lambda^{m-1} \nabla G(v), \tag{8}$$

and there exist positive constants c, c_1 and c_2 such that for all $v \in \mathbb{R}^N$

$$|\nabla G(v)| \leq c|v|^{m-1}, \tag{9}$$

$$c_1|v|^m \leq mG(v) = (\nabla G(v), v) \leq c_2|v|^m. \tag{10}$$

From these properties we also obtain $G(0) = 0$ and $\nabla G(0) = 0$, while from (4) and (6) it also follows that $f(0) = 0$, $Q(t, u, 0) = 0$. Hence $u \equiv 0$ is a solution of (1).

Our results concern decay properties for the energy function

$$E(u(t)) = G(u'(t)) + q^m(t)F(u(t))$$

naturally associated to (1).

The proofs below make use of the main tool of [1], that is the Liouville transformation given by the change of independent variables $t \mapsto y$

$$y = y(t) = \int_T^t q(s)\,ds \qquad t \in J, \tag{11}$$

which transforms the system (1) into

$$(\nabla G(\dot{u}))^{\cdot} + \tilde{Q}(y, u, \dot{u}) + f(u) = 0, \qquad y \in \tilde{J} = [0, \infty), \tag{12}$$

where $\dot{}$ denotes differentiation with respect to y and, for all $(y, u, w) \in \tilde{J} \times \mathbb{R}^N \times \mathbb{R}^N$,

$$\tilde{Q}(y, u, w) = \frac{1}{q}\left\{\frac{Q(t, u, qw)}{q^{m-1}} + (m-1)\frac{q'}{q}\nabla G(w)\right\}. \tag{13}$$

The new system (12) has exactly the form (1), with $q \equiv 1$ and Q replaced by $\tilde{Q}$. For this system the assumptions (H1–2) still hold, while the appropriate conditions on $\tilde{Q}$ replacing (6) and (7) are given by the following lemma (see [1]).

Lemma 1. *For all $y \in \tilde{J}$, $u, w \in \mathbb{R}^N$ we have*

$$(\tilde{Q}(y, u, w), w) \geq \frac{S(t)}{q(t)}|w|^m, \qquad |\tilde{Q}(y, u, w)| \leq \frac{D(t)}{q(t)}|w|^{m-1},$$

where

$$S(t) = \sigma(t) + (m-1)\frac{c_1(q'(t))^+ - c_2(q'(t))^-}{q(t)}, \qquad D(t) = \delta(t) + c(m-1)\frac{|q'(t)|}{q(t)},$$

where σ, δ are the functions in (6) *and* (7), *c, c_1 and c_2 are the constants in* (9) *and* (10) *and we denote $x^+ = \max\{x, 0\}$, $x^- = -\min\{x, 0\}$ for all $x \in \mathbb{R}$.*

The energy of the transformed system (12) is given by $\tilde{E}(u(y)) = G(\dot{u}(y)) + F(u(y))$ and, by (8),

$$E(u(t)) = q^m(t)\tilde{E}(u(y)), \tag{14}$$

where y and t are related by (11).

3. Main results

Set

$$\kappa = \frac{mp}{mc + p(mc_2 - c_1)}. \tag{15}$$

Throughout the paper Const. denotes suitable positive constants, depending on all the data, on u and also on the auxiliary function k which will be introduced later. Our main result is the following.

Theorem 1. *Let $k \in W^{1,1}_{\mathrm{loc}}(J) \cap BV(J)$, $k \geq 0$, be such that*

$$k(t)q(t) \leq \kappa S(t) \qquad \text{for all } t \in J. \tag{16}$$

If $\|k\|_{L^\infty(J)}$ is sufficiently small, then

$$E(u(t)) \leq \text{Const.}\, q^m(t) \exp\left(\text{Const.} \int_T^t Dk\, ds - \int_T^t kq ds\right) \qquad \text{for } t \in J. \tag{17}$$

Proof. We shall apply [6, Theorem 4.1] in the particular case considered in [6, Section 8] to the transformed system (12). At first the basic assumptions (L1–3) of [6] are satisfied. Indeed (L1) is nothing but (H2), and (L2) follows by (9) and (10) with

$$p_0 = (mc_1 - c_2)/m \qquad p_1 = (mc_2 - c_1)/m, \qquad p_2 = mc^{m'}/(mc_2 - c_1). \tag{18}$$

Next a stronger form of (L3) can be written as follows: *there are non–negative measurable functions $\tilde{\sigma} = \tilde{\sigma}(y)$ and $\tilde{\delta} = \tilde{\delta}(y)$ on $\tilde{J}$ such that for all $y \in \tilde{J}$, $u, y \in \mathbb{R}^N$*

$$\tilde{\sigma}(y)|v|^m \leq (\tilde{Q}(y,u,v), v), \qquad |\tilde{Q}(y,u,v)| \leq \tilde{\delta}(y)|v|^{m-1},$$

which is satisfied with

$$\tilde{\sigma}(y) = S(t)/q(t) \qquad \text{and} \qquad \tilde{\delta}(y) = D(t)/q(t), \tag{19}$$

where y and t are related by (11), and $\tilde{\sigma} \geq 0$ because of (16). Moreover k_0 takes the form κ given by (15) thanks to (18). The specific assumption (4.1) of [6, Theorem 4.1] becomes: *there exists $\tilde{k} \in W^{1,1}_{\mathrm{loc}}(\tilde{J}) \cap BV(\tilde{J})$, $\tilde{k} \geq 0$ such that*

$$\tilde{k}(y) \leq \kappa\tilde{\sigma}(y) \qquad \text{on } \tilde{J}. \tag{20}$$

Now we make the choice $\tilde{k}(y) = k(t)$, with y and t related as usual by (11). Then clearly $\tilde{k} \in W^{1,1}_{\mathrm{loc}}(\tilde{J})$. Moreover $\tilde{k} \in BV(\tilde{J})$ because $k \in BV(J)$ and

$$\int_0^\infty |\dot{\tilde{k}}|\, dy = \int_T^\infty |k'/q| q\, dt = \int_T^\infty |k'|\, dt.$$

Next $\tilde{k} \geq 0$, while (20) follows by (16) and (19). Then, by [6, Theorem 4.1]

$$\tilde{E}(u(y)) \leq \text{Const.} \exp\left(\text{Const.} \int_0^t \tilde{\delta}\tilde{k}\, ds - \int_0^t \tilde{k} ds\right) \qquad \text{for } y \in \tilde{J},$$

and then, by (14) and (19) we obtain (17). □

Corollary 1. *Let $q(t) = t^{\nu/m}$, $\sigma(t) = \beta t^\theta$, $\delta(t) = \beta_1 t^{\theta_1}$, with $m + \nu > 0$, $-1 \leq \theta \leq \theta_1 < \nu/m$ and $0 < \beta \leq \beta_1$. Then the following estimates hold as $t \to \infty$.*

(i) *If* $-1 < \theta \leq \theta_1 < \nu/m$ *then*

$$E(u(t)) \leq \text{Const.}\exp\left(-\lambda t^{1+\theta}\right) \qquad \textit{for} \quad \lambda < \lambda_1 = \beta\kappa/(1+\theta).$$

(ii) *If* $-1 = \theta \leq \theta_1 < \nu/m$ *and* $m\beta + (c_1\nu^+ - c_2\nu^-)(m-1) > 0$ *then*

$$E(u(t)) \leq \text{Const.}t^{-\lambda_2}, \qquad \textit{where} \quad \lambda_2 = \kappa[\beta + (c_1\nu^+ - c_2\nu^-)(m-1)/m] - \nu.$$

Proof. In this case (16) becomes

$$k(t) \leq \kappa t^{-\nu/m}\left[\beta t^{\theta} + (c_1\nu^+ - c_2\nu^-)\frac{m-1}{m}t^{-1}\right]. \tag{21}$$

In the case (i) we choose $k(t) = \overline{k}t^{\theta-\nu/m}$, with $\overline{k} < \beta\kappa$, then clearly (21) holds for T large enough, $k \in W^{1,1}_{\text{loc}}(J) \cap BV(J)$, $k \geq 0$ and

$$\max_{[T,\infty)} k \leq \kappa\left[\beta T^{\theta-\nu/m} + c_2\frac{m-1}{m}|\nu|T^{-1-\nu/m}\right],$$

so still choosen T large enough we can apply Theorem 1 and get

$$E(u(t)) \leq \text{Const.}\, t^{\nu}\exp\left(\text{Const.}\int_T^t\left[\beta_1 s^{\theta_1} + c\frac{m-1}{m}|\nu|s^{-1}\right]s^{\theta-\nu/m}ds - \int_T^t \overline{k}s^{\theta}ds\right)$$

for $t \in J$, which yields (i).

In the case (ii) then we choose $k(t) = \kappa[\beta + (c_1\nu^+ - c_2\nu^-)(m-1)/m]t^{-1-\nu/m}$. Still, choosen T large enough we can apply Theorem 1 and get (ii). □

Remark 1. The estimate contained in Corollary 1 are better than those obtained in the case $m = 2$, $\theta \leq 0$, $\nu \leq 0$, in [8], even when both apply.

4. Applications

4.1. Nonlinear networks

At first we consider the linear system

$$Lu''(t) + T(t)u'(t) + R(t)u'(t) + q^2(t)Su(t) = e(t), \tag{22}$$

where S and L are symmetric positive–definite $N \times N$ matrix, T is a continuous skew–symmetric $N \times N$ matrix valued function, R is a continuous symmetric positive–definite matrix valued function and $e \in C(J)$. We assume q verifying (H3).

The system (22) has a clear physical meaning. It describes a linear network of capacitors, resistences and induttances, where u' denotes the corrent vector. The matrix T appears when the network contains gyrators (see [7]). Here the resistive, gyrator–type and capacitative terms are time–dependent, in an essential way. The resistence of the elements of the circuit may change, for example, by effect of known temperature changements, or the capacity may change by effect of external influences.

We set $\lambda_1 = \lambda_1(t)$ and $\lambda_N = \lambda_N(t)$ to be respectively the smallest and the largest eigenvalue of $R(t)$, which are continuous functions being $\lambda_N(t) = \|R(t)\|$, and $\lambda_1(t) = \|[R(t)]^{-1}\|^{-1}$, where $\|\cdot\|$ denotes the Euclidean operator norm. We also denote $\tilde{\lambda}_N(t) = \lambda_N(t) + \|T(t)\|$ and l_1, l_N to be the smallest and the largest eigenvalue of the matrix L.

As a consequence of Theorem 1 we obtain the following

Theorem 2. *Suppose that there exists* $k \in W^{1,1}_{\mathrm{loc}}(J) \cap BV(J)$, $k \geq 0$, *such that*

$$k(t)q^2(t) \leq 2(3l_N - l_1)^{-1}[\lambda_1(t) + l_1(q'(t))^+ - l_N(q'(t))^-] \qquad \text{for all } t \in J. \tag{23}$$

If $\|k\|_{L^\infty(J)}$ *is sufficiently small then, for every pairs* u_1, u_2 *of solutions of* (22) *we have*

$$E(u_1(t) - u_2(t)) \leq \mathrm{Const.}q^2(t)\exp\left[\mathrm{Const.}\int_T^t(\tilde{\lambda}_N + c|q'|/q)k\,ds - \int_T^t kq\,ds\right]. \tag{24}$$

In particular, if

$$q(t) = t^{\nu/2}, \qquad \lambda_1(t) = \beta t^\theta, \qquad \tilde{\lambda}_N(t) = \beta_1 t^{\theta_1},$$

with $\nu > -2$, $-1 < \theta \leq \theta_1 < \nu/2$, $0 < \beta \leq \beta_1$ *then*

$$E(u_1(t) - u_2(t)) \leq \mathrm{Const.}\exp(-\lambda t^{1+\theta}), \tag{25}$$

and then $|u_1(t) - u_2(t)|$, $|u_1'(t) - u_2'(t)| \leq \mathrm{Const.}\exp(-\frac{1}{2}\lambda t^{1+\theta})$, *where* $\lambda < \lambda_1 = 2\beta/[(3l_N - l_1)(1+\theta)]$.

Remark 2. Theorem 2 extends to non–autonomous linear networks the exponential decay property stated by Duffin in the autonomous case in which $\lambda_1(t)$, $\lambda_N(t)$, $q(t)$ and $e(t)$ are constants, (see [7, Stability Property 1, pag. 430]). On the other hand the autonomous case is outside our treatment.

Proof. By the linearity of (22) the function $\tilde{u} = u_1 - u_2$ satisfies the homogeneous system

$$L\tilde{u}''(t) + T(t)\tilde{u}'(t) + R(t)\tilde{u}'(t) + q^2(t)S\tilde{u}(t) = 0. \tag{26}$$

Now we shall apply Theorem 1 to (26). Assumptions (H1–2) hold with $m = p = 2$, and in (9) and (10) we have $c_1 = l_1$, $c = c_2 = l_N$, so $\kappa = 2/(3l_N - l_1)$, while (H3) was directly assumed. To verify (H4) we note that

$$Q(t,u,v) = T(t)v + R(t)v, \qquad \text{and} \quad (Q(t,u,v),v) = (R(t)v,v),$$

so (H4) holds with $\sigma(t) = \lambda_1(t)$ and $\delta(t) = \tilde{\lambda}_N(t)$. Then, by Theorem 1, (24) follows, while (25) is a consequence of Corollary 1. □

The statement of Theorem 2 can be extended, as stated in the introduction, to *nonlinear networks* of the type (3) where V is a nonlinar continuous damping term. When $V = V(u')$, $R(t) = R = \mathrm{Const.}$, Duffin introduced in [9] the concept of *quasilinear replacement* as follows: V *is a quasilinear replacement of the linear term* Ru', *or simply of* R, *if there is a a positive constant* C *and a symmetric* $N \times N$ *matrix* $U = U(v,w)$ *such that for all* $v, w, z \in \mathbb{R}^N$

$$V(v) - V(w) = U(v - w),$$
$$\frac{1}{C}(Rz,z) \leq (Uz,z) \leq C(Rz,z).$$

The physical interpretation of a quasilinear replacement in an electrical network is the replacement of constant linear resistors by variable nonlinear resistors whose differential resistence lies between positive limits (a proof of this fact is given in [9]).

Following this strategy we consider quasilinear replacements of linear time–dependent resistences as $R(t)$ in (22).

Definition 1. We say that $V = V(t,v)$ is a *quasilinear replacement* of $R = R(t)$ if there are positive constant C, D and a symmetric $N \times N$ matrix $U = U(t,v,w)$ such that

$$V(t,v) - V(t,w) = U(v-w) \tag{27}$$

$$\frac{1}{C}(R(t)z,z) \leq (U(t,v,w)z,z) \leq C(R(t)z,z), \tag{28}$$

$$|U(t,v,w)z| \leq D|R(t)z| \tag{29}$$

for all $t \in J$ and $z, v, w \in \mathbb{R}^N$.

Remark 3. Assumptions (28)–(29) are physically justified in [2, Remark 5.2]. See also [9, pag. 834].

Now we are able to state the following

Theorem 3. *Suppose thet $V(t,v)$ is a quasilinear replacemnt of $R(t)v$. Then the same conclusions of Theorem 2 hold for every pairs u_1, u_2 of solutions of* (3), *with λ_1 replaced by $C\lambda_1$ and β replaced by βC^{-1}.*

Proof. Let u_1 and u_2 be two solutions of (3). Then $\tilde{u} = u_1 - u_2$ solves

$$L\tilde{u}'' + T(t)\tilde{u}' + U(t, u_1'(t), u_2'(t))\tilde{u}' + q^2(t)S\tilde{u} = 0. \tag{30}$$

Now we apply Theorem 1 to (30), as in the proof of Theorem 2. The only differences are in the verification of (H4). We note that, being in this case

$$Q = Q(t,v) = T(t)v + U(t, u_1'(t), u_2'(t))v, \tag{31}$$

we have

$$(Q(t,v), v) = -(U(t, u_1'(t), u_2'(t))v, v),$$

and then (7) holds with $\sigma(t) = \lambda_1(t)/C$. Moreover, by (31) and (29) we obtain

$$|Q(t,v)| \leq (\|T(t)\| + \text{Const.}\|R(t)\|)|v|,$$

so (6) holds with $\delta(t) = \text{Const.}\tilde{\lambda}_N(t)$. $\square$

4.2. A canonical example

Consider the system

$$(|u'|^{m-2}u')' + h(t)|u'|^{m-2}u' + t^{\nu}|u|^{p-2}u = 0, \tag{32}$$

where $u : [T, \infty) \to \mathbb{R}^N$, $m > 1$, $m + \nu > 0$, $1 < p \leq m$. In this case, if

$$\beta t^{\theta} \leq h(t) \leq \beta_1 t^{\theta_1}, \qquad -1 \leq \theta \leq \theta_1 < \nu/m, \qquad 0 < \beta \leq \beta_1,$$

we can apply Corollary 1 and give the following estimates as $t \to \infty$.

(i) If $-1 < \theta \leq \theta_1 < \nu/m$ then

$$E(u(t)) \leq \text{Const.}\exp\left(-\lambda t^{1+\theta}\right) \qquad \text{for} \quad \lambda < \lambda_1 = \frac{mp\beta}{[m+(m-1)p](1+\theta)}.$$

(ii) If $-1 = \theta \leq \theta_1 < \nu/m$ and $m\beta + (m-1)\nu > 0$ then

$$E(u(t)) \leq \text{Const.}t^{-\lambda_2}, \qquad \lambda_2 = \frac{mp}{m+(m-1)p}[\beta + (c_1\nu^+ - c_2\nu^-)(m-1)/m] - \nu,$$

so

$$|u(t)| \leq \text{Const.}t^{-\lambda_3}, \qquad \text{and} \quad |u'(t)| \leq \text{Const.}t^{-\lambda_4},$$

where

$$\lambda_3 = \frac{m\beta + (m-1)\nu}{m+(m-1)p}, \qquad \lambda_4 = \frac{\beta p - \nu}{m+(m-1)p}.$$

When $m = p = 2$, $h(t) = \beta t^\theta$ and $N = 1$ the estimate (i) was proved in [4] in the case $\nu = 0$, and is a consequence of the sharp asymptotic behavior given in [5] in the more general case $\nu > -2$. Actually it was proved for $\lambda = \lambda_1$ in [4] and in [5] when $\nu \geq 0$. The estimate (ii) is exactly the estimate given in [3, Theorem 7] in the case in which $|u(t)| \to 0$ as $t \to \infty$, as it is clear from the form of λ_3 and λ_4.

Acknoledgment

The author wish to thanks Prof. Patrizia Pucci for carefully reading the manuscript and also the italian *Gruppo Nazionale per l'Analisi Funzionale e le Sue Applicazioni* for the financial support.

References

1. P. Pucci and J. Serrin, Arch. Rational Mech. Anal., 132, pp. 207–232 (1995).
2. M. C. Bossi and E. Vitillaro. Preprint (1996).
3. P. Pucci and J. Serrin. Asymptotic Analysis, 4, pp. 97–160 (1991).
4. P. Pucci and J. Serrin. Rend. Sem. Mat. Univ. Torino, Fascicolo Speciale, pp. 121–129 (1989).
5. W. A. Harris, P. Pucci, and J. Serrin. Diff. and Integral Equations, 6, pp. 1201–1215 (1993).
6. M. C. Salvatori and E. Vitillaro. Diff. and Integral Equations, in press.
7. R. J. Duffin. J. Math. Anal. Appl., 23, pp. 428–439 (1968).
8. M. Nakao. J. Math. Soc. Japan., 30, pp. 747–762 (1978).
9. R. J. Duffin. Bull. Amer. Math. Soc. (N.S.), 52, pp. 833–838 (1946).

7th Int. Coll. on Differential Equations, pp. 449-454
D. Bainov (Ed)

On a Priori Bound for Nonconvolutional Integral Equations

Akira Yanagiya
Senior High School
of Waseda University
3-31-1,Kamishakuzii,Nerimaku,
Tokyo,162,Japan.
Tel. : 81-3-5991-4151
Fax : 81-3-3928-4110
E-mail : yappy@cfi.waseda.ac.jp

1 Introduction and Theorems

In this paper we consider a priori bound which is defined below for the nonconvolutional integral equation,

$$x(t)+\int_0^t b(t,s)g(x(s))ds=f(t)\quad,\quad 0\le t<t^*. \tag{1.1}$$

In the case for the convolution Levin had obtained the results about a priori bound in his paper [1].

Definition 1 (a priori bound) [1] *A locally bounded function*

$$\Gamma:[0,t^*)\to[0,\infty) \tag{1.2}$$

is called an a priori bound for (1.1) if every pair $(\hat{t},\varphi)$ *that satisfies*

$$\begin{aligned} \hat{t} &\in (0,t^*], \\ \varphi &\in C([0,\hat{t}),R^n)\text{ is a solution of (1.1) on }[0,\hat{t}), \end{aligned} \tag{1.3}$$

must also satisfy

$$|\varphi(t)|=\left(\sum_{i=1}^n \varphi_i^2(t)\right)^{1/2}\le\Gamma(t)\text{ on }[0,\hat{t}). \tag{1.4}$$

Theorem 1 *Let g, b and f satisfy the conditions*

(G1) $g \in C(R^1, R^1), \sup_{\{x|xg(x)\leq 0\}} |g(x)| = K_1 < \infty$,

(B1) $b(t,s) \in C(D, R^1)$, *where* $D = \{(t,s)|0 \leq s \leq t < t^*\}$,

(B2) $b_t(t,s) \in C(D, R^1)$,*where* $b_t(t,s) = \frac{\partial}{\partial t} b(t,s)$

(B3) $b(t,t) > 0$, *on* $[0, t^*)$

(f1) $f \in C^1([0,t^*), R^1)$

Then (1.1) has an a priori bound Γ_1, given by (2.2) below.

Theorem 2 *Let g, b and f satisfiy the conditions*

(G2) $g : R^1 \to R^1$ *is nondecreasing,*

(B4) $b(t,s) : D \to R^1, b(\cdot, s) \in AC_{loc}([s,t^*), R^1)$, *for each fixed $s \in [0,t^*)$, and $b(t,s)$ is measurable in s for each fixed $t \in [0,t^*)$,where AC_{loc} denotes the space of locally absolutely continuous functions.*

(B5) *For all T with $0 \leq T < t^*$ there exists a constant $k_1 = k_1(T) > 0$ such that*

$$\int_0^t |b(t,s)|ds \leq k_1 \text{ for } t \in [0,T],$$

and $k_1 \to 0$ as $T \to 0$,

(B6) *For all T with $0 \leq T < t^*$ there exists a constant $k_2 = k_2(T) > 0$ such that*

$$\operatorname*{ess\,sup}_{0\leq s\leq t\leq T} |b_t(t,s)| \leq k_2,$$

(B7) $\frac{1}{b(t,t)} \in L^\infty_{loc}([0,t^*), R^1), b(t,t) > 0$ *for* $t \in [0,t^*)$

(F2) $f \in AC_{loc}([0,t^*), R^1), f' \in L^\infty_{loc}([0,t^*), R^1)$.

Then (1.1) has an a priori bound $\tilde{\Gamma}_1$, given by below.

2 Proof of Theorem 1

On $[0,t^*)$ set

$$\begin{aligned}
\alpha(t) &= \sup_{0\leq s\leq t} \frac{|b_t(t,s)|}{b(t,t)}, \\
\alpha_1(t) &= \sup_{0\leq s\leq t} \alpha(s), \\
\alpha_2(t) &= \max\left(K_1, \sup_{|x|\leq|f(0)} |g(x)|, \sup_{0\leq s\leq t} \frac{|f'(s)|}{b(s,s)}\right), \\
\alpha_3(t) &= \alpha_1(t) \int_0^t \alpha_2(\tau) \exp\left(\int_\tau^t \alpha_1(s) ds\right) d\tau + \alpha_2(t),
\end{aligned} \tag{2.1}$$

and

$$\Gamma_1(t) = \int_0^t |b(t,s)|\alpha_3(s)ds + \sup_{0\le s\le t} |f(s)|. \tag{2.2}$$

Let $(\hat{t}, \varphi)$ satisfy (1.3) with $n = 1$. Then

$$\varphi(t) + \int_0^t b(t,s)g(\varphi(s))ds = f(t) \quad , \quad 0 \le t < \hat{t}. \tag{2.3}$$

It will be shown that

$$|g(\varphi(t))| \le \alpha_1(t) \int_0^t |g(\varphi(s))|ds + \alpha_2(t) \quad , \quad 0 \le t < \hat{t}. \tag{2.4}$$

From (2.1) , (2.4) and Gronwall inequality, it follows that

$$|g(\varphi(t))| \le \alpha_3(t) \quad , \quad 0 \le t < \hat{t}, \tag{2.5}$$

which together with (2.2) and (2.3) implies (1.4) for $\Gamma = \Gamma_1$ and hence completes the proof.

It is evident from (2.3) and (B2) that $\varphi \in C^1([0,\hat{t}), R^1)$ and that

$$\varphi'(t) + b(t,t)g(\varphi(t)) = -\int_0^t b_t(t,s)g(\varphi(s))ds + f'(t) \quad , \quad 0 \le t < \hat{t}. \tag{2.6}$$

Define the following eight disjoint subsets of $[0, \hat{t})$:

$$\begin{aligned}
S_1 &= \{t|\varphi(t) \ge 0, g(\varphi(t)) < 0\}\\
S_2 &= \{t|\varphi(t) \ge 0, g(\varphi(t)) \ge 0, \varphi'(t) \ge 0\}\\
S_3 &= \{t|\varphi(t) \ge 0, g(\varphi(t)) \ge 0, \varphi'(t) < 0, f(0) \ge \varphi(t)\}\\
S_4 &= \{t|\varphi(t) \ge 0, g(\varphi(t)) \ge 0, \varphi'(t) < 0, f(0) < \varphi(t)\}\\
S_5 &= \{t|\varphi(t) < 0, g(\varphi(t)) > 0\}\\
S_6 &= \{t|\varphi(t) < 0, g(\varphi(t)) \le 0, \varphi'(t) \le 0\}\\
S_7 &= \{t|\varphi(t) < 0, g(\varphi(t)) \le 0, \varphi'(t) > 0, f(0) \le \varphi(t)\}\\
S_8 &= \{t|\varphi(t) < 0, g(\varphi(t)) \le 0, \varphi'(t) > 0, f(0) > \varphi(t)\}.
\end{aligned} \tag{2.7}$$

Clearly

$$[0, \hat{t}) = \bigcup_{k=1}^{8} S_k. \tag{2.8}$$

Inequality (2.4) will be established for an arbitrary $t_0 \in [0, \hat{t})$. Each case, $t_0 \in S_k$, is treated separately. Since the reasoning required for $k = 5, 6, 7, 8$ parallels that employed for $k = 1, 2, 3, 4$, respectively, only the latter cases are considered here.

Let $t_0 \in S_1$. Then (G1), (2.1) and (2.7) imply

$$|g(\varphi(t_0))| \le K_1 \le \alpha(t_0) \le \alpha_1(t_0) \int_0^{t_0} |g(\varphi(s))|ds + \alpha_2(t_0). \tag{2.9}$$

Let $t_0 \in S_2$. Then (B2), (B3), (2.1), (2.6) and (2.7) imply

$$\begin{aligned}
|g(\varphi(t_0))| &= g(\varphi(t_0))\\
&\le \frac{1}{b(t_0,t_0)}\{\varphi'(t_0) + b(t_0,t_0)g(\varphi(t_0))\}\\
&= -\frac{1}{b(t_0,t_0)}\int_0^{t_0} b_t(t_0,s)g(\varphi(s))ds + \frac{1}{b(t_0,t_0)}f'(t_0)\\
&\le \alpha_1(t_0)\int_0^{t_0} |g(\varphi(s))|ds + \alpha_2(t_0).
\end{aligned} \tag{2.10}$$

Let $t_0 \in S_3$.Even without invoking the condition $\varphi'(t_0) < 0$ of (2.7), it follows from (2.1) that

$$|g(\varphi(t_0))| \le \max_{|x|\le|f(0)|} |g(x)| \le \alpha_2(t_0) \le \alpha_1(t_0)\int_0^{t_0} |g(\varphi(t))|dt + \alpha_2(t_0). \tag{2.11}$$

Let $t_0 \in S_4$. Since $\varphi(0) = f(0)$, it is evident from (2.7) that there exists a unique $t_1 \in (0, t_0)$ such that

$$\varphi(t_1) = \varphi(t_0) \text{ and } \varphi(t) > \varphi(t_0) \text{ on } (t_1, t_0). \tag{2.12}$$

Since $\varphi \in C^1([0,\hat{t}), R^1)$, (2.12) implies

$$\varphi^1(t_1) \ge 0. \tag{2.13}$$

Reasoning similar to that employed for $t_0 \in S_2$, but now also using (2.12), (2.13) and the monotonicity of $\alpha_1(t)$ and $\alpha_2(t)$ yields

$$\begin{aligned} |g(\varphi(t_0))| &= g(\varphi(t_0)) = g(\varphi(t_1)) \\ &\le \frac{1}{b(t_1,t_1)}\{\varphi'(t_1) + b(t_1,t_1)g(\varphi(t_1))\} \\ &= -\frac{1}{b(t_1,t_1)}\int_0^{t_1} b_t(t_1,s)g(\varphi(s))ds + \frac{1}{b(t_1,t_1}f'(t_1) \\ &\le \alpha_1(t_1)\int_0^{t_1} |g(\varphi(s))|ds + \alpha_2(t_1) \\ &\le \alpha_1(t_0)\int_0^{t_0} |g(\varphi(s))|ds + \alpha_2(t_0). \end{aligned} \tag{2.14}$$

Then (2.9), (2.10), (2.11) and (2.14) established (2.4) for $t_0 \in S_k$ ($k = 1,2,3,4$), respectively. This, together with similar reasoning for $k = 5,6,7,8$ completes the proof as noted above.

3 Proof of Theorem 2

Analoguos to (2.1) and (2.2), on $[0, t^*)$ set

$$\begin{aligned} \tilde{\alpha}(t) &= \operatorname*{ess\,sup}_{0\le s\le t} \frac{|b_t(t,s)|}{b(t,t)}, \\ \tilde{\alpha}_1(t) &= \sup_{0\le s\le t} \alpha(s), \\ \tilde{\alpha}_2(t) &= \max(|g(0)|, |g(f(0))|, \operatorname{ess\,sup} \frac{|f'(s)|}{b(t,t)}), \\ \tilde{\alpha}_3(t) &= \tilde{\alpha}_1(t)\int_0^t \tilde{\alpha}_2(\tau)\exp\{\int_\tau^t \tilde{\alpha}_1(s)ds\}d\tau + \tilde{\alpha}_2(t), \\ \tilde{\Gamma}_1(t) &= \int_0^t |b(t,s)|\tilde{\alpha}_3(s)ds + \sup_{0\le s\le t} |f(s)|. \end{aligned} \tag{3.1}$$

Let $(\hat{t}, \varphi)$ satisfy (1.3) with $n = 1$. Then (2.3) holds. Instead of (2.4), it will now be shown that

$$|g(\varphi(t))| \le \tilde{\alpha}_1(t)\int_0^t |g(\varphi(s))|ds + \tilde{\alpha}_2(t) \quad \text{a.e. on } [0, t^*). \tag{3.2}$$

As in §2, however, (2.3), (3.1), (3.2) and the Grownwall inequality yield (1.4), for $\Gamma = \tilde{\Gamma}_1$, which completes the proof.

From (2.3) and the hypothesis it follows that

(3.3) $$\varphi \in AC_{loc}([0,\hat{t}), R^1)$$

and that

(3.4) $$\varphi'(t) + b(t,t)g(\varphi(t)) = -\int_0^t b_t(t,s)g(\varphi(s))ds + f'(t) \quad \text{a.e. on } [0,\hat{t}).$$

Let

(3.5) $$\begin{aligned} A &= \{t \in [0,\hat{t}) | \varphi'(t) \text{ exists and (3.4) holds}\}, \\ \tilde{S}_k &= A \bigcap S_k \quad , \quad k = 1,\ldots,8, \end{aligned}$$

where the S_k are defined by (2.7). Then the S_k are disjoint and, instead of (2.8),

(3.6) $$\text{Meas}\{[0,\hat{t}) \cap \bigcap_{k=1}^{8} \tilde{S}_k^c\} = 0.$$

Inequality (3.2) will be established on a.e.$\bigcup_{k=1}^{8} \tilde{S}_k$, which in view of (3.6), establishes (3.2) a.e. on $[0,\hat{t})$.

Let $t_0 \in \tilde{S}_1$. Then (G2), (3.1) and (3.5) imply

$$\begin{aligned} |g(\varphi(t_0))| &\le |g(0)| \le \tilde{\alpha}_2(t_0) \\ &\le \tilde{\alpha}_1(t_0)\int_0^{t_0} |g(\varphi(s))|ds + \tilde{\alpha}_2(t_0) \end{aligned}$$

Let $t_0 \in \tilde{S}_2 \bigcap B$ where

(3.7) $$B = \{t \in [0,t^*) | |f'(t)| \le \operatorname*{ess\,sup}_{0 \le s \le t} |f'(s)|\}.$$

It is an elemantary consequence of (F2) that Meas$(B) = t^*$. The same reasoning employed in (2.10) shows that (3.2) holds for this t_0.

Let $t_0 \in \tilde{S}_3$. Then (G2), (3.1) and (3.5) imply

$$\begin{aligned} |g(\varphi(t_0))| &= g(\varphi(t_0)) \le g(f(0)) \\ &\le \tilde{\alpha}_1(t_0)\int_0^{t_0} |g(\varphi(s))|ds + \tilde{\alpha}_2(t_0). \end{aligned}$$

Let $t_0 \in \tilde{S}_4$. Thus as before, (3.5) implies the existance of a unique $t_1 \in (0,t_0)$ such that (2.12) holds. From (2.12) and (3.3) it follows that

(3.8) $$0 = \int_{t_1}^{t_0} \varphi'(s)ds.$$

From (2.12), (3.3) and (3.8), there exists a set of positive measure $C \subset [t_1,t_0]$ such that

(3.9) $$\varphi'(t_2) \text{ exists}, \varphi'(t_2) \ge 0, \varphi(t_2) \ge \varphi(t_0) \text{ for } t_2 \in C.$$

Since Meas$(B) = t^*$ (recall (3.7)), there exists a $t_2 \in B \bigcap C$. Reasoning similar to (2.14), but now invoking (3.9) and (G2), implies

$$\begin{aligned} |g(\varphi(t_0))| &= g(\varphi(t_0)) \le g(\varphi(t_2)) \\ &\le \frac{1}{b(t_2,t_2)}\{\varphi'(t_2) + b(t_2,t_2)g(\varphi(t_2))\} \\ &= -\frac{1}{b(t_2,t_2)}\int_0^{t_2} b_t(t_2,s)g(\varphi(s))ds + \frac{|f'(t_2)|}{b(t_2,t_2)} \end{aligned}$$

$$\leq \tilde{\alpha}_1(t_2)\int_0^{t_2}|g(\varphi(s))|ds + \tilde{\alpha}_2(t_2)$$
$$\leq \tilde{\alpha}_1(t_0)\int_0^{t_0}|g(\varphi(s))|ds + \tilde{\alpha}_2(t_0).$$

Analogous arguments hold for $\tilde{S}_k$ ($k = 5, 6, 7, 8$). Thus (3.2) is established, which completes the proof.

Reference

[1] J.J. Levin : Some a priori bounds for nonlinear Volterra equations, SIAM J. Math. Anal. 7(1976), pp.872-897

7th Int. Coll. on Differential Equations, pp. 455-468
D. Bainov and V. Covachev (Eds)

A POINTWISE ERROR ESTIMATE FOR DISTRIBUTED PARAMETER IDENTIFICATION

XUE–CHENG TAI AND PEKKA NEITTAANMÄKI

University of Bergen, Department of Mathematics,
Alleg. 55, 5007, Bergen, Norway
and
University of Jyväskylä, Department of Mathematics,
40100 Jyväskylä, Finland.

ABSTRACT. A new error estimate is derived for the numerical identification of a distributed parameter $a(x)$ in a two point boundary value problem, for the case where the finite element method and the fit-to-data output-least-squares technique are used for the identification. With a special weighted norm, we get a pointwise estimate. From this error estimate, we try to dram a few remarks about the approximate identification of the distributed parameters, which were confirmed by numerical experiments.

1. Introduction

We consider the two point boundary value problem

$$\begin{cases} -\dfrac{d}{dx}\left(a(x)\dfrac{du}{dx}\right) = f(x), & x \in (0,1) \\ -a\left.\dfrac{du}{dx}\right|_{x=0} = g_1, \quad a\left.\dfrac{du}{dx}\right|_{x=1} = g_2. \end{cases} \tag{1.1}$$

We assume that u, f, g_1 and g_2 satisfy the usual compatibility conditions, i.e. $\int_0^1 u dx = c_0$ and $\int_0^1 f dx + g_1 + g_2 = 0$, which guarantee that the solution u of (1.1) exists and is unique. Our problem is that we have an observation of the value of $u(x)$ on $[0,1]$ and we want to recover the parameter $a(x)$ in the system (1.1). This problem is usually called an inverse problem or a distributed parameter identification problem.

In order to recover $a(x)$, a frequently used technique is to solve the following optimization problem

$$(P) \qquad \begin{cases} \text{Find } a^* \in M \text{ such that } J(a^*) = \underset{b\in M}{\text{minimize}} J(b)\ , \text{ where } J(b) = \\ \int_0^1 |u(b) - z|^2\, dx \text{ and } u(b) \text{ is the solution of (1.1) with parameter } b. \end{cases}$$

Supported by the Academy of Finland

Above, $z(x)$ is a given observation and M is the admissible parameter set that contains some a priori information about $a(x)$. For some other alternative approaches, see Kunisch and Tai [**9**] and Kunisch and Banks [1]. In order to solve (P) in practice some numerical methods have to be used. In this paper, we apply finite element methods of order $r+1$ ($r \geq 2$), and solve the following discretized optimization problem

$$(P_h) \begin{cases} \text{Find } a_h \in M_h \text{ such that } J_h(a_h) = \underset{b_h \in M_h}{\text{minimize}} J_h(b_h) , \\ \text{where } J_h(b_h) = \int_0^1 |u_h(b_h) - z|^2 \, dx \text{ and } u_h(b_h) \in S_h^{r+1} \text{ satisfies} \\ \int_0^1 b_h \frac{du_h(b_h)}{dx} \frac{dv_h}{dx} \, dx = \int_0^1 f v_h \, dx + g_1 v_h(0) + g_2 v_h(1) \qquad \forall v_h \in S_h^{r+1} , \\ \int_0^1 u_h(b_h) dx = \int_0^1 z dx . \end{cases}$$

Here $M_h = M \cap S_h^{r+1}$ and S_h^{r+1} is a finite element space with order $r+1$ and mesh size h. We shall prove that the obtained parameter a_h of (P_h) satisfies the following error estimate

$$\begin{aligned} &\|(a(x) - a_h(x))(|u'(x)| + h|u''(x)| + \cdots + h^r|u^{(r+1)}(x)|)\|_{L^\infty([0,1])} \\ &\leq C h^{-\frac{1}{2}}(h^{r+1} + \varepsilon/h) . \end{aligned} \tag{1.2}$$

Here ε is the observation error, i.e. $\|z - u(a)\|_{L^2([0,1])} \leq \varepsilon$ and $u', u'' \cdots u^{(r+1)}$ denote the derivatives.

This estimate gives an improvement of the estimates in our paper [**11**], where the following is proved,

$$\begin{aligned} &\left(\int_0^1 |a(x) - a_h(x)|^2 \left(|u'|^2 + |hu''|^2 + \cdots + |h^r u^{(r+1)})|^2\right) dx \right)^{1/2} \\ &\leq C(h^{r+1} + \varepsilon/h) , \end{aligned} \tag{1.3}$$

which is a same kind of estimate as (1.2), but has the L^2–norm for the error. Moreover estimate (1.3) is sharp.

From estimates (1.2) and (1.3) we can draw the following remarks about the approximation in the identification problems. Our numerical tests in a preliminary report [**13**] confirm these remarks.

1. When the observation error is present in the identification problem, we cannot get an accurate identified parameter by just decreasing the mesh size, because then the term $\frac{\varepsilon}{h}$ in the estimates will become dominant. So the maximum error between the identified parameter and the actual parameter may increase as the mesh size decreases. In the numerical tests in [**13**], we do find that the maximum error increases as the mesh size decreases, see [**13**] Example 3.4.
2. As a small mesh size is not preferable in the approximations, in order to improve the accuracy, higher order finite element spaces (p-methods) should be used in the identification. This is especially clear for $\varepsilon = 0$. This was also observed in numerical Example 3.5 of [**13**].

3. When $u_x = 0, u_{xx} = 0\ldots$ at some points, i.e. u is locally approximately constant at these points, then poor approximation for the parameter can be produced at these points. From the error estimate, we can guarantee the convergence if the order of the finite element space is higher than the order of the derivatives that vanish. See Example 3.1–3.3 of [**13**].

For the computing of the distributed parameter identification problem, there are quite a lot of results [**1**], but for their error estimates, there are only a few papers, for example [**5**] [**8**] [**11**]. In an earlier paper [**5**], under the condition $\nabla u \cdot \bar{v} \geq k > 0$ for some direction $\bar{v}$, the estimate

$$\left(\int_\Omega |a - a_h|^2\, dx\right)^{1/2} \leq C(h^r + \varepsilon/h^2)$$

was proved for two dimensional problems. Moreover in [**8**], the estimate

$$\left(\int_0^1 |a - a_h|^2 |u_x|^2\, dx\right)^{1/2} \leq C(h^{r+1} + \varepsilon/h)$$

was proved for one dimensional problems. However the condition $\nabla u \cdot \bar{\geq} k \geq 0$ is restrictive for some problems and with such a condition we can prove that the inverse problem is not illposed, which means that mapping from the solution to the parameter is continuous from H^1 to L^2. Without removing the illposedness of the inverse problem, paper [**5**] show how the derivative of the state u effect the stability and convergence of the identification, but only the first order derivative is taken into account there. As in paper [**13**], if the first order derivative does not satisfy a similar condition as in [**3**], then the second order derivative must satisfy some condition to get convergence and stability. This means the higher order derivatives also play a roll in the stability of the inverse problems. So in this article we consider the affects of all the derivatives on the convergence of the inverse problems under a very weak condition. Moreover our estimate is a pointwise one, not as in the other papers, only the L^2–norm error estimates were obtained.

Our analysis is presented on the very simple one dimensional problem. There are two reasons for this. First, the extension of the results to multidimensional problem is not straight forward and we are still unable to prove the similar results for multidimensional problems. Second, the analysis itself is complicated enough and we feel it is also of interests to present this special case. For the simple identification of (1.1), the exist some other simpler methods to identify the parameter $a(x)$, but they are difficult to be extended to multidimensional problems.

2. Preliminaries

In our paper, we use the standard Sobolev spaces $W^{k,\infty}$, $W^{k,p}$ $(1 \leq p < \infty)$,

H^k and spaces C^k of continuous functions. The norms are

$$\|u\|_{k,p,I} = \left\{ \sum_{|j|=0}^{k} \|D^j u\|_{L^p(I)}^p \right\}^{1/p} \qquad 1 \le p < \infty$$

$$\|u\|_{k,\infty,I} = \sum_{|j|=0}^{k} \|D^j u\|_{L^\infty(I)} .$$

For $p = 2$ we also use notations

$$\|u\|_{k,I} = \|u\|_{k,2,I} .$$

Here I is the considered domain. We do not mention about the domain if $I = [0,1]$. We use notations $\frac{du}{dx}$, u_x to indicate the first order derivative, $\frac{d^2u}{dx^2}$, u_{xx}, $u^{(2)}$, ... etc. to indicate the higher order derivatives.

We divide $[0,1]$ into subintervals and denote this subdivision by Δ_h,

$$\Delta_h = \{e_i \mid e_i = (x_{i-1}, x_i),\ i = 1, 2, \ldots, N,\ 0 = x_0 < x_1 < \ldots < x_N = 1\} .$$

We assume $\{\Delta_h\}$ is a quasi-uniform family of subdivisions, i.e. with $h_i = x_i - x_{i-1}$, and there exists a constant $C > 0$ such that $\frac{\max h_i}{\min h_i} \le C$. The mesh size is $h = \max h_i$. The finite element space is the following

$$S_h^r = \{v \in C^{r-1}([0,1]) \mid \quad v|_{e_i} \in P_r \qquad \forall e_i \in \Delta_h\} ,$$

where P_r is the space of polynomials of degree r. So, we should use the spline functions for the finite element space. The set M of admissible controls is assumed to be

$$M = \{b \mid 0 < \lambda^{-1} < b(x) < \lambda \text{ a.e. in } (0,1),\ \|b\|_{r+1,\infty} < C\}$$

with some constants λ and C. Here and also later, C will always indicate a positive generic constant which is independent of h.

Throughout the paper, we assume that the state u and parameter a in (1.1) satisfy the conditions

$$(A) \qquad \begin{aligned} &u \in W^{r+2,\infty},\ a \in W^{r+2,\infty},\ 0 < \lambda^{-1} < a(x) < \lambda , \\ &\|u\|_{r+2,\infty} < C,\ \|a\|_{r+2,\infty} < C \end{aligned}$$

and the finite element spaces are taken as $S^h = S_h^{r+1}$.

Furthermore, we will use parameters and functions which are listed in Table 2.1.

We shall now define these parameters and functions. We define $a(x)$ as the true parameter in (1.1) which we intend to identify and define πa as its L^2-projection into S^h. Let a_h be the identified parameter, i.e. it is the solution of (P_h) (which may not be unique, but we take one of the obtained solutions [**10**]). Moreover, we denote by u, u^π and w the solutions of (1.1) corresponding to these parameters.

parameter	corresponding state	FE-approximation
a	u	u_h
πa	u^π	u_h^π
a_h	w	w_h

Table 2.1.

Above we mean the weak solutions. So this means u, u^π, w are the solutions of the following variational equations

$$\int_0^1 a u_x v_x \, dx = \int_0^1 f v \, dx + g_1 v(0) + g_2 v(1) \qquad \forall v \in H^1 , \tag{2.1}$$

$$\int_0^1 \pi a u_x^\pi v_x \, dx = \int_0^1 f v \, dx + g_1 v(0) + g_2 v(1) \qquad \forall v \in H^1 , \tag{2.2}$$

$$\int_0^1 a_h w_x v_x \, dx = \int_0^1 f v \, dx + g_1 v(0) + g_2 v(1) \qquad \forall v \in H^1 . \tag{2.3}$$

In order to guarantee the uniqueness of the solutions, we also require

$$\int_0^1 u^\pi dx = \int_0^1 w dx = \int_0^1 u dx = c_0 .$$

The properties of u are given in (A). Because $a_h \in S^h \subset C^r$, and $a_h \in M$, we have $a_h \in W^{r+1,\infty}$. If we also assume that f is regular enough, then we can prove as in [6] that $u^\pi, w \in W^{r+2,\infty}$ and that they satisfy

$$\begin{cases} -(\pi a u_x^\pi)_x = f \quad \text{in } (0,1) \\ -a u_x^\pi \big|_{x=0} = g_1 , \quad a u_x^\pi \big|_{x=1} = g_2 , \end{cases} \tag{2.4}$$

$$\begin{cases} -(a_h w_x)_x = f \quad \text{in } (0,1) \\ -a_h w_x \big|_{x=0} = g_1 , \quad a_h w_x \big|_{x=1} = g_2 . \end{cases} \tag{2.5}$$

Remark 2.1. In fact we can take $a_h \in S_h^r$ and $u_h \in S_h^{r+1}$. But here we take both a_h and u_h from S_h^{r+1}. This is only used to guarantee $a_h \in W^{r+1,\infty}$.

Let u_h, u_h^π, w_h be the corresponding finite element solutions to u, u^π and w, then we can have the following FE-error estimates [2] [4] [7]

$$\|u^\pi - u_h^\pi\|_{\ell,\infty} \le C h^{r+2-\ell} , \qquad 0 \le \ell \le r+1 , \tag{2.6}$$

$$\|w - w_h\|_{\ell,\infty} \le C h^{r+2-\ell} , \qquad 0 \le \ell \le r+1 , \tag{2.7}$$

$$\|u - u_h\|_{\ell,\infty} \le C h^{r+2-\ell} , \qquad 0 \le \ell \le r+1 . \tag{2.8}$$

As πa is the L^2–projection of a, so the following error estimate is standard

$$\|a - \pi a\|_{\ell,\infty} \le C \|a\|_{r+2,\infty} h^{r+2-\ell} , \qquad 0 \le \ell \le r+1 .$$

Therefore for h small enough, $\pi a \in M$.

In the following proofs, the inverse inequalities [**3**, p. 140–142]

$$\forall v_h \in S^h,\ e_i \in \Delta_h, \qquad \|v_h\|_{0,\infty,e_i} \le \frac{C}{h^{1/2}}\|v_h\|_{0,e_i}\,, \tag{2.9}$$

$$\forall v_h \in S^h,\ e_i \in \Delta_h, \qquad \|v_h\|_{1,\infty,e_i} \le \frac{C}{h}\|v_h\|_{0,\infty,e_i} \tag{2.10}$$

are frequently used.

Furthermore, we need some error estimates for u, w, u^π. We state them in the following two lemmas

Lemma 2.1. *Under the condition (A), we have the error estimate*

$$\|u - u^\pi\|_{\ell,\infty} \le C h^{r+3-\ell} \quad 0 \le \ell \le r+1\,. \tag{2.11}$$

The constant C is independent of h, but depends on u and a.

Proof. Substracting (2.2) from (2.1), we get

$$\int_0^1 a(u-u^\pi)_x v_x\,dx = \int_0^1 (\pi a - a)u_x^\pi v_x\,dx \qquad \forall v \in H^1. \tag{2.12}$$

We choose $v = \int_0^x (a(u-u^\pi)_x)^{2p-1}\,dx$. By (2.12) and the Hölder inequality we obtain

$$\begin{aligned}\int_0^1 |a(u-u^\pi)_x|^{2p}\,dx &= \int_0^1 (\pi a - a)u_x^\pi (a(u-u^\pi)_x)^{2p-1}\,dx \\ \le \Big(\int_0^1 |a(u-u^\pi)_x|^{2p}\,dx\Big)^{\frac{2p-1}{2p}} &\Big(\int_0^1 |(\pi a - a)u_x^\pi|^{2p}\,dx\Big)^{\frac{1}{2p}}\,.\end{aligned} \tag{2.13}$$

As $p \to \infty$, we get

$$\|a(u-u^\pi)_x\|_{0,\infty} \le \|(\pi a - a)u_x^\pi\|_{0,\infty} \le C h^{r+2}\|u_x^\pi\|_{0,\infty}\|a\|_{r+2,\infty}\,. \tag{2.14}$$

Next we use the duality method to prove the estimate (2.11) for $\ell = 0$. For a $g \in L^1$, $\int_0^1 g\,dx = 0$, let G ($\int_0^1 G\,dx = 0$) be the solution of

$$\begin{cases} -(aG_x)_x = g \\ aG_x\big|_{x=0,1} = 0\,. \end{cases} \tag{2.15}$$

As πa is the L^2-projection of a into S^h, we get from (2.1) and (2.2) that for all $v_h \in S^h$

$$\begin{aligned}(u-u^\pi, g) &= \int_0^1 aG_x(u-u^\pi)_x\,dx \\ &= \int_0^1 (\pi a - a)(u_x^\pi G_x - v_h)\,dx \\ &\le \|\pi a - a\|_{0,\infty} \inf_{v_h\in S^h}\|u_x^\pi G_x - v_h\|_{0,1} \\ &\le \|\pi a - a\|_{0,\infty} C h \|u_x^\pi G_x\|_{1,1} \\ &\le C h \|\pi a - a\|_{0,\infty}\|u^\pi\|_{2,\infty}\|G\|_{2,1}\,.\end{aligned} \tag{2.16}$$

For the one dimensional problem (2.15), by taking the test function $v_x = \operatorname{sign} G_x$ (the sign of G_x) we can easily prove that $\|G\|_{2,1} \le C\|g\|_{0,1}$. As $\int_0^1 (u-u^\pi)\,dx = 0$, we find by (2.16) that for any $\varepsilon > 0$, there exists a function $g \in L^1, \int_0^1 g\,dx = 0$, such that

$$\|u-u^\pi\|_{0,\infty} - \varepsilon \le 2\frac{(u-u^\pi, g)}{\|g\|_{0,1}} \le Ch\|a-\pi a\|_{0,\infty}\|u^\pi\|_{2,1} \le Ch^{r+3}\|u^\pi\|_{2,1}\|a\|_{r+2,\infty} . \tag{2.17}$$

Because $\varepsilon > 0$ is arbitrary, we see

$$\|u-u^\pi\|_{0,\infty} \le Ch^{r+3}\|u^\pi\|_{2,1}\|a\|_{r+2,\infty} . \tag{2.18}$$

This and (2.14) prove (2.11) for $\ell = 0, 1$. By using the equality

$$u - u^\pi = u - u_h + u_h - u_h^\pi - (u^\pi - u_h^\pi)$$

and applying inverse inequalities to the term $u_h - u_h^\pi$ we can prove (2.11) for $\ell > 1$. □

Remark 2.2. We notice that in the following proofs, instead of (2.11) we only use the inequality

$$\|u-u^\pi\|_{\ell,\infty} \le Ch^{r+2-\ell} , \quad 0 \le \ell \le r+1 . \tag{2.19}$$

This also indicates that $a_h \in S_h^r$ is already sufficient for the needed convergence as in (1.2).

Lemma 2.2. *Suppose that the condition (A) is valid and that*

$$\|u - z\|_0 \le \varepsilon . \tag{2.20}$$

Then for h small enough, we have

$$\|w-u\|_{\ell,\infty} \le C\big(h^{r+2-\ell} + \frac{\varepsilon}{h^\ell}\big)h^{-\frac{1}{2}} , \quad 1 \le \ell \le r+1 . \tag{2.21}$$

Proof. As a_h is the solution of (P_h), we know that

$$\|w_h - z\|_0 \le \|u_h^\pi - z\|_0. \tag{2.22}$$

Therefore from (2.6) and (2.11), we see

$$\begin{aligned} \|w_h - u\|_0 &\le \|w_h - z\|_0 + \|u - z\|_0 \\ &\le \|u_h^\pi - z\|_0 + \varepsilon \\ &\le \|u_h^\pi - u\|_0 + \|u - z\|_0 + \varepsilon \\ &\le \|u_h^\pi - u\|_0 + 2\varepsilon \\ &\le C\left(h^{r+2} + \varepsilon\right) \end{aligned} \tag{2.23}$$

with $C > 2$. Finally, we can prove (2.21) from the inverse inequalities (2.9) and (2.10). For $\ell = 1$, wee see

$$\begin{aligned} &\|(w-u)_x\|_{0,\infty} \le \|(w-w_h)_x\|_{0,\infty} + \|(w_h-u_h)_x\|_{0,\infty} + \|(u_h-u)_x\|_{0,\infty} \\ &\le Ch^{r+1} + Ch^{-\frac{1}{2}}h^{-1}\|w_h - u_h\|_0 \le C\big(h^{r+\frac{1}{2}} + \varepsilon h^{-\frac{3}{2}}\big) . \end{aligned} \tag{2.24}$$

Here we also use the L^∞–error estimate for w and u as in (2.7) and (2.8). This proves (2.21) for $\ell = 1$. For $\ell \ge 2$, we can also prove (2.21) by (2.7) and (2.8) and by using the inverse inequalities samilar as in Lemmma 2.1. □

3. Error estimate for the identified parameter

In this section we will prove the following theorem

Theorem 3.1. *Suppose that the condition (A) is valid and that*

$$(B) \qquad |u_x| + |u_{xx}| + \cdots + |u^{(l+1)}| > k_0 > 0 \text{ in } [0,1], \quad 0 \le l \le r-1 ,$$

then for h small enough, the following error estimate holds

$$\|(a-a_h)(|u_x|+h|u_{xx}|+\cdots+h^{\ell}|u^{(\ell+1)}|)\|_{L^\infty(e_i)} \le Ch^{-1/2}(h^{r+1}+\frac{\varepsilon}{h}), \quad 0 \le \ell \le r-1 .$$

Here e_i is any of the elements and ε is the observation error.

Before we prove Theorem 3.1, we shall first give a technical lemma.

Lemma 3.1. *Assume that $\{\Delta_h\}$ is a quasi–uniform family of subdivisions of $(0,1)$, and that u is a C^2-function satisfying*

$$|u_x| + |u_{xx}| > k_0 > 0 \text{ in } [0,1] . \tag{3.1}$$

When h is small enough, Δ_h can be divided into two disjoint classes Δ_h^1 and Δ_h^2

$$\Delta_h^1 = \{e_i \mid e_i \in \Delta_h, \ |u_x| > k_1 \quad \text{on } e_i\} ,$$
$$\Delta_h^2 = \{e_i \mid e_i \in \Delta_h, \ |u_{xx}| > k_1 \quad \text{on } e_i\} .$$

Here $k_1 = \frac{k_0}{4}$. This means that in any element, either $|u_x| > k_1$ uniformly or $|u_{xx}| > k_1$ uniformly.

Proof. In order to verify Lemma 3.1, we first find that by (3.1) at any point, either $|u_x| > k_0/2$, or $|u_{xx}| > k_0/2$. As u_x and u_{xx} are continuous we can assume that in any element either $|u_x| > k_0/4$, or $|u_{xx}| > k_0/4$. This is just the meaning of this Lemma. A detailed proof can be found in [**11**]. □

Remark 3.1. The condition (B) in theorem 3.1 can be regarded as a generalization of the condition in [**3**]. Also as we know, if $u_x = 0$ at some subintervals in $(0,1)$, then we cannot identify the parameter $a(x)$ in these intervals, therefore we try to avoid this situation. If u is regular enough, then from the Taylor expansion theorem, we see that if u_x is not identically zero, then there must exist an integer $l \ge 0$ such that condition (B) is valid. From this we see that (B) is a natural condition for the identification problems. □

Proof of Theorem 3.1. The proof will be carried out in several steps.
a) Suppose that condition (A) is satisfied, then

$$\|(a_h - a)u_x\|_{0,\infty} \le C(h^{r+1} + \frac{\varepsilon}{h})h^{-1/2} . \tag{3.2}$$

In order to prove this estimate, we first note that by (2.3) and (2.1)

$$\int_0^1 (a_h - a)u_x v_x \, dx = \int_0^1 a_h(u-w)_x v_x \, dx \quad \forall \, v \in H^1 . \tag{3.3}$$

In (3.3) we choose $v(x) = \int_0^x (a_h - a)^{2p+1} u_x^{2p+1}\, ds$, using the Hölder inequality as in Lemma 2.1, we can prove estimate (3.2).
b) Next we shall show that if the condition (A) is satisfied and if

$$|u_x| + |u_{xx}| > k_0 > 0 \quad \text{in } [0,1] \,, \tag{3.4}$$

then, for h small enough, we have

$$\|(a - a_h)u_{xx}\|_{0,\infty,e_i} \le C h^{-1/2}(h^r + \frac{\varepsilon}{h^2}) \,. \tag{3.5}$$

As from (2.11), we can assume that for h small enough u^π also satisfies (3.4) and (A). So we can divide Δ_h into two disjoint classes

$$|u_x^\pi| > k_1 = \frac{k_0}{4} \quad \text{in the elements inside } I_1 \equiv \bigcup_{e_i \in \Delta_h^1} e_i, \tag{3.6}$$

$$|u_{xx}^\pi| > k_1 = \frac{k_0}{4} \quad \text{in the elements inside } I_2 \equiv \bigcup_{e_i \in \Delta_h^2} e_i. \tag{3.7}$$

Substracting (2.4) from (2.5), we get

$$(\pi a - a_h)_x u_x^\pi + (\pi a - a_h) u_{xx}^\pi = (a_h(w - u^\pi)_x)_x \,. \tag{3.8}$$

For simplicity we define

$$z = \pi a - a_h \,, \quad f^* = (a_h(w - u^\pi)_x)_x \,. \tag{3.9}$$

From the definition of f^*, (2.21) and (2.11) we obtain

$$\|f^*\|_{0,\infty} \le C h^{-1/2}(h^r + \frac{\varepsilon}{h^2}) \,. \tag{3.10}$$

Then (3.8) means

$$z_x u_x^\pi + z u_{xx}^\pi = f^* \,. \tag{3.11}$$

From the estimate (3.2) we see that for any element $e_i \subset I_1$,

$$k_1 \|a - a_h\|_{0,\infty,e_i} \le C h^{-1/2}(h^{r+1} + \frac{\varepsilon}{h}) \,.$$

Therefore,

$$\|(a - a_h)u_{xx}\|_{0,\infty,e_i} \le C k_1^{-1} \|u_{xx}\|_{0,\infty,e_i} h^{-1/2}(h^{r+1} + \frac{\varepsilon}{h}) \,.$$

So (3.5) is true. From this we see that we only need to prove (3.5) for the elements in I_2, i.e., for those elements in which $|u_x|$ may be very small.

Multiplying both sides of (3.11) with $z u_{xx}^\pi$, we get

$$|z u_{xx}^\pi|^2 = -z_x u_x^\pi z u_{xx}^\pi + f^* z u_{xx}^\pi \,. \tag{3.12}$$

Taking the L^∞-norm on any element $e_i \in I_2$, we get

$$\|zu^\pi_{xx}\|^2_{0,\infty,e_i} \le \|zu^\pi_x\|_{0,\infty,e_i}\|z_x u^\pi_{xx}\|_{0,\infty,e_i} + \|f^*\|_{0,\infty,e_i}\|zu^\pi_{xx}\|_{0,\infty,e_i} . \tag{3.13}$$

From the inverse inequality (2.10) we see that

$$\begin{aligned}\|z_x u^\pi_{xx}\|_{0,\infty,e_i} &\le \|u^\pi_{xx}\|_{0,\infty}\|z_x\|_{0,\infty,e_i}\\ &\le \|u^\pi_{xx}\|_{0,\infty} C h^{-1}\|z\|_{0,\infty,e_i}\\ &\le \|u^\pi_{xx}\|_{0,\infty} C k_1^{-1} h^{-1}\|zu^\pi_{xx}\|_{0,\infty,e_i} .\end{aligned} \tag{3.14}$$

Substituting this back into (3.13), we obtain

$$\|zu^\pi_{xx}\|_{0,\infty,e_i} \le C(u^\pi_{xx}, k_1)h^{-1}\|zu^\pi_x\|_{0,\infty,e_i} + \|f^*\|_{0,\infty,e_i} . \tag{3.15}$$

Then from (2.11), (3.2) and (3.10) we get the result of (3.5).
c) We suppose now that (A) is satisfied, and

$$|u_x| + |u_{xx}| + |u_{xxx}| > k_0 \text{ in } [0,1] . \tag{3.16}$$

We shall prove that, for h small enough,

$$\big\|(a_h - a)(h^{-1}|u_{xx}| + |u_{xxx}|)\big\|_{0,\infty,e_i} \le C\big(h^{r-1} + \frac{\varepsilon}{h^3}\big)h^{-1/2}. \tag{3.17}$$

As u satisfies (3.16), a similar proof as in Lemma 3.1 can be carried out to show that Δ_h can be divided into three disjoint classes

$$|u_x| > \frac{k_0}{6} = k_1 \text{ in any element inside } I_1, \tag{3.18}$$

$$|u_{xx}| > \frac{k_0}{6} = k_1 \text{ in any element inside } I_2, \tag{3.19}$$

$$|u_{xxx}| > \frac{k_0}{6} = k_1 \text{ in any element inside } I_3. \tag{3.20}$$

By differentiating both sides of (3.8), we again obtain

$$(\pi a - a_h)_{xx}u^\pi_x + 2(\pi a - a_h)_x u^\pi_{xx} + (\pi a - a_h)u^\pi_{xxx} = [(a_h(w - u^\pi)_x)_x]_x . \tag{3.21}$$

From here we get

$$\begin{aligned}&(\pi a - a_h)_{xx}u_{hx} + 2(\pi a - a_h)_x u_{hxx} + (\pi a - a_h)u_{hxxx}\\ &= (a_h(w - u^\pi)_x)_{xx} + (\pi a - a_h)_{xx}(u_h - u^\pi)_x\\ &\quad + 2(\pi a - a_h)_x(u_h - u^\pi)_{xx} + (\pi a - a_h)(u_h - u^\pi)_{xxx} .\end{aligned} \tag{3.22}$$

Furthermore, for simplicity, we also define as above

$$z = \pi a - a_h , \quad f^* = \text{right hand side of (3.22)}. \tag{3.23}$$

Then (3.22) takes the form

$$z_{xx}u_{hx} + 2z_x u_{hxx} + z u_{hxxx} = f^* . \tag{3.24}$$

From estimates (2.11), (2.21) and (2.8) we can see

$$\|f^*\|_{0,\infty} \le Ch^{-1/2}(h^{r-1} + \frac{\varepsilon}{h^3}) . \tag{3.25}$$

This constant depends on $\|\pi a\|_{2,\infty}$ and $\|a_h\|_{2,\infty}$.

From the proof of a) and b) we see that for any element in I_1 or I_2 estimates (3.2) and (3.5) are true. As $|u_{xx}| > k_1$ in I_2 and $|u_x| > k_1$ in I_1, we get from (3.2) and (3.5) that (3.17) is also true in these elements. Therefore, we only need to prove (3.17) for the elements in I_3, i.e., in those elements where $|u_x|$, $|u_{xx}|$ may be very small.

Multiply both sides of (3.24) with zu_{hxxx} and take the L^∞-norm on an element $e_i \subset I_3$, then we see that

$$\|zu_{hxxx}\|^2_{0,\infty,e_i} \le \|z_{xx}u_{hx}zu_{hxxx}\|_{0,\infty,e_i} + 2\|z_x u_{hxx} z u_{hxxx}\|_{0,\infty,e_i} + \|f^* z u_{hxxx}\|_{0,\infty,e_i} .$$

For h small enough, (3.18)–(3.20) are applicable to u_h. As in b), by using the inverse inequality, we arrive at

$$\begin{aligned}
&\|zu_{hxxx}\|^2_{0,\infty,e_i} \\
&\le \|zu_{hx}\|_{0,\infty,e_i}\|z_{xx}u_{hxxx}\|_{0,\infty,e_i} \\
&\quad + 2\|zu_{hxx}\|_{0,\infty,e_i}\|z_x u_{hxxx}\|_{0,\infty,e_i} + \|f^*\|_{0,\infty,e_i}\|zu_{hxxx}\|_{0,\infty,e_i} \\
&\le \|zu_{hx}\|_{0,\infty,e_i} Ch^{-2}\|zu_{hxxx}\|_{0,\infty,e_i} \\
&\quad + \|zu_{hxx}\|_{0,\infty,e_i} Ch^{-1}\|zu_{hxxx}\|_{0,\infty,e_i} + \|f^*\|_{0,\infty,e_i}\|zu_{hxxx}\|_{0,\infty,e_i} .
\end{aligned} \tag{3.26}$$

From the Cauchy inequality and (3.26), we obtain

$$\|zu_{hxxx}\|^2_{0,\infty,e_i} \le Ch^{-4}\|zu_{hx}\|^2_{0,\infty,e_i} + Ch^{-2}\|zu_{hxx}\|^2_{0,\infty,e_i} + \|f^*\|^2_{0,\infty,e_i} . \tag{3.27}$$

Condition (3.4) is not true in I_3, so we cannot use (3.5) to estimate $\|zu_{hxx}\|_{0,\infty,e_i}$. Then we turn to use (3.8) again. From (3.8) we can see

$$\begin{aligned}
(\pi a - a_h)_x u_{hx} + (\pi a - a_h) u_{hxx} &= (a_h(w - u^\pi)_x)_x \\
&+ (\pi a - a_h)_x(u_h - u^\pi)_x + (\pi a - a_h)(u_h - u^\pi)_{xx} .
\end{aligned} \tag{3.28}$$

We define g^* as the right-hand side of (3.28). Then we have

$$z_x u_{hx} + z u_{hxx} = g^* . \tag{3.29}$$

From the estimates (2.8), (2.11), (2.21),(3.28) and (3.29), we see that

$$\|g^*\|_{0,\infty,e_i} \le Ch^{-1/2}(h^r + \frac{\varepsilon}{h^2}) . \tag{3.30}$$

Multiplying both sides of (3.29) by zu_{hxx}, we get

$$\begin{aligned}|zu_{hxx}|^2 &= -z_x u_{hx} z u_{hxx} + g^* z u_{hxx} \\ &= -\frac{1}{2} u_{hx} u_{hxx} (z^2)_x + g^* z u_{hxx} \\ &= -\frac{1}{2}(z^2 u_{hx} u_{hxx})_x + \frac{1}{2}(z^2 u_{hxx}^2 + z^2 u_{hx} u_{hxxx}) + g^* z u_{hxx} \,.\end{aligned}$$

By using the Cauchy inequality to the term $g^* z u_{hxx}$, we see

$$\frac{1}{4}|zu_{hxx}|^2 \le \frac{1}{2}|(z^2 u_{hx} u_{hxx})_x| + \frac{1}{2}|z^2 u_{hx} u_{hxxx}| + |g^*|^2 \,. \tag{3.31}$$

As $z^2 u_{hx} u_{hxx} \in S_h^{4r+1}$, the inverse inequality is also valid. Therefore, for any $\mu > 0$ we find that the first term of the right–hand side of (3.31) satisfies

$$\begin{aligned}&\|(z^2 u_{hx} u_{hxx})_x\|_{0,\infty,e_i} \\ &\le Ch^{-1}\|z^2 u_{hx} u_{hxx}\|_{0,\infty,e_i} \\ &\le \mu\|zu_{hxx}\|^2_{0,\infty,e_i} + \frac{1}{4\mu}C^2 h^{-2}\|zu_{hx}\|^2_{0,\infty,e_i} \,.\end{aligned} \tag{3.32}$$

Taking $\mu = 1/4$ in (3.32), substituting it back to (3.31) and also using the Cauchy inequality, we find that for any $\mu^* > 0$

$$\begin{aligned}\frac{1}{8}\|zu_{hxx}\|^2_{0,\infty,e_i} &\le Ch^{-2}\|zu_{hx}\|^2_{0,\infty,e_i} + \frac{1}{2}\|z^2 u_{hx} u_{hxxx}\|_{0,\infty,e_i} + \|g^*\|^2_{0,\infty,e_i} \\ &\le Ch^{-2}\|zu_{hx}\|^2_{0,\infty,e_i} + \mu^* h^2\|zu_{hxxx}\|^2_{0,\infty,e_i} \\ &\quad + \frac{1}{8\mu^* h^2}\|zu_{hx}\|^2_{0,\infty,e_i} + \|g^*\|^2_{0,\infty,e_i} \,.\end{aligned} \tag{3.33}$$

Take $\mu^* = 1/(2C)$, where C is the constant in (3.27). By using (3.33) and (3.27), we get

$$\begin{aligned}\|zu_{hxxx}\|^2_{0,\infty,e_i} &\le Ch^{-4}\|zu_{hx}\|^2_{0,\infty,e_i} + \frac{1}{2}\|zu_{hxxx}\|^2_{0,\infty,e_i} \\ &\quad + Ch^{-4}\|zu_{hx}\|^2_{0,\infty,e_i} + Ch^{-2}\|g^*\|^2_{0,\infty,e_i} + \|f^*\|^2_{0,\infty,e_i} \,.\end{aligned} \tag{3.34}$$

This means

$$\frac{1}{2}\|zu_{hxxx}\|^2_{0,\infty,e_i} \le Ch^{-4}\|zu_{hx}\|^2_{0,\infty,e_i} + Ch^{-2}\|g^*\|^2_{0,\infty,e_i} + \|f^*\|^2_{0,\infty,e_i} \,. \tag{3.35}$$

If we take $\mu^* = 1$ in (3.33), we see that

$$\|zh^{-1}u_{hxx}\|^2_{0,\infty,e_i} \le Ch^{-4}\|zu_{hx}\|^2_{0,\infty,e_i} + C\|zu_{hxxx}\|^2_{0,\infty,e_i} + Ch^{-2}\|g^*\|^2_{0,\infty,e_i} \,. \tag{3.36}$$

Therefore, combining (3.35) and (3.36) gives

$$\begin{aligned}&\|z\left(h^{-1}|u_{hxx}| + |u_{hxxx}|\right)\|^2_{0,\infty,e_i} \\ &\le Ch^{-4}\|zu_{hx}\|^2_{0,\infty,e_i} + Ch^{-2}\|g^*\|^2_{0,\infty,e_i} + \|f^*\|^2_{0,\infty,e_i} \,.\end{aligned} \tag{3.37}$$

The estimates for g^* and f^* are given in (3.25) and (3.30). From estimates (2.8), (2.11), (2.21) and (3.2) we see that the estimate (3.17) is valid.

d) With the techniques used in part b) and c), we can prove the result of Theorem 3.1. By differentiating (3.21) another time and using the inverse inequalities, we can bound the error with weighted forth order derivative by the errors with weighted lower order derivatives as in (3.27). Then repeatly use the techniques as in (3.28) – (3.36) we can eliminate the errors with weighted lower order derivatives and bound them by some other known errors and then prove the theorem for $\ell = 4$. In this same way, we can prove the theorem for $\ell \geq 4$. □

Remark 3.2. In the above proof, we take the finite element spaces for $a_h(x)$ and $u_h(x)$ both as S_h^{r+1}. This is necessary in proving (2.11) and (2.6)–(2.8). By using a different technique (see [**5**] and Remarks 2.1, 2.2), we could also prove that if we take $a_h(x) \in S_h^r$ and $u_h(x) \in S_h^{r+1}$, the same error estimate as in Theorem 3.1 is valid.

If we use a same finite element space $S_h^{(r+1)}$ for both $a_h(x)$ and $u_h(x)$, but use a smaller mesh size for $u_h(x)$, for example, let h_1 be the mesh size for $a_h(x)$, h_2 be the mesh size for $u_h(x)$ and $h_2 < h_1$, then from the proofs here, we can see the following estimate is true

$$\|(a_h - a)u_x\|_{0,\infty} \leq C h_2^{-1/2}(h_2^{r+1} + h_1^{r+3}h_2^{-1} + \varepsilon/h_2) \ .$$

References

1. Banks, H.T. and Kunisch, K., *Estimation techniques for distributed parameter systems*, Birkhäuser (1989).
2. de Boor, C., *A bound on the L^∞–norm of L^2–approximation by splines in terms of a global mesh ratio*, Math. Comp. **30** (1978), 765–771.
3. Ciarlet, P.G., *The Finite Element Method for Elliptic Problems*, North–Holland, Amsterdam, New York, Oxford, 1978.
4. Douglas, J., Dupont, T. and Wahlbin, L., *Optimal L^∞–error estimates for Galerkin approximation to solutions of two–point boundary value problems*, Math. Comput. **29** (1975), 475–483.
5. Falk, R.S., *Error estimates for the numerical identification of a variable coefficient*, Math. Comp. **40** (1983), 537–546.
6. Gilbarg, D. and Trudinger, N.S., *Elliptic Partial Differential Equations of Second Order*, Springer–Verlag, Berlin, Heidelberg, 1977.
7. Křížek, M. and Neittaanmäki, P, *Finite element approximation of variational problems with applications*, vol. 50, Pitman Monographs in Applied and Pure Mathematics, Longman Scientific, 1990.
8. Kunisch, K., *Rate of convergence for the estimation coefficient in a two point boundary value problem*, Preprint (1987).
9. Kunisch K and Tai X. C., *Sequential and parallel splitting methods for bilinear control problems in Hilbert spaces*, SIAM J. Numer. Anal. (to appear, see also Preprint, Technische Universität Graz, July 1993).
10. Kunisch, K. and White, L., *Identifiability under approximation for elliptic boundary value problem*, SIAM J. Contr. Optim. **25** (1987), 279–297.
11. Tai, X.C. and Neittaanmäki, P., *Error estimates for the numerical identification of distributed parameters*, J. Comp. Math., supplementary issue (1992), 66-78.

12. Tai, X.C. and Neittaanmäki, P., *A FE-splitting-up method and its application to distributed parameter identification of parabolic type*, the Proceeding of the International Symposium on approximation, optimization and computing (ISAOE '89), Dalian, China, July 3–7,1989.
13. Tai, X.C. and Neittaanmäki, P., *On the numerical solution of the distributed parameter parameter identification problem*, Control and estimation of distributed parameter systems, Proceeding of the international conference on control and estimation of distributed parameter systems, Vorau, July 8-14, 1990. International series of numerical mathematics, Vol. 100, Birkhauser Verlag Basel, 1991, pp. 317-329.

Bei Fragen zur Produktsicherheit wenden Sie sich bitte an:
If you have any questions regarding product safety,
please contact:

Walter de Gruyter GmbH
Genthiner Straße 13
10785 Berlin
productsafety@degruyterbrill.com